Probability and Bayesian Modeling

CHAPMAN & HALL/CRC
Texts in Statistical Science Series

Recently Published Titles

A Computational Approach to Statistical Learning
Taylor Arnold, Michael Kane, and Bryan W. Lewis

Theory of Spatial Statistics
A Concise Introduction
M.N.M van Lieshout

Bayesian Statistical Methods
Brian J. Reich, Sujit K. Ghosh

Time Series
A Data Analysis Approach Using R
Robert H. Shumway, David S. Stoffer

The Analysis of Time Series
An Introduction, Seventh Edition
Chris Chatfield, Haipeng Xing

Probability and Statistics for Data Science
Math + R + Data
Norman Matloff

Sampling
Design and Analysis, Second Edition
Sharon L. Lohr

Practical Multivariate Analysis, Sixth Edition
Abdelmonem Afifi, Susanne May, Robin A. Donatello, Virginia A. Clark

Time Series: A First Course with Bootstrap Starter
Tucker S. McElroy, Dimitris N. Politis

Probability and Bayesian Modeling
Jim Albert and Jingchen Hu

For more information about this series, please visit: https://www.crcpress.com/go/textsseries

Probability and Bayesian Modeling

Jim Albert
Jingchen Hu

CRC Press is an imprint of the
Taylor & Francis Group, an **informa** business
A CHAPMAN & HALL BOOK

CRC Press
Taylor & Francis Group
6000 Broken Sound Parkway NW, Suite 300
Boca Raton, FL 33487-2742

Printed on acid-free paper

International Standard Book Number-13: 978-1-138-49256-1 (Hardback)

Visit the Taylor & Francis Web site at
http://www.taylorandfrancis.com

and the CRC Press Web site at
http://www.crcpress.com

Contents

Preface

The Traditional Introduction to Statistics

A traditional introduction to statistical thinking and methods is the two-semester probability and statistics course offered in mathematics and statistics departments. This traditional course provides an introduction to calculus-based probability and statistical inference. The first half of the course is an introduction to probability including discrete, continuous, and multivariate distributions. The chapters on functions of random variables and sampling distributions naturally lead into statistical inference including point estimates and hypothesis testing, regression models, design of experiments, and ANOVA models.

Although this traditional course remains popular, there seems to be little discussion in this course on the application of the inferential material in modern statistical practice. Although there are benefits in discussing methods of estimation such as maximum likelihood, and optimal inference such as a best hypothesis test, the students learn little about statistical computation and simulation-based inferential methods. As stated in Cobb (2015), there appears to be a disconnect between the statistical content we teach and statistical practice.

Developing a New Course

The development of any new statistics course should be consistent with current thinking of faculty dedicated to teaching statistics at the undergraduate level. Cobb (2015) argues that we need to deeply rethink our undergraduate statistics curriculum from the ground up. Towards this general goal, Cobb (2015) proposes "five imperatives" that can help the process of creating this new curriculum. These imperatives are to: (1) flatten prerequisites, (2) seek depth in understanding fundamental concepts, (3) embrace computation in statistics, (4) exploit the use of context to motivate statistical concepts, and (5) implement research-based learning.

Why Bayes?

There are good reasons for introducing the Bayesian perspective at the calculus-based undergraduate level. First, many people believe that the Bayesian approach provides a more intuitive and straightforward introduction than the frequentist approach to statistical inference. Given that the students are learning probability, Bayes provides a useful way of using probability to update beliefs from data. Second, given the large growth of Bayesian applied work in recent years, it is desirable to introduce the undergraduate students to some modern Bayesian applications of statistical methodology. The timing of a Bayesian course is right given the ready availability of Bayesian instructional material and increasing amounts of Bayesian computational resources.

We propose that Cobb's five imperatives can be implemented through a Bayesian statistics course. Simulation provides an attractive "flattened prerequisites" strategy in performing inference. In a Bayesian inferential calculation, one avoids the integration issue by simulating a large number of values from the posterior distribution and summarizing this simulated sample. Moreover, by teaching fundamentals of Bayesian inference of conjugate models together with simulation-based inference, students gain a deeper understanding of Bayesian thinking. Familiarity with simulation methods in the conjugate case prepares students for the use of simulation algorithms later for more advanced Bayesian models.

One advantage of a Bayes perspective is the opportunity to input expert opinion by the prior distribution which allows students to "exploit context" beyond a traditional statistical analysis. This text introduces strategies for constructing priors when one has substantial prior information and when one has little prior knowledge.

To further "exploit context", we introduce one particular Bayesian success story: the use of hierarchical modeling to simultaneously estimate parameters from several groups. In many applied statistical analyses, a common problem is to combine estimates from several groups, often with certain groups having limited amounts of available data. Through interesting applications, we introduce hierarchical modeling as an effective way to achieve partial pooling of the separate estimates.

Thanks to a number of general-purpose software programs available for Bayesian MCMC computation (e.g. openBUGS, JAGS, Nimble, and Stan), students are able to learn and apply more advanced Bayesian models for complex problems. We believe it is important to introduce the students to at least one of these programs which "flattens the prerequisite" of computational experience and "embraces computation". The main task in the use of these programs is the specification of a script defining the Bayesian model, and the Bayesian fitting is implemented by a single function that inputs the model description, the data and prior parameters, and any tuning parameters of the algorithm. By writing the script defining the full Bayesian model, we believe the students get a deeper understanding of the sampling and prior components of the model. Moreover, the use of this software for sophisticated models such as hierarchical models lowers the bar for students implementing these methods. The focus of the students' work is not the computation but rather the summarization and interpretation of the MCMC output. Students interested in the nuts and bolts of the MCMC algorithms can further their learning through directed research or independent study.

Last, we believe all aspects of a Bayesian analysis are communicated best through interesting case studies. In a good case study, one describes the background of the study and the inferential or predictive problems of interest. In a Bayesian applied analysis in particular, one learns about the construction of the prior to represent expert opinion, the development of the likelihood, and the use of the posterior distribution to address the questions of interest. We therefore propose the inclusion of fully-developed case studies in a Bayesian course for students' learning and practice. Based on our teaching experience, having students work on a course project is the best way for them to learn, resonating with Cobb's "teach through research".

Audience and Structure of this Text

This text is intended for students with a background in calculus but not necessarily any experience in programming. Chapters 1 through 6 resemble the material in a traditional probability course, including foundations, conditional probability, discrete and continuous distributions, and joint distributions. Simulation-based approximations are introduced throughout these chapters to get students exposed to new and complementary ways to understand probability and probability distributions, as well as programming in R.

Although there are applications of Bayes' rule in the probability chapters, the main Bayesian inferential material begins in Chapters 7 and 8 with a discussion of inferential and prediction methods for a single binomial proportion and a single normal mean. The foundational elements of Bayesian inference are described in these two chapters, including the construction of a subjective prior, the computation of the likelihood and posterior distributions, and the summarization of the posterior for different types of inference. Exact posterior distributions based on conjugacy, and approximation based on Monte Carlo simulation, are introduced and compared. Predictive distributions are described both for predicting future data and also for implementing model checking.

Chapters 9 through 13 are heavily dependent on simulation algorithms. Chapter 9 provides an overview of Markov Chain Monte Carlo (MCMC) algorithms with a focus on Gibbs sampling and Metropolis-Hastings algorithms. We also introduce the Just Another Gibbs Sampler (JAGS) software, enabling students to gain a deeper understanding of the sampling and prior components of a Bayesian model and stay focused on summarization and interpretation of the MCMC output for communicating their findings.

Chapter 10 describes the fundamentals of hierarchical modeling where one wishes to combine observations from related groups. Chapters 11 and 12 illustrate Bayesian inference, prediction, and model checking for linear and logistic regression models. Chapter 13 describes several interesting case studies motivated by some historical Bayesian studies and our own research. JAGS is the main software in these chapters for implementing the MCMC inference.

For the interested reader, there is a wealth of good texts describing Bayesian modeling at different levels and directed to various audiences. Berry (1996) is a nice presentation of Bayesian thinking for an introductory statistics class, and Gelman, et al (2013) and Hoff (2009) are good descriptions of Bayesian methodology at a graduate level.

Resources

The following website hosts the datasets and R scripts for all chapters and maintains a current errata list:

`https://monika76five.github.io/ProbBayes/`

A special R package, `ProbBayes` (Albert (2019)), containing all of the datasets and special functions for the text, is available on GitHub. The package can be installed by the `install_github()` function from the `devtools` package.

```
library(devtools)
install_github("bayesball/ProbBayes")
```

Teaching material, including lecture slides and videos, homework and labs of an undergraduate Bayesian statistics course taught at one of the authors' institutions, is available at:

`https://github.com/monika76five/BayesianStatistics`

Acknowledgments:

The authors are very grateful to Dalene Stangl, who played an important role in our collaboration, and our editor, John Kimmel, who provided us with timely reviews that led to significant improvements of the manuscript. We wish to thank our partners, Anne and Hao, for their patience and encouragement as we were working on the manuscript. Jingchen Hu also thanks the group of students taking Bayesian Statistics at Vassar College during Spring 2019, who tried out some of the material.

1

Probability: A Measurement of Uncertainty

1.1 Introduction

The magazine *Discover* once had a special issue on "Life at Risk." In an article, Jeffrey Kluger describes the risks of making it through one day:

> Imagine my relief when I made it out of bed alive last Monday morning. It was touch and go there for a while, but I managed to scrape through. Getting up was not the only death-defying act I performed that day. There was shaving, for example; that was no walk in the park. Then there was showering, followed by leaving the house and walking to work and spending eight hours at the office. By the time I finished my day – a day that also included eating lunch, exercising, going out to dinner, and going home – I counted myself lucky to have survived in one piece.

Is this writer unusually fearful? No. He has read mortality studies and concludes "there is not a single thing you can do in an ordinary day – sleeping included – that isn't risky enough to be the last thing you ever do." In *The Book of Risks* by Larry Laudan, we learn that

- 1 out of 2 million people will die from falling out of bed.
- 1 out of 400 will be injured falling out of bed.
- 1 out of 77 adults over 35 will have a heart attack this year.
- The average American faces a 1 in 13 risk of suffering some kind of injury in home that necessitates medical attention.
- 1 out of 7000 will experience a shaving injury requiring medical attention.
- The average American faces a 1 out of 14 risk of having property stolen this year.
- 1 out of 32 risk of being the victim of some violent crime.
- The annual odds of dying in any kind of motor vehicle accident is 1 in 5800.

Where do these reported odds come from? They are simply probabilities calculated from the counts of reported accidents. Since all of these accidents are possible, that means that there is a risk to the average American that an accident will happen to him or her. But fortunately, you need not worry – many of these reported risks are too small to really take seriously or prod you to change style of living.

Uncertainty

Everywhere we are surrounded by uncertainty. If you think about it, there are a number of things that we are unsure about, like

- What is the high temperature next Monday?
- How many inches of snow will our town get next January?
- What's your final grade in this class?
- Will you be living in the same state twenty years from now?
- Who will win the U.S. presidential election in 2024?
- Is there life on Mars?

A probability is simply a number between 0 and 1 that measures the uncertainty of a particular event. Although many events are uncertain, one possesses different degrees of belief about the truth of an uncertain event. For example, most of us are pretty certain of the statement "the sun will rise tomorrow", and pretty sure that the statement "the moon is made of green cheese" is false. One thinks of a probability scale from 0 to 1.

One typically would give the statement "the sun will rise tomorrow" a probability close to 1, and the statement "the moon is made of green cheese" a probability close to 0. It is harder to assign probabilities to uncertain events that have probabilities between 0 and 1. In this chapter, we first get some experience in assigning probabilities. Then three general ways of thinking about probabilities will be described.

1.2 The Classical View of a Probability

Suppose that one observes some phenomena (say, the rolls of two dice) where the outcome is random. Suppose one writes down the list of all possible outcomes, and one believes that each outcome in the list has the same probability. Then the probability of each outcome will be

$$Prob(\text{Outcome}) = \frac{1}{\text{Number of outcomes}}. \tag{1.1}$$

Let's illustrate this classical view of probability by a simple example. Suppose one has a bowl with 4 white and 2 black balls

and two balls from the bowl are drawn at random. It is assumed that the balls are drawn without replacement which means that one doesn't place a ball back into the bowl after it has been selected. What are possible outcomes? There are different ways of writing down the possible outcomes, depending if one decides to distinguish the balls of the same color.

WAY 1: If one doesn't distinguish between balls of the same color, then there are three possible outcomes – essentially one chooses 0 black, 1 black, or 2 black balls.

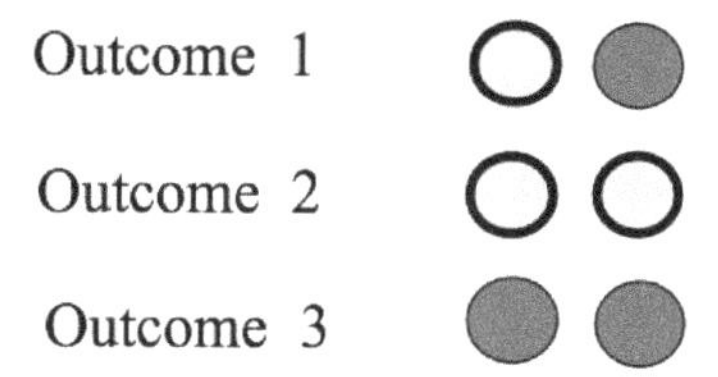

WAY 2: If one does distinguish between the balls of the same color, label the balls in the bowl and then write down 15 distinct outcomes of the experiment of choosing two balls.

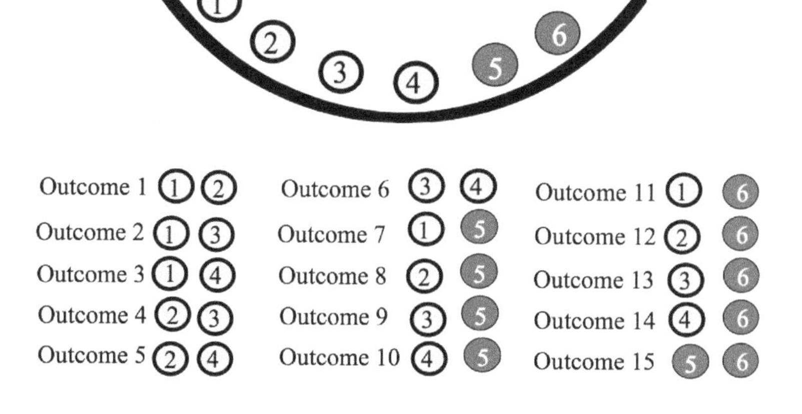

Which is the more appropriate way of listing outcomes?

To apply the classical view of probability, one has to assume that the outcomes are all equally likely. In the first list of three outcomes, one can't assume that they are equally likely. Since there are more white than black balls in the basket, it is more likely to choose two white balls than to choose two black balls. So it is incorrect to say that the probability of each one of the three possible outcomes is 1/3. That is, the probabilities of choosing 0 black, 1 black, and 2 blacks are not equal to 1/3, 1/3, and 1/3.

On the other hand, since one will choosing two balls at random from the basket, it makes sense that the 15 outcomes in the second listing (where we assumed the balls distinguishable) are equally likely. So one applies the classical notion and assigns a probability of 1/15 to each of the possible outcomes. In particular, the probability of choosing two black balls (which is one of the 15 outcomes) is equal to 1/15.

1.3 The Frequency View of a Probability

The classical view of probability is helpful only when we can construct a list of outcomes of the experiment in such a way where the outcomes are equally likely. The frequency interpretation of probability can be used in cases where outcomes are equally likely or not equally likely. This view of probability is appropriate in the situation where one is able to repeat the random experiment many times under the same conditions.

Getting out of jail in Monopoly

Suppose someone is playing the popular game Monopoly and she lands in jail. To get out of jail on the next turn, she either pays $50 or rolls "doubles" when she rolls two fair dice. Doubles means that the faces on the two dice are the same. If it is relatively unlikely to roll doubles, then the player may elect to roll two dice instead of paying $50 to get out of jail.

What is the probability of rolling doubles when she rolls two dice?

In this situation, the frequency notion can be applied to approximate the probability of rolling doubles. Imagine rolling two dice many times under similar conditions. Each time two dice are rolled, we observe whether she rolls doubles or not. Then the probability of doubles is approximated by the relative frequency

$$Prob(\text{doubles}) \approx \frac{\text{Number of doubles}}{\text{Number of experiments}}.$$

R Rolling two dice

The following R code can be used to simulate the rolling of two dice. The `two_rolls()` function simulates rolls of a pair of dice and the `replicate()`

function repeats this process 1000 times and stores the outcomes in the variable many_rolls.

```
two_rolls <- function(){
  sample(1:6, size = 2, replace = TRUE)
}
many_rolls <- replicate(1000, two_rolls())
```

The results of the first 50 experiments are shown in Table 1.1. For each experiment, one records a match (YES) or no match (NO) in the two numbers that are rolled.

TABLE 1.1
The results of the first 50 experiments of rolling two dice.

Die 1	Die 2	Match?	Die 1	Die 2	Match?
3	3	YES	1	6	NO
2	2	YES	2	6	NO
4	6	NO	3	6	NO
6	4	NO	3	1	NO
6	6	YES	6	6	YES
4	5	NO	6	6	YES
4	1	NO	1	5	NO
4	1	NO	1	4	NO
1	2	NO	2	2	YES
5	1	NO	1	3	NO
1	1	YES	5	3	NO
2	6	NO	2	6	NO
3	6	NO	3	5	NO
5	1	NO	3	5	NO
5	3	NO	1	6	NO
3	4	NO	2	5	NO
3	3	YES	2	2	YES
5	5	YES	2	3	NO
4	3	NO	1	5	NO
1	3	NO	2	1	NO
3	2	NO	2	5	NO
5	2	NO	3	1	NO
6	2	NO	2	2	YES
2	6	NO	5	6	NO
1	3	NO	2	3	NO

We see 11 matches (YES results) in the table so

$$Prob(match) \approx 11/50 = 0.22.$$

Let's now roll the two dice 10,000 times with R – this time, 1662 matches are observed, so

$$Prob(match) \approx 1662/10000 = 0.1662.$$

Is 0.1662 the actual probability of getting doubles? No, it is still only an approximation to the actual probability. However, as one continues to roll dice, the relative frequency

$$\text{(number of doubles)/(number of experiments)}$$

will approach the actual probability

$$Prob(\text{doubles}).$$

Here the actual probability of rolling doubles is

$$Prob(\text{doubles}) = 1/6,$$

which is very close to the relative frequency of doubles that we obtained by rolling the dice 10,000 times.

In this example, one can show that are $6 \times 6 = 36$ equally likely ways of rolling two distinguishable dice and there are exactly six ways of rolling doubles. So using the classical viewpoint, the probability of doubles is $6/36 = 1/6$.

1.4 The Subjective View of a Probability

Two ways of thinking about probabilities have been described.

- The classical view. This is a useful way of thinking about probabilities when one lists all possible outcomes in such a way that each outcome is equally likely.
- The frequency view. In the situation when one repeats a random experiment many times under similar conditions, one approximates a probability of an event by the relative frequency that the event occurs.

What if one can't apply these two interpretations of probability? That is, what if the outcomes of the experiment are not equally likely, and it is not feasible or possible to repeat the experiment many times under similar conditions?

In this case, one can rely on a third view of probabilities, the subjective view. This interpretation is arguably the most general way of thinking about a probability, since it can be used in a wide variety of situations.

Suppose one is interested in the probability of the event: "Her team will win the conference title in basketball next season."

One can't use the classical or frequency views to compute this probability. Why? Suppose there are eight teams in the conference. Each team is a possible winner of the conference, but these teams are not equally likely to win – some teams are stronger than the rest. So the classical approach won't help in obtaining this probability.

The event of her team winning the conference next year is essentially a one-time event. Certainly, her team will have the opportunity to win this conference in future years, but the players on her team and their opponents will change and it won't be the same basketball competition. So one can't repeat this experiment under similar conditions, and so the frequency view is not helpful in this case.

What is a subjective probability in this case? The probability

$$Prob(\text{Her team will win the conference in basketball next season})$$

represents the person's belief in the likelihood that her team will win the basketball conference next season. If she believes that her school will have a great team next year and will win most of their conference games, she would give this probability a value close to 1. On the other hand, if she thinks that her school will have a relatively weak team, her probability of this event would be a small number close to 0. Essentially, this probability is a numerical statement about the person's confidence in the truth of this event.

There are two important aspects of a subjective probability.

1. A subjective probability is personal. One person's belief about her team winning the basketball conference is likely different from another person's belief about the team winning the conference since the two people have different information. Perhaps the second person is not interested in basketball and knows little about the teams and the first person is very knowledgeable about college basketball. That means that beliefs about the truth of this event can be different for different people and so the probabilities for these two would also be different.

2. A subjective probability depends on one's current information or knowledge about the event in question. Maybe the first person originally thinks that this probability is 0.7 since her school had a good team last year. But when she learns that many of the star players from last season have graduated, this may change her knowledge about the team, and she may now assign this probability a smaller number.

Measuring probabilities subjectively

Although one is used to expressing one's opinions about uncertain events, using words like likely, probably, rare, sure, maybe, one typically is not used to assigning probabilities to quantify one's beliefs about these events. To make any kind of measurement, one needs a tool like a scale or ruler. Likewise, one needs tools to help us assign probabilities subjectively. Next, a special tool, called a calibration experiment, will be introduced that will help to determine one's subjective probabilities.

A calibration experiment

Consider the event W: "a woman will be President of the United States in the next 20 years".

A college student is interested in his subjective probability of W. This probability is hard to specify precisely since he hasn't had much practice doing it. We describe a simple procedure that will help in measuring this probability.

First consider the following calibration experiment – this is an experiment where the probabilities of outcomes are clear. One has a collection of balls, 5 red and 5 white in a box and one ball is selected at random.

Let B denote the event that the student observes a red ball. Since each of the ten balls is equally likely to be selected, we think he would agree that $Prob(B) = 5/10 = 0.5$.

Now consider the following two bets:

- BET 1 – If W occurs (a women is president in the next 20 years), the student wins \$100. Otherwise, the student wins nothing.
- BET 2 – If B occurs (a red ball is observed in the above experiment), the student wins \$100. Otherwise, the student wins nothing.

Based on the bet that the student prefers, one can determine an interval that contains his $Prob(W)$:

(a) If the student prefers BET 1, then his $Prob(W)$ must be larger than $Prob(B) = 0.5$ – that is, his $Prob(W)$ must fall between 0.5 and 1.

(b) If the student prefers BET 2, then his $Prob(W)$ must be smaller than $Prob(B) = 0.5$ – that is, his probability of W must fall between 0 and 0.5.

What the student does next depends on his answer to part (b).

- If his $Prob(W)$ falls in the interval (0, 0.5), then consider the "balls in box" experiment with 2 red and 8 white balls and he is interested in the probability of choosing a red ball.

- If instead his $Prob(W)$ falls in the interval $(0.5, 1)$, then consider the "balls in box" experiment with 8 red and 2 white balls and he is interested in the probability of choosing a red ball.

Let's suppose that the student believes $Prob(W)$ falls in the interval (0.5, 1). Then he would make a judgment between the two bets

- BET 1 – If W occurs (a women is president in the next 20 years), he wins \$100. Otherwise, he wins nothing.

- BET 2 – If B occurs (observe a red ball with a box with 8 red and 2 white balls), he wins \$100. Otherwise, he wins nothing.

The student decides to prefer BET 2, which means that his probability $Prob(W)$ is smaller than 0.8. Based on the information on the two comparisons, the student now believes that $Prob(W)$ falls between 0.5 and 0.8.

In practice, the student will continue to compare BET 1 and BET 2, where the box has a different number of red and white balls. By a number of comparisons, he will get an accurate measurement at his probability of W.

1.5 The Sample Space

A sample space lists all possible outcomes of a random experiment. There are different ways to write down the sample space, depending on how one thinks about outcomes. Let's illustrate the variety of sample spaces by the simple experiment "roll two fair dice."

Each die is the usual six-sided object that we are familiar with, with a number 1, 2, 3, 4, 5, or 6 on each side. Fair dice implies that each die is constructed such that the six possible numbers are equally likely to come up when rolled.

What can happen when you roll two dice? The collection of all outcomes that are possible is the sample space. But there are different ways of representing the sample space depending on what "outcome" we are considering.

Roll two fair, indistinguishable dice

First, suppose you are interested in the sum of the numbers on the two dice. This would be of interest to a gambler playing the casino game craps. What are the possible sums? After some thought, it should be clear that the smallest possible sum is 2 (if you roll two ones) and the largest possible sum is 12 (with two sixes). Also every whole number between 2 and 12 is a possible sum. So the sample space, denoted by S, would be

$$S = \{2, 3, 4, 5, 6, 7, 8, 9, 10, 11, 12\}.$$

Suppose instead you wish to record the rolls on each of the two dice. One possible outcome would be

(4 on one die, 3 on the other die)

or more simply (4, 3). What are the possible outcomes? Table 1.2 displays the 21 possibilities.

TABLE 1.2
The possible outcomes of rolling two fair, indistinguishable dice.

(1, 1),	(1, 2),	(1, 3),	(1, 4),	(1, 5),	(1, 6)
	(2, 2),	(2, 3),	(2, 4),	(2, 5),	(2, 6)
		(3, 3),	(3, 4),	(3, 5),	(3, 6)
			(4, 4),	(4, 5),	(4, 6)
				(5, 5),	(5, 6)
					(6, 6)

Notice that one is not distinguishing between the two dice in this list. For example, the outcome (2, 3) was written only once, although there are two ways for this to happen – either the first die is 2 and the second die is 3, or the other way around.

Roll two fair, distinguishable dice

Suppose we want to distinguish two dice. Perhaps one die is red and one die is white. We are considering all possible rolls of both dice. We illustrate two ways of showing the sample space in this case.

One way of representing possible rolls of two distinct dice is by a tree diagram shown in Figure 1.1. On the left side of the diagram, the six possible rolls of the red die are represented by six branches of a tree. Then,on the right side, the six possible rolls of the white die are represented by by six smaller branches coming out of each roll of the red die. A single branch on the left and a single branch on the right represent one possible outcome of this experiment.

There are alternative ways for representing the outcomes of this experiment of rolling two distinct dice. Suppose one writes down an outcome by the ordered pair

(roll on white die, roll on red die).

Then, the possible outcomes are listed in Table 1.3.

Since these are ordered pairs, the order of the numbers does matter. The outcome (5, 1) (5 on the red, 1 on the white) is different from the outcome (1, 5) (1 on the red die and 5 on the white die).

Two representations of the sample space of possible rolls of two dice have been illustrated. These representations differ by how one records the outcome

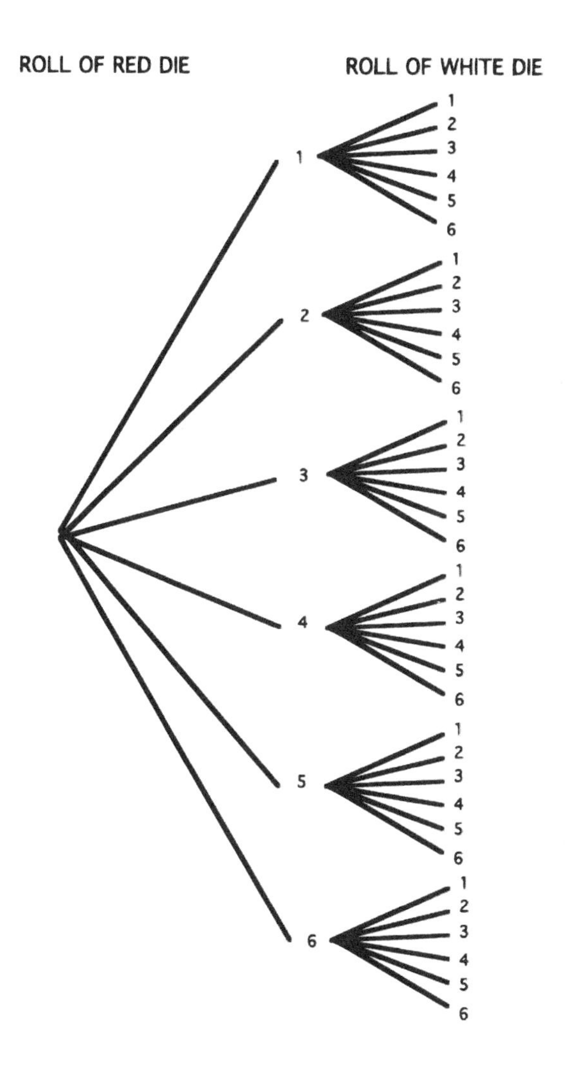

FIGURE 1.1
Tree diagram representation of the rolls of two dice.

TABLE 1.3
The possible outcomes of rolling two fair, distinguishable dice.

(1, 1),	(1, 2),	(1, 3),	(1, 4),	(1, 5),	(1, 6)
(2, 1),	(2, 2),	(2, 3),	(2, 4),	(2, 5),	(2, 6)
(3, 1),	(3, 2),	(3, 3),	(3, 4),	(3, 5),	(3, 6)
(4, 1),	(4, 2),	(4, 3),	(4, 4),	(4, 5),	(4, 6)
(5, 1),	(5, 2),	(5, 3),	(5, 4),	(5, 5),	(5, 6)
(6, 1),	(6, 2),	(6, 3),	(6, 4),	(6, 5),	(6, 6)

of rolling two dice. One either (1) records the sum of the two dice, (2) records the individual rolls, not distinguishing the two dice, or (3) records the individual rolls, distinguishing the two dice.

Which one is the best sample space to use? Actually, all of the sample spaces shown above are correct. Each sample space represents all possible outcomes of the experiment of rolling two dice and we cannot say that one sample space is better than another sample space. We will see that in particular situations some sample spaces are more convenient than other sample spaces when one wishes to assign probabilities. In the current case of rolling two fair dice, the sample space with distinguishable dice is desirable from the viewpoint of computing probabilities since the outcomes are equally likely.

When recording sample spaces, we can use whatever method we like. We could use a tree diagram or table or list the outcomes. The important issue is displaying all possible outcomes in S.

1.6 Assigning Probabilities

When one has a random experiment, the first step is to list all of the possible outcomes in the sample space. The next step is to assign numbers, called probabilities, to the different outcomes that reflect the likelihoods that these outcomes can occur.

To illustrate different assignments of probabilities, suppose a school girl goes to an ice cream parlor and plans to order a single-dip ice cream cone. This particular parlor has four different ice cream flavors. Which flavor will the school girl order?

First, one writes down the sample space in Table 1.4 – the possible flavors that the school girl can order. Probabilities will be assigned to these four possible outcomes that reflect a person's beliefs about her likes and dislikes.

TABLE 1.4
Writing down the sample space: step 1.

Flavor	Vanilla	Chocolate	Butter Pecan	Maple Walnut
Probability				

Can our probabilities be any numbers? Not exactly. Here are some basic facts (or laws) about probabilities:

- Any probability that is assigned must fall between 0 and 1.
- The sum of the probabilities across all outcomes must be equal to 1.
- An outcome will be assigned a probability of 0 if one is sure that that outcome will never occur.
- Likewise, if one assigns a probability of 1 to an event, then that event must occur all the time.

With these facts in mind, consider some possible probability assignments for the flavor of ice cream that this school girl will order.

Scenario 1

Suppose that the school girl likes to be surprised. She has brought a hat in which she has placed many slips of paper – 10 slips are labeled "vanilla", 10 slips are labeled "chocolate", and 10 slips are "butter pecan", and 10 are "maple walnut". She makes her ice cream choice by choosing a slip at random. In this case, each flavor would have a probability of $10/40 = 1/4$ (See Table 1.5).

TABLE 1.5
Writing down the sample space: step 2, scenario 1.

Flavor	Vanilla	Chocolate	Butter Pecan	Maple Walnut
Probability	1/4	1/4	1/4	1/4

Scenario 2

Let's consider a different set of probabilities based on different assumptions about the school girl's taste preferences. She knows that she really doesn't like "plain" flavors like vanilla or chocolate, and she really likes ice creams with nut flavors. In this case, we would assign "vanilla" and "chocolate" each a probability of 0, and assign the two other flavors probabilities that sum to one.

Table 1.6 displays one possible assignment.

TABLE 1.6
Writing down the sample space: step 1, scenario 2.

Flavor	Vanilla	Chocolate	Butter Pecan	Maple Walnut
Probability	0	0	0.7	0.3

Another possible assignment of probabilities that is consistent with these assumptions is displayed in Table 1.7.

TABLE 1.7
Writing down the sample space: step 2, scenario 2.

Flavor	Vanilla	Chocolate	Butter Pecan	Maple Walnut
Probability	0	0	0.2	0.8

Scenario 3

Let's consider an alternative probability assignment from a different person's viewpoint. The worker at the ice cream shop has no idea what flavor the school girl will order. But the worker has been working at the shop all day and she has kept a record of how many cones of each type have been ordered – of 50 cones ordered, 10 are vanilla, 14 are chocolate, 20 are butter pecan, and 6 are maple walnut. If she believes that the school girl has similar tastes to the previous customers, then it would be reasonable to apply the frequency viewpoint to assign the probabilities as displayed in Table 1.8.

TABLE 1.8
Writing down the sample space: step 2, scenario 3.

Flavor	Vanilla	Chocolate	Butter Pecan	Maple Walnut
Probability	10/50	14/50	20/50	6/50

Each of the above probability assignments used a different viewpoint of probability as described in previous sections. The first assignment used the classical viewpoint where each of the forty slips of paper had the same probability of being selected. The second assignment was an illustration of the subjective view where one's assignment was based on one's opinion about the favorite flavors of one's daughter. The last assignment was based on the frequency viewpoint where the probabilities were estimated from the observed flavor preferences of 50 previous customers.

1.7 Events and Event Operations

In this chapter, probability has been discussed in an informal way. Numbers called probabilities are assigned to outcomes in the sample space such that the sum of the numbers over all outcomes is equal to one. In this section, we look at probability from a more formal viewpoint. One defines probability as a function on events that satisfies three basic laws or axioms. Then all of the important facts about probabilities, including some facts that have been used above, can be derived once these three basic axioms are defined.

Suppose that the sample space for our random experiment is S. An event, represented by a capital letter such as A, is a subset of S. Events, like sets, can be combined in various ways described as follows.

- $A \cap B$ is the event that both A and B occur (the intersection of the two events).
- $A \cup B$ is the event that either A or B occur (the union of the two events).
- $\bar{A}$ (or A^c) is the event that A does not occur (the complement of the event A).

To illustrate these event operations, suppose one chooses a student at random from a class and records the month when she or he was born. The student could have been born during 12 possible months and the sample space S is the list of these months:

$$S = \{\text{January, February, March, April, May, June, July, August, September, October, November, December}\}.$$

Define the events L that the student is born during the last half of the year and F that the student is born during a month that is four letters long.

$$L = \{\text{July, August, September, October, November, December}\}.$$

$$F = \{\text{June, July}\}.$$

Various event operations can be illustrated using these events.

- $L \cap F$ is the event that the student is born during the last half of the year AND is born in a four-letter month = {July}.

- $L \cup F$, in contrast, is the event that the student is EITHER born during the last half of the year OR born in a four-letter month = {June, July, August, September, October, November, December}.

- $\bar{L}$ (or L^c) is the event that the student is NOT born during the last half of the year = {January, February, March, April, May, June}

1.8 The Three Probability Axioms

Now that a sample space S and events are defined, probabilities are defined to be numbers assigned to the events. There are three basic laws or axioms that define probabilities:

- **Axiom 1**: For any event A, $P(A) \geq 0$. That is, all probabilities are non-negative values.

- **Axiom 2**: $P(S) = 1$. That is, the probability that you observe something in the sample space is one.

- **Axiom 3:** Suppose one has a sequence of events $A_1, A_2, A_3, ...$ that are mutually exclusive, which means that for any two events in the sequence, say A_2 and A_3 , the intersection of the two events is the empty set (i.e. $A_2 \cap A_3 = \emptyset$). Then one finds the probability of the union of the events by adding the individual event probabilities:

$$P(A_1 \cup A_2 \cup A_3 \cup ...) = P(A_1) + P(A_2) + P(A_3) + ... \tag{1.2}$$

Given the three basic axioms, some additional facts about probabilities can be proven. These additional facts are called properties – these are not axioms, but rather additional facts that are derived knowing the axioms. Below several familiar properties about probabilities are stated and we prove how each property follows logically from the axioms.

Property 1: If A is a subset of B, that is $A \subset B$, then $P(A) \leq P(B)$.

This property states that if one has two events, such that one event is a subset of the other event, then the probability of the first set cannot exceed the probability of the second. This fact may seem pretty obvious, but how can one prove this from the axioms?

Proof: The proof begins with a Venn diagram where a set A is a subset of set B. (See Figure 1.2.)

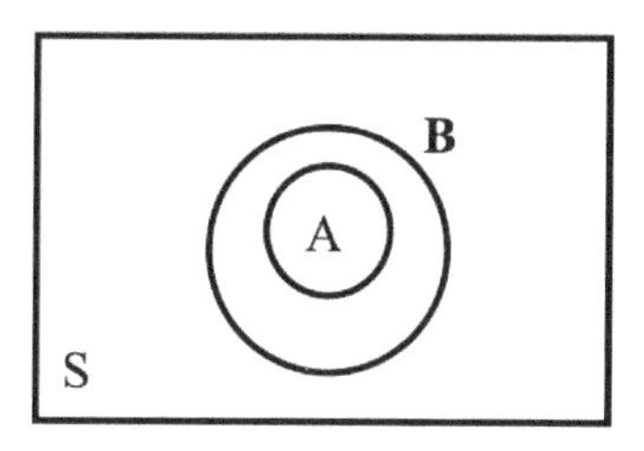

FIGURE 1.2
Two events where one is a subset of the other.

Note that the larger set B can be written as the union of A and $\bar{A} \cap B$, that is,

$$B = A \cup (\bar{A} \cap B) \tag{1.3}$$

Note that A and $\bar{A} \cap B$ are mutually exclusive (i.e. they have no overlap). So one can apply Axiom 3 and write

$$P(B) = P(A) + P(\bar{A} \cap B) \tag{1.4}$$

Also, by Axiom 1, the probability of any event is nonnegative. So the probability of B is equal to the probability of A plus a nonnegative number. So this implies

$$P(B) \geq P(A) \tag{1.5}$$

which is what we wish to prove.

Property 2: $P(A) \leq 1$.

This is pretty obvious – probabilities certainly cannot be larger than 1. But how can this property be shown given our known facts including the axioms and Property 1 that was just proved?

Proof: Actually this property is a consequence of Property 1. Consider the two events A and the sample space S. Obviously A is a subset of the sample space – that is,

$$A \subset S \tag{1.6}$$

So applying Property 1,

$$P(A) \leq P(S) = 1. \tag{1.7}$$

It is known that $P(S) = 1$ from the second Axiom 2. So we have proved our result.

1.9 The Complement and Addition Properties

There are two additional properties of probabilities that are useful in computation. Both of these properties will be stated without proof, but an outline of the proofs will be given in the end-of chapter exercises. The first property, called the complement property, states that the probability of the complement of an event is simply one minus the probability of the event.

Complement property: For an event A,

$$P(\bar{A}) = 1 - P(A). \tag{1.8}$$

The second property, called the addition property, gives a formula for the probability of the union of two events.

Addition property: For two events A and B,

$$P(A \cup B) = P(A) + P(B) - P(A \cap B). \tag{1.9}$$

Both of these properties are best illustrated by an example. Let's revisit the example where one was interested in the birth month of a student selected from a class. As before, let L represent the event that the student is born during the last half of the year and F denote the event that the student is born during a month that is four letters long.

There are 12 possible outcomes for the birth month. One can assume that each month is equally likely to occur, but actually in the U.S. population, the numbers of births during the different months do vary. Using data from the births in the U.S. in 1978, Table 1.9 displays the following probabilities for the months. We see that August is the most likely birth month with a probability of 0.091 and February (the shortest month) has the smallest probability of 0.075.

TABLE 1.9
Probability table of birth months in the U.S. in 1978.

Month	Jan	Feb	Mar	Apr	May	June
Prob	0.081	0.075	0.083	0.076	0.082	0.081
Month	July	Aug	Sept	Oct	Nov	Dec
Prob	0.088	0.091	0.088	0.087	0.082	0.085

Using this probability table, one finds ...

1. $P(L)$ = P(July, August, September, October, November, December) = 0.088 +0.091+0.088+0.098+0.082+0.085 = 0.521.

2. $P(F) = P(\text{June, July}) = 0.081 + 0.088 = 0.169.$

Now we are ready to illustrate the two probability properties.

What is the probability the student is not born during the last half of the year? This can be found by summing the probabilities of the first six months of the year. It is easier to compute this probability by noting that the event of interest is the complement of the event L, and the complement property can be applied to find the probability.

$$P(\bar{L}) = 1 - P(L).$$

What is the probability the student is either born during the last six months of the year *or* a month four letters long? In Figure 1.3, the sample space S is displayed consisting of the twelve possible birth months, and the events F and L are shown by circling the relevant outcomes. The event $F \cup L$ is the union of the two circled events.

Applying the addition property, one finds the probability of $F \cup L$ by adding the probabilities of F and L and subtracting the probability of the intersection event $F \cap L$:

$$\begin{aligned} P(F \cup L) &= P(A) + P(L) - P(F \cap L) \\ &= 0.521 + 0.169 - 0.088 \\ &= 0.602 \end{aligned}$$

Looking at Figure 1.3, the formula should make sense. When one adds the probabilities of the events F and L, one adds the probability of the month July twice, and to get the correct answer, one needs to subtract the outcome (July) common to both F and L.

Special Note: Is it possible to simply add the probabilities of two events, say A and B, to get the probability of the union $A \cup B$? Suppose the sets A and B are mutually exclusive which means they have no outcomes in common. In this special case,$A \cap B = \emptyset$, $P(A \cap B) = 0$ and $P(A \cup B) = P(A) + P(B)$. For example, suppose one is interested in probability that the student is born in the last half the year (event L) or in May (event M). Here, it is not possible to be born in the last half of the year and in May so $L \cap M = \emptyset$. In this case, $P(L \cup M) = P(L) + P(M) = 0.521 + 0.082 = 0.603$.

1.10 Exercises

1. **Probability Viewpoints**

   In the following problems, indicate if the given probability is found using the classical viewpoint, the frequency viewpoint, or the subjective viewpoint.

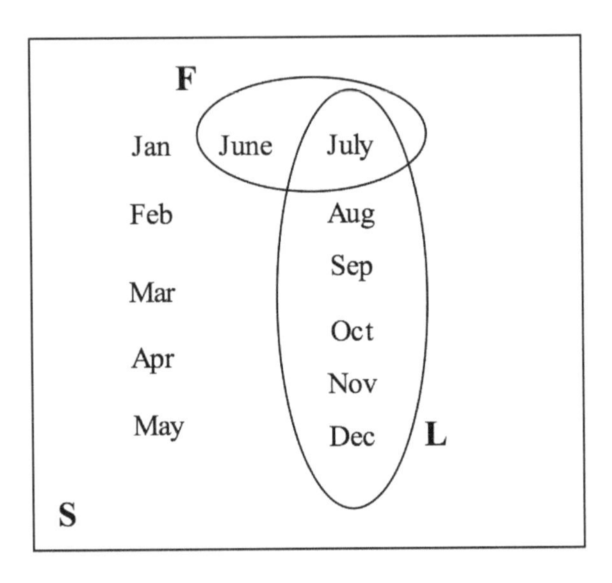

FIGURE 1.3
Representation of two sets F and L in birthday example.

(a) Joe is doing well in school this semester – he is 90% sure that will receive As in all his classes.

(b) Two hundred raffle tickets are sold and one ticket is a winner. Someone purchased one ticket and the probability that her ticket is the winner is 1/200.

(c) Suppose that 30% of all college women are playing an intercollegiate sport. If we contact one college woman at random, the chance that she plays a sport is 0.3.

(d) Two Polish statisticians in 2002 were questioning if the new Belgium Euro coin was indeed fair. They had their students flip the Belgium Euro 250 times, and 140 came up heads.

(e) Many people are afraid of flying. But over the decade 1987-96, the death risk per flight on a US domestic jet has been 1 in 7 million.

(f) In a roulette wheel, there are 38 slots numbered 0, 00, 1, ..., 36. There are 18 ways of spinning an odd number, so the probability of spinning an odd is 18/38.

2. **Probability Viewpoints**

   In the following problems, indicate if the given probability is found using the classical viewpoint, the frequency viewpoint, or the subjective viewpoint.

(a) The probability that the spinner lands in the region A is 1/4.

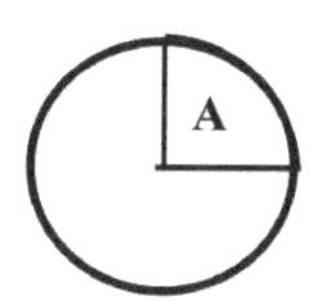

(b) The meteorologist states that the probability of rain tomorrow is 0.5. You think it is more likely to rain and you think the chance of rain is 3/4.

(c) A football fan is 100% certain that his high school football team will win their game on Friday.

(d) Jennifer attends a party, where a prize is given to the person holding a raffle ticket with a specific number. If there are eight people at the party, the chance that Jennifer wins the prize is 1/8.

(e) What is the chance that you will pass an English class? You learn that the professor passes 70% of the students and you think you are typical in ability among those attending the class.

(f) If you toss a plastic cup in the air, what is the probability that it lands with the open side up? You toss the cup 50 times and it lands open side up 32 times, so you approximate the probability by 32/50

3. **Equally Likely Outcomes**

For the following experiments, a list of possible outcomes is given. Decide if one can assume that the outcomes are equally likely. If the equally likely assumption is not appropriate, explain which outcomes are more likely than others.

(a) A bowl contains six marbles of which two are red, three are white, and one is black. One marble is selected at random from the bowl and the color is observed.

Outcomes: {red, white, black}

(b) You observe the gender of a baby born today at your local hospital.

Outcomes: {male, female}

(c) Your school's football team is playing the top rated school in the country.

Outcomes: {your team wins, your team loses}

(d) A bag contains 50 slips of paper, Ten slips are assigned to each category numbered 1 through 5. You choose a slip at random from the bag and notice the number on the slip.

Outcomes: {1, 2, 3, 4, 5}

4. **Equally Likely Outcomes**

For the following experiments, a list of possible outcomes is given. Decide if one can assume that the outcomes are equally likely. If the equally likely assumption is not appropriate, explain which outcomes are more likely than others.

(a) You wait at a bus stop for a bus. From experience, you know that you wait, on average, 8 minutes for this bus to arrive.

Outcomes: {wait less than 10 minutes, wait more than 10 minutes}

(b) You roll two dice and observe the sum of the numbers.

Outcomes: {2, 3, 4, 5, 6, 7, 8, 9, 10, 11, 12}

(c) You get a grade for an English course in college.

Outcomes: {A, B, C, D, F}

(d) You interview a person at random at your college and ask for his or her age.

Outcomes: {17 to 20 years, 21 to 25 years, over 25 years}

5. **Flipping a Coin**

Suppose you flip a fair coin until you observe heads. You repeat this experiment many times, keeping track of the number of flips it takes to observe heads. Here are the numbers of flips for 30 experiments.

1	3	1	2	1	1	2	6	1	2
1	1	1	1	3	2	1	1	2	1
5	2	1	7	3	3	3	1	2	3

(a) Approximate the probability that it takes you exactly two flips to observe heads.

(b) Approximate the probability that it takes more than two flips to observe heads.

(c) What is the most likely number of flips?

6. **Driving to Work**

You drive to work 20 days, keeping track of the commuting time (in minutes) for each trip. Here are the twenty measurements.

25.4,	27.8,	26.8,	24.1,	24.5,	23.0,	27.5,	24.3,	28.4,	29.0
29.4,	24.9,	26.3,	23.5,	28.3,	27.8,	29.4,	25.7,	24.3,	24.2

(a) Approximate the probability that it takes you under 25 minutes to drive to work.

(b) Approximate the probability it takes between 25 and 28 minutes to drive to work.

(c) Suppose one day it takes you 23 minutes to get to work. Would you consider this unusual? Why?

7. **A Person Sent to Mars**

Consider your subjective probability $P(M)$ where M is the event that the United States will send a person to Mars in the next twenty years.

(a) Let B denote the event that you select a red ball from a box of five red and five white balls. Consider the two bets

- BET 1 – If M occurs (United States will send a person to Mars in the next 20 years), you win \$100. Otherwise, you win nothing.
- BET 2 – If B occurs (you observe a red ball in the above experiment), you win \$100. Otherwise, you win nothing.

Circle the bet that you prefer.

(b) Let B represent choosing red from a box of 7 red and 3 white balls. Again compare BET 1 with BET 2 – which bet do you prefer?

(c) Let B represent choosing red from a box of 3 red and 7 white balls. Again compare BET 1 with BET 2 – which bet do you prefer?

(d) Based on your answers to (a), (b), (c), circle the interval of values that contain your subjective probability $P(M)$.

8. **In What State Will You Live in the Future?**

Consider your subjective probability $P(S)$ where S is the event that at age 60 you will be living in the same state as you currently live.

(a) Let B denote the event that you select a red ball from a box of five red and five white balls. Consider the two bets

- BET 1 – If S occurs (you live in the same state at age 60), you win \$100. Otherwise, you win nothing.
- BET 2 – If B occurs (you observe a red ball in the above experiment), you win \$100. Otherwise, you win nothing.

Circle the bet that you prefer.

(b) Let B represent choosing red from a box of 7 red and 3 white balls. Again compare BET 1 with BET 2 – which bet do you prefer?

(c) Let B represent choosing red from a box of 3 red and 7 white balls. Again compare BET 1 with BET 2 – which bet do you prefer?

(d) Based on your answers to (a), (b), (c), circle the interval of values that contain your subjective probability $P(S)$.

9. **Frequency of Vowels in Huckleberry Finn**

Suppose you choose a page at random from the book *Huckleberry Finn* by Mark Twain and find the first vowel on the page.

(a) If you believe it is equally likely to find any one of the five possible vowels, fill in the probabilities of the vowels below.

Vowel	a	e	i	o	u
Probability					

(b) Based on your knowledge about the relative use of the different vowels, assign probabilities to the vowels.

Vowel	a	e	i	o	u
Probability					

(c) Do you think it is appropriate to apply the classical viewpoint to probability in this example? (Compare your answers to parts a and b.)

(d) On each of the first fifty pages of Huckleberry Finn, your author found the first five vowels. Here is a table of frequencies of the five vowels:

Vowel	a	e	i	o	u
Frequency	61	63	34	70	22
Probability					

Use this data to find approximate probabilities for the vowels.

10. **Purchasing Boxes of Cereal**

Suppose a cereal box contains one of four different posters denoted A, B, C, and D. You purchase four boxes of cereal and you count the number of posters (among A, B, C, D) that you do not have. The possible number of "missing posters" is 0, 1, 2, and 3.

(a) Assign probabilities if you believe the outcomes are equally likely.

Number of missing posters	0	1	2	3
Probability				

(b) Assign probabilities if you believe that the outcomes 0 and 1 are most likely to happen.

Number of missing posters	0	1	2	3
Probability				

(c) Suppose you purchase many groups of four cereals, and for each purchase, you record the number of missing posters. The number of missing posters for 20 purchases is displayed below. For example, in the first purchase, you had 1 missing poster, in the second purchase, you also had 1 missing poster, and so on.

```
1, 1, 1, 2, 1, 1, 0, 0, 2, 1,
2, 1, 3, 1, 2, 1, 0, 1, 1, 1
```

Using these data, assign probabilities.

Number of missing posters	0	1	2	3
Probability				

(d) Based on your work in part c, is it reasonable to assume that the four outcomes are equally likely? Why?

11. **Writing Sample Spaces**

For the following random experiments, give an appropriate sample space for the random experiment. You can use any method (a list, a tree diagram, a two-way table) to represent the possible outcomes.

(a) You simultaneously toss a coin and roll a die.

(b) Construct a word from the five letters a, a, e, e, s.

(c) Suppose a person lives at point 0 and each second she randomly takes a step to the right or a step to the left. You observe the person's location after four steps.

(d) In the first round of next year's baseball playoff, the two teams, say the Phillies and the Diamondbacks play in a best-of-five series where the first team to win three games wins the playoff.

(e) A couple decides to have children until a boy is born.

(f) A roulette game is played with a wheel with 38 slots numbered 0, 00, 1, ..., 36. Suppose you place a $10 bet that an even number (not 0) will come up in the wheel. The wheel is spun.

(g) Suppose three batters, Marlon, Jimmy, and Bobby, come to bat during one inning of a baseball game. Each batter can either get a hit, walk, or get out.

12. **Writing Sample Spaces**

For the following random experiments, give an appropriate sample space for the random experiment. You can use any method (a list, a tree diagram, a two-way table) to represent the possible outcomes.

(a) You toss three coins.

(b) You spin the spinner (shown below) three times.

(c) When you are buying a car, you have a choice of three colors, two different engine sizes, and whether or not to have a CD player. You make each choice completely at random and go to the dealership to pick up your new car.

(d) Five horses, Lucky, Best Girl, Stripes, Solid, and Jokester compete in a race. You record the horses that win, place, and show (finish first, second, and third) in the race.

(e) You and a friend each think of a whole number between 0 and 9.

(f) On your computer, you have a playlist of 4 songs denoted by a, b, c, d. You play them in a random order.

(g) Suppose a basketball player takes a "one-and-one" foul shot. (This means that he attempts one shot and if the first shot is successful, he gets to attempt a second shot.)

13. **Writing Sample Spaces**

For the following random experiments, give an appropriate sample space for the random experiment. You can use any method (a list, a tree diagram, a two-way table) to represent the possible outcomes.

(a) Your school plays four football games in a month.

(b) You call a "random" household in your city and record the number of hours that the TV was on that day.

(c) You talk to an Ohio resident who has recently received her college degree. How many years did she go to college?

(d) The political party of our next elected U.S. President.

(e) The age of our next President when he or she is inaugurated.

(f) The year a human will next land on the moon.

14. **Writing Sample Spaces**

For the following random experiments, give an appropriate sample space for the random experiment. You can use any method (a list, a tree diagram, a two-way table) to represent the possible outcomes.

(a) The time you arrive at your first class on Monday that begins at 8:30 AM.

(b) You throw a ball in the air and record how high it is thrown (in feet).

(c) Your cost of textbooks next semester.

(d) The number of children you will have.
(e) You take a five question true/false test.
(f) You drive on the major street in your town and pass through four traffic lights.

15. **Probability Assignments**

Give reasonable assignments of probabilities based on the given information.

(a) In the United States, there were 4.058 million babies born in the year 2000 and 1.98 million were girls. Assign probabilities to the possible genders of your next child.

Gender	Boy	Girl
Probability		

(b) Next year, your school will be playing your neighboring school in football. Your neighboring school is a strong favorite to win the game.

Winner of Game	Your school	Your neighboring school
Probability		

(c) You have an unusual die that shows 1 on two sides, 2 on two sides, and 3 and 4 on the remaining two sides.

Roll	1	2	3	4	5	6
Probability						

16. **Probability Assignments**

Based on the given information, decide if the stated probabilities are reasonable. If they are not, explain how they should be changed.

(a) Suppose you play two games of chess with a chess master. You can either win 0 games, 1 game, or 2 games, so the probability of each outcome is equal to 1/3.
(b) Suppose 10% of cars in a car show are Corvettes and you know that red is the most popular Corvette color. So the chance that a randomly chosen car is a red Corvette must be larger than 10.
(c) In a Florida community, you are told that 30% of the residents play golf, 20% play tennis, and 40% of the residents play golf and tennis.
(d) Suppose you are told that 10% of the students in a particular class get A, 20% get B, 20% get C, and 20% get D. That means that 30% must fail the class.

17. **Finding the Right Key**

Suppose your key chain has five keys, one of which will open up your front door of your apartment. One night, you randomly try keys until the right one is found.

Here are the possible numbers of keys you will try until you get the right one:

1 key, 2 keys, 3 keys, 4 keys, 5 keys

(a) Circle the outcome that you think is most likely to occur.
1 key, 2 keys, 3 keys, 4 keys, 5 keys

(b) Circle the outcome that you think is least likely to occur.
1 key, 2 keys, 3 keys, 4 keys, 5 keys

(c) Based on your answers to parts a and b, assign probabilities to the six possible outcomes.

Outcome	1 key	2 keys	3 keys	4 keys	5 keys
Probability					

18. **Playing Roulette**

One night in Reno, you play roulette five times. Each game you bet \$5 – if you win, you win \$10; otherwise, you lose your \$5. You start the evening with \$25. Here are the possible amounts of money you will have after playing the five games.

\$0, \$10, \$20, \$30, \$40, \$50 .

(a) Circle the outcome that you think is most likely to occur.
\$0, \$10, \$20, \$30, \$40, \$50

(b) Circle the outcome that you think is least likely to occur.
\$0, \$10, \$20, \$30, \$40, \$50

(c) Based on your answers to parts a and b, assign probabilities to the six possible outcomes.

Outcome	\$0	\$10	\$20	\$30	\$40	\$50
Probability						

19. **Cost of Your Next Car**

Consider the cost of the next new car you will purchase in the future. There are five possibilities:

- Cheapest: the car will cost less than \$5000.
- Cheaper: the car will cost between \$5000 and \$10,000.
- Moderate: the car will cost between \$10,000 and \$20,000.
- Expensive: the car will cost between \$20,000 and \$30,000.

- Really expensive: the car will cost over $30,0000.

(a) Circle the outcome that you think is most likely to occur.

cheapest, cheaper, moderate, expensive, really expensive

(b) Circle the outcome that you think is least likely to occur.

cheapest, cheaper, moderate, expensive, really expensive

(c) Based on your answers to parts a and b, assign probabilities to the five possible outcomes.

Outcome	cheapest	cheaper	moderate	expensive	really expensive
Probability					

20. **Flipping a Coin**

Suppose you flip a coin twice. There are four possible outcomes (H stands for heads and T stands for tails).

$$HH, HT, TH, TT$$

(a) Circle the outcome that you think is most likely to occur.

$$HH, HT, TH, TT$$

(b) Circle the outcome that you think is least likely to occur.

$$HH, HT, TH, TT$$

(c) Based on your answers to parts a and b, assign probabilities to the four possible outcomes.

Outcome	HH	HT	TH	TT
Probability				

21. **Playing Songs in Your iPod**

Suppose you play three songs, one each by Jewell (J), Madonna (M), and Plumb (P) in a random order.

(a) Write down all possible ordering of the three songs.

(b) Let M = event that the Madonna song is played first and B = event that the Madonna song is played before the Jewell song. Find $P(M)$ and $P(B)$.

(c) Write down the outcomes in the event $M \cap B$ and find the probability $P(M \cap B)$.

(d) By use of the complement property, find $P(\bar{B})$.

(e) By use of the addition property, find $P(M \cup B)$.

TABLE 1.10
Table of grade level and gender.

	Freshmen	Sophomores	Juniors	Seniors	TOTAL
Male	25	30	24	19	98
Female	20	32	28	15	95
TOTAL	45	62	52	34	193

22. **Student of the Day**

    Suppose that students at a local high school are distributed by grade level and gender in Table 1.10.

    Suppose that a student is chosen at random from the school to be the "student of the day". Let F = event that student is a freshman, J = event that student is a junior, and M = event that student is a male.

    (a) Find the probability $P(\bar{F})$.
    (b) Are events F and J mutually exclusive. Why?
    (c) Find $P(F \cup J)$.
    (d) Find $P(F \cap M)$.
    (e) Find $P(F \cup M)$.

23. **Proving Properties of Probabilities**

    Given the three probability axioms and the properties already proved, prove the complement property $P(\bar{A}) = 1 - P(A)$. An outline of the proof is written below.

    (a) Write the sample space S as the union of the sets A and $\bar{A}$.
    (b) Apply Axiom 3.
    (c) Apply Axiom 2.

24. **Proving Properties of Probabilities**

    Given the three probability axioms and the properties already proved, prove the addition property $P(A \cup B) = P(A) + P(B) - P(A \cap B)$. A Venn diagram and an outline of the proof are written below.

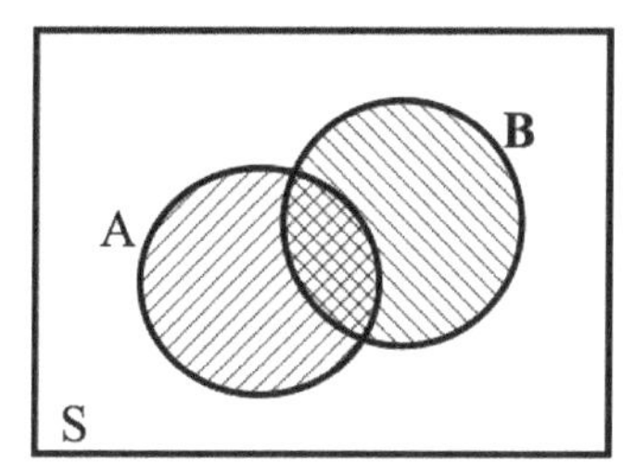

(a) Write the set $A \cup B$ as the union of three sets that are mutually exclusive.

(b) Apply Axiom 2 to write $P(A \cup B)$ as the sum of three terms.

(c) Write the set A as the union of two mutually exclusive sets.

(d) Apply Axiom 2 to write $P(A)$ as the sum of two terms.

(e) By writing the set B as the union of two mutually exclusive sets and applying Axiom 2, write $P(B)$ as the sum of two terms.

(f) By making appropriate substitutions to the expression in part b, one obtains the desired result.

2

Counting Methods

2.1 Introduction: Rolling Dice, Yahtzee, and Roulette

Dice are one of the oldest randomization devices known to man. Egyptian tombs, dated from 2000 BC, were found containing dice and there is some evidence of dice in archaeological excavations dating back to 6000 BC. It is interesting to note that dice appeared to be invented independently by many ancient cultures across the world. In ancient times, the result of a die throw was not just considered luck, but determined by gods. So casting dice was often used as a way of making decisions such as choosing rulers or dividing inheritances. The Roman goddess, Fortuna, daughter of Zeus was believed to bring good or bad luck to individuals.

The Game of Yahtzee

In the 19th and 20th centuries, standard six-sided dice became a basic component of many commercial board games that were developed. One of the most current popular games is Yahtzee that is played with five dice. The Hasbro game company (`http://www.hasbro.com`) presents the history of the game. Yahtzee was invented by a wealthy Canadian couple to play aboard their yacht. This "yacht" game was popular among the couple's friends, who wanted copies of the game for themselves. The couple approached Mr. Edwin Lowe, who made a fortune selling bingo games, about marketing the game. Mr. Lowe's initial attempts to sell the game of Yahtzee by placing ads were not successful. Lowe thought that the game had to be played to be appreciated and he hosted a number of Yahtzee parties and the game became very successful. The Milton Bradley company acquired the E. S. Lowe Company and Yahtzee in 1973 and currently more than 50 million games are sold annually.

The Casino Game of Roulette

Roulette is one of the most popular casino games. The name roulette is derived from the French word meaning small wheel. Although the origin of the game is not clear, it became very popular during the 18th century when Prince Charles introduced gambling to Monaco to alleviate the country's financial problems.

The game was brought to America in the early part of the 19th century and is currently featured in all casinos. In addition, roulette is a popular game among people who like to game online.

The American version of the game discussed in this book varies slightly from the European version. The American roulette wheel contains 38 pockets, numbers 1 through 36 plus zero plus double zero. The wheel is spun and a small metal ball comes to rest in one of the 38 pockets.

Players will place chips on particular locations on a roulette table, predicting where the ball will land when after the wheel is spun and the ball comes to a stop. The dealer places a mark on the winning number. The players who have bet on the winning number are rewarded while the players who bet on losing numbers lose their chips to the casino.

2.2 Equally Likely Outcomes

Assume one writes the sample space in such a way that the outcomes are equally likely. Then, applying the classical interpretation, the probability of each outcome will be

$$Prob(\text{Outcome}) = \frac{1}{\text{Number of outcomes}}. \tag{2.1}$$

If one is interested in the probability of some event, then the probability is given by

$$Prob(\text{Event}) = \frac{\text{Number of outcomes in event}}{\text{Number of outcomes}}. \tag{2.2}$$

This simple formula should be used with caution. To illustrate the use (and misuse) of this formula, suppose one has a box containing five balls of which three are red, one is blue, and one is white. One selects three balls without replacement from the box – what is the probability that all red balls are chosen?

Let's consider two representations of the sample space of this experiment.

Sample space 1: Suppose one does not distinguish between balls of the same color and does not care about the order in which the balls are selected. Then if R, B, W denote choosing a red, blue, and white ball respectively, then there are four possible outcomes:

$$S_1 = \{(R, R, R), (R, R, B), (R, R, W), (R, B, W)\}.$$

If these outcomes in S_1 are assumed equally likely, then the probability of choosing all red balls is

$$Prob(\text{all reds}) = \frac{1}{4}.$$

Sample space 2: Suppose instead that one distinguishes the balls of the same color, so the balls in the box are denoted by $R1, R2, R3, B, W$. Then one writes down ten possible outcomes

$$S_2 = \{(R1, R2, R3), (R1, R2, B), (R1, R2, W), (R1, R3, B), (R1, R3, W), \\ (R2, R3, B), (R2, R3, W), (R1, B, W), (R2, B, W), (R3, B, W)\}.$$

If one assumes these outcomes are equally likely, then the probability of choosing all reds is

$$Prob(\text{all reds}) = \frac{1}{10}.$$

If one compares the answers, one sees an obvious problem since one obtains two different answers for the probability of choosing all reds. What is going on? The problem is that the outcomes in the first sample space S_1 are not equally likely. In particular, the chance of choosing three reds (R, R, R) is smaller than the chance of choosing a red, blue and white (R, B, W) – there is only one way of selecting three reds, but there are three ways of selecting exactly one red. On the other hand, the outcomes in sample space S_2 are equally likely since one was careful to distinguish the five balls in the box, and it is reasonable that any three of the five balls has the same chance of being selected.

From this example, a couple of things have been learned. First, when one writes down a sample space, one should think carefully about the assumption that outcomes are equally likely. Second, when one has an experiment with duplicate items (like three red balls), it may be preferable to distinguish the items when one writes down the sample space and computes probabilities.

R Sampling From a Box

One simulates this experiment on R by first creating a vector `box` with the ball colors, and then using the `sample()` function to sample three balls from the vector. The argument `size = 3` indicates that a sample of 3 is chosen, and the argument `replace = FALSE` ensures that the sampling is done without replacement. In this particular simulation, one observes a red, blue, and red ball in our sample.

```
box <- c("red", "red", "red", "blue", "white")
sample(box, size = 3, replace = FALSE)
[1] "red"  "blue" "red"
```

2.3 The Multiplication Counting Rule

To apply the equally likely recipe for computing probabilities, one needs some methods for counting the number of outcomes in the sample space and the number of outcomes in the event. Here we illustrate a basic counting rule called the multiplication rule.

Suppose you are dining at your favorite restaurant. Your dinner consists of an appetizer, an entrée, and a dessert. You can either choose soup, fruit cup, or quesadillas for your appetizer, you have the choice of chicken, beef, fish, or lamb for your entrée, and you can have either pie or ice cream for your dessert. We first use a tree diagram to write down all of your possible dinners, in Figure 2.1. The first set of branches shows the appetizers, the next set of branches the entrées, and the last set of branches the desserts.

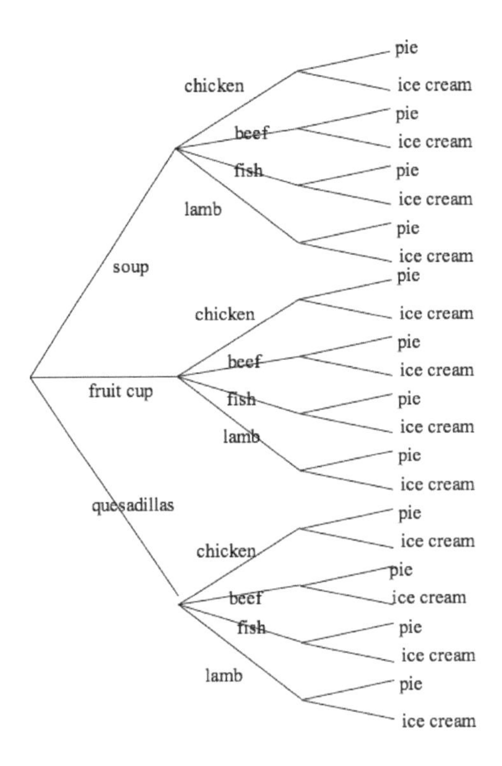

FIGURE 2.1
Tree diagram of possible dinners.

Note that there are 3 possible appetizers, 4 possible entrées, and 2 possible desserts. For each appetizer, there are 4 possible entrées, and so there are $3 \times 4 = 12$ possible choices of appetizer and entrée. Using similar reasoning, for each combination of appetizer and entrée, there are 2 possible desserts, and so the total number of complete dinners would be

$$\text{Number of dinners} = 3 \times 4 \times 2 = 24.$$

The above dining example illustrates a general counting rule that we call the multiplication rule.

Multiplication Rule: Suppose one is performing a task that consists of k steps. One performs the first step in n_1 ways, the second step in n_2 ways, the third step in n_3 ways, and so on. Then the number of ways of completing the task, denoted by n, is the product of the different ways of doing the k steps, or

$$n = n_1 \times n_2 \times ... \times n_k. \tag{2.3}$$

2.4 Permutations

Suppose one places six songs, Song A, Song B, Song C, Song D, Song E, and Song F in one's playlist on the streaming service. The songs are played in a random order and one listens to the first three songs. How many different selections of three songs can one hear? In this example, one is assuming that the order that the songs are played is important. So hearing the selections

Song A, Song B, Song C

in that order will be considered different from hearing the selections in the sequence

Song C, Song B, Song A.

An outcome such as this is called a permutation or arrangement of 3 out of the 6 songs. One represents possible permutations by a set of three blanks, where songs are placed in the blanks.

1st Song	2nd Song	3rd Song

One computes the number of permutations as follows:

1. First, it is known that 6 possible songs can be played first. One places this number in the first blank above.

6 1st Song	2nd Song	3rd Song

2. If one places a particular song, say Song A, in the first slot, there are 5 possible songs in the second position. One places this number in the second blank.

6	5	
1st Song	2nd Song	3rd Song

By use of the multiplication rule, there are $6 \times 5 = 30$ ways of placing two songs in the first two slots.

3. Continuing in the same way, one sees that there are 4 ways of putting a song in the third slot and completing the list of three songs.

6	5	4
1st Song	2nd Song	3rd Song

Again using the multiplication rule, we see that the number of possible permutations of six songs in the three positions is

$$6 \times 5 \times 4 = 120.$$

A second basic counting rule has just been illustrated.

Permutations Rule: If one has n objects (all distinguishable), then the number of ways to arrange r of them, called the number of permutations, is

$$_nP_r = n \times (n-1) \times ... \times (n-r). \tag{2.4}$$

In this example, $n = 6$ and $r = 3$, and If three songs are played in one's playlist, each of the 120 possible permutations will be equally likely to occur. So the probability of any single permutation, say

$$\text{Song } A, \text{ Song } D, \text{ Song } B$$

is equal to 1/120.

Suppose one listens to all six songs on the playlist. How many possible orders are there? In this case, one is interested in finding the number of ways of arranging the entire set of 6 objects. Here $n = 6$ and $r = 6$ and, applying the permutation rule formula, the number of permutations is

$$_6P_6 = n! = 6 \times 5 \times 4 \times ... \times 1 = 720.$$

One uses the special symbol $n!$, pronounced "n factorial", to denote the product of the integers from 1 to n. So the number of ways of arranging n distinct objects is

$$_nP_n = n! = n \times (n-1) \times (n-2) \times ... \times 1. \tag{2.5}$$

R Simulating a Permutation

To illustrate simulating a permutation, define a function `permutation()` with arguments `d` and `Size`. Inside the function, the `sample()` function takes a sample of size `Size` without replacement from the vector `d` and the `str_flatten()`

function creates a single string with the arrangement of the `Size` values. To use this function, a vector `songs` is defined containing the names of the six songs. One applies the `permutation()` function with arguments `songs` and 3 and the simulated arrangement of songs is F, D, and E.

```
permutation <- function(d, Size){
  str_flatten(sample(d, size=Size),
              collapse = " ")
}
songs <- c("Song A", "Song B", "Song C", "Song D",
           "Song E", "Song F")
permutation(songs, Size = 3)
```

```
[1] "Song F Song D Song E"
```

2.5 Combinations

Suppose one has a box with five balls – three are white and two are black. One first shakes up the box and then removes two balls without replacement, i.e. once one takes a ball out, one does not return it to the box before the second ball is taken out.

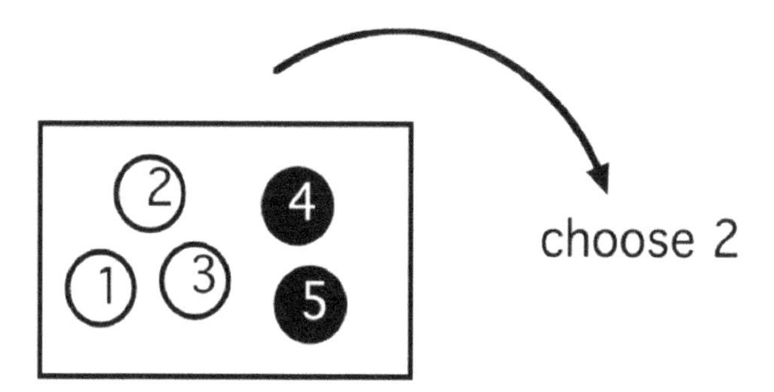

To make it easier to talk about outcomes, the five balls have been labelled from 1 to 5. Remember one is choosing two balls from the box and an outcome is the numbers of the two balls that one selects. When one lists possible outcomes, one should decide if it matters how one orders the selection of balls. That is, if one chooses ball 1 and then ball 2, is that different than choosing ball 2 and then ball 1?

One could say that order is important – so choosing ball 1 then ball 2 is a different outcome from ball 2 then ball 1. But in this type of selection problem, it is common practice not to consider the order of the selection. Then all that matters is the collection of two balls that we select. In this case, one calls the resulting outcome a combination.

When order does not matter, there are 10 possible pairs of balls that one can select. These outcomes or combinations are written below – this list represents a sample space for this random experiment.

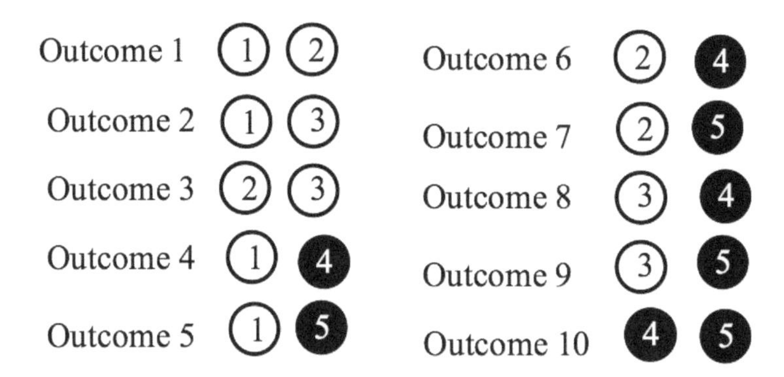

There is a simple formula for counting the number of outcomes in this situation.

Combinations Rule: Suppose one has n objects and one wishes to take a subset of size r from the group of objects without regards to order. Then the number of subsets or combinations is given by the formula

$$\text{number of combinations} = \binom{n}{r} = \frac{n!}{r!(n-r)!}. \tag{2.6}$$

where $k!$ stands for k factorial $k! = k \times (k-1) \times (k-2) \times ... \times 1$. You might have seen another notation $\binom{n}{r}$ when people talk about combinations. This notation is pronounced "n choose r", and it is the same as $\binom{n}{r}$.

Let's try the formula in our example to see if it agrees with our number. In our setting, one has $n = 5$ balls and one is selecting a subset of size $r = 2$ from the box of balls. Using $n = 5$ and $r = 2$ in the formula, one obtains

$$\binom{5}{2} = \frac{5!}{2!(5-2)!} = \frac{5 \times 4 \times 3 \times 2 \times 1}{[2 \times 1] \times [3 \times 2 \times 1]} = \frac{120}{12} = 10.$$

that agrees with our earlier answer of 10 outcomes in the sample space.

R Simulating Combinations

To illustrate combinations, define a vector `Numbers` containing the integers from 1 to 5. The R function `combn()` generates all combinations of a set of a specific size. The matrix `all_combo` is displayed which contains all combinations of size 2 from `Numbers`.

```
Numbers <- c(1, 2, 3, 4, 5)
all_combo <- t(combn(Numbers, 2))
all_combo
      [,1] [,2]
 [1,]    1    2
 [2,]    1    3
 [3,]    1    4
 [4,]    1    5
 [5,]    2    3
 [6,]    2    4
 [7,]    2    5
 [8,]    3    4
 [9,]    3    5
[10,]    4    5
```

Below the function `sample()` is used to simulate random rows of the matrix `all_combo` and a frequency table of the ten possible combinations is displayed. Note that the frequencies for the ten possible combinations are similar since these outcomes are equally likely.

```
N <- nrow(all_combo)
df <- data.frame(Iter = 1:500,
                 Balls = all_combo[sample(N, size = 500,
                                          replace = TRUE), ])
df %>% group_by(Balls.1, Balls.2) %>% count()

   Balls.1 Balls.2     n
     <dbl>   <dbl> <int>
 1       1       2    57
 2       1       3    45
 3       1       4    46
 4       1       5    51
 5       2       3    41
 6       2       4    53
 7       2       5    46
 8       3       4    51
 9       3       5    60
10       4       5    50
```

Number of subsets

Suppose one has a group of n objects and one is interested in the total number of subsets of this group. Then this total number is

$$2^n = \binom{n}{0} + \binom{n}{1} + ... + \binom{n}{n}. \tag{2.7}$$

The formula 2^n is found by noticing there are two possibilities for each object – either the object is in the subset or it is not – and then applying the multiplication rule. The right hand side of the equation is derived by first counting the number of subsets of size 0, of size 1, of size 2, and so on, and then adding all of these subset numbers to get the total number.

Counting the number of pizzas

To illustrate the combinations rule, consider a situation where one is interested in ordering a pizza and there are six possible toppings. How many toppings can there be in the pizza? Since there are six possible toppings, one can either have 0, 1, 2, 3, 4, 5, or 6 toppings on our pizza. Using combinations rule formula,

(a) There are $\binom{6}{0}$ pizzas that have no toppings.

(b) There are $\binom{6}{1}$ pizzas that have exactly one topping.

(c) There are $\binom{6}{2}$ pizzas that have two toppings.

To compute the total number of different pizzas, one continues in this fashion and the total number of possible pizzas is

$$N = \binom{6}{0} + \binom{6}{1} + \binom{6}{2} + \binom{6}{3} + \binom{6}{4} + \binom{6}{5} + \binom{6}{6}.$$

The reader can confirm that $N = 2^6 = 64$.

2.6 Arrangements of Non-Distinct Objects

First let's use a simple example to review the two basic counting rules that we have discussed. Suppose one is making up silly words from the letters "a", "b", "c", "d", "e", "f", like

bacedf, decabf, eabcfd

How many silly words can one make up? Here one has $n = 6$ objects. Using the permutation rule, the number of possible permutations is

$$6! = 6 \times 5 \times 4 \times ... \times 1.$$

To illustrate the second counting rule, suppose one has six letters "a", "b", "c", "d", "e", "f", and one is going to choose three of the letters to construct a three-letter word. One cannot choose the same letter twice and the order in which one chooses the letters is not important. In this case, one is interested in the number of combinations – applying our combination rule with $n = 6$ and $k = 3$, the number of ways of choosing three letters from six is equal to

$$\binom{6}{3} = \frac{6!}{3!\,3!}.$$

Now, consider a different arrangement problem. Suppose one randomly arranges the four triangles and five squares as shown below.

What is the chance that the first and last locations are occupied by triangles? This is an arrangement problem with one difference – the objects are not all distinct – one cannot distinguish the four triangles or the five squares. So one cannot use the earlier permutations rule that assumes the objects are distinguishable. How can one count the number of possible arrangements? It turns out that the combinations rule is useful here. (Surprising, but true.)

To think about possible arrangements, suppose one writes down a list of nine slots and an arrangement is constructed by placing the triangles and the squares in the nine slots. It is helpful to label the slots with the numbers 1 through 9.

1 2 3 4 5 6 7 8 9

One constructs an arrangement in two steps. First, place the four triangles in four slots, and then place the squares in the remaining slots. How many ways can one put the triangles in the slots? First note that one can specify a placement by the numbers of the slots that are used. For example, one could place the triangles in slots 1, 3, 4, and 8.

△ △ △ △

1 2 3 4 5 6 7 8 9

Or one could place the four triangles in slots 2, 5, 7, and 8.

△ △ △ △

1 2 3 4 5 6 7 8 9

One specifies an arrangement by choosing four locations from the slot locations $\{1, 2, 3, 4, 5, 6, 7, 8, 9\}$. How many ways can this be done? One knows that the number of ways of selecting four objects (here labels of locations) from a group of nine objects is

$$\binom{9}{4} = \frac{9!}{4!(9-4)!} = 126.$$

So there are 126 ways of choosing the four locations for the triangles. Once the triangles have been placed, one finishes the arrangement by putting in the squares. But there is only one way of doing this. For example, if one places triangles in slots 2, 5, 6, 7, then the squares must go in slots 1, 3, 4, 8, 9. So applying the multiplication rule, the number of ways of arranging four triangles and five squares is $126 \times 1 = 126$.

A new counting rule has been derived:

Permutations Rule for Non-Distinct Objects: The number of permutations of n non-distinct objects where r are of one type and $n - r$ are of a second type is

$$\binom{n}{r} = \frac{n!}{r!(n-r)!}. \tag{2.8}$$

Recall the question of interest: Suppose four triangles and five squares are randomly arranged. What is the chance that the first and last locations are occupied by triangles?

It has already been shown that there are 126 ways of mixing up four triangles and five squares. Each possible arrangement is equally likely and has a chance of $1/126$ of occurring.

To find the probability, one needs to count the number of ways of arranging the triangles and squares so that the first and last positions are filled with triangles.

△								△
1	2	3	4	5	6	7	8	9

If one places triangles in slots 1 and 9 (and there is only one way of doing that), then one is free to arrange the remaining two triangles and five squares in slots $\{2, 3, 4, 5, 6, 7, 8, 9\}$. By use of the new arrangements formula, the number of ways of doing this is

$$\binom{7}{2} = \frac{7!}{2!(7-2)!} = 21$$

and so the probability the first and last slots are filled with triangles is equal to $21/126$.

R Simulating Arrangements of Non-Distinct Objects

The function `permutation()` is again used to simulate a permutation in this non-distinct object case. A vector `objects` is defined containing three x's and two o's, and a single random permutation is generated by using `permutation()` with arguments `objects` and 5. The `replicate()` function is used to repeat this experiment 1000 times and the frequencies of the different arrangements are displayed. Here the total number of arrangements is $\binom{5}{2} = 10$ and as expected, each of the 10 possible arrangements occurs with approximately the same frequency.

```
permutation <- function(d, Size){
  str_flatten(sample(d, size=Size),
              collapse = " ")
}
objects <- c('x', 'x', 'x', 'o', 'o')
df <- data.frame(Iter = 1:1000,
              Arrangement = replicate(1000,
                 permutation(objects, 5)))
df %>% group_by(Arrangement) %>% count()
```

```
   Arrangement     n
   <fct>       <int>
 1 o o x x x      99
 2 o x o x x      99
 3 o x x o x     109
 4 o x x x o     121
 5 x o o x x      91
 6 x o x o x     109
 7 x o x x o      83
 8 x x o o x     102
 9 x x o x o      82
10 x x x o o     105
```

Which Rule to Use?

Three important counting rules have been described, the permutations rule for distinct objects, the combinations rule, and the permutations rule for non-distinct objects. How can one decide which rule to apply in a given problem? Here are some tips to help one find the right rule.

1. **Do We Care About Order?** If an outcome consists of a collection of objects, does the order in which one lists the objects matter? If order does matter, then a permutations rule may be appropriate. If the order of the objects does not matter, such as choosing a subset from a larger group, then a combinations rule is probably more suitable.

2. **Are the Objects Distinguishable?** There are two permutation rules, one that applies when all of the objects are distinguishable, and the second where there are two types of objects and one cannot distinguish between the objects of each type.
3. **When In Doubt?** If the first two tips do not seem helpful, it may benefit to start writing down a few outcomes in the sample space. When one looks at different outcomes, one should recognize if order is important and if the objects are distinguishable.

2.7 Playing Yahtzee

Yahtzee is a popular game played with five dice. The game is similar to the card game poker – in both games, one is trying to achieve desirable patterns in the dice faces or cards, and some types of patterns are similar in the two games. In this section, some of the dice patterns in the first roll in Yahtzee are described and the problem of determining the chances of several of the patterns are considered.

Outcomes of one roll of five dice

When a player rolls five dice in the game Yahtzee, the most valuable result is when all of the five dice show the same number such as

$$2, 2, 2, 2, 2.$$

This is called a "Yahtzee" and the player scores 50 points with this pattern. A second valuable pattern is a "four-of-a-kind' where you observe one number appearing four times, such as

$$3, 4, 3, 3, 3.$$

Table 2.1 gives all of the possible patterns when you roll five dice in Yahtzee. When one plays the game, some of these patterns are worth a particular number of points and these points are given in the right column.

Total number of outcomes

As in the case of two dice, it is useful to distinguish the five dice when one counts outcomes. One can represent an outcome by placing a value of individual die rolls (1 through 6) in the six slots.

____	____	____	____	____
die 1	die 2	die 3	die 4	die 5

TABLE 2.1
Possible patterns of rolling five dice in Yahtzee.

Pattern	Sample of pattern	Point value
Yahtzee	4, 4, 4, 4, 4	50
Four-of-a-kind	6, 6, 6, 4, 6	
Large straight	2, 6, 4, 5, 3	40
Small straight	4, 2, 1, 3, 2	30
Full house	5, 1, 1, 5, 1	25
Three-of-a-kind	2, 2, 3, 4, 2	
Two pair	6, 3, 3, 6, 2	
One pair	4, 3, 4, 1, 5	
Nothing	1, 3, 2, 5, 6	

So two possible outcomes are

$$2, 3, 4, 5, 5 \text{ and } 3, 2, 4, 5, 5.$$

Each die has 6 possibilities and so, applying the multiplication rule, the total number of outcomes in the rolls of five dice is

$$6 \times 6 \times 6 \times 6 \times 6 = 7776.$$

Since all of the outcomes are equally likely, we assign a probability of 1/7776 to each outcome.

Probability of a Yahtzee

One represents the Yahtzee roll as the outcome

$$x, x, x, x, x$$

where x denotes an arbitrary roll of one die. There are six possible choices for x, and so the number of possible Yahtzees is 6.

Since each outcome has probability 1/7776, the probability of a Yahtzee is

$$Prob(\text{Yahtzee}) = \frac{6}{7776}.$$

Probability of four-of-a-kind

In the pattern "four of a kind", one wants to have one number appear four times and a second number appear once. In other words, one is interested in counting outcomes of the form

$$x, x, x, x, y$$

where the four x's and the single y can be in different orders. To apply the multiplication rule, think of writing down a possible "four-of-a-kind" in three steps.

- Step 1: Choose the number for x (the number that appears four times).
- Step 2: Next choose the number for the singleton y.
- Step 3: Mix up the orders of the four x's and the one y.

For example, one chooses the outcome 5, 5, 5, 3, 5 by (1) choosing 5 to be the number that appears four times, (2) choosing 3 as the number that appears once, and then arranging the digits 5, 5, 5, 5, 3 to get 5, 5, 5, 3, 5.

Next the number of ways of doing each of the three steps is counted.

- Step 1: There are 6 ways of choosing x.
- Step 2: Once x has been chosen, there are 5 ways of choosing the value for y.
- Step 3: Last, once x and y have been selected, there are $\binom{5}{4} = 5$ ways of mixing up the x's and y's.

To find the number of four-of-a-kinds, one uses the multiplication rule using the number of ways of doing each of the three steps:

$$\text{Number of ways} = 6 \times 5 \times 5 = 150.$$

The corresponding probability of four-of-a-kind is

$$Prob(\text{four}-\text{of}-\text{a}-\text{kind}) = \frac{150}{7776}.$$

R Simulating Yahtzee

Some of the Yahtzee probabilities are conveniently approximated by simulation. In the following, the function `four_kind()` uses the `sample()` function to simulate the rolls of five dice. By tabulating the roll outcomes (using the `table()` function), one checks if a four-of-a-kind is observed – if so, the string "4 kind" is returned, otherwise a "nothing" is returned.

```
four_kind <- function(){
  rolls <- sample(6, size = 5, replace = TRUE)
  ifelse(max(table(rolls) == 4),
         "4 kind", "nothing")
}
```

This Yahtzee experiment is simulated 1000 times by use of the `replicate()` function. One sees below that one observed four-of-a-kind 20 times, so the approximated probability of four-of-a-kind is 16/1000 = 0.016. This agrees closely with the exact probability of 150/7776 = 0.0193.

```
df <- data.frame(Iter = 1:1000,
          Result = replicate(1000, four_kind()))
df %>% group_by(Result) %>% count()
  Result      n
  <fct>   <int>
1 4 kind     16
2 nothing   984
```

2.8 Exercises

1. **Constructing a Word**

   Suppose you select three letters at random from {a, b, c, d, e, f} to form a word.

   (a) How many possible words are there?
   (b) What is the probability the word you choose is "fad"?
   (c) What is the probability the word you choose contains the letter "a"?
   (d) What is the chance that the first letter in the word is "a"?
   (e) What is the probability that the word contains the letters "d", "e", and "f"?

2. **Running a Race**

   There are seven runners in a race – three runners are from Team A and four runners are from Team B.

   (a) Suppose you record which runners finish first, second, and third. Count the number of possible outcomes of this race.
   (b) If the runners all have the same ability, then each of the outcomes in (a) is equally likely. Find the probability that Team A runners finish first, second, and third.
   (c) Find the probability that the first runner across the finish line is from Team A.

3. **Rolling Dice**

   Suppose you roll three fair dice.

   (a) How many possible outcomes are there?
   (b) Find the probability you roll three sixes.
   (c) Find the probability that all three dice show the same number.
   (d) Find the probability that the sum of the dice is equal to 10.

4. **Ordering Hash Browns**

   When you order Waffle House's world famous hash browns, you can order them scattered (on the grill), smothered (with onions), chunked (with ham), topped (with chili), diced (with tomatoes), and peppered (with peppers). How many ways can you order 5 hash browns at Waffle House?

5. **Selecting Balls from a Box**

   A box contains 5 balls – 2 are white, 2 are black, and one is green. You choose two balls out of the box at random without replacement.

   (a) Write down all possible outcomes of this experiment. (Assume that the order in which you select the balls is important.)
   (b) Find the probability that you choose two white balls.
   (c) Find the probability you choose two balls of the same color.
   (d) Find the probability you choose a white ball second.

6. **Dividing into Teams**

   Suppose that ten boys are randomly divided into two teams of equal size. Find the probability that the three tallest boys are on the same team.

7. **Choosing Numbers**

   Suppose you choose three numbers from the set {1, 2, 3, 4, 5, 6, 7, 8} without replacement.

   (a) How many possible choices can you make?
   (b) What is the probability you choose exactly two even numbers?
   (c) What is the probability the three numbers add up to 10?

8. **Choosing People**

   Suppose you choose two people from three married couples.

   (a) How many selections can you make?
   (b) What is the probability the two people you choose are married to each other?
   (c) What is the probability that the two people are of the same gender?

9. **Football Plays**

   Suppose a football team has five basic plays, and they will randomly choose a play on each down.

   (a) On three downs, find the probability that the team runs the same play on each down.
   (b) Find the probability the team runs three different plays on the three downs.

10. **Playing the Lottery**

    In a lottery game, you make a random guess at the winning three-digit number (each digit can be 0, 1, 2, 3, 4, 5, 6, 7, 8, 9). You win \$200 if your guess matches the winning number, \$20 if your guess matches in exactly two positions and \$2 if your guess matches in exactly one position. Find the probabilities of winning \$200, winning \$20, and winning \$2.

11. **Dining at a Restaurant**

    Suppose you are dining at a Chinese restaurant with the menu given below. You decide to order a combination meal where you get to order one soup or appetizer, one entrée (seafood, beef, or poultry), and a side dish (either fried rice or noodles).

SOUP	**POULTRY**
HOT AND SOUR SOUP	KUNG PAO CHICKEN
WONTON SOUP	HUNAN CHICKEN
EGG DROP SOUP	CHICKEN WITH DOUBLE NUTS
APPETIZERS	CHICKEN WITH GARLIC SAUCE
EGG ROLL	CURRY CHICKEN
BARBECUED SPARERIBS	
FRIED CHICKEN STRIPS	**FRIED RICE**
BUTTERFLY SHRIMP	CHICKEN FRIED RICE
CRAB RANGOON	BEEF FRIED RICE
	SHRIMP FRIED RICE
SEAFOOD	PORK FRIED RICE
SHRIMP WITH GARLIC SAUCE	THREE DELIGHT FRIED RICE
CURRY SHRIMP	VEGETABLE FRIED RICE
KUNG PAO SCALLOPS	
FLOWER SHRIMP	**NOODLES/RICE**
SHRIMP WITH PEA PODS	PAN FRIED NOODLES
	MOO SHU PANCAKE
BEEF	CHOW MEIN NOODLES
KUNG PAO BEEF	STEAMED RICE
HUNAN BEEF	
SZECHUAN STYLE BEEF	
ORANGE BEEF (HOT & SPICY)	

    (a) How many possible combination meals can you order?
    (b) If you are able to go to this restaurant every day, approximately how many years could you dine there and order different combination meals?
    (c) Suppose that you are allergic to seafood (this includes crab, shrimp, and scallops). How many different combination meals can you order?

(d) Suppose your friend orders two different entrées completely at random. How many possible dinners can she order? What is the probability the two entrées chosen contain the same meat?

12. **Ordering Pizza**

    If you buy a pizza from Papa John's, you can you order the following toppings: ham, bacon, pepperoni, Italian sausage, sausage, beef, anchovies, extra cheese, baby portabella mushrooms, onions, black olives, Roma tomatoes, green peppers, jalapeño peppers, banana peppers, pineapple, grilled chicken.

    (a) If you have the option of choosing two toppings, how many different two topping pizzas can you order?
    (b) Suppose you want your two toppings to be some meat and some peppers. How many two-topping pizzas are of this type?
    (c) If you order a "random" two-topping pizza, what is the chance that it will have peppers?
    (d) If you are able to order at most four toppings, how many different pizzas can you order?

13. **Mixed Letters**

    You randomly mix up the letters "s", "t", "a", "t", "s".

    (a) Find the probability the arrangement spells the word "stats".
    (b) Find the probability the arrangement starts and ends with "s".

14. **Arranging CDs**

    Suppose you have three Taylor Swift CDs and three Lady Gaga CDs sitting on a shelf as follows. We assume that you can't distinguish the CDs of a given artist.

    $$T, T, T, L, L, L$$

    The CDs are knocked off of the shelf and you place them back on the shelf completely at random.

    (a) What is the probability that the mixed-up CDs remain in the same order?
    (b) What is the probability that the first and last CDs on the shelf are both Lady Gaga music?
    (c) What is the probability that the Jewel CDs stay together on the shelf?

15. **Playing a Lottery Game**

    The Minnesota State Lottery has a game called Daily 3. A three digit number is chosen randomly from the set 000, 001, ... , 999 and you win by guessing correctly certain characteristics of this three digit number. The lottery website lists the following possible plays such as First Digit, Front Pair, etc. Find the probability of winning for each play.

First Digit: Pick one number. To win, match the first number drawn.

Front Pair: Pick 2 numbers. To win, match the first 2 numbers drawn in exact order

Straight: Pick 3 numbers. To win, match all 3 numbers drawn in exact order.

3-Way Box: Pick 3 numbers, 2 that are the same. To win, match all three numbers drawn in any order.

6-Way Box: Pick 3 different numbers. To win, match all 3 numbers drawn in any order.

16. **Booking a Flight**

    Suppose you are booking a flight to San Francisco on Orbitz. To save money, you agree to either leave Monday, Tuesday, or Wednesday, and return on either Friday, Saturday, or Sunday. Assume that Orbitz randomly assigns you a day to leave and randomly assigns you a day to return.

    (a) What is the probability you leave on Tuesday and return on Saturday?
    (b) What is the chance that your trip will be exactly three days long?
    (c) What is the most likely trip length in days?
    (d) Do you think that the assumptions about Orbitz are reasonable? Explain.

17. **Assigning Grades**

    A math class of ten students takes an exam.

    (a) If the instructor decides to give exam grades of A to two randomly selected students, how many ways can this be done?
    (b) Of the remaining eight students, three will receive B's and the remaining will receive C's. How many ways can this be done?
    (c) If the instructor assigns at random, two A's, three B's and five C's to the ten students, how many ways can this be done?
    (d) Under this grading method, what is the probability that Jim (the best student in the class) gets an A?

18. **Choosing Officers**

    A club consisting of 8 members has to choose three officers.

    (a) How many ways can this be done?

(b) Suppose that the club needs to choose a president, a vice-president, and a treasurer. How many ways can this be done?

(c) If the club consists of 4 men and 4 women and the officers are chosen at random, find the probability the three officers are all of the same gender.

(d) Find the probability the president and the vice-president are different genders.

19. **Playing Yahtzee**

Find the number of ways and the corresponding probabilities of getting all of the following patterns in Yahtzee. Here are some hints for the different patterns.

Four of a kind: The pattern here is x, x, x, x, y, where x is the number that appears four times and y is the number that appears once.

Small straight: This roll will either include the numbers 1, 2, 3, 4, the numbers 2, 3, 4, 5, or the numbers 3, 4, 5, 6. If the numbers 1, 2, 3, 4 are the small straight, then the remaining number can not be 5 (otherwise it would be a large straight).

Full house: The pattern here is x, x, x, y, y, where x is the number that appears three times and y is the number that appears twice.

Three of a kind: The pattern here is x, x, x, y, z, where x is the number that appears three times, and y and z are the numbers that appear only once.

One pair: The pattern here is x, x, w, y z, where x is the number that appears two times, and w, y and z are the numbers that appear only once.

Nothing: This is the most difficult number to count directly. Once the number of each of the remaining patterns is found, then the number of "nothings" can be found by subtracting the total number of other patterns from the total number of rolls (7776).

R Exercises

20. **Sampling Letters**

The built-in vector `letters` contains the 26 lower-case letters of the alphabet.

(a) Using the `sample()` function, take a sample of 10 letters without replacement from `letters`.

(b) Using the `sample()` function, take a sample of 10 letters *with replacement* from `letters`.

21. **Sampling Letters (continued)**

    (a) Write a function to take a sample of 10 letters without replacement from `letters`.

    (b) Add a line in the function so that the function returns the number of vowels in the sample. (If the sample is stored in the vector `y`, then the line of code

    ```
    sum(y %in% c("a", "e", "i", "o", "u"))
    ```

    will count the number of vowels in the sample.)

    (c) Using the function `replicate()`, take 50 samples, storing the number of vowels from the samples in the vector `n_vowels`.

    (d) Approximate the probability that there are two vowels in your sample.

22. **Simulating Permutations**

    Suppose a license plate in a particular state consists of two letters followed by a number (for example, "CD9" and "EE0" are two possible license plates).

    (a) Write a function to simulate a random license plate.

    (b) Using the `replicate()` function to simulate 50 random license plates.

    (c) From the simulated plates, approximate the probability there is at least one vowel in the license plate.

23. **Simulating Yahtzee**

    (a) Write a function to roll five dice and record by a 1 or 0 if one observes a large straight.

    (b) Use the `replicate()` function and the function found in part (a) to approximate the probability of rolling a large straight.

    (c) By changing the function in part (a) and using the `replicate()` function, approximate the probability of rolling a small straight.

    (d) In a similar fashion, approximate the probability of rolling a full house.

3

Conditional Probability

3.1 Introduction: The Three Card Problem

Suppose one has three cards – one card is blue on both sides, one card is pink on both sides, and one card is blue on one side and pink on the other side. Suppose one chooses a card and places it down showing "blue". What is the chance that the other side is also blue?

This is an illustration of a famous conditional probability problem. One is given certain information – here the information is that one side of the card is blue – and one wishes to determine the probability that the other side is blue.

Most people think that this probability is 1/2, but actually this is wrong. The correct answer is demonstrated by simulating this experiment many times. One can do this simulation by hand, but we will illustrate this using an R script.

Suppose one thinks of this experiments are first choosing a card, and then choosing a side from the card. There are three possible cards, which we call "Blue", "Pink" and "mixed". For the blue card, there are two blue sides; for the pink card, there are two pink sides, and the "mixed" card has a blue side and a pink side.

R Conditional Probabilities by Simulation

We illustrate using R to perform this simulation. A data frame `df` with two variables `Card` and `Side` is defined. The `sample()` function randomly chooses a card and a side by choosing a random row from the data frame. This experiment was repeated 1000 times and the `table()` function is used to classify the outcomes by card and side.

```
df <- data.frame(Card = c("Blue", "Blue",
                          "Pink", "Pink",
                          "Mixed", "Mixed"),
                 Side = c("Blue", "Blue",
                          "Pink", "Pink",
                          "Blue", "Pink"))
cards <- df[sample(6, size = 1000, replace = TRUE), ]
table(cards$Card, cards$Side)
```

```
      Blue Pink
Blue   326    0
Mixed  173  152
Pink     0  349
```

One observed side is blue and we are interested in the probability of the event "card is blue". In this experiment, the blue side was observed 326 + 173 = 499 times – of these, the card was blue 326 times. So the probability the other side is blue is approximated by 326/499 which is close to 2/3. This example illustrates that one's intuition can be faulty in figuring out probabilities of the conditional type.

Selecting Slips of Paper

To illustrate the conditional nature of probabilities, suppose one has a box that has 6 slips of paper – the slips are labeled with the numbers 2, 4, 6, 8, 10, and 12. One selects two slips at random from the box. It is assumed that one is sampling without replacement and the order that one selects the slips is not important. Then one lists all of the possible outcomes. Note that since two numbers are chosen from six, the total number of outcomes will be ${}_6C_2 = 15$.

$$S = \{(2, 4), (2, 6), (2, 8), (2, 10), (2, 12), (4, 6), (4, 8), (4, 10), (4, 12)\ (6, 8), (6, 10), (6, 12), (8, 10), (8, 12), (10, 12)\}.$$

Suppose one is interested in the probability the sum of the numbers on the two slips is 14 or higher. Assuming that the 15 outcomes listed above are equally likely, one sees there are 9 outcomes where the sum is 14 or higher and so

$$Prob(\text{sum 14 or higher}) = \frac{9}{15}.$$

Next, suppose one is given some new information about this experiment – both of the numbers on the slips are single digits. Given this information, one now has only six possible outcomes. This new sample space is called the reduced sample space based on the new information.

$$S = \{(2, 4), (2, 6), (2, 8), (4, 6), (4, 8), (6, 8)\}$$

One evaluates the probability $Prob$(sum is 14 or higher) given that both of the slip numbers are single digits. Since there is only one way of obtaining a sum of 14 or higher in our new sample space, one sees

$$Prob(\text{sum 14 or higher}) = \frac{1}{6}.$$

Notation: Suppose that E is our event of interest and H is our new information. Then one writes the probability of E given the new information H as

$Prob(E \mid H)$, where the vertical line "|" means "conditional on" or given the new information. Here it was found

$$Prob(\text{sum is 14 or higher} \mid \text{both slip numbers are single digits}).$$

How does the probability of "14 or higher" change given the new information? Initially, the probability of 14 and higher was pretty high (9/15), but given the new information, the probability dropped to 1/6. Does this make sense? Yes. If one is told that both numbers are single digits, then one has drawn small numbers and that would tend to make the sum of the digits small.

Independent Events

One says that events A and B are independent if the knowledge of event A does not change the probability of the event B. Using symbols

$$P(B \mid A) = P(B). \tag{3.1}$$

Rolls of Two Dice

To illustrate the concept of independence, consider an example where one rolls a red die and a white die. Consider the following three events:

- S = the sum of the two rolls is 7
- E = the red die is an even number
- D = the rolls of the two dice are different

Are events S and E independent?

1. First one finds the probability one rolls a sum equal to 7, that is, $P(S)$. There are 36 outcomes and 6 outcomes results in a sum of 7, so $P(S) = 6/36$.
2. Next, one finds $P(S \mid E)$. Given that the red die is an even number (event E), note that there are 18 outcomes where E occurs. Of these 18 outcomes, there are 3 outcomes where the sum is equal to 7. So $P(S \mid E) = 3/18$.
3. Note $P(S \mid E) = P(S)$, so events S and E are independent. Knowing the red die is even does not change one's probability of rolling a 7.

Are events S and D independent?

To see if these two events are independent, one computes $P(S \mid D)$ and checks if $P(S \mid D) = P(S)$. One can show that $P(S \mid D) = 6/30$. This probability is not equal to $P(S)$ so S and D are not independent events.

R Conditional Probabilities by Simulation

One can demonstrate conditional probability by the use of the `filter()` function in the `dplyr` package. To illustrate, a data frame `df` is constructed with simulated rolls of two dice – the associated variables are `Roll_1` and `Roll_2`.

```
df <- data.frame(Roll_1 = sample(6, size = 1000,
                                 replace = TRUE),
                 Roll_2 = sample(6, size = 1000,
                                 replace = TRUE))
```

The `mutate()` function is used to define a new variable `Sum` that is the sum of the two rolls. Suppose one is told that the roll of the first die is greater than 3 – how does that information change the probabilities for `Sum`? In the following script, the `filter()` function is used to restrict die rolls to only the ones where `Roll_1 > 3`. Then the frequencies and corresponding approximate probabilities of different sums are found on these "restricted" die rolls. For example, one sees that the probability $Prob(\text{Sum} = 10 | \text{Roll\_1} > 3) \approx 0.164$.

```
df %>%
  mutate(Sum = Roll_1 + Roll_2) %>%
  filter(Roll_1 > 3) %>%
  group_by(Sum) %>%
  summarize(Count = n()) %>%
  mutate(Probability = Count / sum(Count))

    Sum Count Probability
  <int> <int>       <dbl>
1     5    20      0.0405
2     6    55      0.111
3     7    85      0.172
4     8    78      0.158
5     9    89      0.180
6    10    81      0.164
7    11    58      0.117
8    12    28      0.0567
```

3.2 In Everyday Life

Generally one's beliefs about uncertain events can change when new information is obtained. Conditional probability provides a way to precisely say how one's beliefs change. Let's illustrate this with a simple example.

Suppose one is interested in estimating the population of Philadelphia, Pennsylvania in the current year. Consider three possible events:

- A = Philadelphia's population is under one million
- B = Philadelphia's population is between one and two million
- C = Philadelphia's population is over two million

If one knows little about Philadelphia, then one probably is not very knowledgeable about its population. So initially the probabilities are assigned shown in Table 3.1.

TABLE 3.1
Probabilities of events about Philadelphia's population, $P(Event \mid I)$.

Event	$P(Event \mid I)$
under one million	0.3
between one and two million	0.3
over two million	0.4
TOTAL	1.0

One is assigning approximately the same probability to each of the three events, indicating that they are all equally likely in his or her mind. These can be viewed as conditional probabilities since they are conditional on one's initial information – these probabilities are denoted by $P(Event \mid I)$, where I denotes one's initial information. Now suppose some new information is provided about Philadelphia's population. One is not told the current population, but is told that in 1990, Philadelphia was the fifth largest city in the country, and the population of the sixth largest city, San Diego, was 1.1 million in 1990. So this tells one that in 1990, the population of Philadelphia had to exceed 1.1 million. Now one might not be sure about how the population of Philadelphia has changed between 1990 and 2020, but it probably has not changed a significant amount. So one thinks that

- The population of Philadelphia is most likely to be between 1 and 2 million.
- It is very unlikely that Philadelphia's population is over 2 million.
- There is a small chance that Philadelphia's population is under 1 million.

One revises his or her probabilities that reflect these beliefs as shown in Table 3.2. These probabilities are denoted as $P(Event \mid N)$, which are probabilities of these population events conditional on the newer information N, in Table 3.2.

Now, additional information is provided. To find the current population of Philadelphia, one looks up the census estimated figures and the population of

TABLE 3.2
Probabilities of events about Philadelphia's population, $P(Event \mid N)$.

Event	$P(Event \mid N)$
under one million	0.2
between one and two million	0.78
over two million	0.02
TOTAL	1.0

Philadelphia's population was reported to be 1,567,872 in 2016. Even though the census number is a few years old, one doesn't think that the population has changed much – definitely not enough to put in a new category of the table. So one's probabilities will change again as shown in Table 3.3. We call these probabilities of events conditional on additional information A.

TABLE 3.3
Probabilities of events about Philadelphia's population, $P(Event \mid A)$.

Event	$P(Event \mid A)$
under one million	0
between one and two million	1
over two million	0
TOTAL	1.0

All of us actually make many judgments every day based on uncertainty. For example, we make decisions about the weather based on information such as the weather report, how it looks outside, and advice from friends. We make decisions about who we think will win a sports event based on what we read in the paper, our knowledge of the teams' strengths, and discussion with friends. Conditional probability is simply a way of quantifying our beliefs about uncertain events given information.

3.3 In a Two-Way Table

It can be easier to think about, and compute conditional probabilities when they are found from observed counts in a two-way table.

In Table 3.4, high school athletes in 14 sports are classified with respect to their sport and their gender. These numbers are recorded in thousands, so the 454 entry in the Baseball/Softball – Male cell means that 454,000 males played baseball or softball this year.

TABLE 3.4
Counts of high school athletes, by sport and gender.

	Male	Female	TOTAL
Baseball/Softball	454	373	827
Basketball	541	456	997
Cross Country	192	163	355
Football	1048	1	1049
Gymnastics	2	21	23
Golf	163	62	225
Ice Hockey	35	7	42
Lacrosse	50	39	89
Soccer	345	301	646
Swimming	95	141	236
Tennis	145	163	308
Track and Field	550	462	1012
Volleyball	39	397	436
Wrestling	240	4	244
TOTAL	3899	2590	6489

Suppose one chooses a high school athlete at random who is involved in one of these 14 sports. Consider several events

- F = athlete chosen is female
- S = athlete is a swimmer
- V = athlete plays volleyball

What is the probability that the athlete is female? Of the 6489 (thousand) athletes, 2590 were female, so the probability is

$$P(F) = 2590/6489 = 0.3991$$

Likewise, the probability that the randomly chosen athlete is a swimmer is

$$P(S) = 236/6489 = 0.0364.$$

and the probability he or she plays volleyball is

$$P(V) = 436/6489 = 0.0672.$$

Next, consider the computation of some conditional probabilities. What is the probability a volleyball player is female? In other words, conditional on the fact that the athlete plays volleyball, what is the chance that the athlete is female:

$$P(F \mid V).$$

To find this probability, restrict attention only to the volleyball players in the table.

	Male	Female	TOTAL
Volleyball	39	397	436

Of the 436 (thousand) volleyball players, 397 are female, so

$$P(F \mid V) = 397/436 = 0.9106.$$

What is the probability a woman athlete is a swimmer? In other words, if one knows that the athlete is female, what is the (conditional) probability she is a swimmer, or $P(S \mid F)$?

Here since one is given the information that the athlete is female, one restricts attention to the "Female" column of counts. There are a total of 2590 (thousand) women who play one of these sports; of these, 141 are swimmers. So

$$P(S \mid F) = 141/2590 = 0.0544.$$

Are events F and V independent? One can check this several ways. Above it was found that the probability a randomly chosen athlete is a volleyball player is $P(V) = 0.0672$. Suppose one is told that the athlete is a female (F). Will that change the probability that she is a volleyball player? Of the 2590 women, 397 are volleyball players, and so $P(V \mid F) = 397/2590 = 0.1533$, Note that $P(V)$ is different from $P(V \mid F)$, that means that the knowledge the athlete is female has increased one's probability that the athlete is a volleyball player. So the two events are not independent.

R Conditional Probabilities in a Two-Way Table

Suppose one has two spinners, each that will record a 1, 2, 3, or 4 with equal probabilities. Suppose the smaller of the two spins is 2 – what is the probability that the larger spin is equal to 4? One can answer this question by use of a simulation experiment. First one constructs a data frame – by two uses of the `sample()` function, 1000 random spins of the first spinner are stored in `Spin_1` and 1000 spins of the second spinner in `Spin_2`.

```
df <- data.frame(Spin_1 = sample(4, size = 1000,
                                 replace = TRUE),
                 Spin_2 = sample(4, size = 1000,
                                 replace = TRUE))
```

By use of the `mutate()` function, one computes the smaller and larger of the two spins and stores the result in the respective variables `Min` and `Max`. Then one finds a frequency table of the simulated values of `Min` and `Max`.

```
df %>%
  mutate(Min = pmin(Spin_1, Spin_2),
         Max = pmax(Spin_1, Spin_2)) %>%
  group_by(Min, Max) %>%
  summarize(n = n()) %>%
  spread(Max, n)
```

```
    Min   `1`   `2`   `3`   `4`
  <int> <int> <int> <int> <int>
1     1    58   127   119   129
2     2    NA    67   127   123
3     3    NA    NA    63   122
4     4    NA    NA    NA    65
```

Since one is told that the smaller of the two spins is equal to 2, one restricts attention to the row where `Min = 2`. One observes that `Max` is equal to 2, 3, 4 with frequencies 67, 127, and 123. So

$$P(\text{Max spin} = 4 \mid \text{Min spin} = 2) = \frac{123}{67 + 127 + 123} = 0.388.$$

3.4 Definition and the Multiplication Rule

In this chapter, conditional probabilities have been computed by considering a reduced sample space. There is a formal definition of conditional probability that is useful in computing probabilities of complicated events.

Suppose one has two events A and B where the probability of event B is positive, that is $P(B) > 0$. Then the probability of A given B is defined as the quotient

$$P(A \mid B) = \frac{P(A \cap B)}{P(B)}. \tag{3.2}$$

How many boys?

To illustrate this conditional probability definition, suppose a couple has four children. One is told that this couple has at least one boy. What is the chance that they have two boys?

If one lets L be the event "at least one boy" and B be the event "have two boys", one wishes to find $P(B \mid L)$.

Suppose one represents the genders of the four children (from youngest to oldest) as a sequence of four letters. For example, the sequence $BBGG$ means that the first two children were boys and the last two were girls. If we represent outcomes this way, there are 16 possible outcomes of four births:

BBBB	BGBB	GBBB	GGBB
BBBG	BGBG	GBBG	GGBG
BBGB	BGGB	GBGB	GGGB
BBGG	BGGG	GBGG	GGGG

If one assumes that boys and girls are equally likely (is this really true?), then each of the outcomes is equally likely and each outcome is assigned a probability of 1/16. Applying the definition of conditional probability, one has

$$P(B \mid L) = \frac{P(B \cap L)}{P(L)}.$$

There are 15 outcomes in the set L, and 6 outcomes where both events B and L occur. So using the definition

$$P(B \mid L) = \frac{6/16}{15/16} = \frac{6}{15}.$$

The Multiplication Rule

If one takes the conditional probability definition and multiplies both sides of the equation by $P(B)$, one obtains the multiplication rule

$$P(A \cap B) = P(B)P(A \mid B). \tag{3.3}$$

Choosing balls from a random bowl

The multiplication rule is especially useful for experiments that can be divided into stages. Suppose one has two bowls – Bowl 1 is filled with one white and 5 black balls, and Bowl 2 has 4 white and 2 black balls. One first spins the spinner below that determines which bowl to select, and then selects one ball from the bowl. What is the chance the ball one selects is white?

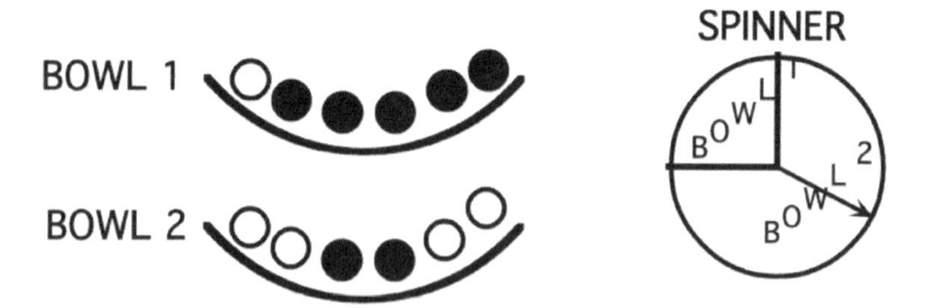

One can demonstrate the multiplication rule by the tree diagram in Figure 3.1. The first set of branches corresponds to the spinner result (choose Bowl 1 or choose Bowl 2) and the second set of branches corresponds to the ball selection.

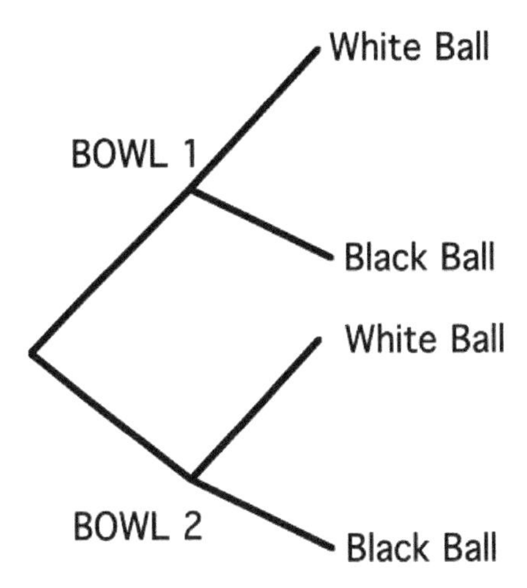

FIGURE 3.1
Tree diagram of choosing balls from a random bowl, part 1.

One places numbers on the diagram corresponding to the probabilities that are given in the problem, shown in Figure 3.2. Since one quarter of the spinner region is "Bowl 1", the chance of choosing Bowl 1 is $1/4$ and so the chance of choosing Bowl 2 is $3/4$ – these probabilities are placed at the first set of branches. Also one knows that if Bowl 1 is selected, the chances of choosing a white ball and a black ball are respectively $1/6$ and $5/6$. These conditional probabilities, $P(\text{white} \mid \text{Bowl } 1)$ and $P(\text{black} \mid \text{Bowl } 2)$, are placed at the top set of branches at the second level. Also, if one selects Bowl 2, the conditional probabilities of selecting a white ball and a black ball are given by $P(\text{white} \mid \text{Bowl } 2) = 4/6$ and $P(\text{black} \mid \text{Bowl } 2) = 2/6$ – these probabilities are placed at the bottom set of branches.

Now that the probabilities are assigned on the tree, one uses the multiplication rule to compute the probabilities of interest:

- What is the probability of selecting Bowl 1 and selecting a white ball? By the multiplication rule

$$\begin{aligned} P(\text{Bowl}\,1 \cap \text{white ball}) &= P(\text{Bowl}\,1)P(\text{white ball} \mid \text{Bowl}\,1) \\ &= \frac{1}{4} \times \frac{1}{6} = \frac{1}{24}. \end{aligned}$$

  One is just multiplying probabilities along the top branch of the tree.

- What is the probability of selecting a white ball? One sees from the tree that there are two ways of selecting a white depending on which bowl is selected. One can either (1) select Bowl 1 and choose a white ball or (2)

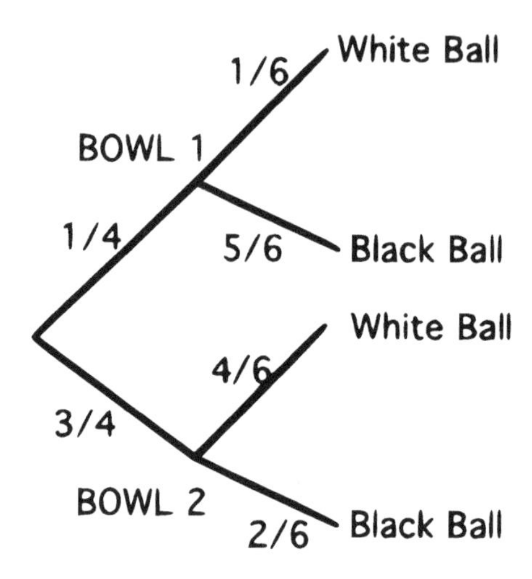

FIGURE 3.2
Tree diagram of choosing balls from a random bowl, part 2.

select Bowl 2 and choose a white ball. One finds the probability of each of the two outcomes and add the probabilities to get the answer.

$$
\begin{aligned}
P(\text{white ball}) &= P(\text{Bowl 1} \cap \text{white ball}) + P(\text{Bowl 2} \cap \text{white ball}) \\
&= \frac{1}{4} \times \frac{1}{6} + \frac{3}{4} \times \frac{4}{6} = \frac{13}{24}.
\end{aligned}
$$

Simulating choosing balls from a random bowl

One simulate this balls and bowl experiment on R. Using the `sample()` function, one simulates 1000 choices of the bowl where the probabilities of choosing Bowl 1 and Bowl 2 are 1/4 and 3/4 and places the bowl choices in variable `Bowl`. In a similar fashion, one simulates 1000 ball selections from Bowl 1 (variable `Color_1`) and 1000 selections from Bowl 2 (variable `Color_2`). Last, by use of a single `ifelse()` function, one lets the ball color be equal to `Color_1` if Bowl 1 is selection, or `Color_2` if Bowl 2 is selected.

```
Bowl <- sample(1:2, size = 1000, replace = TRUE,
               prob = c(1, 3) / 4)
Color_1 <- sample(c("white", "black"), size = 1000,
                  replace = TRUE,
                  prob = c(1, 5) / 6)
Color_2 <- sample(c("white", "black"), size = 1000,
                  replace = TRUE,
                  prob = c(4, 2) / 6)
```

```
Color <- ifelse(Bowl == 1, Color_1, Color_2)
```

By use of the `table()` function, one categorizes all simulations by the values of `Bowl` and `Color`.

```
table(Bowl, Color)
```

```
    Color
Bowl black white
   1   197    41
   2   265   497
```

The probability that Bowl 1 was selected and a white ball was chosen is approximately equal to $41/1000 = 0.41$. The chance of choosing a white ball is approximated by $(41+497)/1000 = 0.538$.

3.5 The Multiplication Rule under Independence

When two events A and B are independent, then the multiplication rule takes the simple form

$$P(A \cap B) = P(A) \times P(B). \tag{3.4}$$

Moreover, if one has a sequence of independent events, say $A_1, A_2, \cdots, A_k$, then the probability that all events happen simultaneously is the product of the probabilities of the individual events

$$P(A_1 \cap A_2 \cap \cdots \cap A_k) = P(A_1) \times P(A_2) \times \cdots \times P(A_k). \tag{3.5}$$

By use of the assumption of independent events and multiplying, one finds probabilities of sophisticated events. We illustrate this in several examples.

Blood Types of Couples

Americans have the blood types O, A, B, and AB with respectively proportions 0.45, 0.40, 0.11, and 0.04. Suppose two people in this group are married.

1. **What is the probability that the man has blood type O and the woman has blood type A?** Let O_M denote the event that the man has O blood type and A_W the event that the woman has A blood type. Since these two people are not related, it is reasonable to assume that O_M and A_W are independent events. Applying

the multiplication rule, the probability the couple have these two specific blood types is

$$\begin{aligned} P(O_M \cap A_W) &= P(O_M) \times P(A_W) \\ &= (0.45) \times (0.40) = 0.18. \end{aligned}$$

2. **What is the probability the couple have O and A blood types?** This is a different question from the first since we have no indication which person has which blood type. Either the man has blood type O and the woman has blood type A, or the other way around. So the probability of interest is

$$\begin{aligned} P(\text{two have A, O types}) &= P((O_M \cap A_W) \cup (O_W \cap A_M)) \\ &= P(O_M \cap A_W) + P(O_W \cap A_M). \end{aligned}$$

One adds the probabilities since $O_M \cap A_W$ and $O_W \cap A_M$ are different outcomes. One uses the multiplication rule with the independence assumption to find the probability:

$$\begin{aligned} P(\text{two have A, O types}) &= P((O_M \cap A_W) \cup (O_W \cap A_M)) \\ &= P(O_M \cap A_W) + P(O_W \cap A_M) \\ &= P(O_M) \times P(A_W) + P(O_W) \times P(A_M) \\ &= (0.45) \times (0.40) + (0.45) \times (0.40) \\ &= 0.36. \end{aligned}$$

3. **What is the probability the man and the woman have the same blood type?** This is a more general question than the earlier parts since one hasn't specified the blood types – one is just interested in the event that the two people have the same type. There are four possible ways for this to happen: they can both have type O, they both have type A, they have type B, or they have type AB. One first finds the probability of each possible outcome and then sums the outcome probabilities to obtain the probability of interest. One obtains

$$\begin{aligned} P(\text{same type}) = &\; P((O_M \cap O_W) \cup (A_M \cap A_W) \cup \\ &(B_M \cap B_W) \cup (AB_W \cap AB_M)) \\ = &\; (0.45)^2 + (0.40)^2 + (0.11)^2 + (0.04)^2 \\ = &\; 0.3762. \end{aligned}$$

4. **What is the probability the couple have different blood types?** One way of doing this problem is to consider all of the

ways to have different blood types – the two people could have blood types O and A, types O and B, and so on, and add the probabilities of the different outcomes. But it is simpler to note that the event "having different blood types" is the complement of the event "have the same blood type". Then using the complement property of probability,

$$\begin{aligned} P(\text{different type}) &= 1 - P(\text{same type}) \\ &= 1 - 0.3762 \\ &= 0.6238. \end{aligned}$$

A Five-Game Playoff

Suppose two baseball teams play in a "best of five" playoff series, where the first team to win three games wins the series. Suppose the Yankees play the Angels and one believes that the probability the Yankees will win a single game is 0.6. If the results of the games are assumed independent, what is the probability the Yankees win the series?

This is a more sophisticated problem than the first example, since there are numerous outcomes of this series of games. The first thing to note is that the playoff can last three games, four games, or five games. In listing outcomes, one lets Y and A denote respectively the single-game outcomes "Yankees win" and "Angels win". Then a series result is represented by a sequence of letters. For example, $YYAY$ means that the Yankees won the first two games, the Angels won the third game, and the Yankees won the fourth game and the series. Using this notation, all of the possible outcomes of the five-game series are written below.

Three games	Four games	Five games
$\underline{YYY}$	$\underline{YYAY}$, $AAYA$	$\underline{YYAAY}$, $AAYYA$
AAA	$\underline{YAYY}$, $AYAA$	$\underline{YAYAY}$, $AYAYA$
	$\underline{AYYY}$, $YAAA$	$\underline{YAAYY}$, $AYYAA$
		$\underline{AYYAY}$, $YAAYA$
		$\underline{AYAYY}$, $YAYAA$
		$\underline{AAYYY}$, $YYAAA$

One is interested in the probability the Yankees win the series. All of the outcomes above where the Yankees win are underlined. By the assumption of independence, one finds the probability of a specific outcome – for example, the probability of the outcome $YYAY$ as

$$\begin{aligned} P(YYAY) &= (0.6) \times (0.6) \times (0.4) \times (0.6) \\ &= 0.0864. \end{aligned}$$

One finds the probability that the Yankees win the series by finding the probabilities of each type of Yankees win and adding the outcome probabilities. The probability of each outcome is written down in Table 3.5.

TABLE 3.5
Table of probabilities of all Yankees winning outcomes.

Three games	Four games	Five games
$P(YYY) = 0.216$	$P(YYAY) = 0.0864$	$P(YYAAY) = 0.0346$
	$P(YAYY) = 0.0864$	$P(YAYAY) = 0.0346$
	$P(AYYY) = .0864$	$P(YAAYY) = 0.0346$
		$P(AYYAY) = 0.0346$
		$P(AYAYY) = 0.0346$
		$P(AAYYY) = 0.0346$

So the probability of interest is given by

$$\begin{aligned} P(\text{Yankees win series}) &= P(YYY, YYAY, YAYY, ...) \\ &= 0.216 + 3 \times 0.864 + 6 \times 0.0346 \\ &= 0.683. \end{aligned}$$

Playing Craps

One of the most popular casino games is craps. Here we describe a basic version of the game, and we will use the multiplication rule together with the use of conditional probabilities to find the probability of winnings.

This game is based on the roll of two dice. One begins by rolling the dice: if the sum of the dice is 7 or 11, the player wins, and if the sum is 2, 3, or 12, the player loses. If any other sum of dice is rolled (that is, 4, 5, 6, 8, 9, 10), this sum is called the "point". The player continues rolling two dice until either his point or a 7 are observed – he wins if he sees his point and loses if he observes a 7. What is the probability of winning at this game?

(a) On the first roll, the player can win by rolling the sum of 7 or 11, or lose by rolling the sum of 2, 3, or 12. The probabilities of these five outcomes are placed in Table 3.6.

(b) If the player rolls initially a sum 4, 5, 6, 8, 9 or 10, he keeps rolling. The probabilities of rolling these sums (of two dice) are placed in the P(Roll) column of Table 3.7.

(c) Suppose the player initially rolls 4 and this becomes his or her point. Now the player keep rolling until the point of 4 (player wins) or a 7 (player loses) are observed. All of the other sums of two dice are not important. In this case, there are only the following nine possible outcomes.

TABLE 3.6
Probabilities of outcomes with first roll of sum of 7, 11, 2, 3, or 12.

First roll	Probability	Outcome
7	6/36	Win
11	2/36	Win
2	1/36	Lose
3	2/36	Lose
12	1/36	Lose

TABLE 3.7
Probabilities of outcomes with first roll of sum of 4, 5, 6, 8, 9 or 10, part 1.

First Roll	P(Roll)	Second Roll	Outcome	P(Win | Roll)
4	3/36			
4	3/36			
5	4/36			
5	4/36			
6	5/36			
6	5/36			
8	5/36			
8	5/36			
9	4/36			
9	4/36			
10	3/36			
10	3/36			

$$(1, 3), (1, 6), (2, 2), (2, 5), (3, 1), (3, 4), (4, 3), (5, 2), (6, 1)$$

Of these nine outcomes, the player wins (point of 4) in three of them – so the conditional probability $P(\text{Win} \mid \text{First Roll is } 4) = 3/9$. This value is placed in the $P(\text{Win} \mid \text{Roll})$ column. Using a similar method, one computes $P(\text{Win} \mid \text{First Roll})$ if the first roll is 5, if the first roll is 6, ..., the first roll is 10. The secondary roll, the outcome (Win or Lose), and conditional win probabilities are placed in the $P(\text{Win} \mid \text{Roll})$ column in Table 3.8.

(d) Using the multiplication rule, the probability of rolling a 4 first and then winning is given by

$$P(\text{Roll} = 4 \cap \text{Win}) = P(\text{Roll} = 4)\ P(\text{Win} \mid \text{Roll} = 4).$$

Using a similar calculation, the probabilities $P(\text{Roll} = 5 \cap \text{Win})$, $P(\text{Roll} = 6 \cap \text{Win})$, $P(\text{Roll} = 8 \cap \text{Win})$, $P(\text{Roll} = 9 \cap \text{Win})$, P(Roll

TABLE 3.8
Probabilities of outcomes with first roll of sum of 4, 5, 6, 8, 9 or 10, part 2.

First Roll	P(Roll)	Second Roll	Outcome	P(Win | Roll)
4	3/36	4	Win	3/9
4	3/36	7	Lose	
5	4/36	5	Win	4/10
5	4/36	7	Lose	
6	5/36	6	Win	5/11
6	5/36	7	Lose	
8	5/36	8	Win	5/11
8	5/36	7	Lose	
9	4/36	9	Win	4/10
9	4/36	7	Lose	
10	3/36	10	Win	3/9
10	3/36	7	Lose	

$= 10 \cap$ Win) are found by multiplying entries in the P(Roll) and P (Win | Roll) columns of Table 3.8.

(e) The probability the player wins at craps is the following sum

$$\begin{aligned}
P(\text{Win}) = & P(\text{Roll} = 7) + P(\text{Roll} = 11) + P(\text{Roll} = 4 \cap \text{Win}) \\
& + P(\text{Roll} = 5 \cap \text{Win}) + P(\text{Roll} = 6 \cap \text{Win}) \\
& + P(\text{Roll} = 8 \cap \text{Win}) + P(\text{Roll} = 9 \cap \text{Win}) \\
& + P(\text{Roll} = 10 \cap \text{Win}) \\
& = \frac{6}{36} + \frac{2}{36} + \left(\frac{3}{36}\right)\left(\frac{3}{9}\right) + \left(\frac{4}{36}\right)\left(\frac{4}{10}\right) + \\
& \left(\frac{5}{36}\right)\left(\frac{5}{11}\right) + \left(\frac{5}{36}\right)\left(\frac{5}{11}\right) + \left(\frac{4}{36}\right)\left(\frac{4}{10}\right) + \\
& \left(\frac{3}{36}\right)\left(\frac{3}{9}\right) \\
& = 0.493.
\end{aligned}$$

Is craps a fair game? In other words, who has the advantage in this game: the player or the casino? Since the probability of the player winning at craps is 0.493, it is not a fair game. But the advantage to the casino is relatively small.

3.6 Learning Using Bayes' Rule

We have seen that probabilities are conditional in that one's opinion about an event is dependent on our current state of knowledge. As we gain new information, our probabilities can change. Bayes' rule provides a mechanism for changing our probabilities when we obtain new data.

Suppose that you are given a blood test for a rare disease. The proportion of people who currently have this disease is 0.1. The blood test comes back with two results: positive, which is some indication that you may have the disease, or negative. It is possible that the test will give the wrong result. If you have the disease, it will give a negative reading with probability 0.2. Likewise, it will give a false positive result with probability 0.2. Suppose that you have a blood test and the result is positive. Should you be concerned that you have the disease?

In this example, you are uncertain if you have the rare disease. There are two possible alternatives: you have the disease, or you don't have the disease. Before you have a blood test, you assign probabilities to "have disease" and "don't have disease" that reflect the plausibility of these two models. You think that your chance of having the disease is similar to the chance of a randomly selected person from the population. Thus you assign the event "have disease" a probability of 0.1 By the complement property, this implies that the event "don't have disease" has a probability of 1- 0.1 = 0.9.

The new information that one obtains to learn about the different models is called data. In this example, the data is the result of the blood test. Here the two possible data results are a positive result $(+)$ or a negative result $(-)$. One is given the probabilities of the observations for each model. If one "has the disease," the probability of a $+$ observation is 0.8 and the probability of a $-$ observation is 0.2. Since these are conditional probabilities, one writes

$$P(+ \mid \text{disease}) = 0.8, \;\; P(- \mid \text{disease}) = 0.2.$$

Likewise, if the result is "don't have the disease", the probabilities of the outcomes are 0.2 and 0.8, respectively. Using symbols, one has

$$P(+ \mid \text{no disease}) = 0.2, \;\; P(- \mid \text{no disease}) = 0.8.$$

Suppose you take the blood test and the result is positive $(+)$ – what is the chance you really have the disease? We are interested in computing the conditional probability

$$P(\text{disease} \mid +).$$

This should not be confused with the earlier probability $P(+ \mid \text{disease})$ that is the probability of getting a positive result if you have the disease. Here the focus is on the so-called inverse probability – the probability of having the disease given a positive blood test result.

We describe the computation of this inverse probability using two methods. They are essentially two ways of viewing the same calculation.

Method 1: Using a tree diagram

A person either has or does not have the disease, and given the person's disease state, he or she either gets a positive or negative test result. One represents the outcomes by a tree diagram where the first set of branches corresponds to the disease states and the second set of branches corresponds to the blood test results. The branches of the tree are labelled by the given probabilities, shown in Figure 3.3.

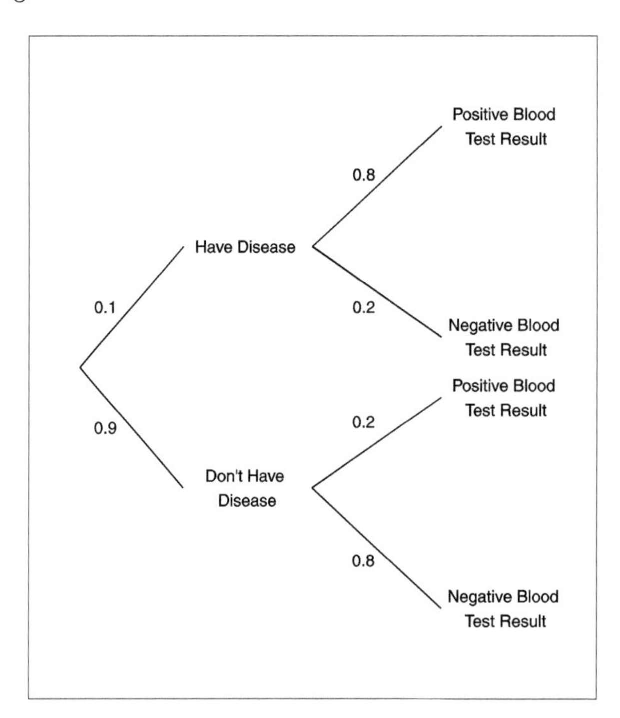

FIGURE 3.3
Tree diagram of the disease problem.

By the definition of conditional probability,

$$P(\text{disease} \mid +) = \frac{P(\text{disease} \cap +)}{P(+)}.$$

One finds the numerator $P(\text{disease} \cap +)$ by use of the multiplication rule:

$$\begin{aligned} P(\text{disease} \cap +) &= P(\text{disease})P(+ \mid \text{disease}) \\ &= 0.1 \times 0.8 = 0.08. \end{aligned}$$

In the tree diagram, one is multiplying probabilities along the disease/+ branch to find this probability.

To find the denominator $P(+)$, note that there are two ways of getting a positive blood test result – either the person has the disease and gets a positive blood test result, or the person doesn't have the disease and gets a positive result. These two outcomes are the disease/+ and no disease/+ branches of the tree. One finds the probability by using the multiplication rule to find the probability of each outcome, and then summing the outcome probabilities:

$$\begin{aligned} P(+) &= P(\text{disease} \cap +) + P(\text{no disease} \cap +) \\ &= P(\text{disease})P(+ \mid \text{disease}) + P(\text{no disease})P(+ \mid \text{no disease}) \\ &= 0.1 \times 0.8 + 0.9 \times 0.2 \\ &= 0.26. \end{aligned}$$

So the probability of having the disease, given a positive blood test result is

$$P(\text{disease} \mid +) = \frac{P(\text{disease} \cap +)}{P(+)} = \frac{0.08}{0.26} = 0.31.$$

As one would expect, the new probability of having the disease (0.31) is larger than the initial probability of having the disease (0.1) since a positive blood test was observed.

Method 2: Using a Bayes' box

There is an alternative way of computing the inverse probability based on a two-way table that classifies people by the disease status and the blood test result. This is an attractive method since it based on expected counts rather than probabilities.

Suppose there are 1000 people in the community – one places "1000' in the lower right corner of Table 3.9.

TABLE 3.9
Bayes' box procedure, step 1.

		Blood test result		
		+	−	TOTAL
Disease	Have disease			
status	Don't have disease			
TOTAL				1000

One knows that the chance of getting the disease is 10% – so one expects 10% of the 1000 = 100 people to have the disease and the remaining 900 people to be disease-free. One places these numbers in the right column corresponding to "Disease status", in Table 3.10.

TABLE 3.10
Bayes' box procedure, step 2.

		Blood test result		
		+	−	TOTAL
Disease	Have disease			100
status	Don't have disease			900
TOTAL				1000

One knows the test will err with probability 0.2. So if 100 people have the disease, one expects 20% of 100 = 20 to have a negative test result and 80 will have a positive result – one places these counts in the first row of the table. Likewise, if 900 people are disease-free, then 20% of 900 = 180 will have an incorrect positive result and the remaining 720 will have a negative result – one places these in the second row of Table 3.11.

TABLE 3.11
Bayes' box procedure, step 3.

		Blood test result		
		+	−	TOTAL
Disease	Have disease	80	20	100
status	Don't have disease	180	720	900
TOTAL				1000

Now one is ready to compute the probability of interest $P(\text{disease} \mid +)$ from the table of counts. Since one is conditioning on the event +, one restricts attention to the + column of the table – 260 people had positive test result. Of these 260, 80 actually had the disease, so

$$P(\text{disease} \mid +) = \frac{80}{260} = 0.31.$$

Note that, as expected, one obtains the same answer for the inverse probability.

3.7 R Example: Learning about a Spinner

The `ProbBayes` package is designed to illustrate Bayesian thinking. This package is used here to learn about the identity of an unknown spinner. It is supposed that each spinner is divided in several regions and the outcomes of the spins are the integers 1, 2, ... and so on.

R A spinner is constructed by specifying a vector of areas of the spinner regions. For example one spinner is defined with five outcomes with corresponding areas 2, 1, 2, 1, 2. The `spinner_plot()` function will produce the spinner as displayed in Figure 3.4.

```
library(ProbBayes)
areas <- c(2, 1, 2, 1, 2)
spinner_plot(areas)
```

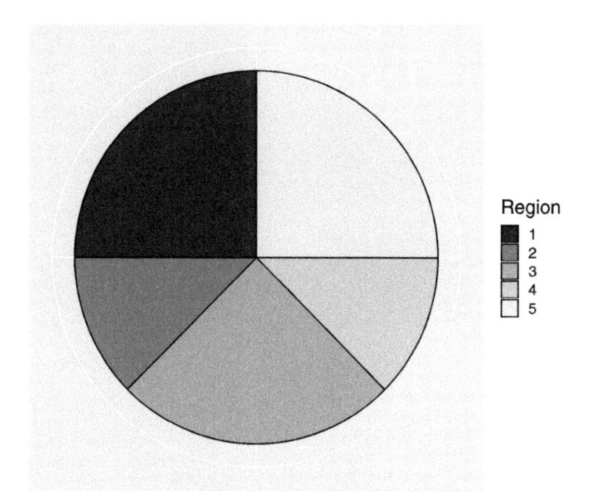

FIGURE 3.4
A spinner with five outcomes 1, 2, 3, 4, 5 and corresponding areas 2, 1, 2, 1 and 2.

One figures out the probability distribution for the spins from knowing the areas of the five outcomes. Each region area is divided by the sum of the areas, obtaining the probabilities as displayed using the function `spinner_probs()`. This data frame of probabilities is stored in the R object `p_dist`.

```
(p_dist <- spinner_probs(areas))
  Region  Prob
1      1 0.250
2      2 0.125
3      3 0.250
4      4 0.125
5      5 0.250
```

To illustrate Bayes' rule, suppose there are four spinners, A, B, C, D defined by the vectors `s_reg_A`, `s_reg_B`, `s_reg_C`, and `s_reg_D` pictured in Figure 3.5.

```
s_reg_A <- c(2, 2, 2, 2)
s_reg_B <- c(4, 1, 1, 2)
s_reg_C <- c(2, 4, 2)
s_reg_D <- c(1, 3, 3, 1)
```

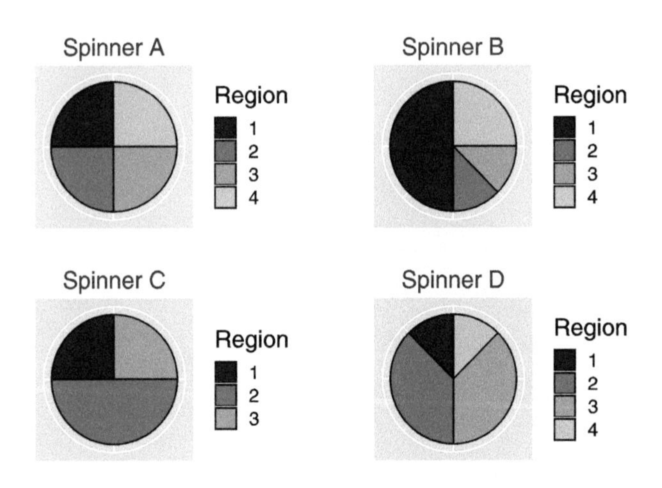

FIGURE 3.5
Four possible spinners in the Bayes' rule example.

A box contains four spinners, one of each type. A friend selects one and holds it behind a curtain. Which spinner is she holding?

The identity of her spinner is called a *model.* There are four possible models, the friend could be holding Spinner A, or Spinner B, or Spinner C, or Spinner D. In R, a data frame is created with a single variable `Model` and the names of these spinners are placed in that column.

```
(bayes_table <- data.frame(Model=c("Spinner A", "Spinner B",
                                   "Spinner C", "Spinner D")))
      Model
1 Spinner A
2 Spinner B
3 Spinner C
4 Spinner D
```

One does not know what spinner this person is holding. But one can assign probabilities to each model that reflect her opinion about the likelihood of these four spinners. There is no reason to think that any of the spinners are more or less likely to be chosen so the same probability of 1/4 is assigned to each model. These probabilities are called the person's *prior* since they reflect her beliefs before observing any data. It is called a uniform prior since the probabilities are spread uniformly over the four models. In the data frame, a new column `Prior` is added with the values 1/4, 1/4, 1/4, 1/4.

```
bayes_table$Prior <- rep(1/4, 4)
bayes_table
      Model Prior
1 Spinner A  0.25
2 Spinner B  0.25
3 Spinner C  0.25
4 Spinner D  0.25
```

The `prob_plot()` function graphs the prior distribution (see Figure 3.6).

```
prob_plot(bayes_table)
```

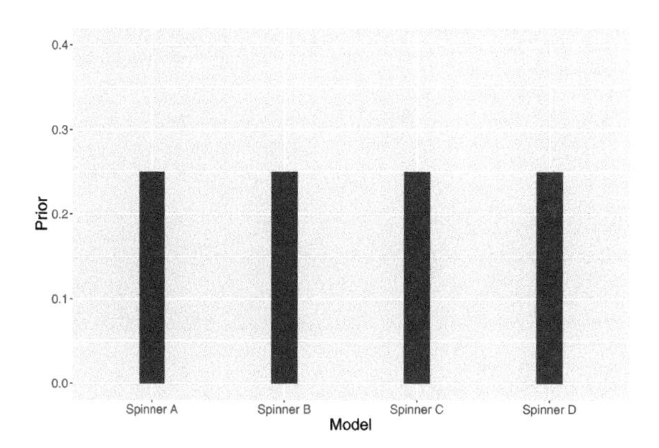

FIGURE 3.6
Prior on the four spinners.

Next, our friend will spin the unknown spinner once – it turns out to land in Region 1. The next step is to compute the *likelihoods* – these are the probabilities of observing a spin in Region 1 for each of the four spinners. In other words, the likelihood is the conditional probability

$$Prob(\text{Region1} \mid \text{Model}),$$

where model is one of the four spinners.

One figures out these likelihoods by looking at the spinners. For example, look at Spinner A. Region 1 is one quarter of the total area for Spinner A, so the likelihood for Spinner A is one fourth, or

$$Prob(\text{Region1} \mid \text{Spinner } A) = 1/4.$$

Looking at Spinner B, Region 1 is one half of the total area so its likelihood is one half. In a similar fashion, one determines the likelihood for Spinner C is one fourth and the likelihood for Spinner D is one eighth. These likelihoods are added to our `bayes_table`.

```
bayes_table$Likelihood <- c(1/4, 1/2,  1/4, 1/8)
bayes_table
      Model Prior Likelihood
1 Spinner A  0.25      0.250
2 Spinner B  0.25      0.500
3 Spinner C  0.25      0.250
4 Spinner D  0.25      0.125
```

Once the prior probabilities and the likelihoods are found, it is straightforward to compute the posterior probabilities. Basically, Bayes' rule says that the posterior probability of a model is proportional to the product of the prior probability and the likelihood. That is,

$$Prob(\text{model} \mid \text{data}) \propto Prob(\text{model}) \times Prob(\text{data} \mid \text{model})$$

Bayesians use the phrase "turn the Bayesian crank" to reflect the straightforward way of computing posterior probabilities using Bayes' rule.

An R function `bayesian_crank()` takes as input a data frame with variables `Prior` and `Likelihood` and outputs a data frame with new columns `Product` and `Posterior`. This function is applied for our example where we observe "Region 1" outcome.

```
(bayesian_crank(bayes_table) -> bayes_table)
      Model Prior Likelihood Product Posterior
1 Spinner A  0.25      0.250 0.06250 0.2222222
2 Spinner B  0.25      0.500 0.12500 0.4444444
3 Spinner C  0.25      0.250 0.06250 0.2222222
4 Spinner D  0.25      0.125 0.03125 0.1111111
```

For each possible model, the prior probability is multiplied by its likelihood. After finding the four products, these are changed to probabilities by dividing each product by the sum of the products. These are called *posterior* probabilities since they reflect our new opinion about the identity of the spinner after observing the spin in Region 1.

By using the `prior_post_plot()` function, the prior and posterior distributions are graphically compared for our spinners, in Figure 3.7.

```
prior_post_plot(bayes_table)
```

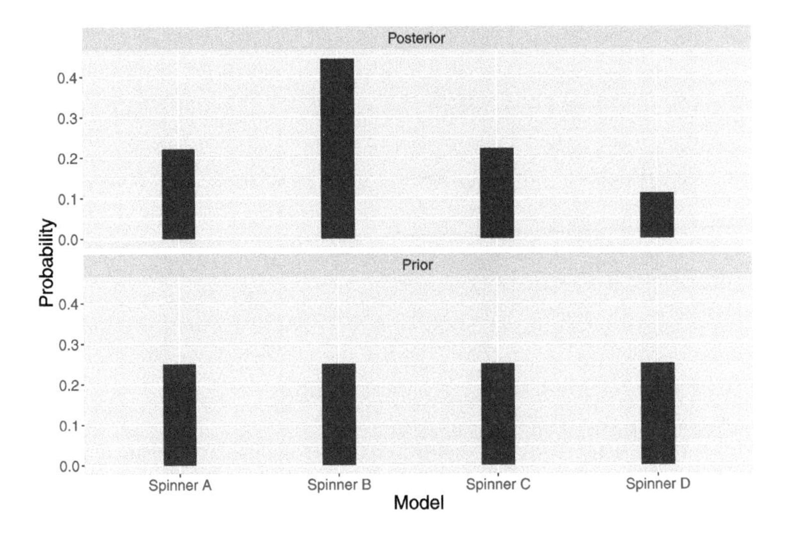

FIGURE 3.7
Prior and posterior distributions on the four spinners.

These calculations can be viewed from a learning perspective. Initially, one had no reason to favor any spinner and each of the four spinners was given the same prior probability of 0.25. Now after observing one spin in Region 1, the person's opinions have changed. Now the most likely spinner behind the curtain is Spinner B since it has a posterior probability of 0.44. In contrast, it is unlikely that Spinner D has been spun since its new probability is only 0.11.

3.8 Exercises

1. **Flipping Coins**

   Suppose you flip a fair coin four times. The 16 possible outcomes of this experiment are shown below.

$HHHH$	$HHHT$	$HHTT$	$HHTH$
$HTHH$	$HTHT$	$HTTT$	$HTTH$
$THHH$	$THHT$	$THTT$	$THTH$
$TTHH$	$TTHT$	$TTTT$	$TTTH$

(a) Let A denote the event that you flip exactly three heads. Find the probability of A.

(b) Suppose you are given the information N that at least two heads are flipped. Circle the possible outcomes in the reduced sample space based on knowing that event N is true.

(c) Using the reduced sample space, find the conditional probability $P(A \mid N)$.

(d) Compare $P(A)$ computed in part a with $P(A \mid N)$ computed in part (c). Based on this comparison, are events A and N independent? Why?

2. **Choosing a Committee**

Suppose you randomly choose three people from the group {Sue, Ellen, Jill, Bob, Joe, John} to be on a committee. Below we have listed all possible committees of size three:

{Sue, Ellen, Jill}	{Sue, Ellen, Bob}	{Sue, Ellen, Joe}
{Sue, Jill, Bob}	{Sue, Jill, Joe}	{Sue, Jill, John}
{Sue, Bob, John}	{Sue, Joe, John}	{Ellen, Jill, Bob}
{Ellen, Jill, John}	{Ellen, Bob, Joe}	{Ellen, Bob, John}
{Jill, Bob, Joe}	{Jill, Bob, John}	{Jill, Joe, John}
{Sue, Ellen, John}	{Sue, Bob, Joe}	{Ellen, Jill, Joe}
{Ellen, Joe, John}	{Bob, Joe, John}	

(a) Find the probability of the event A that exactly two women are in the committee (Sue, Ellen, and Jill are women; Bob, Joe, and John are men).

(b) Suppose you are told that Jill is on the committee– call this event J. Circle the possible outcomes in the reduced sample space if we know that J is true.

(c) Compute the conditional probability $P(A \mid J)$.

(d) Based on your computations in parts (a) and (c), are events A and J independent?

(e) Let F denote the event that more women are on the committee than men. Find $P(F)$.

(f) Suppose you are given the information S that all three people on the committee are of the same gender. Find $P(F \mid S)$.

(g) Based on your computations in parts (e) and (f), are events F and S independent?

3. **Arranging Letters**

Suppose you randomly arrange the letters a, s, s, t, t. You used a computer to do this arranging 200 times and below lists all of the possible "words" that came up. There were 30 distinct arrangements.

asstt	astst	astts	atsst	atsts	attss
sastt	satst	satts	ssatt	sstat	sstta
stast	stats	stsat	ststa	sttas	sttsa
tasst	tasts	tatss	tsast	tsats	tssat
tssta	tstas	tstsa	ttass	ttsas	ttssa

(a) Assuming each possible arrangement is equally likely, what is the probability that the word formed is "stats"?

(b) What is the probability that the word formed begins and ends with an "s"?

(c) Suppose you are told that the word formed starts with "s" – write down all of the possible words in the reduced sample space.

(d) Given that the word begins with "s", what is the probability the word is "stats"?

4. **Rolling Two Dice**

Suppose two dice are rolled.

(a) Suppose you are told that the sum of the dice is equal to 7. Write down the six possible outcomes.

(b) Given the sum of the dice is equal to 7, find the probability the largest die roll is 6.

(c) Suppose you are told that the two dice have different numbers. Write down the possible outcomes.

(d) If the two dice have different numbers, what is the probability the largest die roll is 6?

5. **Choosing Sport Balls**

Suppose you have a bin in your garage with three sports balls – four are footballs, three are basketballs, and two are tennis balls. Suppose you take three balls from the bin – you count the number of footballs and the number of basketballs. The first time this is done, the following balls were selected:

basketball, basketball, football,

so the number of footballs selected was 1 and the number of basketballs selected was 2. We repeat this sampling experiment 1000 times, each time recording the number of footballs and basketballs we select. Table 3.12 summarizes the results of the 1000 experiments.

Let F_1 denote that event that you have chosen exactly one football and B_1 the event that you chose exactly one basketball from the bin.

TABLE 3.12
Summaries of 1000 experiments of choosing sport balls.

		Number of Basketballs				Total
		0	1	2	3	
Numbers of Footballs	0	13	66	64	34	177
	1	49	198	169	0	416
	2	118	180	0	0	298
	3	109	0	0	0	109
	Total	289	444	233	34	1000

(a) Find $P(F_1)$ and $P(B_1)$.

(b) Find $P(F_1 \mid B_1)$.

(c) Find $P(B_1 \mid F_1)$.

(d) From your calculations above, explain why F_1 and B_1 are not independent events.

6. **Rating Movies**

On the Internet Movie Database (`www.imdb.com`), people are given the opportunity to rate movies on a scale from 1 to 10. Table 3.13 shows the ratings of the movie "Sleepless in Seattle" for men and women who visited the website.

TABLE 3.13
Movie ratings of "Sleepless in Seattle", by gender.

Rating	8,9,10 (High)	5,6,7 (Medium)	1,2,3,4 (Low)	TOTAL
Males	2217	3649	754	6620
Females	1059	835	178	2072
TOTAL	3276	4484	932	8692

(a) Suppose you choose at random a person who is interested in rating this movie on the website. Find the probability that the person gives this movie a high rating between 8 and 10 – that is, $P(H)$.

(b) Find the conditional probabilities $P(H \mid M)$ and $P(H \mid F)$, where M and F are the events that a man and a woman rated the movie, respectively.

(c) Interpret the conditional probabilities in part (b) – does this particular movie appeal to one gender?

(d) Table 3.14 below shows the ratings of the movie "Die Hard" for men and women who visited the website. Answer questions (a), (b), and (c) for this movie.

TABLE 3.14
Movie ratings of "Die Hard", by gender.

Rating	8,9,10 (High)	5,6,7 (Medium)	1,2,3,4 (Low)	TOTAL
Males	16197	6737	882	24016
Females	1720	1243	258	3221
TOTAL	17917	8180	1140	27237

7. **Rating Movies (continued)**

The Internet Movie Database also breaks down the movie ratings by the age of the reviewer. For the movie "Sleepless in Seattle", Table 3.15 classifies the reviewers by age and their rating.

TABLE 3.15
Movie ratings of "Sleepless in Seattle", by age.

Rating	8,9,10 (High)	5,6,7 (Medium)	1,2,3,4 (Low)	TOTAL
under 18	74	76	16	166
18-29	1793	2623	555	4971
30-44	886	1280	272	2438
45+	438	300	60	798
TOTAL	3191	4279	903	8373

(a) Find the probability that a reviewer gives this movie a high rating – that is, find $P(H)$.

(b) Define a "young adult" (YA) as a person between the ages of 18 and 29, and a "senior" (S) as a person 45 or older. Compute $P(H \mid YA)$ and $P(H \mid S)$.

(c) Based on your computations in parts (a) and (b), are "giving a high rating" and "age of rater" independent events? If not, explain how the probability of giving a high rating depends on age.

8. **Family Planning**

Suppose a family plans to have children until they have two boys. Suppose there are two events of interest, $A =$ event that they have at least five children and $B =$ event that the first child born is male. Assuming that each child is equally likely to be a boy or girl, and genders of different children born are independent, then this process of building a family was simulated 1000 times. The results of the simulation arc displayed in Table 3.16.

TABLE 3.16
Simulation results of family planning: two boys.

		Gender of First Born		
		Female	Male	TOTAL
Number of Children	2	0	247	247
	3	125	138	263
	4	126	58	184
	5 or more	250	56	306
	TOTAL	501	499	1000

(a) Use the table to find $P(A)$.

(b) Find $P(A \mid B)$ and decide if events A and B are independent.

(c) Suppose another family plans to continue to have children until they have at least one of each gender. Table 3.17 shows simulated results of 1000 families of this type . Again find $P(A)$, $P(A \mid B)$ and decide if events A and B are independent.

TABLE 3.17
Simulation results of family planning: each gender.

		Gender of First Born		
		Female	Male	TOTAL
Number of Children	2	235	261	496
	3	106	152	258
	4	71	63	134
	5 or more	50	62	112
	TOTAL	462	538	1000

9. **Conditional Nature of Probability** For each of the following problems

   - Make a guess at the probabilities of the three events based on your current knowledge.
   - Ask a friend about this problem. Based on his or her opinion about the event, make new probability assignments.
   - Do some research on the Internet to learn about the right answer to the question. Make new probability assignments based on your new information.

(a) What is the area of Pennsylvania?

Initial probabilities:

Event	under 30,000 sq miles	between 30,000 and 50,000 sq miles	over 50,000 sq miles
Probability			

Probabilities after talking with a friend:

Event	under 30,000 sq miles	between 30,000 and 50,000 sq miles	over 50,000 sq miles
Probability			

Probabilities after doing research on the Internet.

Event	under 30,000 sq miles	between 30,000 and 50,000 sq miles	over 50,000 sq miles
Probability			

(b) Robin Williams has appeared in how many movies?

Probabilities after talking with a friend:

Event	Under 15	Between 16 and 30	Over 30
Probability			

Probabilities after doing research on the Internet.

Event	Under 15	Between 16 and 30	Over 30
Probability			

Probabilities after doing research on the Internet.

Event	Under 15	Between 16 and 30	Over 30
Probability			

10. **Conditional Nature of Probability**

For each of the following problems

- Make a guess at the probabilities of the three events based on your current knowledge.
- Ask a friend about this problem. Based on his or her opinion about the event, make new probability assignments.
- Do some research on the Internet to learn about the right answer to the question. Make new probability assignments based on your new information.

(a) How many plays did Shakespeare write?

Initial probabilities:

Event	Under 30	Between 31 and 50	Over 50
Probability			

Probabilities after talking with a friend:

Event	Under 30	Between 31 and 50	Over 50
Probability			

Probabilities after doing research on the Internet.

Event	Under 30	Between 31 and 50	Over 50
Probability			

(b) What is the average temperature in Melbourne, Australia in June?

Initial probabilities:

Event	Under 40°	Between 40° and 60°	Over 60°
Probability			

Probabilities after talking with a friend:

Event	Under 40°	Between 40° and 60°	Over 60°
Probability			

Probabilities after doing research on the Internet.

Event	Under 40°	Between 40° and 60°	Over 60°
Probability			

11. **Picnic Misery**

Twenty boys went on a picnic. 5 got sunburned, 8 got bitten by mosquitoes, and 10 got home without mishap. What is the probability that the mosquitoes ignored a sunburned boy? What is the probability that a bitten boy was also burned?

12. **A Mall Survey**

Suppose 30 people are surveyed at a local mall. Half of the 10 men surveyed approve the upcoming school levy and a total of 17 people do not approve of the levy. Based on the survey data,

(a) What is the probability a woman is in favor of the levy?

(b) If the person is in favor of the levy, what is the probability the person is a woman?

13. **Drawing Tickets**

    Have 12 tickets numbered from 1 to 12. Two tickets are drawn, one after the other, without replacement.

    (a) Find the probability that both numbers are even.
    (b) Find the probability both numbers are odd.
    (c) Find the probability one number is even and one is odd.

14. **Testing for Steroids**

    Suppose that 20% of all baseball players are currently on steroids. You plan on giving a random player a test, but the test is not perfectly reliable. If the player is truly on steroids, he will test negative (for steroids) with probability 0.1. Likewise, if the player is not on steroids, he will get a positive test result with probability 0.1.

    (a) What is the probability the player is on steroids and will test negative?
    (b) If you give a player a test, what is the probability he will test positive?
    (c) If the test result is positive, what is the probability the player is on steroids?

15. **Preparing for the SAT**

    Suppose a student has a choice of enrolling (or not) in an expensive program to prepare for taking the SAT exam. The chance that she enrolls in this class is 0.3. If she takes the program, the chance that she will do well on the SAT exam is 0.8. On the other hand, if she does not take the prep program, the chance that she will do well on the SAT is only 0.4. Let E denote the event "enrolls in the class" and W denote the event "does well on the SAT exam".

    (a) Find $P(W \mid E)$.
    (b) Find $P(E \cap W)$.
    (c) Find $P(E \mid W)$, that is, the probability that she took the class given that she did well on the test.

16. **Working Off-Campus**

    At a college campus, 33% of the students are freshmen and 25% are seniors. Also, 13% of the freshmen work over 10 hours off-campus, and 37% of the seniors work over 10 hours off-campus.

    (a) Suppose you sample a student who is either a freshman and senior. Find the probability she works over 10 hours off-campus.
    (b) If this person does work over 10 hours off campus, find the probability she is a senior.

17. **Flipping Coins**

    You flip a coin three times. Let A be the event that a head occurs on the first flip and B is the event that (exactly) one head occurs. Are A and B independent?

18. **A Two-headed Coin?**

    One coin in a collection of 65 has two heads. Suppose you choose a coin at random from the collection – you toss it 6 times and observe all heads. What is the probability it was the two-headed coin?

19. **Smoking and Gender**

    Suppose the proportion of female students at your school is 60%. Also you know that 26% of the male students smoke and only 16% of the female students smoke. Suppose you randomly select a student.

    (a) Find the probability the student is a male smoker.
    (b) Find the probability the student smokes.
    (c) If the student smokes, what is the probability the student is female?

20. **Choosing until You Select a Red**

    Suppose you have a box with 4 green and 2 red balls. You select balls from the box one at a time until you get a red, or until you select three balls. If you do not select a red on the first draw, find the probability that you will select three balls.

21. **Mutually Exclusive and Independence**

    Suppose that two events A and B are mutually exclusive. Are they independent events?

22. **Blood Type of Couples**

    Consider the example where Americans have the blood types O, A, B, AB with proportions .45, .40, .11, 04. If two people are married

    (a) Find the probability both people have blood type A.
    (b) Find the probability the couple have A and B blood types.
    (c) Find the probability neither person has an A type.

23. **Five-Game Playoffs**

    Consider the "best of five" playoff series between the Yankees and the Indians. We assume the probability the Yankees win a single game is 0.6.

    (a) Find the probability the Yankees win in three games.
    (b) Find the probability the series lasts exactly three games.
    (c) Find the probability the series lasts five games and no team wins more than one game in a row.

24. **Computer and Video Games**

    The Entertainment Software Association reports that of all computer and video games sold, 53% are rated E (Everyone), 30% are rated T (Teen), and 16% are rated M (Mature). Suppose three customers each purchase a game at a local store. Assume that the software choices for the customers can be regarded as independent events.

    (a) Find the probability that all three customers buy games that are rated E.
    (b) Find the probability that exactly one customer purchases an M rated game.
    (c) Find the probability that the customers purchase games with the same rating.

25. **Washer and Dryer Repair**

    Suppose you purchase a washer and dryer from a particular manufacturer. From reading a consumer magazine, you know that 20% of the washers and 10% of the dryers will need some repair during the warranty period.

    (a) Find the probability that both the dryer and washer will need repair during the warranty period.
    (b) Find the probability that exactly one of the machines will need repair.
    (c) Find the probability that neither machine will need repair.

26. **Basketball Shooting**

    In a basketball game, a player has a "one and one" opportunity at the free-throw line. If she misses the first shot, she is done. If instead she makes the first shot, she will have an opportunity to make a second shot. From past data, you know that the probability this player will make a single free-throw shot is 0.7.

    (a) Find the probability the player only takes a single free-throw.
    (b) Find the probability the player makes two shots.
    (c) Find the probability the player makes the first shot and misses the second.

27. **Playing Roulette**

    You play the game of roulette in Reno. Each game you always bet on "red" and the chance that you win is 18/38. Suppose you play the game four times.

    (a) Find the probability you win in all games.
    (b) Find the probability you win in the first and third games, and lose in the second and fourth games.
    (c) Find the probability you win in exactly two of the four games.

28. **Is a Die Fair?**

    Suppose a friend is about to roll a die. The die either is the usual "fair" type or it is a "special" type that has two sides showing 1, two sides showing 2, and two sides showing 3. You believe that the die is the fair type with probability 0.9. Your friend rolls the die and you observe a 1.

    (a) Find the probability that a 1 is rolled.

    (b) If you observe a 1, what is the probability your friend was rolling the fair die?

29. **How Many Fish?**

    You are interested in learning about the number of fish in the pond in your back yard. It is a small pond, so you do not expect many fish to live in it. In fact, you believe that the number of fish in the pond is equally likely to be 1, 2, 3, or 4. To learn about the number of fish, you will perform a capture-recapture experiment. You first catch one of the fish, tag it, and return it to the pond. After a period of time, you catch another fish and observe that it is tagged and this fish is also tossed back into the pond.

    (a) There are two stages of this experiment. At the first stage you have 1, 2, 3, or 4 fish in the pond, and at the second stage, you observe either a tagged or not-tagged fish. Draw a tree diagram to represent this experiment, and label the branches of the tree with the given probabilities.

    (b) Find the probability of getting a tagged fish.

    (c) If you find a tagged fish, find the probability there was exactly 1 fish in the pond. Also find the probabilities of exactly 2 fish, 3 fish, and 4 fish in the pond.

30. **Shopping at the Mall**

    Suppose that you are shopping in a large mall in a metropolitan area. The people who shop at this mall either live downtown or in the suburbs. Recently a market research firm surveyed mall shoppers — from this survey, they believe that 70% of the shoppers live in the suburbs and 30% live downtown. You know that there is a relationship between a person's political affiliation and where he or she lives. You know that 40% of the adults who live in the suburbs are registered Democrats and 80% of the downtown residents are Democrats.

    (a) If you let T = event that shopper lives downtown, S = event that shopper lives in the suburbs and D = event that shopper is a Democrat, write down the probabilities given in the above paragraph.

(b) Suppose you interview a random shopper. Find the probability that the shopper is a Democrat.

(c) If your shopper is a Democrat, find the probability he or she lives in the suburbs.

31. **What Bag?**

Suppose that you have two bags in your closet. The white bag contains four white balls and the mixed bag contains two white and two black balls. The closet is dark and you just grab one bag out at random and select a ball. The ball you choose can either be white or black.

(a) Suppose there are 1000 hypothetical bags in your closet. By use the Bayes' box shown below, classify the 1000 bags by the type "white" and 'mixed" and the ball color observed.

		Ball color observed		
		White	Black	TOTAL
Bag	White			
type	Mixed			
TOTAL				1000

(b) Using the Bayes' box, find the probability that you observe a white ball.

(c) If you observe a white ball, find the probability that you were selecting from the white bag.

R Exercises

32. **Conditional Probability**

Suppose you have two spinners – one spinner is equally likely to land on the numbers 1, 2, 3, and the second spinner is equally likely to land on 1, 2, 3, 4, 5.

(a) Using two applications of the `sample()` function, create a data frame containing 1000 random spins from the first spinner and 1000 random spins from the second spinner.

(b) Use the `filter()` function to take a subset of the data frame created in part (a), keeping only the rows where the sum of spins is fewer than 5.

(c) Using the output from part (b), approximate the probability the first spin is equal to 1 given that the sum of spins is fewer than 5.

(d) By use of the `filter()` function, approximate the probability the sum of spins is fewer than 5 given that the first spin is equal to 1.

33. **Rolling a Random Die**

Suppose you spin a spinner that is equally likely to land on the values 1, 2, 3, 4. If the spinner lands on 1, then you roll a fair die; otherwise (if the spinner lands on 2, 3, 4), you roll a weighted die where an even roll is twice as likely as an odd roll.

(a) Create a data frame where `Spin` contains 1000 spins from the random spinner, `Die1` contains 1000 rolls from the fair die and `Die2` contains 1000 rolls from the biased die.

(b) By use of the `ifelse()` function, define a new variable `Die` representing the roll of the "random" die, where the outcome depends on the value of the spinner.

(c) Find the probability the random die roll is equal to 3.

4

Discrete Distributions

4.1 Introduction: The Hat Check Problem

Some time ago, it was common for men to wear hats when they went out for dinner. When a man entered a restaurant, he would give his hat to an attendant who would keep the hat in a room until his departure. Suppose the attendant gets confused and returns hats in some random fashion to the departing men. What is the chance that no man receives his personal hat? How many hats, on average, will be returned to the right owners?

This is a famous "matching" probability problem. To start thinking about this problem, it is helpful to start with some simple cases. Suppose only one man checks his hat at the restaurant. Then obviously this man will get his hat back. Then the probability of "no one receives the right hat" is 0, and the average number of hats returned will be equal to 1.

Let n denote the number of men who enter the restaurant. The case $n = 1$ was considered above. What if $n = 2$? If the two men are Barry and Bobby, then there are two possibilities shown in Table 4.1. These two outcomes are equally likely, so the probability of no match is $1/2$. Half the time there will be 2 matches and half the time there will be 0 matches, and so the average number of matches will be 1.

TABLE 4.1
Possibilities of the hat check problem when $n = 2$.

	Barry receives	Bobby receives	# of matching hats
1.	Barry's hat	Bobby's hat	2
2.	Bobby's hat	Barry's hat	0

What if we have $n = 3$ men that we'll call Barry, Bobby, and Jack. Then there are $3! = 6$ ways of returning hats to men, listed in Table 4.2. Again these outcomes are equally likely, so the probability of no match is $2/6$. One can show that the average number of matches is again 1.

TABLE 4.2
Possibilities of the hat check problem when $n = 3$.

	Barry receives	Bobby receives	Jack receives	# of matching hats
1.	Barry's hat	Bobby's hat	Jack's hat	3
2.	Barry's hat	Jack's hat	Bobby's hat	1
3.	Bobby's hat	Barry's hat	Jack's hat	1
4.	Bobby's hat	Jack's hat	Barry's hat	0
5.	Jack's hat	Barry's hat	Bobby's hat	0
6.	Jack's hat	Bobby's hat	Barry's hat	1

What happens if there are a large number of hats checked? It turns out that the probability of no matches is given by

$$Prob(\text{no matches}) = \frac{1}{e},$$

where e is the special irrational number 2.718. Also it is interesting that the average number of matches for any value of n is given by

$$\text{Average number of matches} = 1.$$

The reader will get the opportunity of exploring this famous problem by simulation in the end-of-chapter exercises.

4.2 Random Variable and Probability Distribution

Suppose that Peter and Paul play a simple coin game. A coin is tossed. If the coin lands heads, then Peter receives $2 from Paul; otherwise Peter has to pay $2 to Paul. The game is played for a total of five coin flips. After the five flips, what is Peter's net gain (in dollars)?

The answer depends on the results of the coin flips. There are two possible outcomes of each coin flip (heads or tails) and, by applying the multiplication rule, there are $2^5 = 32$ possibilities for the five flips. The 32 possible outcomes are written below.

$HHHHH$	$HTHHH$	$THHHH$	$TTHHH$
$HHHHT$	$HTHHT$	$THHHT$	$TTHHT$
$HHHTH$	$HTHTH$	$THHTH$	$TTHTH$
$HHHTT$	$HTHTT$	$THHTT$	$TTHTT$
$HHTHH$	$HTTHH$	$THTHH$	$TTTHH$
$HHTHT$	$HTTHT$	$THTHT$	$TTTHT$
$HHTTH$	$HTTTH$	$THTTH$	$TTTTH$
$HHTTT$	$HTTTT$	$THTTT$	$TTTTT$

For each possible outcome of the flips, say $HTHHT$, there will be a corresponding net gain for Peter. For this outcome, Peter won three times and lost twice, so his net gain is $3(2) - 2(2) = 2$ dollars. The net gain is an example of a *random variable* – this is simply a number that is assigned to each outcome of the random experiment.

Generally, a capital letter will be used to represent a random variable – here the capital letter G denotes Peter's gain in this experiment. For each of the 32 outcomes, one can assign a value of G – this is done in Table 4.3.

TABLE 4.3
The 32 outcomes and value of G in the 5 coin flips problem.

$HHHHH, G = 10$	$HTHHH, G = 6$	$THHHH, G = 6$	$TTHHH, G = 2$
$HHHHT, G = 6$	$HTHHT, G = 2$	$THHHT, G = 2$	$TTHHT, G = -2$
$HHHTH, G = 6$	$HTHTH, G = 2$	$THHTH, G = 2$	$TTHTH, G = -2$
$HHHTT, G = 2$	$HTHTT, G = -2$	$THHTT, G = -2$	$TTHTT, G = -6$
$HHTHH, G = 6$	$HTTHH, G = 2$	$THTHH, G = 2$	$TTTHH, G = -2$
$HHTHT, G = 2$	$HTTHT, G = -2$	$THTHT, G = -2$	$TTTHT, G = -6$
$HHTTH, G = 2$	$HTTTH, G = -2$	$THTTH, G = -2$	$TTTTH, G = -6$
$HHTTT, G = -2$	$HTTTT, G = -6$	$THTTT, G = -6$	$TTTTT, G = -10$

It is seen from the table that the possible gains for Peter are -10, -6, -2, 2, 6, and 10 dollars. One is interested in the probability that Peter will get each possible gain. To do this, one puts all of the possible values of the random variable in Table 4.4. Although a capital letter will be used to denote a random variable, a small letter will denote a specific value of the random variable. So g refers to one specific value of the gain G, and $P(G = g)$ refers to the corresponding probability.

TABLE 4.4
Table of gain, number of outcomes, and corresponding probability, step 1.

Gain g (dollars)	Number of outcomes	$P(G = g)$
-10		
-6		
-2		
2		
6		
10		

What is the probability that Peter gains \$6 in this game? Looking at the table of outcomes, one sees that Peter won \$6 in five of the outcomes. Since there are 32 possible outcomes of the five flips, and each outcome has the same probability, one sees that the probability of Peter winning \$6 is $5/32$.

This process is continued for all of the possible values of G. In Table 4.5, one places the number of outcomes for each value and the corresponding probability. This is an example of a probability distribution for G – This is simply a list of all possible values for a random variable together with the associated probabilities.

TABLE 4.5
Table of gain, number of outcomes, and corresponding probability, step 2.

Gain g (dollars)	Number of outcomes	$P(G = g)$
−10	1	1/32
−6	5	5/32
−2	10	10/32
2	10	10/32
6	5	5/32
10	1	1/32

Probability distribution

In general, suppose X is a discrete random variable. This type of random variable only assigns probability to a discrete set of values. In other words, the *support* of X is a set of discrete values. The function $f(x)$ is a probability mass function (pmf) for X if the function satisfies two properties.

(1) $f(x) \geq 0$ for each possible value x of X

(2) $\sum_x f(x) = 1$

The table of values of the gain G and the associated probabilities $f(g) = P(G = g)$ do satisfy these two properties. Each of the assigned probabilities is positive, so property (1) is satisfied. If one sums the assigned probabilities, one finds

$$\sum_g P(G = g) = \frac{1}{32} + \frac{5}{32} + \frac{10}{32} + \frac{10}{32} + \frac{5}{32} + \frac{1}{32} = 1,$$

and so property (2) is satisfied.

A probability distribution is a listing of the values of X together with the associated values of the pmf. One graphically displays this probability distribution with a bar graph. One places all of the values of G on the horizontal axis, marks off a probability scale on the vertical scale, and then draws vertical lines on the graph corresponding to the pmf values.

Figure 4.1 visually shows that it is most likely for Peter to finish with a net gain of +2 or −2 dollars. Also note the symmetry of the graph – the graph looks the same way on either side of 0. This symmetry about 0 indicates that

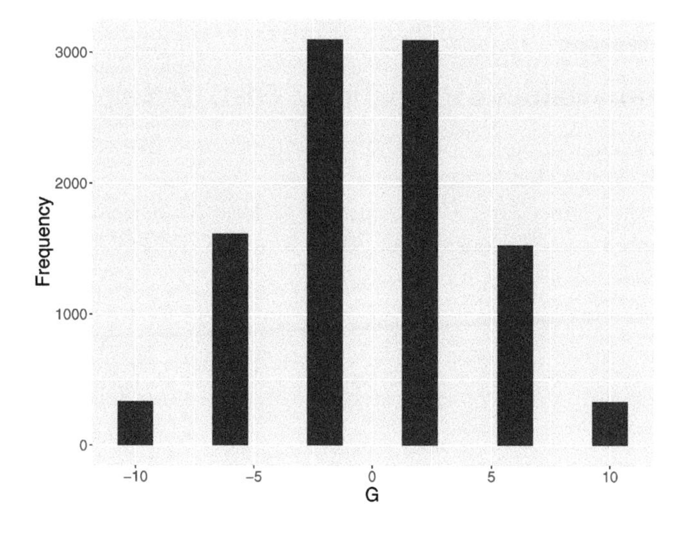

FIGURE 4.1
Probability distribution of the net gains for Peter in the Peter-Paul game.

this game is fair. We will shortly discuss a way of summarizing this probability distribution that confirms that this is indeed a fair game.

R Simulating the Peter-Paul Game

It is straightforward to simulate this game in R. A function `one_play()` is written which will play the game one time. The `sample()` function is used to flip a coin five times and the function returns the net gain for Paul.

```
one_play <- function(){
  flips <- sample(c("H", "T"),
                  size = 5,
                  replace = TRUE)
  2 * sum(flips == "H") -
       2 * sum(flips == "T")
}
```

The `replicate()` function is used to simulate 1000 plays of the game and the net gains for all plays are stored in the vector `G`. If one constructs a bar graph of the net gains, it will resemble the graph of the probability distribution of G showed in Figure 4.1.

```
G <- replicate(1000, one_play())
```

4.3 Summarizing a Probability Distribution

Once we have constructed a probability distribution – like was one above– it is convenient to use this to find probabilities.

What is the chance that Peter will win at least \$5 in this game? Looking at the probability table, ones sees that winning "at least \$5" includes the possible values

$$G = 6 \text{ and } G = 10$$

One finds the probability of interest by adding the probabilities of the individual values.

$$\begin{aligned} P(G \geq 5) &= P(G = 6 \text{ or } G = 10) \\ &= P(G = 6) + P(G = 10) \\ &= \frac{5+1}{32} = \frac{6}{32}. \end{aligned}$$

What is the probability Peter wins money in this game? Peter wins money if the gain G is positive and this corresponds to the values $G = 2, 6, 10$. By adding up the probabilities of these three values, one sees the probability that Peter wins money is

$$\begin{aligned} P(\text{Peter wins}) &= P(G > 0) \\ &= P(G = 2) + P(G = 6) + P(G = 10) \\ &= \frac{10+5+1}{32} = \frac{1}{2}. \end{aligned}$$

It is easy to compute the probability Peter loses money – also $1/2$. Since the probability Peter wins in the game is the same as the probability he loses, the game is clearly fair.

When one has a distribution of data, it is helpful to summarize the data with a single number, such as median or mean, to get some understanding about a typical data value. In a similar fashion, it is helpful to compute an "average" of a probability distribution – this will give us some feeling about typical or representative values of the random variable when one observes it repeated times.

A common measure of "average" is the mean or expected value of X, denoted μ or $E(X)$. The mean (or expected value) is found by

1. Computing the product of a value of X and the corresponding value of the pmf $f(x) = P(X = x)$ for all values of X.

2. Summing the products.

In other words, one finds the mean by the formula

$$\mu = \sum_{x} x f(x). \quad (4.1)$$

The computation of the mean for the Peter-Paul game is illustrated in Table 4.6. For each value of the gain G, the value is multiplied by the associated probability – the products are given in the rightmost column of the table. Then the products are added – one sees that the mean of G is $\mu = 0$.

TABLE 4.6
Calculation of the mean for the Peter-Paul game.

g	$P(G = g)$	$g \times P(G = g)$
-10	1/32	$-10/32$
-6	5/32	$-30/32$
-2	10/32	$-20/32$
2	10/32	20/32
6	5/32	30/32
10	1/32	10/32
SUM	1	0

How does one interpret a mean value of 0? Actually it is interesting to note that $G = 0$ is not a possible outcome of the game – that is, Peter cannot break even when this game is played. But if Peter and Paul play this game a large number of times, then the value $\mu = 0$ represents (approximately) the mean winnings of Peter in all of these games.

R Simulating the Peter-Paul Game (continued)

The functions `sample()` and `replicate()` were earlier illustrated to simulate this game 1000 times in R. Peter's winnings in the different games are stored in the vector `G`. Here is a display of Peter's winnings in the first 100 games:

```
G[1:100]
  [1]   6  -6  -6  -2  -6   2   6  -2  -6  -6 -10  -6
 [13]  -2   2  -2   2  10   6   2  -2  -6   6  -2  -2
 [25]  -2  -2  -2   2  10   2  -2  -2   6  -2   2   2
 [37]   6   2  -2  -6  -6   2  -6  -2   2  -6 -10  -6
 [49]   2   6   6   6   2  -2  -2  -2   2  -6  -2   2
 [61]   2  -2   6  -2   6   6   2   6  -6   6   2   6
 [73]  -6  -2   2   2   6   2   6  -2 -10  -6   2  -6
 [85]   6   2  -2  -2   6  -6  -6  -2 -10  -2 -10  -6
 [97]  -2  10   6  -2
```

One approximates the mean winning μ by finding the sample mean $\bar{G}$ of the winning values in the 1000 simulated games.

```
mean(G)
[1] -0.0748
```

This value is approximately equal to the mean of G, $\mu = 0$. If Peter was able to play this game for a much larger number of games, then one would see that his average winning would be very close to $\mu = 0$.

4.4 Standard Deviation of a Probability Distribution

Consider two dice – one we will call the "fair die" and the other one will be called the "loaded die". The fair die is the familiar one where each possible number (1 through 6) has the same chance of being rolled. The loaded die is designed in a special way that 3's or 4's are relatively likely to occur, and the remaining numbers (1, 2, 5, and 6) are unlikely to occur. Table 4.7 gives the probabilities of the possible rolls for both dice.

TABLE 4.7
Probabilities of the possible rolls for a fair die and a loaded die.

Fair Die		Loaded Die	
Roll	Probability	Roll	Probability
1	1/6	1	1/12
2	1/6	2	1/12
3	1/6	3	1/3
4	1/6	4	1/3
5	1/6	5	1/12
6	1/6	6	1/12

How can one distinguish the fair and loaded dice? An obvious way is to roll each a number of times and see if we can distinguish the patterns of rolls that we get. One first rolls the fair die 20 times with the results

3, 3, 5, 6, 6, 1, 2, 1, 4, 3, 2, 5, 6, 4, 2, 5, 6, 1, 2, 3 (mean 3.5)

Next one rolls the loaded die 20 times with the results

3, 2, 1, 4, 4, 1, 4, 3, 3, 3, 1, 3, 3, 5, 3, 3, 3, 6, 3, 4 (mean 3.1)

Figure 4.2 displays dotplots of 50 rolls from each of the two dice.

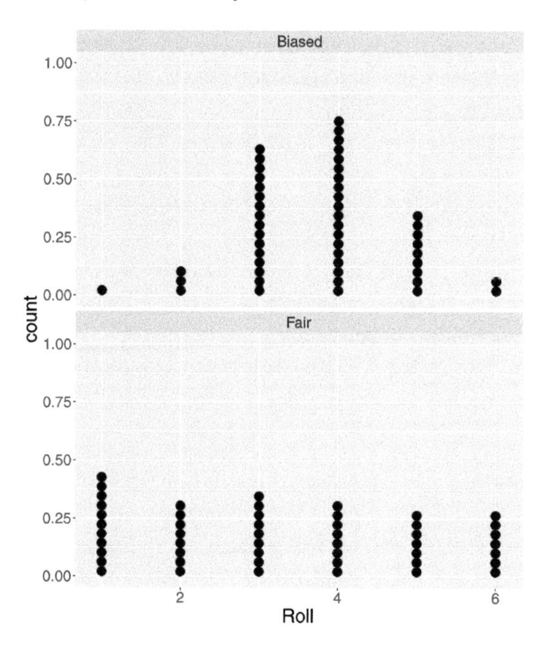

FIGURE 4.2
Dotplots of rolls from fair and loaded dice.

What doe one see? For the fair die, the rolls appear to be evenly spread out among the six possible numbers. In contrast, the rolls for the loaded die tend to concentrate on the values and 3 and 4, and the remaining numbers were less likely to occur.

Can one compute a summary value to contrast the probability distributions for the fair and loaded dice? One summary number for a random variable has already been discussed, the mean μ. This number represents the average outcome for the random variable when one performs the experiment many times.

Suppose the mean is computed for each of the two probability distributions. For the fair die, the mean is given by

$$\begin{aligned} \mu_{FairDie} &= (1)(\frac{1}{6}) + (2)(\frac{1}{6}) + (3)(\frac{1}{6}) + (4)(\frac{1}{6}) + (5)(\frac{1}{6}) + (6)(\frac{1}{6}) \\ &= 3.5, \end{aligned}$$

and for the loaded die the mean is given by

$$\begin{aligned} \mu_{LoadedDie} &= (1)(\frac{1}{12}) + (2)(\frac{1}{12}) + (3)(\frac{1}{3}) + (4)(\frac{1}{3}) + (5)(\frac{1}{12}) + (6)(\frac{1}{12}) \\ &= 3.5. \end{aligned}$$

The means of the two probability distributions are the same – this means that one will tend to get the same average roll when the fair die and the loaded die are rolled many times.

But one knows from our rolling data that the two probability distributions are different. For the loaded die, it is more likely to roll 3's or 4's. In other words, for the loaded die, it is more likely to roll a number close to the mean value $\mu = 3.5$.

The standard deviation of a random variable X, denoted by the Greek letter σ, measures how close the random variable is to the mean μ. It is called a standard deviation since it represents an "average" (or standard) distance (or deviation) from the mean μ. This standard deviation, denoted σ is defined as follows:

$$\sigma = \sqrt{\Sigma_x (x - \mu)^2 P(X = x)}. \tag{4.2}$$

To find the standard deviation σ for a random variable, one first computes (for all values of X) the difference (or deviation) of x from the mean value μ. Next, one squares each of the differences, and finds the average squared deviation by multiplying each squared deviation by the corresponding value of the pmf and summing the products. The standard deviation σ is the square root of the average squared deviation.

Tables 4.8 and 4.9 illustrate the computation of the standard deviation for the roll of the fair die and for the roll of the loaded die, where R denotes the roll random variable.

TABLE 4.8
Computation of the standard deviation $\sigma_{FairDie}$ for the fair die.

r	$r - \mu$	$(r - \mu)^2 \times P(R = r)$
1	$1 - 3.5 = -2.5$	$(-2.5)^2 \times (1/6)$
2	$2 - 3.5 = -1.5$	$(-1.5)^2 \times (1/6)$
3	$3 - 3.5 = -0.5$	$(-0.5)^2 \times (1/6)$
4	$4 - 3.5 = 0.5$	$(0.5)^2 \times (1/6)$
5	$5 - 3.5 = 1.5$	$(1.5)^2 \times (1/6)$
6	$6 - 3.5 = 2.5$	$(2.5)^2 \times (1/6)$
SUM		2.917

$$\sigma_{FairDie} = \sqrt{2.917} = 1.71$$

$$\sigma_{LoadedDie} = \sqrt{1.583} = 1.26$$

It is seen from our calculations that

$$\sigma_{FairDie} = 1.71, \sigma_{LoadedDie} = 1.26$$

TABLE 4.9
Computation of the standard deviation $\sigma_{LoadedDie}$ for the loaded die.

r	$r - \mu$	$(r-\mu)^2 \times P(R=r)$
1	$1 - 3.5 = -2.5$	$(-2.5)^2 \times (1/12)$
2	$2 - 3.5 = -1.5$	$(-1.5)^2 \times (1/12)$
3	$3 - 3.5 = -0.5$	$(-0.5)^2 \times (1/3)$
4	$4 - 3.5 = 0.5$	$(0.5)^2 \times (1/3)$
5	$5 - 3.5 = 1.5$	$(1.5)^2 \times (1/12)$
6	$6 - 3.5 = 2.5$	$(2.5)^2 \times (1/12)$
SUM		1.583

What does this mean? Since the loaded die roll has a smaller standard deviation, this means that the roll of the loaded die tends to be closer to the mean (3.5) than for the fair die. When one rolls the loaded die many times, one will notice a smaller spread or variation in the rolls than when one rolls the fair die many times.

R Simulating Rolls of Fair and Loaded Dice

One illustrates the difference in distributions of rolls of fair and loaded dice by an R simulation. The probabilities of 100 rolls of each of the two types of dice are stored in the vectors `die1` and `die2`. Two applications of the `sample()` function are used to simulated rolls – the rolls for the fair and loaded dice are stored in the vectors `rolls1` and `rolls2`. respectively.

```
die1 <- c(1, 1, 1, 1, 1, 1) / 6
die2 <- c(1, 1, 4, 4, 1, 1) / 12
rolls1 <- sample(1:6, prob = die1,
                 size = 100,
                 replace = TRUE)
rolls2 <- sample(1:6, prob = die2,
                 size = 100,
                 replace = TRUE)
```

One approximates the means and standard deviations for the probability distributions by computing sample means and sample standard deviations of the simulated rolls.

```
c(mean(rolls1), sd(rolls1))
[1] 3.340000 1.585779
c(mean(rolls2), sd(rolls2))
[1] 3.280000 1.246055
```

Note that both types of dice display similar means, but the loaded die displays a smaller standard deviation than the fair die.

Interpreting the standard deviation for a bell-shaped distribution

Once one has computed a standard deviation σ for a random variable, how can one use this summary measure? One use of σ was illustrated in the dice example above. The probabilities for the roll of the loaded die were more concentrated about the mean than the probabilities for the roll of the fair die, and that resulted in a smaller value of σ for the roll of the loaded die.

The standard deviation has an attractive interpretation when the probability distribution of the random variable is bell-shaped. When the probability distribution has the following shape:

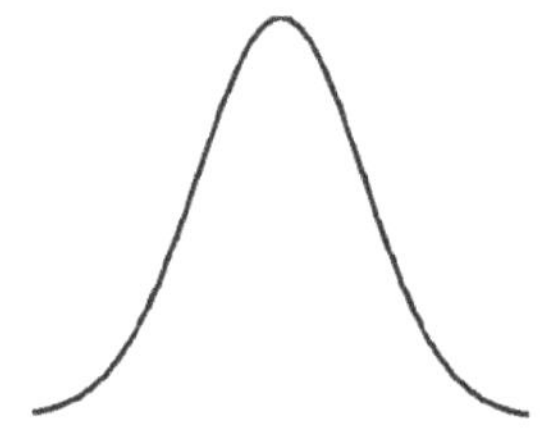

then approximately

- the probability that X falls within one standard deviation of the mean is 0.68.
- the probability that X falls within two standard deviations of the mean is 0.95.

Mathematically, one writes,

- $Prob(\mu - \sigma < X < \mu + \sigma) \approx 0.68$
- $Prob(\mu - 2\sigma < X < \mu + 2\sigma) \approx 0.95$

R Simulating Rolls of Ten Dice

To illustrate this interpretation of the standard deviation, suppose ten fair dice are rolled and the sum of the numbers appearing on the dice is recorded. It is easy to simulate this experiment in R using the following script. The function `roll10()` will roll 10 dice, the function `replicate()` repeats the experiment for 1000 trials, and the variable `sum_rolls` contains the sum of the rolls from the experiments.

```
roll10 <- function(){
  sum(sample(1:6, size = 10, replace = TRUE))
}
sum_rolls <- replicate(1000, roll10())
```

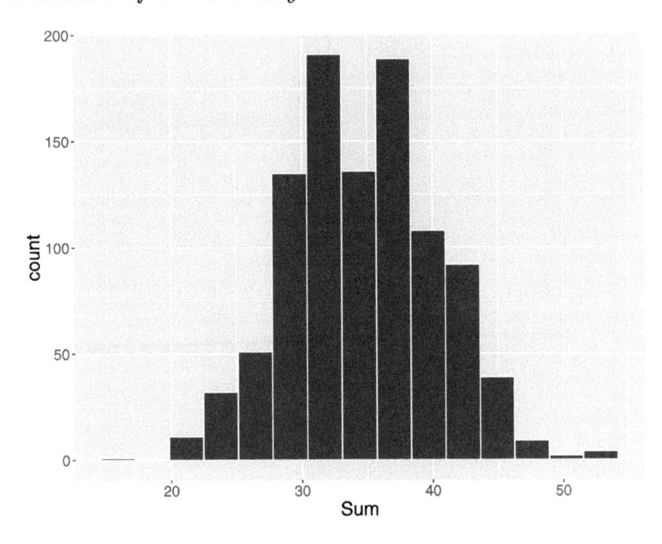

FIGURE 4.3
Histogram of the sum of ten dice in 1000 simulated trials.

A histogram of the results from 1000 trials of this experiment is shown in Figure 4.3.

Note that the shape of this histogram is approximately bell shaped about the value 35. Since this histogram is a reflection of the probability distribution of the sum of the rolls of ten dice, this means that the shape of the probability distribution for the sum will also be bell-shaped.

For this problem, it can be shown (as an end-of-chapter exercise) that the mean and standard deviation for the sum of the rolls of ten fair dice are respectively

$$\mu = 35, \ \sigma = 5.4.$$

Applying our rule, the probability that the sum falls between

$$\mu - \sigma \text{ and } \mu + \sigma, \text{ or } 35 - 5.4 = 29.6 \text{ and } 35 + 5.4 = 40.4$$

is approximately 0.68, and the probability that the sum of the rolls falls between

$$\mu - 2\sigma \text{ and } \mu + 2\sigma, \text{ or } 35 - 2(5.4) = 24.2 \text{ and } 35 + 2\ (5.4) = 45.8$$

is approximately 0.95.

R Simulating Rolls of Ten Dice (continued)

To see if these are accurate probability computations, return to our simulation of this experiment and see how often the sum of the ten rolls fell within the above limits. Recall that the simulation sums were stored in the vector `sum_rolls`. Below the proportions of sums of ten rolls that fall between 29.6 and 40.4, and between 24.2 and 45.8, are computed.

```
sum(sum_rolls > 29.6 & sum_rolls < 40.4) / 1000
[1] 0.702
sum(sum_rolls > 24.2 & sum_rolls < 45.8) / 1000
[1] 0.955
```

One sees that the proportions of values that fall within these limits are 0.702 and 0.955, respectively. Since these proportions are close to the numbers 0.68 and 0.95, we see in this example that this rule is pretty accurate.

4.5 Coin-Tossing Distributions

Introduction: A Galton Board

A Galton board is a physical device for simulating a special type of random experiment. It was named after the famous scientist Sir Francis Galton who lived from 1822 to 1911. Galton is noted for a wide range of achievements in the areas of meteorology, genetics, psychology, and statistics. The Galton board consists of a set of pegs laid out in the configuration shown in Figure 4.4 – one peg is in the top row, two pegs are in the second row, three pegs in the third row, and so on. A ball is placed above the top peg. When the ball is dropped and hits a peg, it is equally likely to fall left or right. We are interested in the location of the ball after striking five pegs – as shown in the figure, the ball can land in locations 0, 1, 2, 3, 4, or 5.

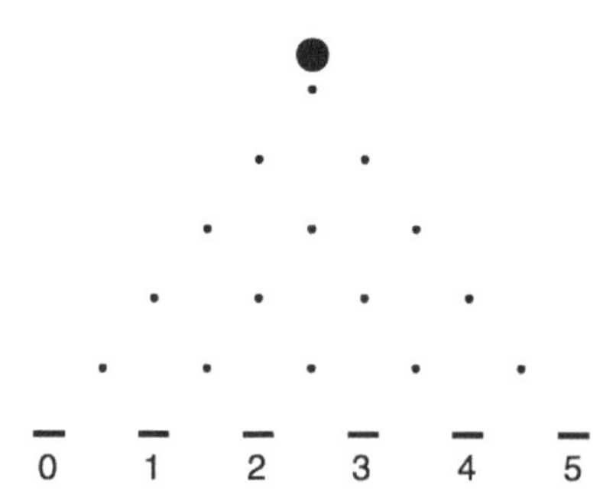

FIGURE 4.4
Illustration of a Galton board.

Figure 4.5 shows the path of four balls that fall through a Galton board. The chances of falling in the locations follow a special probability distribution that has a strong connection with a simple coin-tossing experiment.

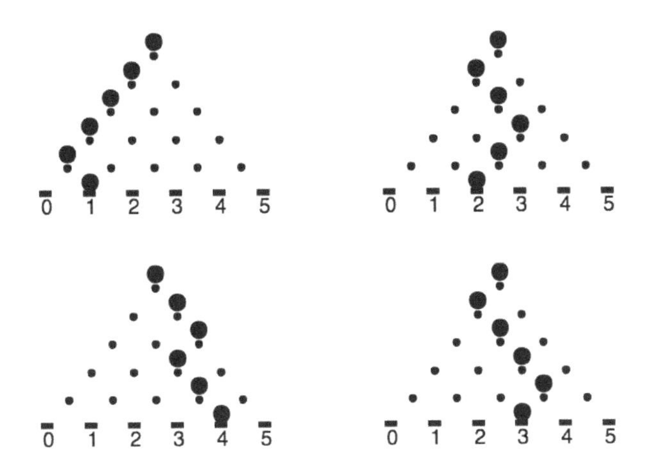

FIGURE 4.5
Illustration of the path of four balls falling through a Galton board.

Consider the following random experiment. One takes a quarter and flip it ten times, recording the number of heads one gets. There are four special characteristics of this simple coin-tossing experiment.

1. One is doing the same thing (flip the coin) ten times. We will call an individual coin flip a trial, and so our experiment consists of ten identical trials.
2. On each trial, there are two possible outcomes, heads or tails.
3. In addition, the probability of flipping heads on any trial is 1/2.
4. The results of different trials are independent. This means that the probability of heads, say, on the fourth flip, does not depend on what happened on the first three flips.

One is interested in the number of heads one gets – this number will be referred to X. In particular, one is interested in the probability of getting five heads, or $Prob(X = 5)$.

In this section, one will see that this binomial probability model applies to many different random phenomena in the real world. Probability computations for the binomial and the closely related negative binomial models will be discussed and the usefulness of these models in representing the variation in real-life experiments will be illustrated.

4.5.1 Binomial probabilities

Let's return to our experiment where a quarter is flipped ten times, recording X, the number of heads. One is interested in the probability of flipping exactly five heads, that is, $Prob(X = 5)$. To compute this probability, one first has to think of possible outcomes in this experiment. Suppose one records if each flip is heads (H) or tails (T). Then one possible outcome with ten flips is

Trial	1	2	3	4	5	6	7	8	9	10
Result	H	H	T	T	H	T	T	H	H	T

Another possible outcome is $TTHHTHTHHH$. The sample space consists of all possible ordered listings of ten letters, where each letter is either an H or a T.

Next, consider computing the probability of a single outcome of ten flips such as the $HHTTHHTHHT$ sequence shown above. The probability of this outcome is written as

$$P(\text{“}H \text{ on toss 1” AND “}H \text{ on toss 2” AND ... AND “}T \text{ on toss 10”}).$$

Using the fact that outcomes on different trials are independent, this probability is written as the product

$$P(H \text{ on toss } 1) \times P(H \text{ on toss } 2) \times ... \times P(T \text{ on toss } 10).$$

Since the probability of heads (or tails) on a given trial is 1/2, one has

$$P(HHTTHHTTHT) = \frac{1}{2} \times \frac{1}{2} \times ... \times \frac{1}{2} = \left(\frac{1}{2}\right)^{10}.$$

Actually, the probability of any outcome (sequence of ten letters with H's or T's) in this experiment is equal to $\left(\frac{1}{2}\right)^{10}$.

Let's return to our original question – what is the probability that one gets exactly five heads? If one thinks of the individual outcomes of the ten trials, then one will see that there are many ways to get five heads. For example, one could observe

$$HHHHHTTTTT \text{ or } HHHHTTTTTH \text{ or } HHHTTTTTHH$$

In each of the three outcomes, note that the number of heads is five. How many outcomes (like the ones shown above) will result in exactly five heads? As before, label the outcomes of the individual flips by the trial number:

Trial	1	2	3	4	5	6	7	8	9	10
Outcome	____	____	____	____	____	____	____	____	____	____

If five heads are observed, then one wishes to place five H's in the ten slots above. In the outcome $HHHHHTTTTT$, the heads occur in trials 1, 2, 3, 4, 5, and in the outcome $HHHTTTTTHH$, the heads occur in trials 1, 2, 3, 9, and 10. If one observes exactly 5 heads, then one must choose five numbers from the possible trial numbers 1, 2, ..., 10 to place the five H's. There are $\binom{10}{5}$ ways of choosing these trial numbers. Note that the order in which one chooses the trial numbers is not important. Since there are ways of getting exactly five heads, and each outcome has probability $\left(\frac{1}{2}\right)^{10}$, one sees that

$$Prob(X = 5) = \binom{10}{5}\left(\frac{1}{2}\right)^{10} = 0.246.$$

From the complement property, one sees that the *Prob*(five heads are *not* tossed) $= 1 - 0.246 = 0.754$. It is interesting to note that although one expects to get five heads when flipping a coin ten times, it is actually much more likely *not* to flip five heads than to flip five heads.

Binomial experiments

Although the coin tossing experiment described above seems pretty artificial, many random experiments share the same basic properties as coin tossing. Consider the following binomial experiment:

1. One repeats the same basic task or trial many times – let the number of trials be denoted by n.
2. On each trial, there are two possible outcomes, which are called "success" or "failure". One could call the two outcomes "black" and "white", or "0" or "1", but they are usually called success and failure.
3. The probability of a success, denoted by p, is the same for each trial.
4. The results of outcomes from different trials are independent.

Here are some examples of binomial experiments.

Example: A sample survey. Suppose the Gallup organization is interested in estimating the proportion of adults in the United States who use the popular auction website eBay. They take a random sample of 100 adults and 45 say that they use eBay. In this story, we see that

1. The results of this survey can be considered to be a sequence of 100 trials where one trial is asking a particular adult if he or she uses eBay.
2. There are two possible responses to the survey question – either the adult says "yes" (he or she uses eBay) or "no" (he or she doesn't use eBay).
3. Suppose the proportion of all adults that use eBay is p. Then the probability that the adult says "yes" will be p.

4. If the sampling is done randomly, then the chance that one person says "yes" will not depend on the answers of the people who were previously asked. This means that the responses of different adults to the question can be regarded as independent events.

Example: A baseball hitter's performance during a game. Suppose you are going to a baseball game and your favorite player comes to bat five times during the game. This particular player is a pretty good hitter and his batting average is about 0.300. You are interested in the number of hits he will get in the game. This can also be considered a binomial experiment:

1. The player will come to bat five times – these five at-bats can be considered the five trials of the experiment ($n = 5$).
2. At each at-bat, there are two outcomes of interest – either the player gets a hit or he doesn't get a hit.
3. Since the player's batting average is 0.300, the probability that he will get a hit in a single at-bat is $p = 0.300$.
4. It is reasonable to assume that the results of the different at-bats are independent. That means that the chance that the player will get a hit in his fifth at-bat will be unrelated to his performance in the first four at-bats. We note that this is a debatable assumption, especially if you believe that a player can have a hot-hand.

Example: Sampling without replacement. Suppose a committee of four will be chosen at random from a group of five women and five men. You are interested in the number of women that will be in the committee. Is this a binomial experiment?

1. If one thinks of selecting this committee one person at a time, then one can think this experiment as four trials (corresponding to selecting the four people).
2. On each trial, there are two possible outcomes – either one selects a woman or a man. At this point, things are looking good – this may be a binomial experiment. But...
3. Is the probability of choosing a woman the same for each trial? For the first pick, the chance of picking a woman is $5/10$. But once this first person has been chosen, the probability of choosing a woman is not $5/10$ – it will be either $4/9$ or $5/9$ depending on the outcome of the first trial. So the probability of a "success" is not the same for all trials, so this violates the third property of a binomial experiment.
4. Likewise, in this experiment, the outcomes of the trials are not independent. The probability of choosing a woman on the fourth trial is dependent on who was selected in the first three trials, so again the binomial assumption is violated.

4.5.2 Binomial computations

A binomial experiment is defined by two numbers

$$n = \text{the number of trials, and}$$

$$p = \text{probability of a "success" on a single trial.}$$

If one recognizes an experiment as being binomial, then all one needs to know is n and p to determine probabilities for the number of successes X. Using the same argument as was made in the coin-tossing example, one can show that the probability of x successes in a binomial experiment is given by

$$P(X = x) = \binom{n}{x} p^x (1 - p)^{n-x}, \ k = x..., n. \tag{4.3}$$

Let's illustrate using this formula for a few examples.

Example: A baseball hitter's performance during a game (revisited). Remember our baseball player with a true batting average of 0.300 is coming to bat five times during a game. What is the probability that he gets exactly two hits? It was shown earlier that this was a binomial experiment. Since the player has five opportunities, the number of trials is $n = 5$. If one regards a success as getting a hit, the probability of success on a single trial is $p = 0.3$. The random variable X is the number of hits of the player during this game. Using the formula, the probability of exactly two hits is

$$P(X = 2) = \binom{5}{2} (0.3)^2 (1 - 0.4)^{5-2} = 0.3087.$$

What is the probability that the player gets at least one hit? To do this problem, one first constructs the collection of binomial probabilities for $n = 5$ trials and probability of success $p = 0.3$. Table 4.10 shows all possible values of X (0, 1, 2, 3, 4, 5) and the associated probabilities found using the binomial formula.

TABLE 4.10
Possible values and associated probabilities for the baseball hitter.

x	$P(X = x)$
0	0.168
1	0.360
2	0.309
3	0.132
4	0.029
5	0.002

One is interested in the probability that the player gets at least one hit or $P(X \geq 1)$. "At least one hit" means that X can be 1, 2, 3, 4, or 5. To find this one simply sums the probabilities of X between 1 and 5:

$$P(X \geq 1) = P(X = 1, 2, 3, 4, 5) = 0.360+0.309+0.132+0.029+0.002 = 0.832.$$

There is a simpler way of doing this computation using the complement property of probability. We note that if the player does not get at least one hit, then he was hitless in the game (that is, $X = 0$). Using the complement property

$$P(X \geq 1) = 1 - P(X = 0) = 1 - 0.168 = 0.832.$$

R Binomial Calculations

By use of the `dbinom()` and `pbinom()` functions in R, one can perform probability calculations for any binomial distribution. In our baseball example the number of hits X is binomial with sample size 5 and probability of success $p = 0.3$. In the following R script a data frame is constructed with the possible values of the number of hits `x`, and the function `dbinom()` with arguments `size` and `prob` used to compute the binomial probabilities:

```
data.frame(x = 0:5) %>%
  mutate(Probability = dbinom(x, size = 5, prob = .3))
  x Probability
1 0     0.16807
2 1     0.36015
3 2     0.30870
4 3     0.13230
5 4     0.02835
6 5     0.00243
```

The function `pbinom()` will compute cumulative probabilities of the form $P(X \leq x)$. For example, to find the probability that number of hits X is 2 or less, $P(X \leq 2)$:

```
pbinom(2, size = 5, prob = .3)
[1] 0.83692
```

One computes the probability $P(X \geq 2)$ by finding the cumulative probability $P(X \leq 1)$, and subtracting the result from 1:

```
1 - pbinom(1, size = 5, prob = .3)
[1] 0.47178
```

R Simulating Binomial Experiments

One conveniently simulates outcomes from binomial experiments by use of the `rbinom()` function. The arguments to this function are the number of simulated draws, the number of binomial trials `size` and the probability of success `prob`. To illustrate, consider the baseball hitter who is coming to bat 5 times in a game where the probability of a hit on each at-bat is 0.3. One simulates the number of hits in 50 games by using arguments 50, `size = 5` and `prob = 0.3`.

```
(hits <- rbinom(50, size = 5, prob = 0.3))
 [1] 3 1 1 1 1 1 1 1 2 2 2 1 3 3 2 1 1 3 1 3 0 1
[23] 3 3 3 0 2 2 2 2 1 1 2 1 0 0 1 2 3 2 1 3 2 3
[45] 2 0 0 1 1 1
```

```
table(hits)
hits
 0  1  2  3
 6 20 13 11
```

By use of the `table()` function, we tally the outcomes. Here this player got exactly one hit in a game in 20 games, so the approximate probability that $X = 1$ is equal to $20/50 = 0.4$.

4.5.3 Mean and standard deviation of a binomial

There are simple formulas for the mean and variance for a binomial random variable. First let X_1 denote the result of the first binomial trial where

$$X_1 = \begin{cases} 1 & \text{if we observe a success} \\ 0 & \text{if we observe a failure} \end{cases}$$

In the end-of-chapter exercises, the reader will be asked to show that the mean and variance of X_1 are given by

$$E(X_1) = p, \;\; Var(X_1) = p(1-p).$$

If $X_1, ..., X_n$ represent the results of the n binomial trials, then the binomial random variable X can be written as

$$X = X_1 + ... + X_n.$$

Using this representation, the mean and variance of X are given by

$$E(X) = E(X_1) + ... + E(X_n), \;\; Var(X) = Var(X_1) + ... + Var(X_n).$$

The result about the variance is a consequence of the fact that the results of different trials of a binomial experiment are independent. Using this result and the previous result on the mean and variance of an individual trial outcome, we obtain

$$E(X) = p + ... + p = np, \tag{4.4}$$

and

$$Var(X) = p(1-p) + ... + p(1-p) = np(1-p). \tag{4.5}$$

To illustrate these formulas, recall the first example where X denoted the number of heads when a fair coin is flipped 10 times. Here the number of trials and probability of success are given by $n = 10$ and $p = 0.5$. The expected number of heads would be

$$E(X) = 10(0.5) = 5$$

and the variance of the number of heads would be

$$V(X) = 10(0.5)(1 - 0.5) = 2.5.$$

R Simulating Binomial Experiments (continued)

In our baseball example, the number of successes X were simulated in 50 binomial experiments where $n = 5$ and $p = 0.3$. The mean and standard deviation of X are given by $\mu = 5(0.3) = 1.5$ and $\sigma = \sqrt{5(.3)(1-.3)} = 1.02$. One approximates the mean and standard deviation by finding the sample mean and standard deviation from the simulated values of X. Below one sees that these approximate values agree closely with the exact values of μ and σ.

```
hits <- rbinom(50, size = 5, prob = 0.3)
mean(hits)
[1] 1.58
sd(hits)
[1] 0.9707981
```

4.5.4 Negative binomial experiments

The 2004 baseball season was exciting since particular players had the opportunity to break single-season records. Let's focus on Ichiro Suzuki of the Seattle Mariners who had the opportunity to break the season record for the most hits that was set by George Sisler in 1920. Sisler's record was 257 hits and Suzuki had 255 hits before the Mariners' game on September 30. Was it likely that Suzuki would tie Sisler's record during this particular game?

One can approximate this process as a coin-tossing experiment. When Suzuki comes to bat, there are two relevant outcomes: either he will get a hit, or he will get an out. Note that other batting plays such as a walk or sacrifice bunt that don't result in a hit or an out are ignored. Assume the probability that he gets a hit on a single at-bat is $p = 0.372$ (his 2004 batting average) and one assumes (for simplicity) that the outcomes on different at-bats are independent.

Suzuki needs two more hits to tie the record. How many at-bats will it take him to get two hits?

This is not a binomial experiment since the number of trials is not fixed. Instead the number of successes (hits) is fixed in advance and the number of trials to achieve this is random. Consider

$$Y = \text{number of at-bats to get two hits.}$$

One is interested in probabilities about the number of bats Y.

It should be obvious that Y has be at least 2 (he needs at least 2 at-bats to get 2 hits), but Y could be 3, 4, 5, etc. Let's find the probability that $Y = 5$.

First we know that the second hit must have occurred in the fifth trial (since Y=5). Also it is known that there must have been one hit and three outs in the first four trials – there are $\binom{4}{1}$ ways of arranging the H's and the O's in these trials.

H

H, 3 O's

Also the probability of each possible outcome is $p^2(1-p)^3$, where p is the probability of a hit. So the probability that it takes 5 trials to observe 2 hits is

$$P(Y = 5) = \binom{4}{1} p^2(1-p)^3.$$

Since $p = 0.372$ in this case, we get

$$P(Y = 5) = \binom{4}{1} 0.372^2(1 - 0.372)^3 = 0.1371.$$

A general negative binomial experiment is described as follows:

- One has a sequence of independent trials where each trial can be a success (S) or a failure.
- The probability of a success on a single trial is p.
- The experiment is continued until one observes r successes, and Y = number of trials one observes.

The probability that it takes y trials to observe r successes is

$$P(Y = y) = \binom{y-1}{r-1} p^r(1-p)^{y-r}, y = r, r+1, r+2, ... \tag{4.6}$$

Let's use this formula in our baseball example where $r = 2$ and $p = 0.372$. Table 4.11 gives the probabilities for the number of at-bats y = 2, 3, ..., 9.

TABLE 4.11
Probability distribution for the number of at-bats for Suzuki to get two additional hits.

y	$P(Y = y)$
2	.1384
3	.1738
4	.1637
5	.1371
6	.1076
7	.0811
8	.0594
9	.0426

Note that it is most likely that Suzuki will only need three at-bats to get his two additional hits, but the probability of three at-bats is only 17%. Actually each of the values 2, 3, 4, 5, and 6 have probabilities exceeding 10%. There is a significant probability that Suzuki will take a large number of bats – by adding the probabilities in Table 4.11, we see that the probability that Y is at most 9 is 0.904, so the probability that Y exceeds 9 is 1 - 0.904 = 0.096.

For a negative binomial experiment where Y is the number of trials needed to observe r successes, one can show that the mean value is

$$E(Y) = \frac{r}{p}. \tag{4.7}$$

For the baseball example, $r = 2$ and $p = 0.372$, so the expected number of at-bats to get two hits would be $E(Y) = 2/0.372 = 5.4$. It is interesting to note that although $Y = 3$ is the most probable value, Suzuki would average over 5 at-bats to get 2 hits in many repetitions of this random experiment.

R Negative Binomial Calculations and Simulations

The R functions `dnbinom()` and `rnbinom()` can be used to compute probabilities and simulate from negative binomial distributions. One small complication is that these functions define the random variable to be the number of failures (instead of the total number of trials) until the r-th success.

To illustrate the use of these functions, consider our baseball example where X is the number of at-bats for Suzuki to get $r = 2$ hits where the probability of a hit on a single at-bat is $p = 0.372$. The probability $P(X = 5)$ is the same as the probability $P(Y = 3)$ where Y is the number of failures until the second success. Using the function `dnbinom()`, one computes $P(Y = 3)$

```
dnbinom(3, size = 2, prob = .372)
[1] 0.137096
```

which is equivalent to the probability that $X = 5$ computed earlier. Also, `rnbinom()` can be used to simulate negative binomial experiments. For example, one can simulate the number of failures until the second success for 10 experiments as follows.

```
rnbinom(10, size = 2, prob = .372)
 [1]  4  1  2  3  1  3  2 15  0  1
```

It is interesting to note that Suzuki had 15 outs until the second success for one of these experiments.

4.6 Exercises

1. **Coin-tossing Game**

   In the Peter-Paul coin-tossing game described in the text, let the random variable X be the number of times Paul is in the lead. For example, if the coin tosses are $HTHHT$, Paul's running winnings are \$-2, 0, \$2, \$4, \$2, and the number of times he is in the lead is $X = 4$.

$HHHHH$	$HTHHH$	$THHHH$	$TTHHH$
$HHHHT$	$HTHHT$	$THHHT$	$TTHHT$
$HHHTH$	$HTHTH$	$THHTH$	$TTHTH$
$HHHTT$	$HTHTT$	$THHTT$	$TTHTT$
$HHTHH$	$HTTHH$	$THTHH$	$TTTHH$
$HHTHT$	$HTTHT$	$THTHT$	$TTTHT$
$HHTTH$	$HTTTH$	$THTTH$	$TTTTH$
$HHTTT$	$HTTTT$	$THTTT$	$TTTTT$

   (a) Find the probability distribution for X.
   (b) Construct a graph of the pmf for X.
   (c) What is the most likely value of X?
   (d) Find the probability that $X > 2$.

2. **Sampling Without Replacement**

   Suppose you choose two coins from a box with two nickels and three quarters. Let X denote the number of nickels you draw.

   (a) Write out all possible 10 outcomes of this experiment.
   (b) Find the probability distribution for X.
   (c) What is the most likely value of X?
   (d) Find the probability that $X > 1$.

3. **Shooting Free Throws**

Suppose you watch your favorite basketball player attempt five free throw shots during a game. You know that the chance that he is successful on a single shot is 0.5, so that the possible sequences of successes (S) and misses (M) shown below are equally likely. Suppose you measure the number of runs X where a run is defined to be a streak of S's or M's. For example, in the sequence $MMSSM$, there are three runs (one run of two misses, one run of two successes, and one run of one miss).

$SSSSS$	$SMSSS$	$MSSSS$	$MMSSS$
$SSSSM$	$SMSSM$	$MSSSM$	$MMSSM$
$SSSMS$	$SMSMS$	$MSSMS$	$MMSMS$
$SSSMM$	$SMSMM$	$MSSMM$	$MMSMM$
$SSMSS$	$SMMSS$	$MSMSS$	$MMMSS$
$SSMSM$	$SMMSM$	$MSMSM$	$MMMSM$
$SSMMS$	$SMMMS$	$MSMMS$	$MMMMS$
$SSMMM$	$SMMMM$	$MSMMM$	$MMMMM$

(a) Find the probability distribution for X.

(b) Construct a graph of the pmf for X.

(c) What is the most likely number of runs in the sequence?

(d) Find the probability that you have at most 2 runs in the sequence.

4. **Rolling Two Dice**

Suppose you roll two dice and you keep track of the larger of the two rolls which we denote by X. For example, if you roll a 4 and a 5, then $X = 5$.

(a) Find the probability distribution for X.

(b) Construct a graph of the pmf for X.

(c) What is the most likely value of X?

(d) Find the probability that X is either 5 or 6.

5. **Spinning a Spinner**

Let X denote the number you get when you spin the spinner shown below.

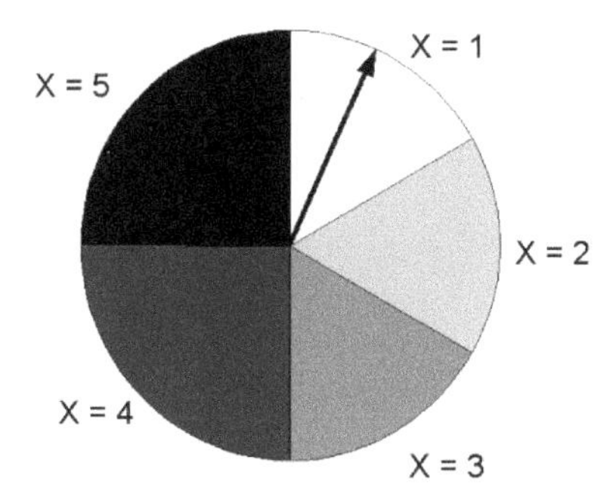

(a) Find the probability distribution for X.

(b) Find the probability that $X \geq 2$.

(c) Find the mean and standard deviation of X.

6. **Rolling Four Dice**

Suppose you are asked to roll four dice and record the sum X. A lazy student thinks this is too much work. As a shortcut, he decides to roll only two dice, record the sum of the dice, and then double the result – call this random variable Y.

The probability distributions of X and Y are shown in Tables 4.12 and 4.13. The distribution of X was obtained by simulating the rolls of four dice for one million trials.

TABLE 4.12
Probability distribution of X.

x	$P(X = x)$	x	$P(X = x)$
4	0.001	15	0.108
5	0.003	16	0.096
6	0.008	17	0.080
7	0.016	18	0.062
8	0.027	19	0.043
9	0.044	20	0.027
10	0.062	21	0.015
11	0.080	22	0.008
12	0.097	23	0.003
13	0.108	24	0.001
14	0.113		

(a) Compute the mean and standard deviation of the probability distributions of X and Y.

TABLE 4.13
Probability distribution of Y.

y	$P(Y = y)$	y	$P(Y = y)$
4	0.028	16	0.139
6	0.056	18	0.111
8	0.083	20	0.083
10	0.111	12	0.056
12	0.139	14	0.028
14	0.167		

(b) Plot the probability distributions of X and Y on the same graph.

(c) Compare and contrast the two probability distributions. How are the distributions similar? How are they different? How would you respond to the lazy student who thinks that doubling a two-dice result is equivalent to finding the sum of four fair dice?

7. **Running a Marathon Race**

Suppose three runners from college A and four runners from college B are participating in a marathon race. Suppose that all seven runners have equal abilities and so all possible orders of finish of the seven runners are equally likely. For example, one possible order of finish is $AAABBBB$ where the three A runners finish first, second, and third. Let X denote the finish position of the best runner from college A.

(a) Find the probability distribution of X.
(b) Find the probability that X is at most 2.
(c) Find the average finish of the best runner from college A.

8. **Choosing a Slip from a Random Box**

Suppose you roll a die. If the die roll is 1 or 2, you choose a slip from box 1; otherwise you choose a slip from box 2. Let Y denote the number on the slip.

(a) Find the probability distribution for Y.
(b) Find the probability that Y is between 2 to 4.

9. **A Random Walk**

Suppose that a person starts at location 0 on the number line and each minute he is equally likely to take a step to the left and to the right. Let Y denote the person's location after four steps.

(a) Find the probability distribution for Y.

(b) Find the probability that he is at least two steps away from his start after four steps.

(c) Suppose there is some gravitational pull towards the 0 (home) location. Then if he is currently at a negative location, the probability he will take a positive step is 0.7, and likewise if he is at a positive location, the probability he takes a negative step is 0.7. If he is at point 0, he is equally likely to take a negative or positive step. Find the probability distribution of Y.

(d) Compare the two probability distributions in parts (a) and (c) using the mean and standard deviation.

10. **Selecting a Prize from a Bag**

Suppose you select a prize (with replacement) from a bag that contains three prizes – one worth \$1, one worth \$5, and one worth \$10. You have three opportunities to select a prize and you get to keep the largest prize of the three you select. Let X denote the value of the prize you keep.

(a) Find the probability distribution of X.

(b) Find the probability you win more than \$1.

(c) Find your expected winning.

11. **Playing Roulette**

Suppose you place a single \$5 bet on three numbers (the Trio Bet) in roulette that has a payoff odds of 11 to 1. Let X denote your payoff. Recall that if you win you receive 11 times your betting amount plus your \$5 bet; if you lose, your payoff is nothing.

(a) Find the probability distribution for X.

(b) Find the mean of X. On average, how much money do you lose in a single \$5 bet?

(c) Consider placing \$5 instead on a Five Number Bet that pays at 6 to 1. Find the probability distribution for the payoff Y for this bet. Compute the mean of Y. How does this average payoff compare with the average payoff for the Trio Bet?

(d) Find the standard deviation of the payoffs for X and Y. Which bet has the larger standard deviation? Interpret what it means to have a large standard deviation.

12. **Sum of Independent Random Variables**

Suppose you have k random variables $X_1, ..., X_k$. Each random variable has a mean μ and a standard deviation σ. Suppose the random variables are independent – this means that the probability that one variable, say takes a value will not be affected by the values of the other random variables. In this case, it can be shown that the mean and standard deviation of the sum $S = X_1 + ... + X_k$ will have mean $E(S) = k\mu$ and standard deviation $SD(S) = \sqrt{k}\sigma$.

(a) It has been shown that if X denotes the roll of a single die, then the mean and standard deviation of X are given by $\mu = 3.5$ and $\sigma = 1.71$. Suppose you roll 10 dice and the outcomes of these dice are represented by $X_1, ..., X_{10}$. Using the above result, find the mean and standard deviation of the sum of these 10 rolls.

(b) Suppose you spin the spinner pictured here five times and record the sum of the five spins S. Find the mean and standard deviation of S. [Hint: First you need to find the mean and standard deviation of X, a single spin of the spinner. Then you can apply the above result.]

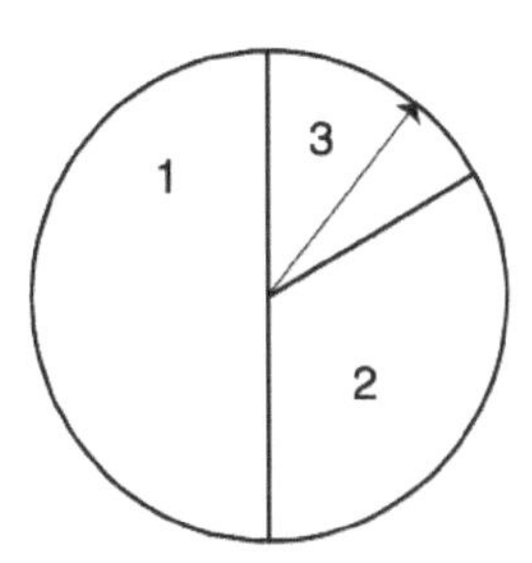

13. **Selecting a Coin from a Box**

Suppose you select a coin from a box containing 3 nickels, 2 dimes and one quarter. Let X represent the value of the coin.

(a) Find the probability distribution of X.

(b) Find the mean and standard deviation of X.

(c) Suppose that your instructor will give you twice the value of the coin that you select, so your profit is $Y = 2X$. Make intelligent guesses at the mean and standard deviation of Y.

(d) Check your guesses by actually computing the mean and standard deviation of Y.

(e) This is an illustration of a general result. If X has mean μ and standard deviation $\sigma$and $Y = cX$ where c is a positive constant, then the mean of Y is equal to ___________ and the standard deviation of Y is equal to ___________.

14. **How Many Tries to Open the Door?**

You have a ring with four keys, one of which will open your door. Suppose you try the keys in a random order until you open the door. Let X denote the number of wrong keys you try before you find the right one. It can be shown that X has the following distribution.

x	$P(X = x)$
0	1/4
1	1/4
2	1/4
3	1/4

(a) Find the mean and standard deviation of X.

(b) Suppose you record instead Y, the total number of keys you try. Note that $Y = X + 1$. Find the probability distribution for Y and the mean and standard deviation.

(c) This is an illustration of a general result. If X has mean μ and standard deviation σ and $Y = X + c$ for some constant c, then the mean of Y is equal to ____________ and the standard deviation of Y is equal to ____________.

15. **The Hat Check Problem**

Consider the hat check problem described in Section 4.1. Consider the special case where $n = 4$ men are checking their hats. If the names of the four men are represented by the initials A, B, C, D, then you can represent the hats given to these four men by the arrangements $ABCD, ABDC$, and so on.

(a) Write down the 24 possible arrangements and find the probability distribution for X, the number of matches.

(b) Find the probability of no matches.

(c) Find the expected number of matches.

16. **Binomial Experiments**

Is each random process described below a binomial experiment? If it is, give values of n and p. Otherwise, explain why it is not binomial.

(a) Roll a die 20 times and count the number of sixes you roll.

(b) There is a room of 10 women and 10 men – you choose five people from the room without replacement and count the number of women you choose.

(c) Same process as part (b) but you sample with replacement instead of without replacement.

(d) You flip a coin repeatedly until you observe 3 heads.

(e) The spinner below is spun 50 times – you count the number of spins in the black region.

17. **Binomial and Negative Binomial Experiments**

Each of the random processes below is a binomial experiment, a negative binomial experiment, or neither. If the process is binomial, give values of n and p, and if the process is negative binomial, give values of r and p.

(a) Suppose that 30% of students at a college regularly commute to school. You sample 15 students and record the number of commuters.

(b) Same scenario as part (a). You continue to sample students until you find two commuters and record the number of students sampled.

(c) Suppose that a restaurant offers apple and orange juice. From past experience, the restaurant knows that 30% of the breakfast customers order apple juice, 50% order orange juice, and 20% order no juice. One morning, the restaurant has 30 customers and the numbers ordering apple juice, orange juice, and no juice are recorded.

(d) Same scenario as part (c). The restaurant only records the number ordering orange juice out of the first 30 customers.

(e) Same scenario as part (c). The restaurant counts the number of customers that order breakfast until exactly three order apple juice.

(f) Same scenario as part (c). Suppose that from past experience, the restaurant knows that 40% of the breakfast bills will exceed $10. Of the first 30 breakfast bills, the number of bills exceeding $10 is observed.

18. **Shooting Free Throws**

Suppose that Michael Jordan makes 80% of his free throws. Assume he takes 10 free shots during one game.

(a) What is the most likely number of shots he will make?

(b) Find the probability that he makes at least 8 shots.

(c) Find the probability he makes more than 5 shots.

19. **Purchasing Audio CDs**

Suppose you know that 20% of the audio CD's sold in China are defective. You travel to China and you purchase 20 CD's on your trip.

(a) What is the probability that at least one CD in your purchase is defective?

(b) What is the probability that between 4 and 7 CD's are defective?

(c) Compute the "average" number of defectives in your purchase.

20. **Rolling Five Dice**

    Suppose you roll five dice and count the number of 1's you get.

    (a) Find the probability you roll exactly two 1's. Perform an exact calculation.
    (b) Find the probability all the dice are 1's. Perform an exact calculation.
    (c) Find the probability you roll at least two 1's. Perform an exact calculation.

21. **Choosing Socks from a Drawer**

    Suppose a drawer contains 10 socks, of which 4 are brown. We select 5 socks from the drawer with replacement.

    (a) Find the probability two of the five selected are brown.
    (b) Find the probability we choose more brown than non-brown.
    (c) How many brown socks do we expect to select?
    (d) Does the answer to part (a) change if we select socks from the drawer without replacement? Explain.

22. **Choosing Socks from a Drawer**

    Suppose that we select socks from the drawer with replacement until we see two that are brown.

    (a) Find the probability that it takes us four selections.
    (b) Find the probability it takes more than 2 selections.
    (c) How many selections do we expect to make?

23. **Sampling Voters**

    In your local town, suppose that 60% of the residents are supportive of a school levy that will be on the ballot in the next election. You take a random sample of 15 residents.

    (a) Find the probability that a majority of the sample support the levy.
    (b) How many residents in the sample do you expect will support the levy?
    (c) If you sample the residents one at a time, find the probability that it will take you five residents to find three that support the levy.

24. **Taking a True/False Test**

    Suppose you take a true/false test with twenty questions and you guess at the answers.

    (a) Find the probability you pass the test assuming that passing is 60% or higher correct.

(b) Find the probability you get a B or higher where B is 80% correct.

(c) If you get an 80% on this test, is it reasonable to assume that you were guessing? Explain.

25. **Bernoulli Experiment**

Let X_1 denote the result of one binomial trial, where $X_1 = 1$ if you observe a success and $X_1 = 0$ if you observe a failure. Find the mean and variance of X_1.

26. **Rolling a Die**

Suppose we roll a die until we observe a 6. This is a special case of a negative binomial experiment where $r = 1$ and $p = 1/6$. When we are interested in the number of trials until the first success, this is a geometric experiment and Y is a geometric random variable.

(a) Find the probability that it takes you 4 rolls to get a 6.

(b) Find the probability that it takes you more than 2 rolls to get a 6.

(c) How many rolls do you need, on average, to get a 6?

27. **Heights of Male Freshmen**

Suppose that one third of male freshmen entering a college are over 6 feet tall. Four men are randomly assigned to a dorm room. Let X denote the number of men in this room that are under 6 feet tall. You can ignore the fact that the actual sampling of men is done without replacement.

(a) Assuming X has a binomial distribution, what is a "success" and give values of n and p.

(b) What is the most likely value of X? What is the probability of this value?

(c) Find the probability that at least three men in this room will be under 6 feet tall.

28. **Basketball Shooting**

Suppose a basketball player is practicing shots from the free-throw line. She hasn't been playing for a while and she becomes more skillful in making shots as she is practicing. Let X represent the number of shots she makes in 50 attempts. Explain why the binomial distribution should not be used in finding probabilities about X.

29. **Collecting Posters from Cereal Boxes**

Suppose that a cereal box contains one of four posters and you are interested in collecting a complete set. You first purchase one box of cereal and find poster #1.

(a) Let X_2 denote the number of boxes you need to purchase to find a different poster than #1. Find the expected value of X_2.

(b) Once you have found your second poster, say #2, let X_3 denote the number of boxes you need to find a different poster than #1 or #2. Find the expected value of X_3.

(c) Once you have collected posters #1, #2, #3, let X_4 denote the number of boxes you need to purchase to get poster #4. Find the expected value of X_4.

(d) How many posters do you need, on average, to get a complete set of four?

30. **Baseball Hitting**

In baseball, it is important for a batter to get "on-base" and batters are rated in terms of their on-base percentage. In the 2004 baseball season, Bobby Abreu of the Philadelphia Phillies had 705 "plate appearances" or opportunities to bat. Suppose we divide his plate appearances into groups of five – we record the number of times Abreu was on-base for plate appearances 1 through 5, for 6 through 10, for 11 through 15, and so on. If we let X denote the number of times on-base for five plate appearances, then we observe the following counts for X:

x	0	1	2	3	4	5	Total
Count	10	29	44	40	15	3	141

To help understand this table, note that the count for $X = 1$ is 29 – this means there were 29 periods where Abreu was on-base exactly one time. The count for $X = 2$ is 44 – this means that for 44 periods Abreu was on-base two times.

Since each outcome is either a success or failure, where success is getting on-base, one wonders if the variation in these data can be explained by a binomial distribution.

x	0	1	2	3	4	5	TOTAL
$P(X = x)$							
Expected							
Count							

(a) Find the probabilities for a binomial distribution with $n = 5$ and $p = 0.443$. This value of p is Abreu's on-base rate for the entire 2004 baseball season. Place these probabilities in the $P(X = x)$ row of the table.

(b) Multiply the probabilities you found in part (a) by 141, the number of periods in the 2004 season. Place these numbers in the Expected Count row of the table. These represent the expected number of times Abreu would have 0, 1, 2, .., 5 times on-base if the probabilities followed a binomial distribution.

(c) Compare the expected counts with the actual observed counts in the first table. Does a binomial distribution provide a good description of these data?

31. **Graphs of Binomial Distributions**

Figure 4.6 shows the binomial distributions with $n = 20$ and $p = 0.5$ (above) and $n = 20$ and $p = 0.2$ (below).

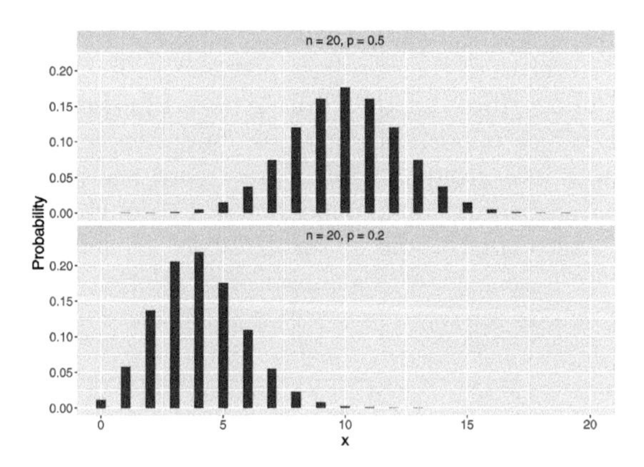

FIGURE 4.6
Histograms of two binomial distributions.

Recall in Section 4.4 that if a probability distribution is approximately bell-shaped, then approximately 68% of the probability falls within one standard deviation of the mean.

(a) For the binomial distribution with $n = 20$ and $p = 0.5$, find the mean μ and standard deviation σ and compute the interval $(\mu - \sigma, \mu + \sigma)$.

(b) Find the exact probability that X falls in the interval $(\mu - \sigma, \mu + \sigma)$.

(c) Repeat parts (a) and (b) for the binomial distribution $n = 20$ and $p = 0.2$.

(d) For which distribution was the 68% rule more accurate? Does that make sense based on the shapes of the two distributions?

32. **Guessing on a Test**

Students in a statistics class were given a five-question baseball trivia quiz. On each question, the students had to choose one of two possible answers. The number correct X was recorded for each student – a count table of the values of X are shown below.

$X = x$	Count	$P(X = x)$	Expected
0	0		
1	3		
2	4		
3	7		
4	6		
5	1		

(a) Suppose the students know little about baseball and so they are guessing on each question. If this is true, find the probability distribution of the number correct X.

(b) Using this distribution, find the probability of each value of X and place these probabilities in the above table.

(c) By multiplying these probabilities by the number of students (21), find the expected number of students for each value of X.

(d) Compare your expected counts with the actual counts – does a binomial distribution seem like a reasonable assumption in this example?

33. **Playing Roulette**

Suppose you play the game roulette 20 times. For each game, you place a Trio Bet on three numbers and you win with probability $3/38$.

(a) Find the probability you win the game exactly two times.

(b) Find the probability that you are winless in the 20 games.

(c) Find the probability you win at least once.

(d) How many games do you expect to win?

34. **The Galton Board**

Consider the Galton board described in Section 4.5. A ball is placed above the first peg and dropped. When it strikes a peg, it is equally likely to fall left or right. The location at the bottom X is equal to the number of times that the ball falls right.

(a) Explain why X has a binomial distribution and give the values of n and p.

(b) Find $P(X = 2)$.

(c) Find the probability the ball falls to the right of the location "1".

(d) Suppose that we change the experiment so that the probability of falling right is equal to $1/4$. Explain how this changes the binomial experiment and find $P(X = 2)$.

35. **Drug Testing**

In a *New York Times* article "Facing Questions, Rodriguez Raises More?" (February 21, 2008), Major League Baseball is said to have a drug-testing policy where 600 tests are randomly given to a group of 1200 professional ballplayers. Alex Rodriguez claimed one season that he received five random tests.

(a) If every player is equally likely to receive a single random blood test, what is the probability that Rodriguez gets tested?
(b) If X represents the number of tests administered to Rodriguez among the 600 tests, then explain why X has a binomial distribution and give the values of n and p.
(c) Compute the probability that Rodriguez receives exactly one test.
(d) Recall Rodriguez's claim that he received five random tests. Compute the probability of this event.
(e) You should find the probability computed in part (d) to be very small. If Rodriguez is indeed telling the truth, what do you think about the randomness of the drug-testing policy?

R Exercises

36. **Peter-Paul Game**

(a) Implement the Peter-Paul game simulation as described in the text, storing 1000 values of the gain variable in the R variable `G`.
(b) Use the simulated values to estimate the probability $P(G > 2)$.
(c) Estimate the standard deviation of G from the simulated values.

37. **The Hat Check Problem (continued)**

Suppose that $n = 10$ men are checking their hats. It would be too tedious to write down all $10! = 3,628,800$ possible arrangements of hats, but it is straightforward to design a simulation experiment for this problem.

(a) Write a function to mix up the integers 1 through 10 and returning the number of matches.
(b) Using the function written in part (a), simulate this experiment 1000 times. Approximate the probability of no matches and the expected number of matches. Compare your answers with the "large sample" answers given in the introduction to this chapter.

38. **A Random Walk (continued)**

Suppose that a person starts at location 0 on the number line and each minute he is equally likely to take a step to the left and to the right. Let Y denote the person's location after four steps.

(a) Write a function to implement one random walk, returning the person's location after four steps.

(b) By use of the `replicate()` function, simulate this random walk for 1000 iterations. Summarize the simulated locations by a mean and standard deviation.

(c) Make an adjustment to your function so that if the person is currently at a negative location, the probability he will take a positive step is 0.7, and likewise if he is at a positive location, the probability he takes a negative step is 0.7. (If he is at point 0, he is equally likely to take a negative or positive step.) Simulate this adjusted random walk 1000 iterations. Compute the mean and standard deviation of this new random walk and compare to the values computed in part (b).

39. **Dice Rolls**

(a) Construct a data frame with variables `roll1`, `roll2`, ..., `roll5`, each containing 1000 simulated rolls of a fair die.

(b) Using the function `pmax()` as shown below, define a new variable `Max` that is equal to the maximum among the five rolls for each of the 1000 iterations.

```
Max <- pmax(roll1, roll2, roll3, roll4, roll5)
```

(c) Estimate the probability that the maximum roll is equal to 6.

(d) Estimate the mean and standard deviation of the maximum roll.

40. **Binomial Experiments**

(a) Suppose 25 percent of the students are commuters. You take a survey of 12 students and count X the number of commuters. Simulate 1000 surveys using the function `rbinom()`, storing the number of commuters in these 1000 samples.

(b) Approximate the probability that exactly 3 people in your sample are commuters.

(c) Compute the sample mean and standard deviation of the simulated values and compare with the exact values of the mean μ and standard deviation σ.

5

Continuous Distributions

5.1 Introduction: A Baseball Spinner Game

The baseball board game All-Star Baseball has been honored as one of the fifty most influential board games of all time according to the Wikipedia Encyclopedia (http://en.wikipedia.org). This game is based on a collection of spinner cards, where one card represents the possible batting accomplishments for a single player. The game is played by placing a card on a spinner and a spin determines the batting result for that player.

A spinner card is constructed by use of the statistics collected for a player during a particular season. To illustrate this process, the table below shows the batting statistics for the famous player Mickey Mantle for the 1956 baseball season. When Mantle comes to bat, that is called a plate appearance (PA) – we see from the table that he had 632 plate appearances this season. There were several events possible when Mantle came to bat – he could get a single (1B), a double (2B), a triple (3B), or a home run (HR). Also he could walk (BB), strike out (SO), or get other type of out.

PA	1B	2B	3B	HR	BB	SO	Other OUTS
632	109	22	5	52	99	112	233

The probability of each type of event can be found by dividing each count by the number of plate appearances. Each probability is converted to an angle on the spinner by multiplying each probability by the total number of degrees (360). From these degree measurements, a spinner is constructed, displayed in Figure 5.1, where the area of each wedge of the circle is proportional to the probability of that event occurring. A single plate appearance of Mickey Mantle can be simulated by spinning the spinner and observing the batting event.

PA	1B	2B	3B	HR	BB	SO	Other OUTS
632	109	22	5	52	99	112	233
Probability	0.172	0.035	0.008	0.082	0.157	0.177	0.369
Degrees in spinner	62	13	3	30	57	64	133

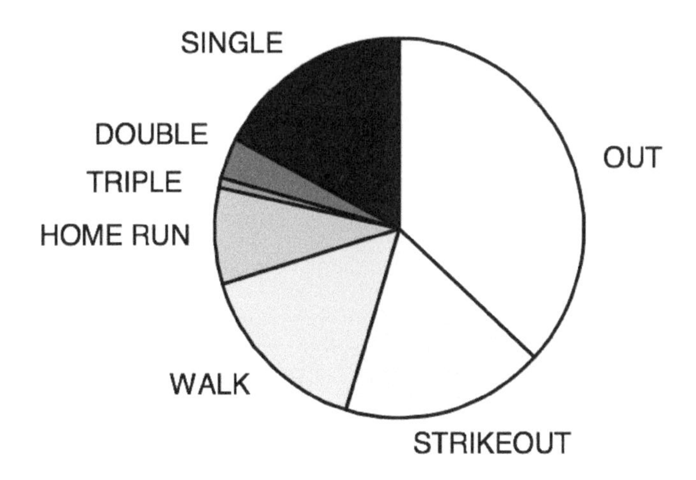

FIGURE 5.1
Spinner constructed based on Mantle's statistics.

The binomial described in Chapter 4 is an example of a discrete random variable which takes on only values in a list, such as $\{0, 1, ..., 10\}$. How can one think about probabilities where the random variable is not discrete? As a simple example, consider the experiment of spinning the spinner in Figure 5.2 where the random variable X is the recorded location. Here X is a continuous random variable that can take on any value between 0 and 100.

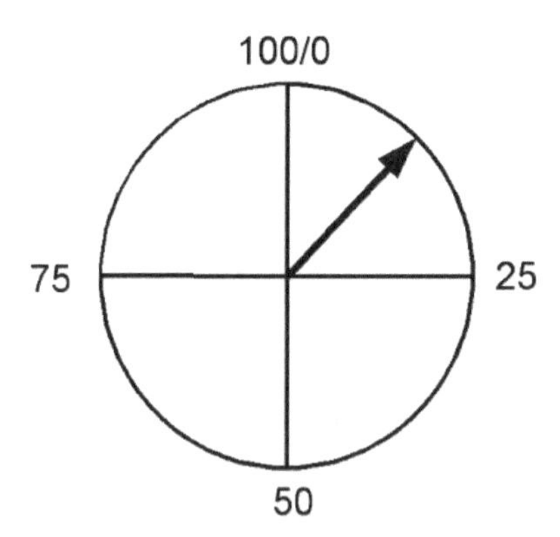

FIGURE 5.2
A spinner with continuous random outcomes.

In this chapter, probabilities for a continuous random variable will be shown to be represented by means of a smooth curve where the probability that X falls in a given interval is equal to an area under the curve. Through a series of examples, we will illustrate probability calculations for this type of random variables.

5.2 The Uniform Distribution

Consider the spinner experiment described in Section 5.1 where the location of the spinner X can be any number between 0 and 100. Our computer simulated spinning this spinner 20 times with the following results (rounded to the nearest tenth):

95.0	23.1	60.7	48.6	89.1	76.2	45.6	1.9	93.5	91.7
82.1	44.5	61.5	79.2	92.2	73.8	17.6	40.6	41.0	89.4

A histogram of these values of X is shown in the Figure 5.3.

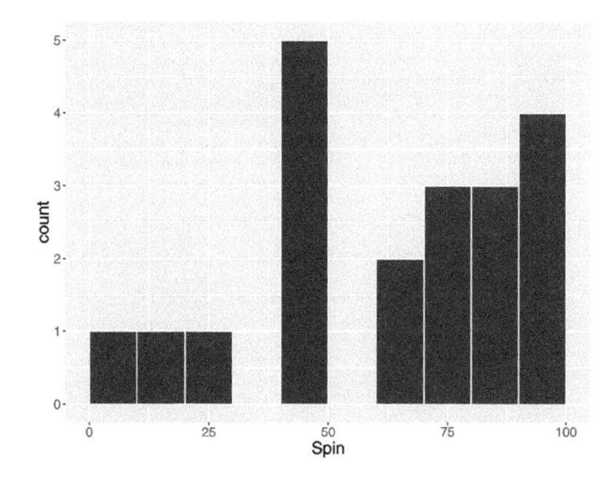

FIGURE 5.3
Histogram of 20 simulated values of a spinner.

Although one thinks that any spin between 0 and 100 is equally likely to occur, there does not appear to be any obvious shape of this histogram. But the spinner was only spun 20 times. Let's try spinning 1000 times– a histogram of the spins is shown in Figure 5.4.

Note that since there is a large sample of values, a small interval width was chosen for each bin in the histogram. Now a clearer shape in the histogram can be seen – although there is variation in the bar heights, the general shape of the histogram seems to be pretty flat or uniform over the entire interval of possible values of X between 0 and 100.

Suppose one was able to spin the spinner a large number of times. If one does this, then the shape of the histogram looks close to the uniform density shown in Figure 5.5.

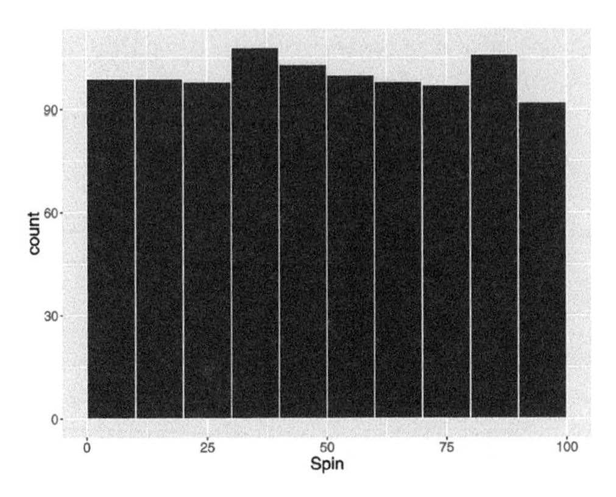

FIGURE 5.4
Histogram of 1000 simulated values of the spinner.

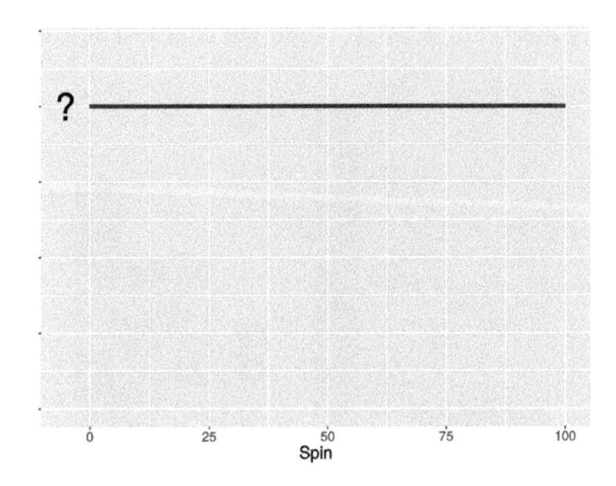

FIGURE 5.5
Shape of the histogram for a large number of simulated values of the spinner.

When the random variable X is continuous, such as the case of the spinner result here, then one represents probabilities by means of a smooth curve that is called a density curve; more formally, a probability density curve. How does one find probabilities? When X is continuous, then probabilities are represented by areas under the density curve.

As a simple example, what is the chance that the spinner result falls between 0 and 100? Since the scale of the spinner is from 0 to 100, one knows that all spins must fall in this interval, so the probability of X landing in (0, 100) is 1. This probability is represented by the total area under the flat line between 0 and 100. Since the area of this rectangle is given by height times base, and the base is equal to 100, the height of this density curve must be $1/100 = 0.01$. This is the value that should replace the "?" in Figure 5.5. In this case, one says that the spinner result has a uniform distribution and the curve is a uniform density.

By means of similar area computations, one finds other probabilities about the spinner location X.

1. What is the probability the spin falls between 20 and 60? That is, what is
$$P(20 < X < 60)?$$
This probability is equal to the shaded area under the uniform density between 20 and 60. See Figure 5.6. Using again the formula for the area of a rectangle, the base is $60 - 20 = 40$ and the height is 0.01, so
$$P(20 < X < 60) = 40(0.01) = 0.4.$$

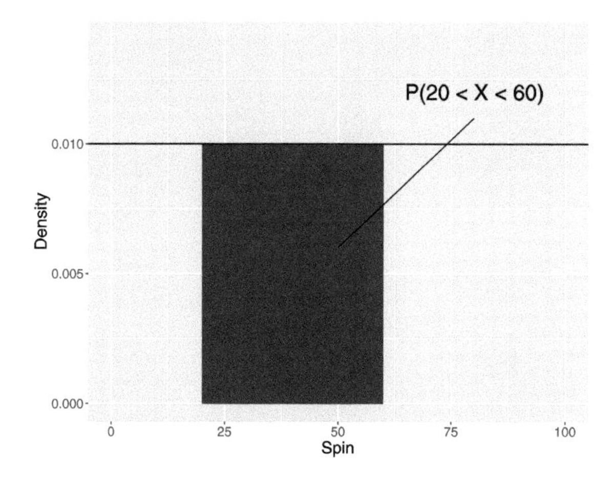

FIGURE 5.6
Illustration of finding the probability of $P(20 < X < 60)$.

2. What is the probability the spin is greater than 80? That is, what is $P(X > 80)$? Figure 5.7 shows the area that needs to be computed to find this probability. Note that the area under the curve only between the values 80 and 100 is shaded, since X cannot be larger than 100. Again by finding the area of the shaded rectangle, we see that $P(X > 80) = 20\ (0.01) = 0.2$.

R Simulating from a Uniform Density

The R function `runif()` is helpful for simulating from a uniform density. The arguments are the number of simulations and the minimum and maximum value of the support of the density. Below 50 values of a random spinner are simulated that fall uniformly on the interval from 0 to 50. The histogram in

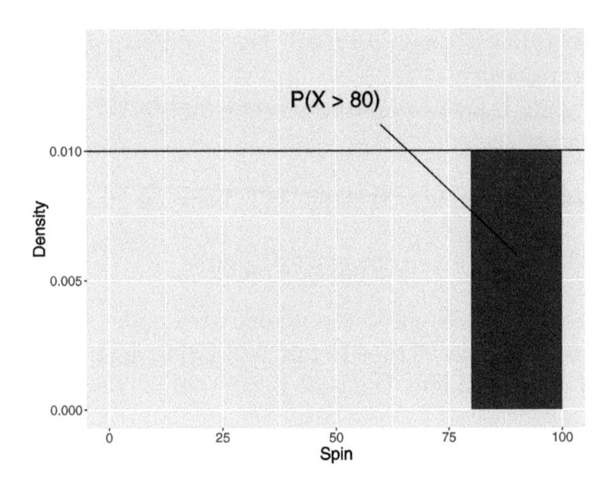

FIGURE 5.7
Illustration of finding the probability of $P(X > 80)$.

Figure 5.8 graphs these simulated spins with the uniform density drawn on top.

```
spins <- runif(50, min = 0, max = 50)
```

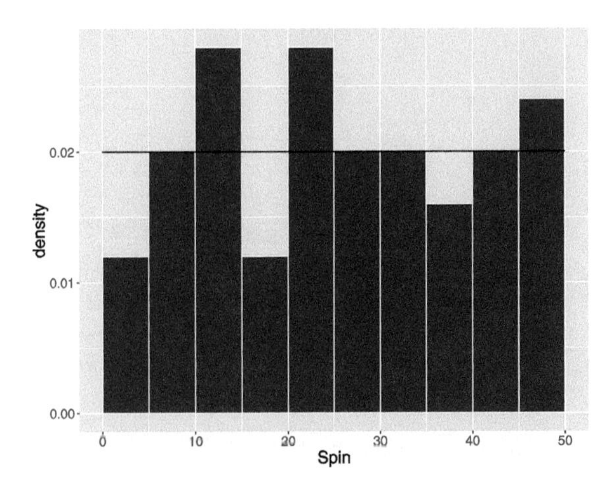

FIGURE 5.8
Histogram of 50 simulated uniform values.

5.3 Probability Density: Waiting for a Bus

Consider a random experiment where a continuous random variable X is observed such as the location of the spinner in Section 5.2. Define the *support* of X to be the set of possible values for X. For example, the support of X for the spinner example is the interval (0, 100). To describe probabilities about X, a density function denoted by $f(x)$ is defined. Any function f will not work – one requires that f satisfy two properties:

Property 1. The probability density f must be **nonnegative** which means that

$$f(x) \geq 0, \text{for all } x. \tag{5.1}$$

Property 2. The total area under the probability density curve f must be equal to 1. Mathematically,

$$\int_{-\infty}^{\infty} f(x)dx = 1. \tag{5.2}$$

To illustrate a probability density, suppose that a professor has a class that meets three times a week. To get to class, the professor walks and waits for a bus to go to school. From past experience, the professor knows that she can wait any time between 0 and 10 minutes for the bus, and she knows that each waiting time between 0 and 10 minutes is equally likely.

For a given week, what's the chance that her longest wait will be under 7 minutes?

Let W denote her longest waiting time for the week. One can show that the density for W is given by

$$f(w) = \frac{3w^2}{1000}, 0 < w < 10.$$

This density for this longest waiting time is shown in Figure 5.9.

Before we go any further, we should check if this is indeed a legitimate probability density:

1. Note from the graph that the density does not take on negative values, so the first property is satisfied.

2. Second, for it to be a probability density, the entire area under the curve must be equal to 1. One can check this by finding the integral of the density between 0 and 10 (the region where the density is positive):

$$\int_0^{10} \frac{3w^2}{1000} dw = \frac{w^3}{1000}\Big|_0^{10} = \frac{10^3}{1000} - \frac{0^3}{1000} = 1.$$

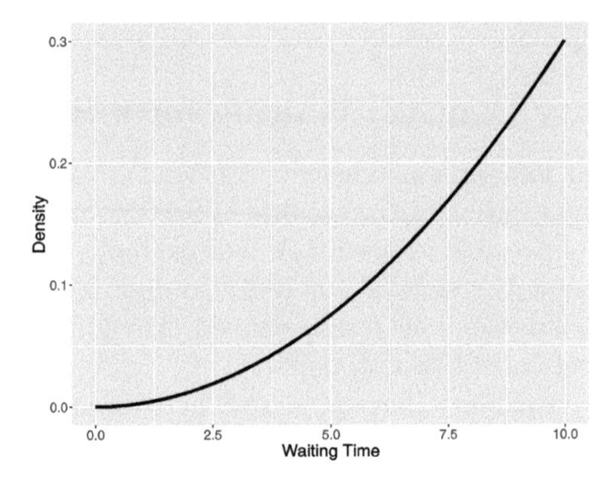

FIGURE 5.9
Density curve for the longest waiting time W.

The entire area under the curve is indeed equal to 1, so f is a legitimate probability density. Now that f is known to be a probability density, one can use it to find probabilities. To find the probability that this longest waiting time is less than 7 minutes, $P(W < 7)$, one wishes to compute the area under the density curve between 0 and 7, as shown in Figure 5.10.

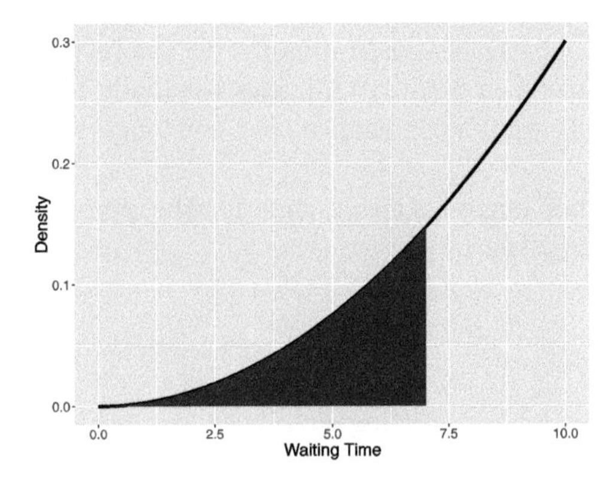

FIGURE 5.10
Density curve for the longest waiting time W, and $P(W < 7)$.

This is equivalent to the integral

$$\int_0^7 \frac{3w^2}{1000} dw$$

and, by evaluating this, one obtains the probability

$$\int_0^7 \frac{3w^2}{1000} dw = \frac{w^3}{1000}\Big|_0^7 = \frac{7^3}{1000} - \frac{0^3}{1000} = 0.343.$$

Suppose one is interested in the probability that the longest waiting time is between 6 and 8 minutes. This is represented by the shaded area in Figure 5.11.

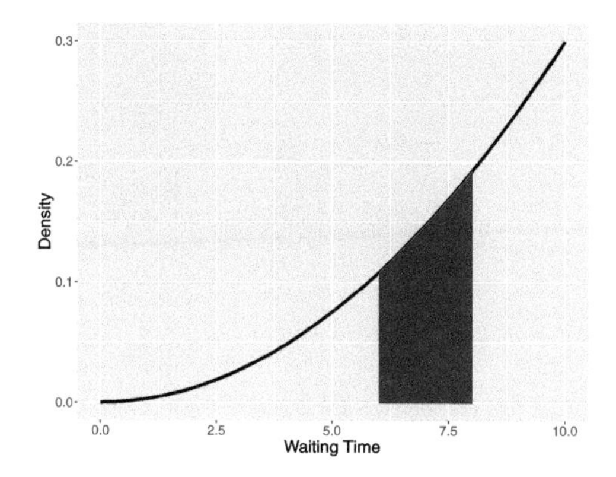

FIGURE 5.11
Density curve for the longest waiting time W, and $P(6 < W < 8)$.

To compute this area, one finds the integral of the density between 6 and 8:

$$\int_6^8 \frac{3w^2}{1000} dw = \frac{w^3}{1000}\bigg|_6^8 = \frac{8^3}{1000} - \frac{6^3}{1000} = 0.296.$$

R Simulating Waiting Times

Recall that the waiting time variable W was defined as the longest waiting time for the week where each of the separate waiting times has a uniform distribution from 0 to 10 minutes. By simulating the process, one simulates values of W. By use of three applications of `runif()` one simulates 1000 waiting times for Monday, Wednesday, and Friday. The `pmax()` function is used to simulate the longest waiting time for each group of waiting times.

```
wait_monday <- runif(1000, min = 0, max = 10)
wait_wednesday <- runif(1000, min = 0, max = 10)
wait_friday <- runif(1000, min = 0, max = 10)
longest_wait <- pmax(wait_monday,
                     wait_wednesday,
                     wait_friday)
```

Figure 5.12 shows 1000 simulated values of W and the density function $3w^2/1000$ is drawn on top. It appears that the histogram is a good match to the actual density function.

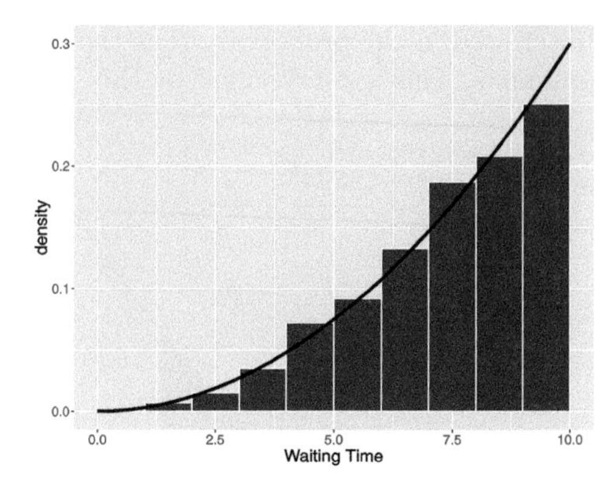

FIGURE 5.12
Histogram of 1000 simulated values of W with the density function drawn on top.

5.4 The Cumulative Distribution Function

To find any probability about the maximum waiting time, one computes an area under the curve that is equivalent to integrating the density curve over a region. But there is a basic function that can be computed at the beginning that will simplify these probability computations.

Choose an arbitrary point x – the cumulative distribution function at x, or cdf for short, is the probability that W is less than or equal to x:

$$F(x) = P(W \leq x) = \int_{-\infty}^{x} f(w)dw. \tag{5.3}$$

Here suppose one chooses a value of x in the interval (0, 10). Then $F(x)$ would be the area under the density curve between 0 and x shown in Figure 5.13.

Writing this area as an integral, one computes $F(x)$ as

$$F(x) = P(W \leq x) = \int_{0}^{x} \frac{3w^2}{1000} dw = \frac{w^3}{1000}\Big|_0^x = \frac{x^3}{1000}.$$

This formula is valid for any value of x in the interval (0, 10).

In fact, $F(x)$ is defined for all values of x on the real line.

- If x is a value smaller than or equal to 0, then we see from the figure that the probability that W is smaller than x is equal to 0. So $F(x) = 0$ for $x \leq 0$.

- On the other hand, if x is greater than or equal to 10, then the probability that W is smaller than x is 1. So $F(x) = 1$ for $x \geq 10$.

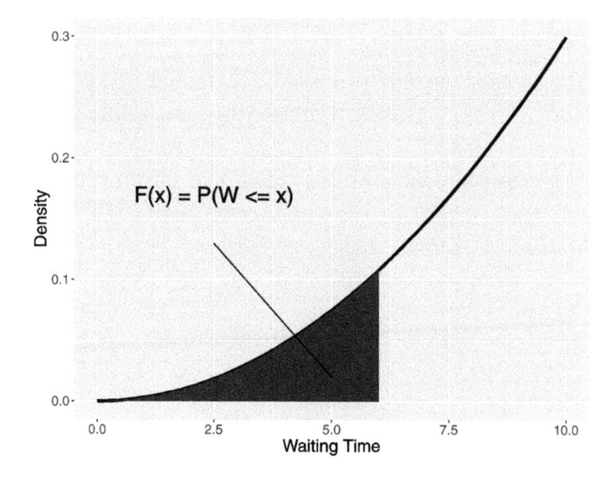

FIGURE 5.13
Illustration of the cumulative density function.

Putting all together, one sees that the cdf F is given by

$$F(x) = \begin{cases} 0, & x \leq 0 \\ x^3/1000, & 0 < x < 10 \\ 1, & x \geq 10, \end{cases}$$

illustrated in Figure 5.14.

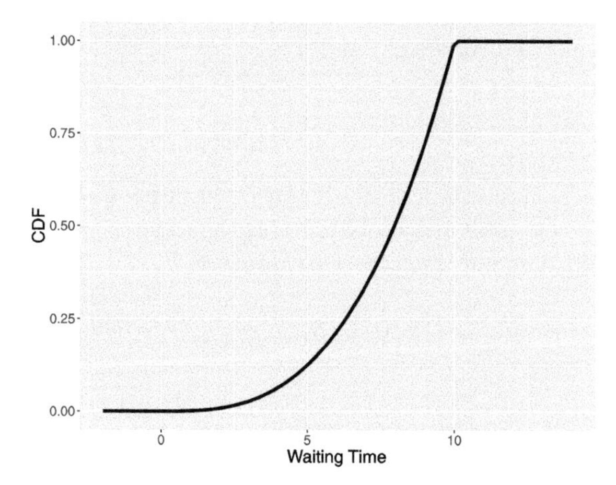

FIGURE 5.14
The cumulative density function, $F(x)$, of the bus waiting example.

Finding probabilities using the CDF

Once we have computed the cdf function F, probabilities are found simply by evaluating F at different points. Fortunately, no additional integration is needed.

For example, to find the probability that the maximum waiting time W is less than equal to 6 minutes, one just computes $F(6) = P(W \leq 6) = 6^3/1000 = 0.216$ which is shown in Figure 5.15.

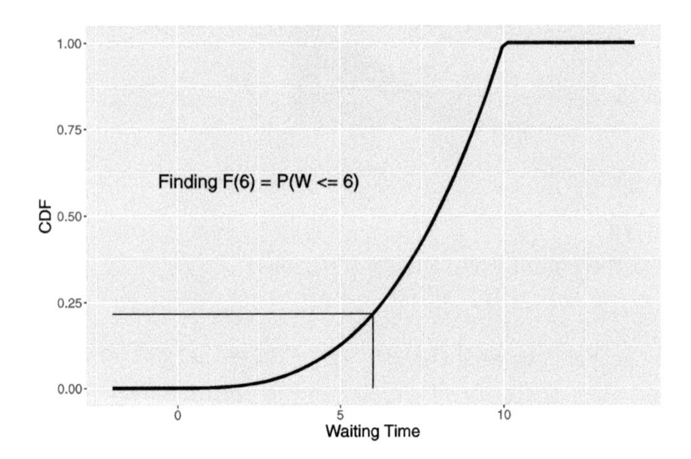

FIGURE 5.15
The cumulative density function $F(x)$ and evaluation of $F(6) = P(W \leq 6)$.

To compute the probability that the maximum waiting time exceeds 8 minutes, first note that "exceeding 8 minutes" is the complement event to "less than or equal to 8 minutes", and so

$$P(W > 8) = 1 - P(W \leq 8) = 1 - F(8) = 1 - \frac{8^3}{1000} = 0.488.$$

Likewise, if one is interested in the chance that the waiting time W falls between 2 and 4, represent the probability as the difference of two "less-than" probabilities, and then subtract the two values of F.

$$P(2 < W < 4) = P(W \leq 4) - P(W \leq 2) = F(4) - F(2) = \frac{4^3}{1000} - \frac{2^3}{1000} = 0.056.$$

R Computing Probabilities by Simulation

For the waiting for a bus example, the variable `longest_wait` contains 1000 simulated values of our longest waiting time. This sample is used to compute approximate probabilities. To illustrate, to find the probability that the longest wait exceeds 8 minutes, one finds the proportion of simulated values of W that exceeds 8.

```
mean(longest_wait > 8)
[1] 0.502
```

In a similar fashion one approximates the probability that a longest waiting time falls between 6 and 10 minutes.

```
mean(longest_wait > 6 & longest_wait < 10)
[1] 0.798
```

5.5 Summarizing a Continuous Random Variable

Mean and standard deviation

One is interested in summarizing a continuous random variable. Natural summaries are given by the mean μ and the standard deviation σ, where these quantities are defined in a similar manner as for a discrete random variable, with the exception that summations are replaced by integrals.

The mean μ, or equivalently the expected value of X, is given by

$$\mu = E(X) = \int_{-\infty}^{\infty} x f(x) dx. \tag{5.4}$$

Just as in the discrete random variable case, there is an attractive interpretation of μ. If one is able to observe a large number of values of X, then μ will be approximately equal to the sample mean $\bar{X}$ of these random values of X.

To define the spread of the values of X, one first computes the average squared deviation about the mean, the variance,

$$\sigma^2 = Var(X) = E(X - \mu)^2 = \int_{-\infty}^{\infty} (x - \mu)^2 f(x) dx. \tag{5.5}$$

The standard deviation of X, σ, is defined to be the square root of the variance.

Let's illustrate the computation of the mean and standard deviation for the bus waiting time problem. Using the definition of f, one gets that the mean is equal to

$$\mu = \int_0^{10} x \left(\frac{3x^2}{1000} \right) dx.$$

Performing the integration, one gets

$$\mu = \int_0^{10} x \left(\frac{3x^2}{1000} \right) dx = \frac{3x^4}{4000}\Big|_0^{10} = \frac{3(10)^4}{1000} = 7.5.$$

On, the average, one expects the longest wait in a week to be 7.5 minutes.

The computation of the variance is a bit more tedious, but straightforward.

$$\sigma^2 = \int_0^{10} (x - \mu)^2 \left(\frac{3x^2}{1000}\right) dx = 3.75.$$

So the standard deviation of X is $\sigma = \sqrt{3.75} = 1.94$.

R Computing the Mean and Standard Deviation by Simulation

Earlier, we demonstrated simulating 1000 values of the longest waiting time W. To check the computations of the mean μ and standard deviation σ, one computes the sample mean and standard deviation of the simulated values.

```
mean(longest_wait)
[1] 7.581979
sd(longest_wait)
[1] 1.878144
```

One sees that these empirical values are close approximations to the exact values $\mu = 7.5$ and $\sigma = 1.94$.

Percentiles

Another useful summary of a continuous random variable is a percentile. The 70th percentile, for example, is the value of X, call it x, such that 70% of the probability is to the left, shown in Figure 5.16. That is, the 70th percentile, call it x_{70}, satisfies the equation

$$P(X \leq x_{70}) = 0.70.$$

Since one recognizes the left hand side of the equation as equivalent to the cdf F (which already has been computed as $x^3/1000$), the equation is written as

$$F(x_{70}) = 0.70,$$

that is,

$$\frac{x_{70}^3}{1000} = 0.70.$$

To find the 70th percentile, the above equation is solved for x_{70} – after some algebra, we get

$$x_{70} = \sqrt[3]{700} = 8.88.$$

This means that approximately 70% of the longest waiting times will be shorter than 8.88 minutes over a duration of many weeks.

R Computing Percentiles by Simulation

For the waiting for a bus example, the variable `longest_wait` contains 1000 simulated values of our longest waiting time. This sample is used to compute

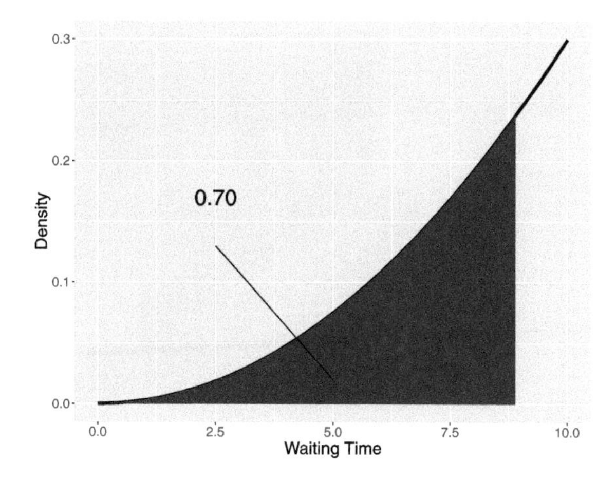

FIGURE 5.16
Illustration of the 70th percentile.

approximate percentiles by computing sample percentiles of the simulated values. For example, by use of the `quantile()` function, one finds that the 10th and 90th percentiles of W are approximately 4.80 and 9.66 minutes.

```
quantile(longest_wait, c(0.1, 0.9))
     10%      90%
4.798759 9.661885
```

The probability a longest waiting time is between 4.79 and 9.66 minutes is approximately 0.80.

5.6 Normal Distribution

Normal probability curve

One of the most popular races in the United States is marathon, a grueling 26-mile run. Most people are familiar with the Boston Marathon that is held in Boston, Massachusetts every April. But other cities in the U.S. hold yearly marathons. Here we look at data collected from Grandma's Marathon that is held in Duluth, Minnesota every June.

In the year 2003, there were 2515 women who completed Grandma's Marathon. The completion times in minutes for all of these women can be downloaded from the marathon's website. A histogram of these times, measured in minutes, is shown in Figure 5.17.

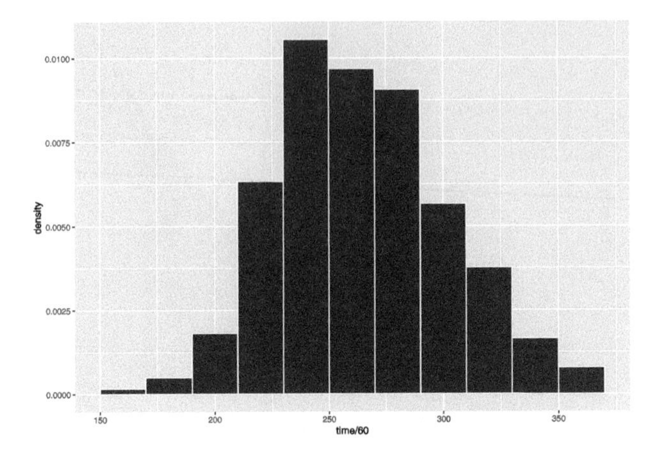

FIGURE 5.17
Histogram of women's completion times in the Grandma's Marathon.

Note that these measured times have a bell shape. Figure 5.18 superimposes a normal curve on top of this histogram. Note that this curve is a pretty good match to the histogram. In fact, data like this marathon time data that are measurements are often well approximated by a normal curve.

A normal density curve has the general form

$$f(x) = \frac{1}{\sqrt{2\pi}\sigma} \exp\left\{-\frac{(x-\mu)^2}{2\sigma^2}\right\}, -\infty < x < \infty. \tag{5.6}$$

This density curve is described by two parameters – the mean μ and the standard deviation σ. The mean μ is the center of the curve. Looking at the normal curve above, one sees that the curve is centered about 270 minutes – actually the mean of the normal curve is $\mu = 274$. The number σ, the standard deviation, describes the spread of the curve. Here the normal curve standard deviation is $\sigma = 43$. If one knows the mean and standard deviation of the normal curve, one can make reasonable predictions where the majority of times of the women runners will fall.

Early use of the Normal curve

The famous normal curve was independently discovered by several scientists. Abraham De Moivre in the 18th century showed that a binomial probability for a large number of trials n could be approximated by a normal curve. Pierre Simon Laplace and Carl Friedrich Gauss also made important discoveries about this curve. By the 19th century, it was believed by some scientists such as Adolphe Quetelet that the normal curve would represent the distribution of any group of homogeneous measurements. To illustrate his thinking,

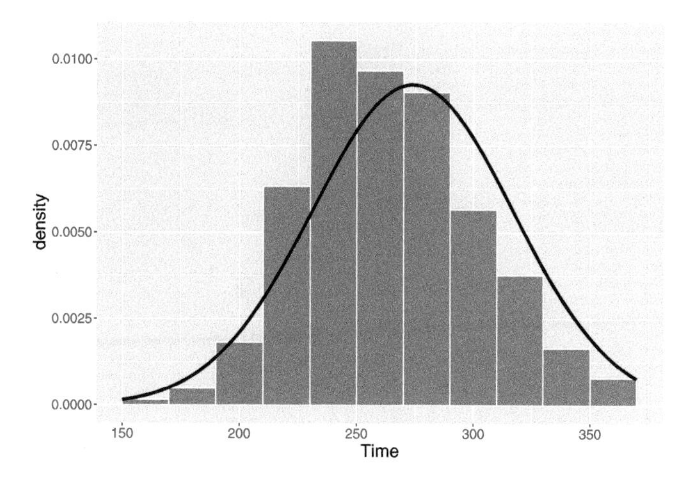

FIGURE 5.18
Histogram of women's completion times in the Grandma's Marathon, with a normal curve on top.

Quetelet considered the frequency measurements for the chest circumference measurements (in inches) for 5738 Scottish soldiers taken from the *Edinburgh Medical and Surgical Journal* (1817). A histogram of the chest measurements is shown in Figure 5.19. Quetelet's beliefs were a bit incorrect – any group of measurements will not necessarily be normal-shaped. However, it is generally true that a distribution of physical measurements from a homogeneous group, say heights of American women or foot lengths of Chinese men will generally have this bell shape.

In the previous sections of this chapter, the notion of a continuous random variable was introduced. Here the normal curve is introduced that is a popular model for representing the distribution of a measurement random variable. Also it will be seen that the normal curve is helpful for computing binomial probabilities and for representing the distributions of means taken from a random sample.

Computing normal probabilities

Suppose that the normal density with μ = 274 minutes and σ= 43 minutes represents the distribution of women racing times. Say one is interested in the probability that a runner completes the race less than 4 hours or 240 minutes. One computes this probability by finding an area under the normal curve. Specifically, as indicated in Figure 5.20, this probability is the area under the curve for all times less than 240 minutes.

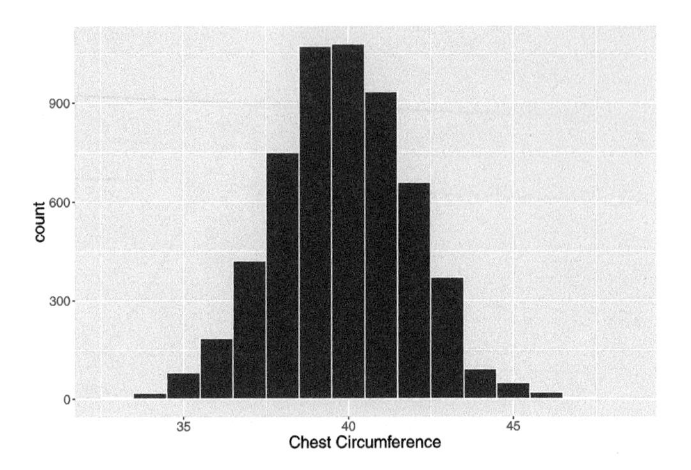

FIGURE 5.19
Histogram of chest circumference measurements of Scottish soldiers.

R Normal Probability Calculations

One expresses this area as the integral

$$P(X \leq 240) = \int_{-\infty}^{240} \frac{1}{\sqrt{2\pi}\sigma} \exp\left\{-\frac{(x-\mu)^2}{2\sigma^2}\right\} dx$$

but unfortunately one cannot integrate this function analytically (as was done for a uniform density) to find the probability. Instead one finds this area by use of the R `pnorm()` function in R. This function is used for three examples, illustrating the computation of three types of areas.

Returning to our example, recall that the marathon times were approximately normally distributed with mean $\mu = 274$ and standard deviation $\sigma = 43$.

1. **Finding a "less than" area.** Suppose one is interested in the probability that a woman marathon runner completes the race in under 240 minutes. That is, one wishes to find $P(X < 240)$ which is the area under the normal curve to the left of 240. The function value `pnorm(x, m, s)` gives the value of the cdf of a normal random variable with mean $\mu = a$ and $\sigma = s$ evaluated at the value x. For our example, the mean and standard deviation are given by 274 and 43, respectively, so the desired probability is given by

```
pnorm(240, 274, 43)
[1] 0.2145602
```

2. **Finding a "between two values" area.** Suppose one is interested in computing the probability that a marathon runner completes a race between two values, such as $P(230 < X < 280)$, shown in Figure 5.21.

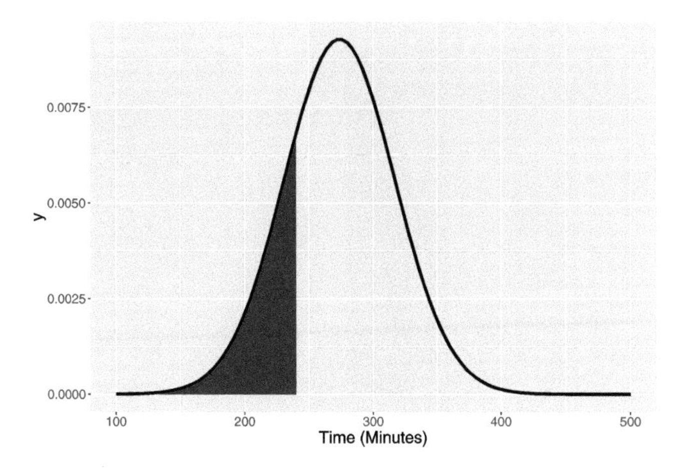

FIGURE 5.20
Normal density with μ = 274 and σ= 43, with illustration of the area under the curve less than 240 (minutes).

One writes this probability as the difference of two "less than" probabilities:

$$\begin{aligned} P(230 < X < 280) &= P(X < 280) - P(X < 230) \\ &= F(280) - F(230), \end{aligned}$$

where $F(x)$ is the cdf of a Normal(274, 43) random variable evaluated at x. Therefore, by use of the `pnorm()` function, this probability is equal to

```
pnorm(280, 274, 43) - pnorm(230, 274, 43)
[1] 0.4023928
```

3. **Finding a "greater than" area.** Last, sometimes one will be interested in the probability that X is greater than some value, such as $P(X > 300)$, the probability a runner takes more than 300 minutes to complete the race, shown in Figure 5.22.

   This probability is found by the complement property of probability, that

$$\begin{aligned} P(X > 300) &= 1 - P(X \leq 300) \\ &= 1 - F(300). \end{aligned}$$

   Therefore, one uses the `pnorm()` function to compute the probability that X is smaller than 300, and then subtract the answer from 1.

```
1 - pnorm(300, 274, 43)
[1] 0.2727054
```

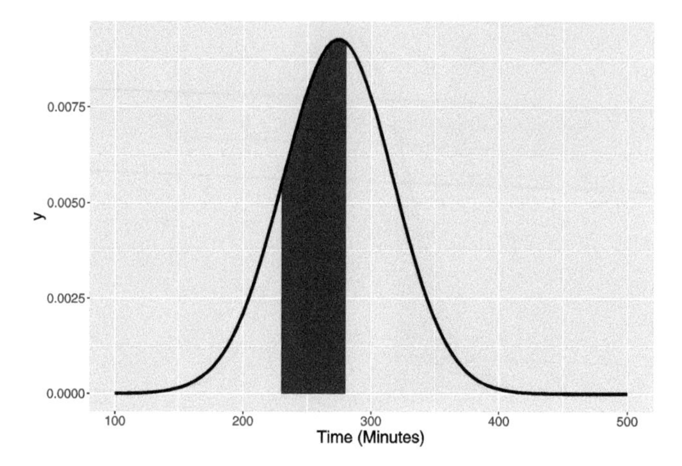

FIGURE 5.21
Normal density with $\mu = 274$ and $\sigma = 43$, with illustration of the area under the curve between 230 and 280 (minutes).

Computing Normal percentiles

In the marathon completion times example, we were interested in computing a probability that was equivalent to finding an area under the normal curve. A different problem is to compute a percentile of the distribution. In the marathon example, suppose that t-shirts will be given away to the runners who get the 25% fastest times. How fast does a runner need to run the race to get a t-shirt?

Here one wishes to compute the 25th percentile of the distribution of times. This is a time, call it x_{25}, such that 25% of all times are smaller than x_{25}. This is shown graphically in Figure 5.23.

Equivalently, we wish to find the value x_{25} such that

$$P(X \leq x_{25}) = F(x_{25}) = 0.25.$$

R Calculating Normal Percentiles

Percentiles of a normal curve are conveniently computed in R by use of the `qnorm()` function. Specifically, `qnorm(p, m, s)` gives the percentile of a Normal(m, s) curve corresponding to a "left area" of `p`. In our example, the value of `p` is 0.25, and so the 25th percentile of the running times (with mean 274 minutes and standard deviation 43 minutes) is computed to be

```
qnorm(0.25, 274, 43)
[1] 244.9969
```

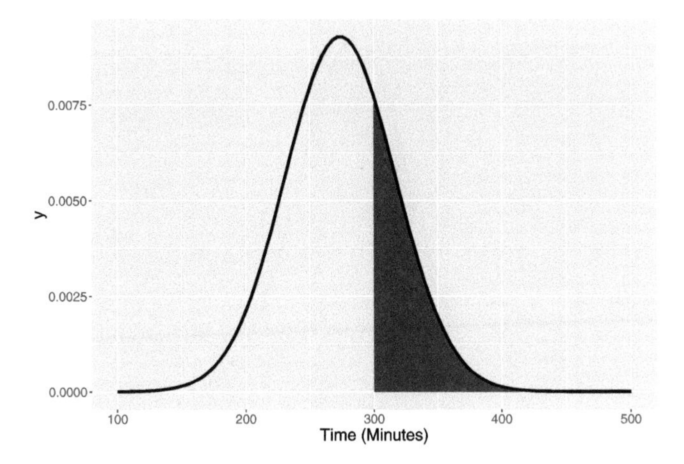

FIGURE 5.22
Normal density with $\mu = 274$ and $\sigma = 43$, with illustration of the area under the curve greater than 300 (minutes).

This means one needs to run faster (fewer than 245.0 minutes) to get a t-shirt in this competition.

Suppose one needs to complete the race faster than 10% of the runners to be invited to run in the race the following year. How fast does one need to run? If one wishes to have a 10% of the times to be larger than one's time, this means that 90% of the times will be smaller than one's time. That is, one wishes to find the 90th percentile, x_{90} of the normal distribution, shown in Figure 5.24.

```
qnorm(0.90, 274, 43)
[1] 329.1067
```

So 329 minutes is the time to beat if one wishes to be invited to participate in next year's race.

5.7 Binomial Probabilities and the Normal Curve

The normal curve is useful for modeling batches of data, especially when one is collecting measurements of some process. But the normal curve actually has a more important justification. We will explore several important results about the pattern of binomial probabilities and sample means and we will find these results useful in our introduction to statistical inference.

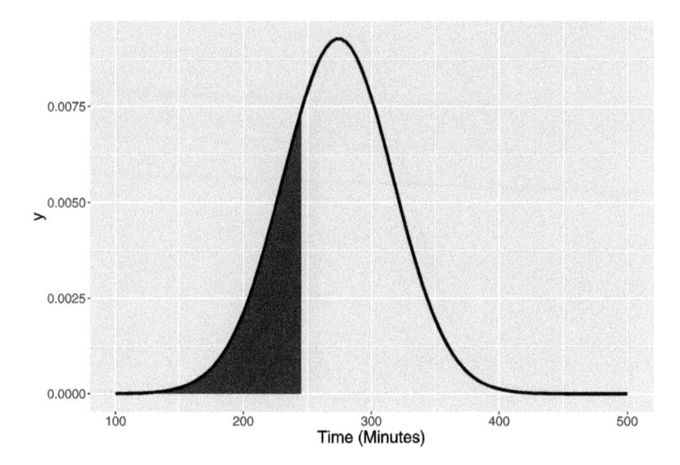

FIGURE 5.23
Normal density with $\mu = 274$ and $\sigma = 43$, with illustration of the 25th percentile.

First, consider different shapes of binomial distributions. Suppose that half of one's student body is female and one plans on taking a sample survey of n students to learn if they are interested in using a new recreational sports complex that is proposed. Let X denote the number of females in the sample. Assuming a random sample is chosen, it is known that X will be distributed binomial with parameters n and $p = 1/2$. What is the shape of the binomial probabilities? Figure 5.25 displays the binomial probabilities for sample sizes $n = 10$, 20, 50, and 100.

What does one notice about these probability graphs? First, note that each distribution is symmetric about the mean $\mu = np$. But, more interesting, the shape of the distribution seems to resemble a normal curve as the number of trials n increases.

Perhaps this pattern happens since one started with a binomial distribution with $p = 0.5$ and one would not see this behavior if a different value of p was used. Suppose that only 10% of all students would use the new facility and let X denote the number of students in your sample who say they would use the facility. The random variable X would be distributed binomial with parameters n and $p = 0.1$. Figure 5.26 shows the probability distributions again for the sample sizes $n = 10$, 20, 50, and 100. As one might expect the shapes of the probabilities for n=10 are not very normal-shaped – the distribution is skewed right. But, note that as n increases, the probabilities become more normal-shaped and the normal curve seems to be a good match for $n = 100$.

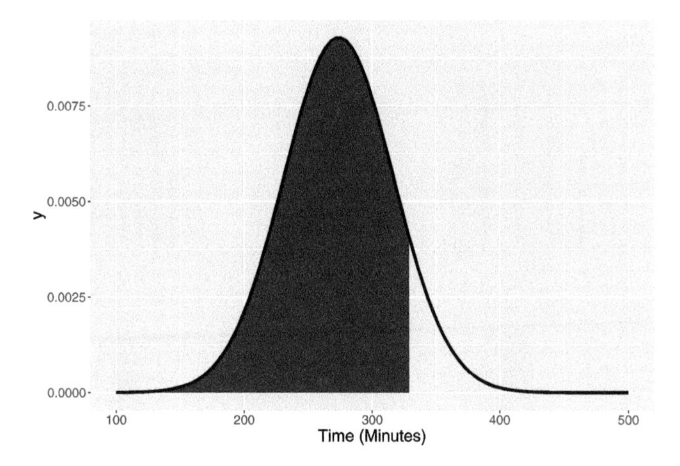

FIGURE 5.24
Normal density with μ = 274 and σ= 43, with illustration of the 90th percentile.

Figures 5.25 and 5.26 illustrate a basic result: if one has a binomial random variable X with n trials and probability of success p, then, as the number of trials n approaches infinity, the distribution of the standardized score

$$Z = \frac{X - np}{\sqrt{np(1-p)}} \tag{5.7}$$

approaches a standard normal random variable, that is a normal distribution with mean 0 and standard deviation 1. This is a very useful result. It means, that for a large number of trials, one can approximate a binomial random variable X by a normal random variable with mean and standard deviation

$$\mu = np, \ \ \sigma = \sqrt{np(1-p)}. \tag{5.8}$$

This approximation result can be illustrated with our student survey example. Suppose that 10% of the student body would use the new recreational sports complex. One takes a random sample of 100 students — what's the probability that 5 or fewer students in the sample would use the new facility?

The random variable X in this problem is the number of students in the sample that would use the facility. This random variable has a binomial distribution with $n = 100$ and $p = 0.1$ that is pictured as a histogram in Figure 5.27. By the approximation result, this distribution is approximated by a normal curve with $\mu = 100(0.1) = 10$ and $\sigma = \sqrt{100(0.1)(0.9)} = 3$. This normal curve is placed on top of the probability histogram in Figure 5.27 – note that it is a pretty good fit to the histogram.

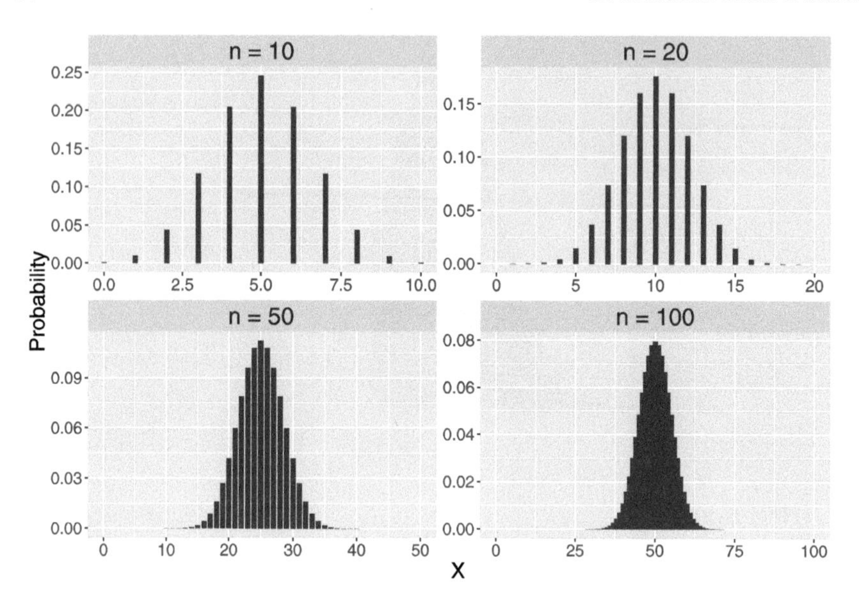

FIGURE 5.25
Binomial probabilities for sample sizes $n = 10$, 20, 50, and 100, and success probability $p = 1/2$.

Binomial Computations Using a Normal Curve

One is interested in the probability that at most 5 students use the facility, that is, $P(X \leq 5)$. This probability is approximated by the area under a Normal(10, 3) curve between $X = 0$ and $X = 5$. Using the R `pnorm()` function, we compute this normal curve area to be

```
pnorm(5, 10, 3) - pnorm(0, 10, 3)
[1] 0.04736129
```

In this case, one can also find this probability exactly by a calculator or computer program that computes binomial probabilities. Using the `pbinom()` function, we find the probability that X is at most 5 is

```
pbinom(5, size = 100, prob = 0.10)
[1] 0.05757689
```

Normal approximation gives a similar answer to the exact binomial computation.

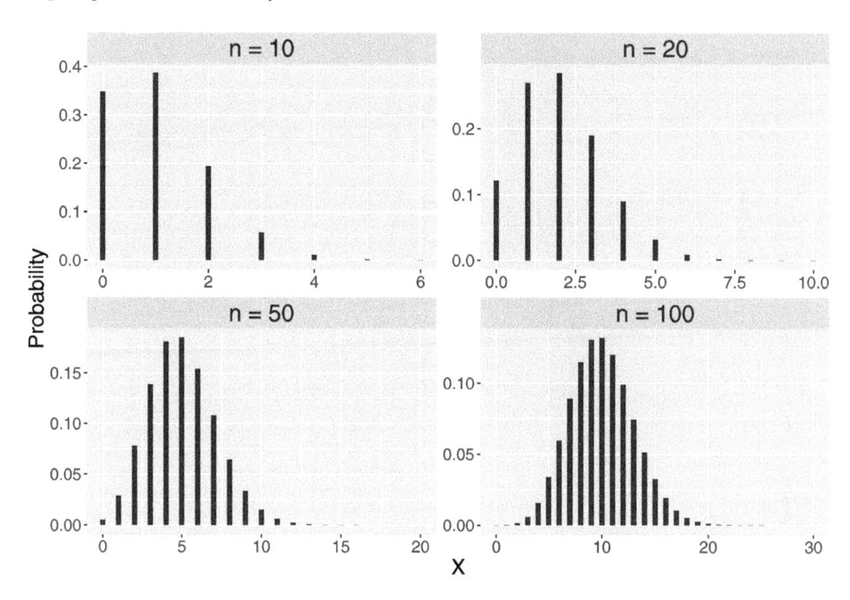

FIGURE 5.26
Binomial probabilities for sample sizes n = 10, 20, 50, and 100, and success probability $p = 0.1$.

5.8 Sampling Distribution of the Mean

We have seen that binomial probabilities are well-approximated by a normal curve when the number of trials is large. There is a more general result about the shape of sample means that are taken from any population.

To begin our discussion about the sampling behavior of means, suppose one has a jar filled with a variety of candies of different weights. One is interested in learning about the mean weight of a candy in the jar. One could obtain the mean weight by measuring the weight for every single candy in the jar, and then finding the mean of these measurements. But that could be a lot of work. Instead of weighing all of the candies, suppose one selects a random sample of 10 candies from the jar and finds the mean of the weights of these 10 candies. What has one learned about the mean weight of all candies from this sample information?

To answer this type of question, one assumes he or she knows the weights of all candies in the jar and examines the pattern of means obtained after taking random samples from the jar.

The group of items (here, candies) of interest is called the population. Assume first that one knows the population – that is, we know exactly the

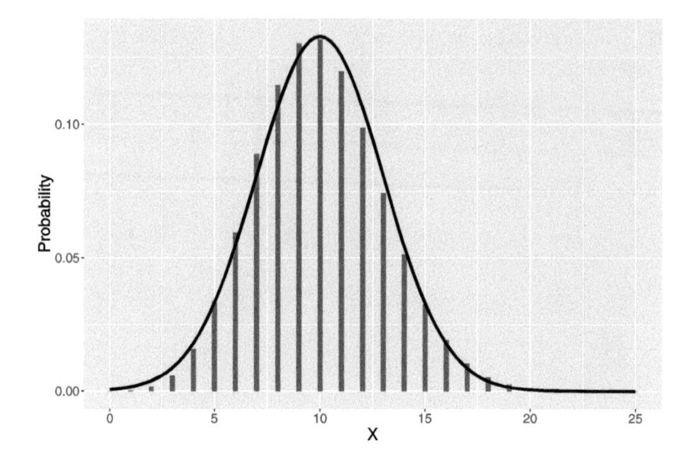

FIGURE 5.27
Histogram of binomial probabilities with the approximated normal curve on top.

weights of all candies in the jar. There are five types of candies – Table 5.1 gives the weight of each type of candy (in grams) and the proportion of candies of that type.

TABLE 5.1
Weights (in grams) and proportions of 5 types of candies.

	Weight	Proportion
fruity square	2	0.15
milk maid	5	0.35
jelly nougat	8	0.20
caramel	14	0.15
candy bars	18	0.15

Let X denote the weight of a randomly selected candy from the jar. Note that X is a discrete random variable with the probability distribution given in Table 5.1. This distribution is summarized by computing a mean μ and a standard deviation σ. The reader can verify in the end-of-chapter exercises that $\mu = 8.4500$ and $\sigma = 5.3617$. So if one was really able to weigh each candy in the jar, one would find the mean weight to be 8.45 gm.

Suppose a random sample of 10 candies is selected with replacement from the jar and the mean is computed. Note that this is called the sample mean $\bar{X}$ to distinguish it from the population mean μ.

R Sampling Candies

This sampling can be simulated using the following R code. The distribution of candies is stored in the vectors `weights` and `proportion`. By use of the `sample()` function, one obtains the following candy weights:

```
weights <- c(2, 5, 8, 14, 18)
proportion <- c(.15, .35, .2, .15, .15)
sample(weights, size = 10, prob = proportion, replace = TRUE)
 [1]  5  8  5 14  5  18  8  18  5  8
```

One computes the sample mean

$$\bar{X} = (5 + 8 + 5 + 14 + 5 + 18 + 8 + 18 + 5 + 8)/10 = 9.4\,\text{gm}.$$

Suppose this process is repeated two more times – in the second sample, one obtains $\bar{X}$= 6.9 gm and in the third sample, one obtains $\bar{X}$= 8.8 gm. The three sample mean values are plotted in Figure 5.28.

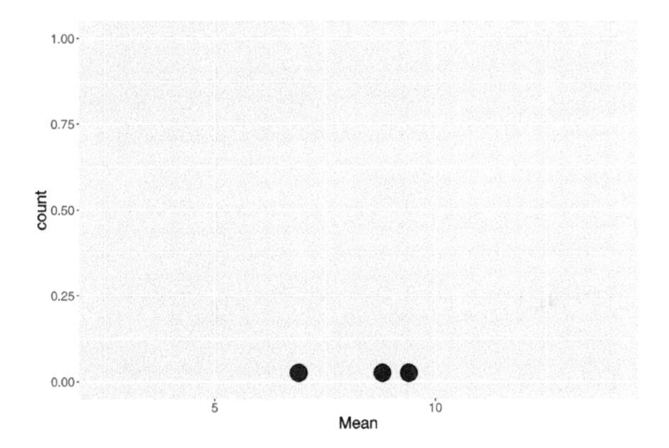

FIGURE 5.28
Graph of 3 sample means from 10 randomly selected candies.

Suppose that one continues to take random samples of 10 candies from the jar and plot the values of the sample means on a graph – one obtains the sampling distribution of the mean $\bar{X}$, shown in Figure 5.29.

Note that there is an interesting pattern of these sample means – they appear to have a normal shape. This motivates an amazing result, called the Central Limit Theorem, about the pattern of sample means. If one takes sample means from any population with mean μ and standard deviation σ, then the sampling distribution of the means (for large enough sample size) will be approximately normally distributed with mean and standard deviation

$$E(\bar{X}) = \mu, \;\; SD(\bar{X}) = \frac{\sigma}{\sqrt{n}}. \tag{5.9}$$

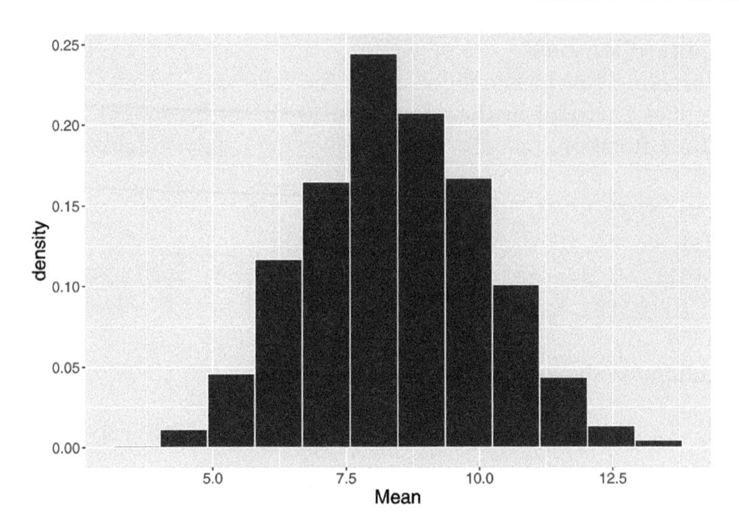

FIGURE 5.29
Histogram of the sampling distribution of the mean $\bar{X}$.

Let's illustrate this result for our candy example. Recall that the population of candy weights had a mean and standard deviation given by $\mu = 8.45$ and $\sigma = 5.36$, respectively. If one takes samples of size $n = 10$, then, by this result, the sample mean $\bar{X}$ will be approximately normally distributed where

$$E(\bar{X}) = 8.45, \;\; SD(\bar{X}) = \frac{5.36}{\sqrt{10}} = 1.69.$$

This normal curve is drawn on top of the histogram of sample means, shown in Figure 5.30.

There are two important points to mention about this result.

1. First the expected value of the sample means, $E(\bar{X})$, is equal to the population mean μ. When one takes a random sample, it is possible that the sample mean $\bar{X}$ is far away from the population mean μ. But, if one takes many random samples, then, on the average, the sample mean will be close to the population mean.
2. Second, note that the spread of the sample means, as measured by the standard deviation, is equal to $\sigma/\sqrt{n}$. Since the spread of the population is σ, note that the spread of the sample means will be smaller than the spread of the population. Moreover, if one takes random samples of a larger size, then the spread of the sample means will decrease.

The second point can be illustrated in the context of our candy example. Above, we selected random samples of size $n = 10$ and computed the sample

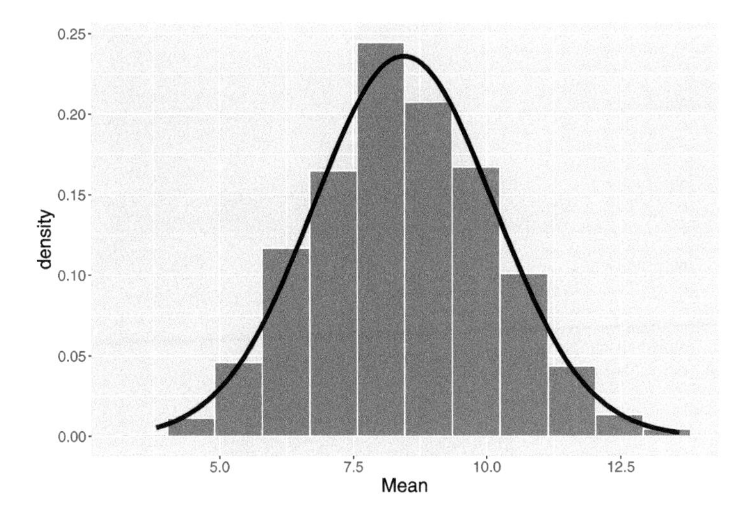

FIGURE 5.30
Histogram of the sampling distribution of the mean $\bar{X}$, with approximated normal curve on top.

means. Suppose instead one selected repeated samples of size $n = 25$ from the candy jar – how does the sampling distribution of means change?

Using R, one can simulate the process of taking samples of size 25 – histograms of the sample means are shown in Figure 5.31. By the Central Limit Theorem, the sample means will be approximately normal-shaped with mean and standard deviation

$$E(\bar{X}) = 8.45, \ \ SD(\bar{X}) = \frac{5.36}{\sqrt{25}} = 1.07.$$

Comparing the $n = 10$ sample means with the $n = 25$ sample means in Figure 5.31, what's the difference? Both sets of sample means are normally distributed with an average equal to the population mean. But the $n = 25$ sample means have a smaller spread – this means that as you take bigger samples, the sample mean $\bar{X}$ is more likely to be close to the population mean μ. The simulation is left as an end-of-chapter exercise.

The Central Limit Theorem works for any population

We illustrate the Central Limit Theorem for a second example where the population has a distinctive non-normal shape. At one university, many of the students' hometowns are within 40 miles of the school. There also are a large number of students whose homes are between 80-120 miles of the university. Given the population of "distances of home" of all students, it is interesting to see what happens when we take random samples from this population.

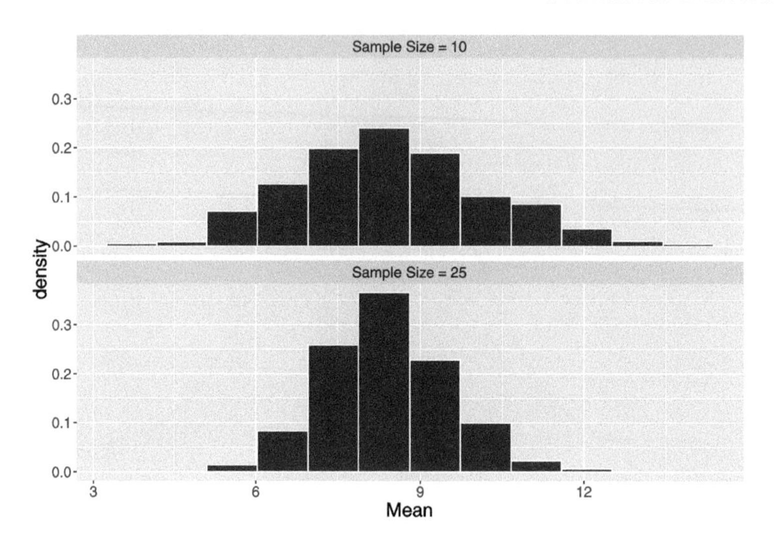

FIGURE 5.31
Histogram of the sampling distribution of the mean $\bar{X}$, with sample sizes $n = 10$ and $n = 25$.

If we let X denote "distance from home", imagine that the population of distances is described by the continuous density curve in Figure 5.32. Two humps can be seen in this density – these correspond to the large number of students whose homes are in the ranges 0 to 40 miles and 70 to 130 miles. Suppose the mean and standard deviation of this population are given by $\mu = 60$ miles and $\sigma = 41.6$ miles, respectively.

Now imagine that one takes a random sample of n students from this population and computes the sample mean from this sample. For example, suppose one takes a random sample of 20 students and collect the distances from home from these students – once one has collected the 20 distances, one computes the sample mean $\bar{X}$. Here are two samples and the values of $\bar{X}$:

```
Sample 1:
  102    22    23    24   114   102   114   102    22    19
   88    31    30   100   111   105     105   17   100    21
   xbar =67.6 mi.

Sample 2:
   12   127    33    34    73    19   111    99    16    20
   22    16    24    62    22    76    91   115   117    93
   xbar =59.1 mi.
```

If this sampling process is repeated many times, what will the distribution of sample means look like? Also, what is the effect of the sample size n? To

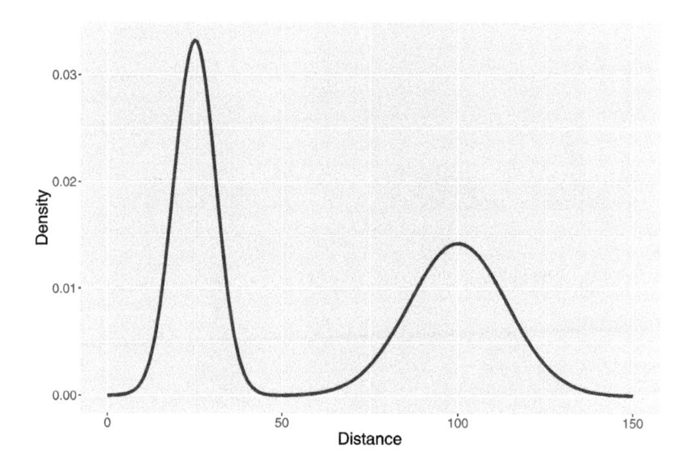

FIGURE 5.32
Density curve of the population of distances.

answer this question, one can let the computer simulate repeated samples of sizes $n = 1, n = 2, n = 5$, and $n = 20$. The histograms in Figure 5.33 show the distributions of sample means for the four sample sizes.

As one might expect, if samples of size 1 are selected, our sample means look just like the original population. If samples of size 2 are selected, then the sample means have a funny three-hump distribution. But, note as one takes samples of larger sizes, the sampling distribution of means looks more like a normal curve. This is what one expects from the Central Limit Theorem result – no matter what the population shape, the distribution of the sample means will be approximately normal if the sample size is large enough.

What is the distribution of the sample means when we take samples of size $n = 20$? One just applies the Central Limit Theorem result. The sample means will be approximately normal with mean and standard deviation

$$E(\bar{X}) = \mu, \ \ SD(\bar{X}) = \frac{\sigma}{\sqrt{n}}. \tag{5.10}$$

Since one knows the mean and standard deviation of the population and the sample size, one just substitute these quantities and obtains

$$E(\bar{X}) = 60, \ \ SD(\bar{X}) = \frac{41.6}{\sqrt{20}} = 9.3.$$

These results can be used to answer some questions.

1. **What is the probability that a student's distance from home is between 40 and 60 miles?**

   Actually this is a difficult question to answer exactly, since one does not know the exact shape of the population. But, looking at

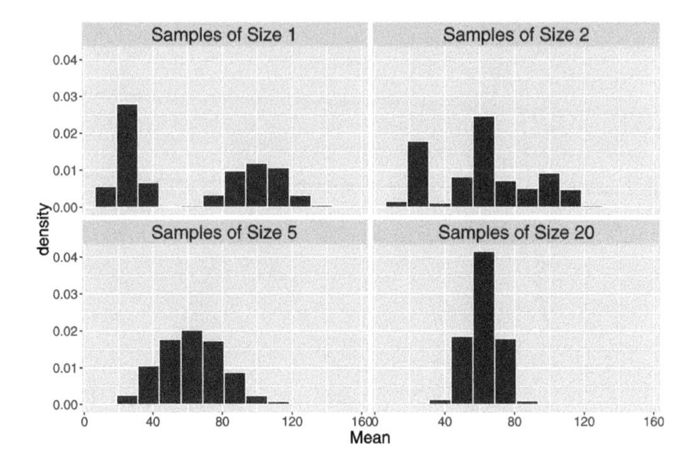

FIGURE 5.33
Histograms of random samples of distances, with sample sizes of $n = 1, n = 2, n = 5$, and $n = 20$.

the graph of the population, one sees that the curve takes on very small values between 40 and 60 miles. So this probability is close to zero – very few students live between 40 and 60 miles from our school.

2. **What is the probability that, if one takes a sample of 20 students, the mean distance from home for these 20 students is between 40 and 60 miles?**

   This is a different question than the first one. This question is asking about the chance that the sample mean falls between 40 and 60 miles. Since the sampling distribution of $\bar{X}$ is approximately normal with mean 60 and standard deviation 9.3, one can compute this by using R. Using the `pnorm()` function, one obtains

```
pnorm(60, 60, 9.3) - pnorm(40, 60, 9.3)
[1] 0.4842436
```

   It is interesting to note that although it is unlikely for students to live between 40 and 60 miles from the school, it is pretty likely for the sample mean for a group of 20 students to fall between 40 and 60 miles.

3. **What is the probability that the mean distance exceeds 100 miles?**

   Here one wants to find the probability that $\bar{X}$ is greater than 100, that is $P(\bar{X} > 100)$. Using R, one computes

```
1 - pnorm(100, 60, 9.3)
[[1] 8.498565e-06
```

This probability is essentially zero, which means that it is highly unlikely that a sample mean of 20 student distances will exceed 100 miles.

5.9 Exercises

1. **Waiting at a ATM Machine**

   You are waiting at your local ATM machine and as usual, you are waiting in a line. Suppose you know that your waiting time can be between 0 to 5 minutes and any value between 0 and 5 minutes is equally likely.

   (a) The graph below shows the density function for X, the waiting time. What is the height of this function?

   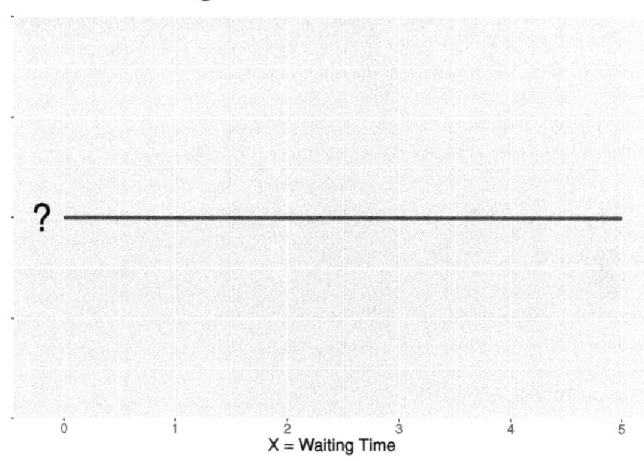

   (b) Find the probability you wait more than 2 minutes.

   (c) Find the probability you wait between 2 and 3 minutes.

2. **Morning Wake-Up**

   Suppose you wake up at a random time in the morning between 6 am and 12 pm.

   (a) Find the probability you wake up before 11 am.

   (b) Find the probability you wake up between 8 and 10 am.

   (c) What is an "average" or typical time you will wake up? Explain how you computed this number.

   (d) Find the standard deviation of the time.

3. **The Median Waiting Time**

In the "waiting for a bus" example in Section 5.3, suppose that you record the median time T (in minutes) that you wait for the bus on the three days. The density function for this median time is given by

$$f(t) = \frac{6t(10-t)}{1000}, \quad 0 < t < 10.$$

(a) Draw a graph of this density function.

(b) Find the probability that the median time is between 5 and 7 minutes.

(c) Find the cdf $F(t)$ for all values of t.

(d) Using the cdf you found in part c, find the probability the median time is over 6 minutes.

(e) Find the 75% percentile of your median waiting time.

4. **The Sum of Two Spins**

Suppose you spin two spinners, where the location of the arrow for each spinner is equally likely to fall between 0 and 10.

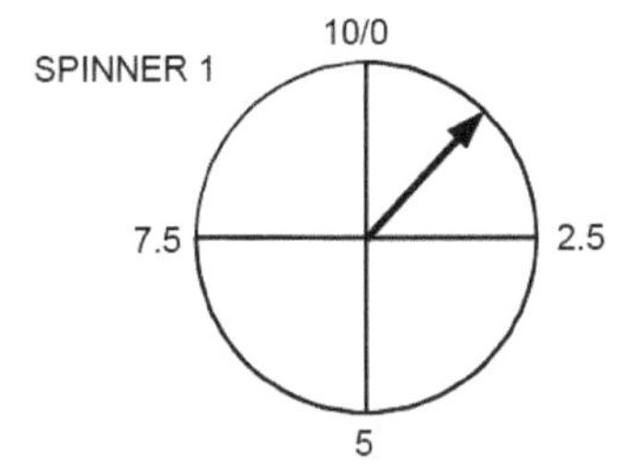

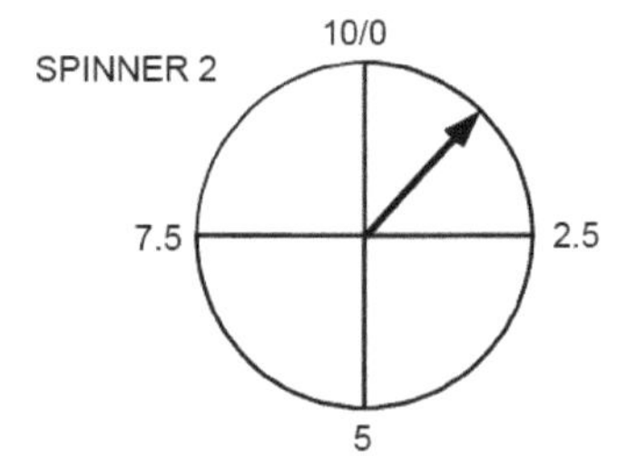

If you let S be the sum of the two spins, it can be shown that the density function of S is given by

$$f(s) = \begin{cases} s/100, & 0 < s \le 10 \\ (20-s)/100, & 10 < s \le 20, \end{cases}$$

and shown by the figure below.

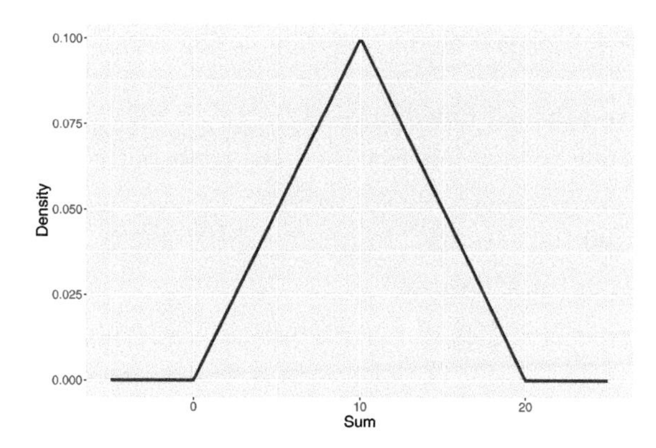

(a) Check that this function satisfies the two properties of a probability density function.

(b) Find the probability the sum of the two spins is smaller than 5.

(c) Find the cdf function F.

(d) Using the cdf function, find the probability the sum of spins falls between 8 and 12.

(e) Using the cdf function, find the probability the sum of spins exceeds 12.

5. **Salaries for Professional Basketball Players**

Let X denote the salary (in millions of dollars) of a professional basketball player. A reasonable density function for X is given by

$$f(x) = \frac{0.15}{x^{1.3}}, \; x \geq 0.1$$

shown by the figure below.

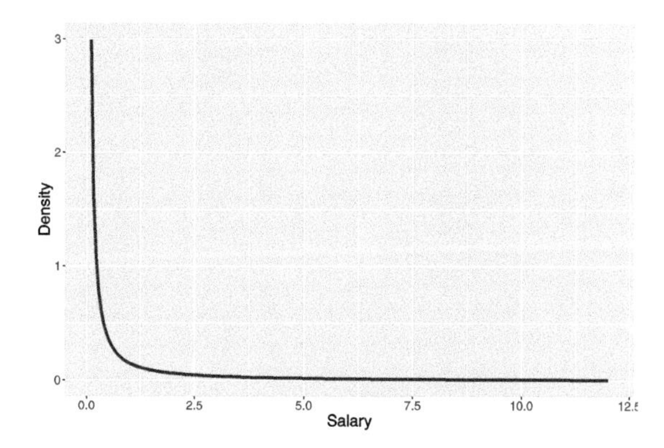

(a) What proportion of basketball players earn more than 1 million dollars?

(b) What proportion of players earn between 1 and 2 million dollars?

(c) Find the cdf function.

(d) Using the cdf function, find the probability a player earns less than one-half a million dollars.

(e) Find the "average" salary of a NBA player.

6. **Grading on a Curve**

Suppose the grades on a math test are distributed according to the curve.

$$f(x) = \frac{x}{5000}, 0 < x < 100.$$

(a) Draw a graph of this density curve.

(b) Find the mean grade on this test.

(c) What proportion of students who take this test get a grade of 90 or higher?

(d) What proportion of students get a C grade, where C is defined to be between 70 and 80?

(e) Is this test harder or easier than the test grades in your statistics class? Explain.

7. **Time to Clean Your Room**

Suppose the time that it takes you to clean your room (in hours) is a random variable X with the cdf function given below. A graph of the cdf is also shown.

$$F(x) = \begin{cases} 0, & x < 0 \\ 0.75(2x^3/3 - x^4/4) & 0 \le x \le 2 \\ 1, & x > 2 \end{cases}$$

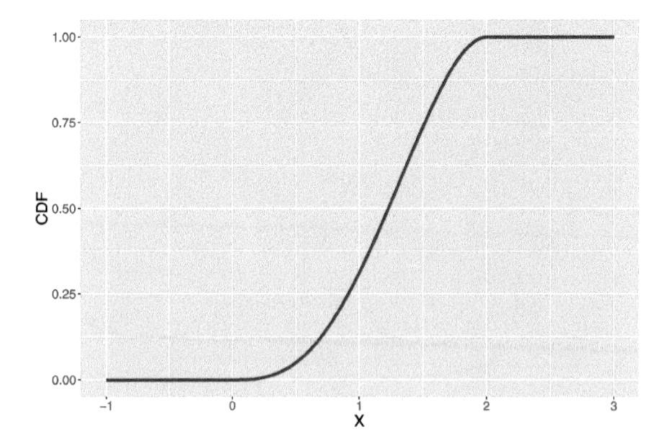

(a) Find the probability you can clean your room in under one hour.

(b) Find the probability it takes you over one and a half hour to clean your room?

(c) Using the graph, find a value M such that it is equally likely that X is smaller than M and X is larger than M. [Hint: M is the 50th percentile of X.]

8. **Time to Complete a Race**

Suppose a group of children are running a race. The times (in minutes) that the children complete the race can be described by the density function

$$f(x) = \frac{4 + (x - 3)^2}{21}, 3 < x < 6.$$

(a) Graph this density function.

(b) Looking at your graph, is it more common to have a slow time (near 6 minutes) or a fast time (near 3 minutes)?

(c) Find the probability a child completes the race in under 4 minutes.

(d) Find the probability that a child's time exceeds 5 1/2 minutes.

(e) Find the median running time.

9. **Spinning a Random Spinner**

Suppose you flip a coin. If the coin lands heads, you spin a spinner that is equally likely to fall at any point in the interval (0, 4). If the coin lands tails, you spin a different spinner that lands at any point in the interval (2, 6). If X denotes your spin, the density function for X is graphed below.

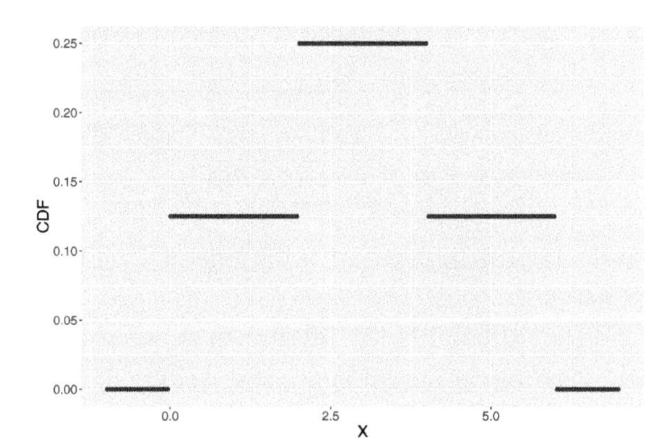

(a) Check that this graphed function is indeed a probability density.

(b) Find the probability that X is greater than 5.

(c) Find the probability that X falls between 1 and 3.

10. **Lifetimes of Light Bulbs**

    Suppose that a company is interested in the amount of time that a particular type of light bulb will last until it burns out. After sampling the lifetimes for a large group of light bulbs, it is decided that the lifetime X (in hours) is well-described by the exponential distribution of the form

$$f(x) = \frac{1}{100}e^{-x/100}, x > 0.$$

    The cdf for X is drawn below.

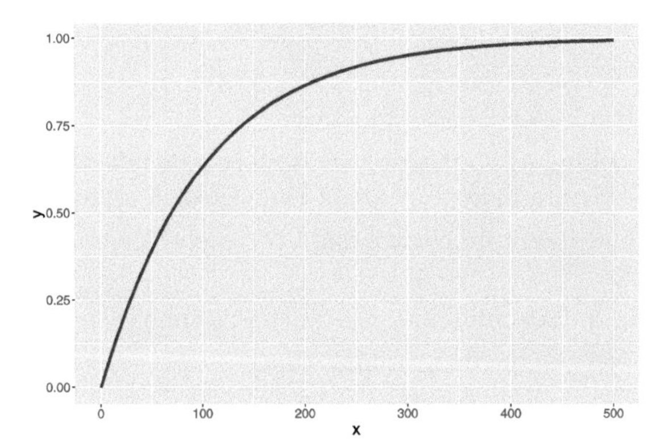

    In addition, the cdf is computed for some values of X in the following table.

x	$F(x)$	x	$F(x)$
0	0	180	0.8347
30	0.2592	210	0.8775
60	0.4512	240	0.9093
90	0.5934	270	0.9328
120	0.6988	300	0.9502
150	0.7769		

    (a) Find the probability that a lifetime of a bulb will be less than 90 hours.
    (b) Find the probability the lifetime is between 120 and 180 hours.
    (c) From the table, approximate the median lifetime.
    (d) Approximate the 95th percentile.

11. **Locations of Dart Throws**

    Suppose you throw a dart at a circular target such that the dart is equally likely to land in any location on the target. The locations for a large number of dart throws are shown in the figure below.

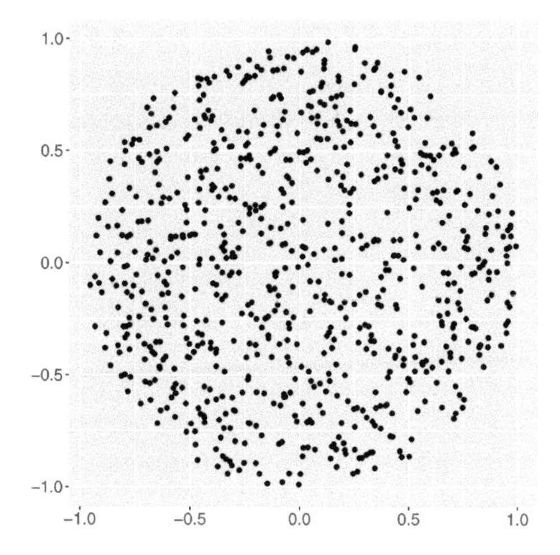

Let X denote the distance of a throw from the bulls eye. It can be shown that the density function of X has the form

$$f(x) = \frac{x}{2}, 0 < x < 2.$$

(a) Find the probability your throw lands within a distance of 1 unit from the target.

(b) Find the probability your throw lands between .5 and 1.5 units from the target.

(c) If you threw the dart many times at the target, find your average distance from the target.

12. **Heights of Men**

Suppose heights of American men are approximately normally distributed with mean 70 inches and standard deviation 4 inches.

(a) What proportion of men is between 68 and 74 inches?

(b) What proportion of men is taller than 6 feet?

(c) Find the 90th percentile of heights.

13. **Test Scores**

Test scores in a precalculus test are approximately normally distributed with mean 75 and standard deviation 10. If you choose a student at random from this class

(a) What is the probability he or she gets an A (over 90)?

(b) What is the probability he or she gets a C (between 70 and 80)?

(c) What is the letter grade of the lower quartile of the scores?

14. **Body Temperatures**

    The normal body temperature was measured for 130 subjects in an article published in the *Journal of the American Medical Association.* These body temperatures are approximately normally distributed with mean $\mu = 98.2$ degrees and standard deviation $\sigma = 0.73$.

    (a) Most people believe that the mean body temperature of healthy individuals is 98.6 degrees, but actually the mean body temperature is less than 98.6. What proportion of healthy individuals have body temperatures less than 98.6?

    (b) Suppose a person has a body temperature of 96 degrees. What is the probability of having a temperature less than or equal to 96 degrees? Based on this computation, would you say that a temperature of 96 degrees is unusual? Why?

    (c) Suppose that a doctor diagnoses a person as sick if his or her body temperature is above the 95th percentile of the temperature of "healthy" individuals. Find this body temperature that will give a sick diagnosis.

15. **Baseball Batting Averages**

    Batting averages of baseball players can be well approximated by a normal curve. The figure below displays the batting averages of players during the 2003 baseball season with at least 300 at-bats (opportunities to hit). The mean and standard deviation of the matching normal curve shown in the figure are $\mu = 0.274$ and $\sigma = 0.027$, respectively.

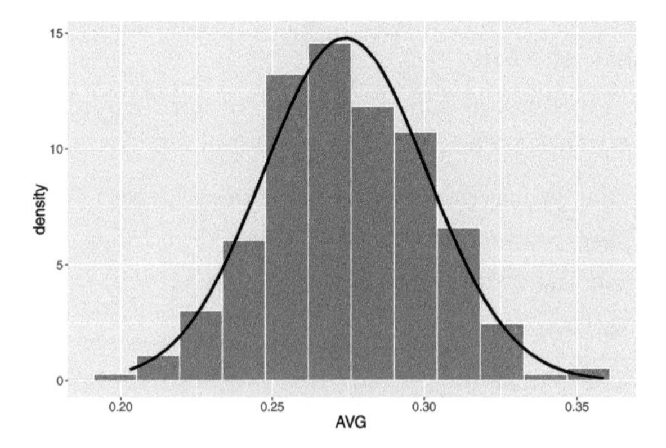

    (a) If you choose a baseball player at random, find the probability his batting average is over 0.300. (This is a useful benchmark for a "good" batting average.)

    (b) Find the probability this player has a batting average between 0.200 and 0.250.

(c) A baseball player is said to hit below the Mendoza line (named for weak-hitting baseball player Minnie Mendoza) if his batting average is under 0.200. Given our model, find the probability that a player hits below the Mendoza line.

(d) Suppose that a player has an incentive clause in his contract that states that he will earn an additional $1 million if his batting average is in the top 15%. How well does the player have to hit to get this additional salary?

16. **Emergency Calls**

Suppose that the AAA reports that the average time it takes to respond to an emergency call on the highway is 25 minutes. Assume that the times to respond to emergency calls are approximately normally distributed with mean 25 minutes and standard deviation 4 minutes.

(a) If your car gets stuck on a highway and you call the AAA for help, find the probability that it will take longer than 30 minutes to get help.

(b) Find the probability that you'll wait between 20 and 30 minutes for help.

(c) Find a time such that you are 90% sure that the wait will be smaller than this number.

17. **Buying a Battery for your iPod**

Suppose you need to buy a new battery for your iPod. Brand A lasts an average of 11 hours and Brand B lasts an average of 12 hours. You plan on using your iPod for 8 hours on a trip and you want to choose the battery that is most likely to last 8 hours (that is, have a life that is least as long as 8 hours).

(a) Based on this information, can you decide which battery to purchase? Why or why not?

(b) Suppose that the battery lives for Brand A are normally distributed with mean 11 hours and standard deviation 1.5 hours, and the battery lives for Brand B are normally distributed with mean 12 hours and standard 2 hours. Compute the probability that each battery will last at least 8 hours.

(c) On the basis of this calculation in part (b), which battery should you purchase?

18. **Lengths of Pregnancies**

It is known that the lengths of completed pregnancies are approximately normally distributed with mean 266 days and standard deviation 16 days.

(a) What is the probability a pregnancy will last more than 270 days?

(b) Find an interval that will contain the middle 50% of the pregnancy lengths.

(c) Suppose a doctor wishes to tell a mother that he is 90% confident that the pregnancy will be shorter than x days. Find the value of x.

19. **Attendances at Baseball Games**

Attendances for home page of the Cleveland Indians for a recent baseball season can be approximated by a normal curve with mean $\mu = 24{,}667$ and standard deviation $\sigma = 6144$.

Consider the attendance for one randomly selected game during the 2006 season.

(a) Find the probability the attendance exceeds 30,000.

(b) Find the probability the attendance is between 20,000 and 30,000.

(c) Suppose that the attendance at one game in the following season is 12,000. Based on the normal curve, compute the probability that the attendance is at most 12,000. Based on this computation, is this attendance unusual? Why?

20. **Coin Flipping**

Suppose you flip a fair coin 1000 times.

(a) How many heads do you expect to get?

(b) Find the probability that the number of heads is between 480 and 520.

(c) Suppose your friend gets 550 heads. What is the probability of getting at least 550 heads? Do you believe that your friend's coin really was fair? Explain.

21. **Use of Online Banking Services**

Suppose that a newspaper article claims that 80% of adults currently use online banking services. You wonder if the proportion of adults who use online banking services in your community, p, is actually this large. You take a sample of 100 adults and 70 tell you they use online banking.

(a) If the newspaper article is accurate, find the probability that 70 or fewer of your sample would use on-line banking.

(b) Based on your computation, is there sufficient evidence to suggest that less than 80% of your community use online banking services? Explain.

22. **Time to Complete a Race**

    Suppose a group of children are running a race. The times (in minutes) that the children complete the race can be described by the density function

$$f(x) = \frac{4 + (x - 3)^2}{21}, 3 < x < 6.$$

A graph of this density is shown below. The mean and standard deviation of this density are given by 4.83 and 0.84 minutes, respectively.

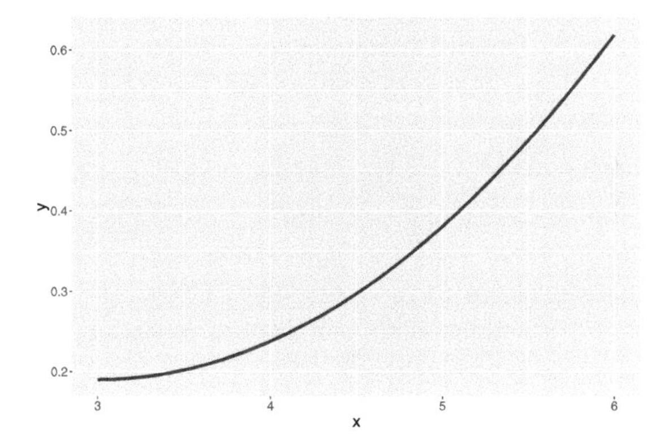

(a) Suppose 25 students run this race and you find the mean completion time. Find the probability that the mean time exceeds 5 minutes.

(b) Find an interval that you are 90% confident contains the mean completion time for the 25 students.

23. **Snowfall Accumulation**

    Your local meteorologist has collected data on snowfall for the past 100 years. Based on these data, you are told that the amount of snowfall in January is approximately normally distributed with mean 15 inches and standard deviation 4 inches.

    (a) Find the probability you get more than 20 inches of snow this year.

    (b) In the next 10 years, find the probability that the average snowfall (for these 10 years) will exceed 20 inches.

24. **Total Waiting Time at a Bank**

    You are waiting to be served at your bank. From past experience, you know that your time to be served has a uniform distribution between 0 and 10 minutes.

    (a) Find the mean and standard deviation of your waiting time.

(b) The Central Limit Theorem can be also stated in terms of the sum of random variables. If the random variables $X_1, ..., X_n$ represent a random sample drawn from a population with mean μ and standard deviation σ, then the sum of random variables $S = \sum_{i=1}^{n} X_i$, for large sample size n, will be approximately normally distributed with mean $n\mu$ and standard deviation $\sqrt{n}\sigma$. Suppose you wait every day at the bank for a period of 30 days. Use the version of the Central Limit Theorem to find the probability that your total waiting time will exceed three hours.

25. **Total Errors in Check Recording**

Suppose you record the amount of a written check to the nearest dollar. It is reasonable to assume that the error between the actual check amount and the written amount has a uniform distribution between -0.50 and $+0.50$.

(a) Find the mean and standard deviation of one error.

(b) Suppose you write 100 checks in a single month and S denotes the total error in recording these checks. Find the probability that S is smaller than \$5. (Use the version of the Central Limit Theorem described in Exercise 5.)

(c) Find an interval of the form $(-c, c)$ so that $P(-c < S < c) = 0.95$.

26. **Distribution of Measurements**

Suppose that a group of measurements is approximately normally distributed with mean μ and standard deviation σ.

(a) Find the probability that a measurement falls within one standard deviation of the mean.

(b) Is it likely that you collect a measurement that is larger than $\mu + 3\sigma$? Explain.

(c) Find an interval that contains the middle 50% of the measurements.

27. **Salaries of Professional Football Players**

Suppose you learn that the mean salary of all professional football players this season is 7 million dollars with a standard deviation of 2 million dollars.

(a) Do you believe that the distribution of salaries is approximately normally distributed? If your answer is no, sketch a plausible distribution for the salaries.

(b) From your graph, find an approximate probability that a salary is smaller than 6 million dollars.

(c) Suppose you take a random sample of 30 salaries. Find the probability that the mean salary for this sample is smaller than 6 million dollars.

28. **Weights of Candies**

In the candy bowl example, the probability distribution of the candy weight X is given in the following table.

	x	$P(X = x)$
fruity square	2	0.15
milk maid	5	0.35
jelly nougat	8	0.20
caramel	14	0.15
candy bars	18	0.15

Verify by calculation that the mean and standard deviation of X are given by $\mu = 8.4500$ and $\sigma = 5.3617$, respectively.

29. **Sleeping Times**

Suppose sleeping times of college students are approximately normally distributed. You are told that 25% of students sleep less than 6.5 hours and 25% of students sleep longer than 8 hours. Given this information, determine the mean and standard deviation of the normal distribution.

R Exercises

30. **A Continuous Spinner**

Suppose you spin a spinner where all values from 0 to 100 are equally likely.

(a) Write down the density function for X, one spin from this spinner.

(b) Use the following command to simulate 1000 values from this uniform distribution and store the values in the vector `spinner`:

```
spinner <- runif(1000, min = 0, max = 100)
```

(c) Construct a histogram of the simulated spins.

(d) Use the simulated spins to approximate the probability $P(X > 70)$.

31. **Simulating a Normal Distribution**

Suppose monthly snowfalls in Rochester, New York are normally distributed with mean 25 inches and standard deviation 10 inches.

(a) Using the `rnorm()` function, simulate snowfalls for 1000 hypothetical months in Rochester.

(b) Construct a graph of these snowfall amounts.

(c) Approximate from the simulated values the probability that a snowfall falls in the interval $(20, 30)$. Compare your answer with the exact probability found using the `pnorm()` function.

(d) From the simulated values, find an interval that contains the middle 80% of the snowfalls. Compare your answer with the exact interval found using the `qnorm()` function.

32. **Waiting for a Bus**

In the example, the amount of time that one waits for a bus has a uniform distribution from 0 to 10 minutes. One waits for a bus on Monday, Wednesday, and Friday and records the minimum of the three waiting times.

(a) Write a program to simulate 1000 values of this minimum waiting time.

(b) One can show that the minimum waiting time Y has density given by

$$f(y) = \frac{3}{1000}(10 - y)^2, \ 0 < y < 10.$$

Compare a histogram of simulated values from (a) with this density function to confirm that you have indeed simulated from the correct distribution.

33. **Weights of Candies (continued)**

Suppose one takes a sample of 10 candies from the distribution of candy weights shown in Exercise 28.

(a) Write a function to take a random sample of 10 candies from the bowl and return the sample mean $\bar{X}$.

(b) Use the `replicate()` function to repeat this process for 1000 iterations – store the sample means in the vector `xbars`.

(c) Construct a histogram of the sample means and comment on its shape. Also find the mean and standard deviation of the sample means.

(d) Repeat this exercise using samples of size $n = 25$. Are there any changes in the mean and standard deviation of the sample means?

34. **Spins and the Central Limit Theorem**

Suppose you are spinning a spinner with equally likely outcomes 1, 2, 3, 4, 5. X represents a single spin from this spinner.

(a) Find the mean μ and standard deviation σ of X.

(b) Write a function to simulate 10 spins from this spinner and compute the sample mean $\bar{X}$.

(c) Simulate 1000 samples of 10 spins, obtaining a vector of sample means.

(d) Construct a histogram of the sample means and comment on its shape. Also find the mean and standard deviation of the sample means.

(e) Check your calculations in part (d) by finding the exact mean and standard deviation of the sample mean $\bar{X}$.

6

Joint Probability Distributions

6.1 Introduction

In Chapters 4 and 5, the focus was on probability distributions for a single random variable. For example, in Chapter 4, the number of successes in a binomial experiment was explored and in Chapter 5, several popular distributions for a continuous random variable were considered. In addition, in introducing the Central Limit Theorem, the approximate distribution of a sample mean $\bar{X}$ was described when a sample of independent observations $X_1, ..., X_n$ is taken from a common distribution.

In this chapter, examples of the general situation will be described where several random variables, e.g. X and Y, are observed. The joint probability mass function (discrete case) and the joint density (continuous case) are used to compute probabilities involving X and Y.

6.2 Joint Probability Mass Function: Sampling from a Box

To begin the discussion of two random variables, we start with a familiar example. Suppose one has a box of ten balls – four are white, three are red, and three are black. One selects five balls out of the box without replacement and counts the number of white and red balls in the sample. What is the probability one observes two white and two red balls in the sample?

This probability can be found using ideas from previous chapters.

1. First, one thinks the total number of ways of selecting five balls with replacement from a box of ten balls. One assumes the balls are distinct and one does not care about the order that one selects the balls, so the total number of outcomes is

$$N = {10 \choose 5} = 252.$$

2. Next, one thinks about the number of ways of selecting two white and two red balls. One does this in steps – first select the white balls, then select the red balls, and then select the one remaining black ball. Note that five balls are selected, so exactly one of the balls must be black. Since the box has four white balls, the number of ways of choose two white is $\binom{4}{2} = 6$. Of the three red balls, one wants to choose two – the number of ways of doing that is $\binom{3}{2} = 3$. Last, the number of ways of choosing the remaining one black ball is $\binom{3}{1} = 3$. So the total number of ways of choosing two white, two red, and one black ball is the product

$$\binom{4}{2} \times \binom{3}{2} \times \binom{3}{1} = 6 \times 3 \times 3 = 54.$$

3. Each one of the $\binom{10}{5} = 252$ possible outcomes of five balls is equally likely to be chosen. Of these outcomes, 54 resulted in two white and two red balls, so the probability of choosing two white and two red balls is

$$P(2\,\text{white and}\,2\,\text{red}) = \frac{54}{252}.$$

Here the probability of choosing a specific number of white and red balls has been found. To do this calculation for other outcomes, it is convenient to define two random variables

$$X = \text{number of red balls selected},\ Y = \text{number of white balls selected}.$$

Based on what was found,

$$P(X = 2, Y = 2) = \frac{54}{252}.$$

Joint probability mass function

Suppose this calculation is done for every possible pair of values of X and Y. The table of probabilities is given in Table 6.1.

TABLE 6.1
Joint pmf for (X, Y) for balls in box example.

	Y = # of White				
X = # of Red	0	1	2	3	4
0	0	0	6/252	12/252	3/252
1	0	12/252	54/252	36/252	3/252
2	3/252	36/252	54/252	12/252	0
3	3/252	12/252	6/252	0	0

This table is called the joint probability mass function (pmf) $f(x, y)$ of (X, Y). As for any probability distribution, one requires that each of the probability values are nonnegative and the sum of the probabilities over all values of X and Y is one. That is, the function $f(x, y)$ satisfies two properties:

(1) $f(x, y) \geq 0$, for all x, y

(2) $\Sigma_{x,y} f(x, y) = 1$

It is clear from Table 6.1 that all of the probabilities are nonnegative and the reader can confirm that the sum of the probabilities is equal to one.

Using Table 6.1, one sees that some particular pairs (x, y) are not possible as $f(x, y) = 0$. For example, $f(0, 1) = 0$ which means that it is not possible to observe 0 red balls and 1 white ball in the sample. Note that five balls were sampled, and if one only observed one red or white ball, that means that one must have sampled $5 - 1 = 4$ black balls which is not possible.

One finds probabilities of any event involving X and Y by summing probabilities from Table 6.1.

1. **What is $P(X = Y)$, the probability that one samples the same number of red and white balls?** By the table, one sees that this is possible only when $X = 1, Y = 1$ or $X = 2, Y = 2$. So the probability

$$P(X = Y) = f(1, 1) + f(2, 2) = \frac{12}{252} + \frac{54}{252} = \frac{66}{252}.$$

2. **What is $P(X > Y)$, the probability one samples more red balls than white balls?** From the table, one identifies the outcomes where $X > Y$, and then sums the corresponding probabilities.

$$\begin{aligned} P(X > Y) &= f(1, 0) + f(2, 0) + f(2, 1) + f(3, 0) + f(3, 1) + f(3, 2) \\ &= \frac{12}{252} + \frac{3}{252} + \frac{36}{252} + \frac{3}{252} + \frac{12}{252} + \frac{6}{252} \\ &= \frac{72}{252} \end{aligned}$$

R Simulating sampling from a box

The variable `box` is a vector containing the colors of the ten balls in the box. The function `one_rep()` simulates drawing five balls from the box and computing the number of red balls and number of white balls.

```
box <- c("white", "white", "white", "white",
         "red", "red", "red",
         "black", "black", "black")
one_rep <- function(){
  balls <- sample(box, size = 5, replace = FALSE)
```

```
  X <- sum(balls == "red")
  Y <- sum(balls == "white")
  c(X, Y)
}
```

Using the `replicate()` function, one simulates this sampling process 1000 times, storing the outcomes in the data frame `results` with variable names `X` and `Y`. Using the `table()` function, one classifies all outcomes with respect to the two variables. By dividing the observed counts by the number of simulations, one obtains approximate probabilities similar to the exact probabilities shown in Table 6.1.

```
results <- data.frame(t(replicate(1000, one_rep())))
names(results) <- c("X", "Y")
table(results$X, results$Y) / 1000
        0     1     2     3     4
  0 0.000 0.000 0.022 0.055 0.011
  1 0.000 0.036 0.214 0.154 0.013
  2 0.009 0.138 0.226 0.037 0.000
  3 0.009 0.048 0.028 0.000 0.000
```

Marginal probability functions

Once a joint probability mass function for (X, Y) has been constructed, one finds probabilities for one of the two variables. In our balls example, suppose one wants to find the probability that exactly three red balls are chosen, that is $P(X = 3)$. This probability is found by summing values of the pmf $f(x, y)$ where $x = 3$ and y can be any possible value of the random variable Y, that is,

$$
\begin{aligned}
P(X = 3) &= \sum_y f(3, y) \\
&= f(3, 0) + f(3, 1) + f(3, 2) \\
&= \frac{3}{252} + \frac{12}{252} + \frac{6}{252} \\
&= \frac{21}{252}.
\end{aligned}
$$

This operation is done for each of the possible values of X – the *marginal* probability mass function of X, $f_X()$ is defined as follows:

$$f_X(x) = \sum_y f(x, y). \tag{6.1}$$

One finds this marginal pmf of X from Table 6.1 by summing the joint probabilities for each row of the table. The marginal pmf is displayed in Table

TABLE 6.2
Marginal pmf for X in the balls example.

x	$f_X(x)$
0	21/252
1	105/252
2	105/252
3	21/252

6.2. Note that a marginal pmf is a legitimate probability function in that the values are nonnegative and the probabilities sum to one.

One can also find the marginal pmf of Y, denoted by $f_Y()$, by a similar operation – for a fixed value of $Y = y$ one sums over all of the possible values of X.

$$f_Y(y) = \sum_x f(x, y). \tag{6.2}$$

For example, if one wants to find $f_Y(2) = P(Y = 2)$ in our example, one sums the joint probabilities in Table 6.1 over the rows in the column where $Y = 2$. One obtains the probability:

$$\begin{aligned} f_Y(2) &= \sum_x f(x, 2) \\ &= f(0, 2) + f(1, 2) + f(2, 2) + f(3, 2) \\ &= \frac{6}{252} + \frac{54}{252} + \frac{54}{252} + \frac{6}{252} \\ &= \frac{120}{252}. \end{aligned}$$

By repeating this exercise for each value of Y, one obtains the marginal pmf displayed in Table 6.3.

TABLE 6.3
Marginal pmf for Y in the balls example.

y	$f_Y(y)$
0	6/252
1	60/252
2	120/252
3	60/252
4	6/252

Conditional probability mass functions

In Chapter 3, the conditional probability of an event A was defined given knowledge of another event B. Moving back to the sampling balls from a box example, suppose one is told that exactly two red balls are sampled, that is $X = 2$ – how does that information change the probabilities about the number of white balls Y?

In this example, one is interested in finding $P(Y = y \mid X = 2)$. Using the definition of conditional probability, one has

$$\begin{aligned} P(Y = y \mid X = 2) &= \frac{P(Y = y, X = 2)}{P(X = 2)}. \\ &= \frac{f(2, y)}{f_X(2)} \end{aligned}$$

For example, the probability of observing two white balls given that we have two red balls is equal to

$$\begin{aligned} P(Y = 2 \mid X = 2) &= \frac{P(Y = 2, X = 2)}{P(X = 2)} \\ &= \frac{f(2, 2)}{f_X(2)} \\ &= \frac{54/252}{105/252} = \frac{54}{105}. \end{aligned}$$

Suppose this calculation is repeated for all possible values of Y – one obtains the values displayed in Table 6.4.

TABLE 6.4
Conditional pmf for Y given $X = s$ in the balls example.

y	$f_{Y\mid X}(y \mid X = 2)$
0	3/105
1	36/105
2	54/105
3	12/105

These probabilities represent the conditional pmf for Y conditional on $X = 2$. This conditional pmf is just like any other probability distribution in that the values are nonnegative and they sum to one. To illustrate using this distribution, suppose one is told that two red balls are selected (that is, $X = 2$) and one wants to find the probability that more than one white ball is chosen. This probability is given by

$$
\begin{aligned}
P(Y > 1 \mid X = 2) &= \Sigma_{y>1} \; f_{Y|X}(y \mid X = 2) \\
&= f_{Y|X}(2 \mid X = 2) + f_{Y|X}(3 \mid X = 2) \\
&= \frac{54}{105} + \frac{12}{105} = \frac{66}{105}.
\end{aligned}
$$

In general, the conditional probability mass function of Y conditional on $X = x$, denoted by $f_{Y\ midX}(y \mid x)$, is defined to be

$$
f_{Y|X}(y \mid x) = \frac{f(x, y)}{f_X(x)}, \text{ if } f_X(x) > 0. \tag{6.3}
$$

R Simulating sampling from a box

Recall that the data frame `results` contains the simulated outcomes for 1000 selections of balls from the box. By filtering on the value `X = 2` and tabulating the values of `Y`, one is simulating from the conditional pmf of Y conditional on $X = 2$. Note that the relative frequencies displayed below are approximately equal to the exact probabilities shown in Table 6.2.

```
results %>%
  filter(X == 2) %>%
  group_by(Y) %>%
  summarize(N = n()) %>%
  mutate(P = N / sum(N))
```

```
      Y     N      P
  <int> <int>  <dbl>
1     0     9 0.0220
2     1   138 0.337
3     2   226 0.551
4     3    37 0.0902
```

6.3 Multinomial Experiments

Suppose one rolls the usual six-sided die where one side shows 1, two sides show 2, and three sides show 3. One rolls this die ten times – what is the chance that one will observe three 1's and five 2's?

This situation resembles the coin-tossing experiment described in Chapter 4. One is repeating the same process, that is rolling the die, repeated times, and one regards the individual die results as independent outcomes. The difference is that the coin-tossing experiment had only two possible outcomes on a single trial, and here there are three outcomes on a single die roll, 1, 2, and 3.

Suppose a random experiment consists of a sequence of n independent trials where there are k possible outcomes on a single trial where $k \geq 2$. Denote the possible outcomes as 1, 2, ..., k, and let $p_1, p_2, ..., p_k$ denote the associated probabilities. If X_1, X_2, ..., X_k denote the number of 1s, 2s, ..., ks observed in the n trials, the vector of outcomes $X = (X_1, X_2, ..., X_n)$ has a multinomial distribution with sample size n and vector of probabilities $p = (p_1, p_2, ..., p_k)$.

In our example, each die roll has $k = 3$ possible outcomes and the associated vector of probabilities is $p = (1/6, 2/6, 3/6)$. The number of observed 1's, 2's, 3's in $n = 10$ trials, $X = (X_1, X_2, X_3)$ has a multinomial distribution with parameters n and p.

By generalizing the arguments made in Chapter 4, one can show that the probability that $X_1 = x_1, ..., X_k = x_k$ has the general form

$$f(x_1, ..., x_k) = \left(\frac{n!}{n_1!...n_k!}\right) \prod_{j=1}^{k} p_j^{x_j}, \tag{6.4}$$

where $x_j = 0, 1, 2, ..., j = 1, ...k$ and $\sum_{j=1}^{n} x_j = n$.

This formula can be used to compute a probability for our example. One has $n = 10$ trials and the outcome three 1's and five 2's is equivalent to the outcome $X_1 = 3, X_2 = 5$. The number of 3's X_3 is not random since we know that $X_1 + X_2 + X_3 = 10$. The probability vector is $p = (1/6, 2/6, 3/6)$. By substituting in the formula, we have

$$P(X_1 = 3, X_2 = 5, X_3 = 2) = \left(\frac{10!}{3!\,5!\,2!}\right) \left(\frac{1}{6}\right)^3 \left(\frac{2}{6}\right)^5 \left(\frac{3}{6}\right)^2.$$

R By use of the `factorial()` function in R, we compute this probability to be 0.012.

```
factorial(10) / (factorial(3) * factorial(5) * factorial(2)) *
+   (1 / 6) ^ 3 * (2 / 6) ^ 5 * (3 / 6) ^ 2
[1] 0.01200274
```

Other probabilities can be found by summing the joint multinomial pmf over sets of interest. For example, suppose one is interested in computing the probability that the number of 1's exceeds the number of 2's in our ten dice rolls. One is interested in the probability $P(X_1 > X_2)$ which is given by

$$P(X_1 > X_2) = \sum_{x_1 > x_2} \left(\frac{10!}{3!\,5!\,2!}\right) \left(\frac{1}{6}\right)^{x_1} \left(\frac{2}{6}\right)^{x_2} \left(\frac{3}{6}\right)^{10-x_1-x_2},$$

where one is summing over all of the outcomes (x_1, x_2) where $x_1 > x_2$.

Marginal distributions

One attractive feature of the multinomial distribution is that the marginal distributions have familiar functional forms. In the dice roll example, suppose one is interested only in X_1, the number of 1's in ten rolls of our die. One obtains the marginal probability distribution of X_1 directly by summing out the other variables from the joint pmf of X_1 and X_2. For example, one finds, say $P(X_1 = 2)$, by summing the joint probability values over all (x_1, x_2) pairs where $x_1 = 2$:

$$P(X_1 = 2) = \sum_{x_2, x_1 + x_2 \leq 10} f(x_1, x_2).$$

In this computation, it is important to recognize that the sum of rolls of 1 and 2, $x_1 + x_2$ cannot exceed the number of trials $n = 10$.

A more intuitive way to obtain a marginal distribution relies on the previous knowledge of binomial distributions. In each die roll, suppose one records if one gets a one or not. Then X_1, the number of ones in n trials, will be binomial distributed with parameters n and $p = 1/6$. Using a similar argument, X_2, the number of twos in n trials, will be binomial with n trials and $p = 2/6$.

Conditional distributions

One applies the knowledge about marginal distributions to compute conditional distributions in the multinomial situation. Suppose that one is given that $X_2 = 3$ in $n = 10$ trials. What can one say about the probabilities of X_1?

One uses the conditional pmf definition to compute the conditional probability $P(X_1 = x \mid X_2 = 3)$. First, it is helpful to think about possible values for X_1. Since one has $n = 10$ rolls of the die and we know that we observe $X_2 = 3$ (three twos), the possible values of X_1 can be 0, 1, ..., 7. For these values, we have

$$P(X_1 = x \mid X_2 = 3) = \frac{P(X_1 = x, X_2 = 3)}{P(X_2 = 3)}.$$

The numerator is the multinomial probability and since X_2 has a marginal binomial distribution, the denominator is a binomial probability. Making the substitutions, one has

$$P(X_1 = x \mid X_2 = 3) = \frac{\left(\frac{10!}{x!\,3!\,(10-x-3)!}\right)\left(\frac{1}{6}\right)^x \left(\frac{2}{6}\right)^3 \left(\frac{3}{6}\right)^{10-x-3}}{\binom{10}{3}\left(\frac{2}{6}\right)^3\left(1-\frac{2}{6}\right)^{10-3}}.$$

After some simplification, one obtains

$$P(X_1 = x \mid X_2 = 3) = \binom{7}{x}\left(\frac{1}{4}\right)^x\left(1-\frac{1}{4}\right)^{7-x}, \; x = 0, ..., 7.$$

which is a binomial distribution with 7 trials and probability of success $1/4$.

An alternative way to figure out the conditional distribution is based on an intuitive argument. One is told there are three 2's in 10 rolls of the die. The results of the remaining $10 - 3 = 7$ trials are unknown where the possible outcomes are 1 and 3 with probabilities proportional to $1/6$ and $3/6$. So X_1 will be binomial with 7 trials and success probability equal to $(1/6)/(1/6+3/6) = 1/4$.

R Simulating Multinomial experiments

The function `sim_die_rolls()` will simulate 10 rolls of the special weighted die. The `sample()` function draws values of 1, 2, 3 with replacement where the respective probabilities are 1/6, 2/6, and 3/6. The outputs are values of X_1, X_2 and X_3.

```
sim_die_rolls <- function(){
        rolls <- sample(1:3, size = 10,
                        replace = TRUE,
                        prob = c(1, 2, 3) / 6)
        c(sum(rolls == 1),
          sum(rolls == 2),
          sum(rolls == 3))
}
```

Using the `replicate()` function, one simulates the Multinomial experiment for 5000 iterations. The outcomes are placed in a data frame with variable names `X1`, `X2` and `X3`.

```
results <- data.frame(t(replicate(5000,
                                sim_die_rolls())))
names(results) <- c("X1", "X2", "X3")
head(results)
```

```
  X1 X2 X3
1  1  4  5
2  0  7  3
3  2  4  4
4  1  4  5
5  0  5  5
6  1  2  7
```

Given this simulated output, one can compute many different probabilities of interest. For example, suppose one is interested in $P(X_1 + X_2 < 5)$. One approximates this probability by simulation by finding the proportion of simulated pairs (`X1, X2`) where `X1 + X2 < 5`.

```
results %>%
  summarize(P = sum(X1 + X2 < 5) / 5000)
       P
1 0.3774
```

Suppose one is interested in finding the mean of the distribution of X_1 conditional on $X_2 = 3$. The `filter()` function is used to choose only the Multinomial results where `X2 = 3` and the `summarize()` function finds the mean of `X1` among these results. One estimates $E(X_1 \mid X_2 = 3) \approx 1.79193$. Note that it was found earlier that the conditional distribution of X_1 conditional on $X_2 = 3$ is binomial$(7, 1/4)$ with mean $7(1/4)$ which is consistent with the simulation-based calculation.

```
results %>%
  filter(X2 == 3) %>%
  summarize(X1_M = mean(X1))
      X1_M
1 1.79193
```

6.4 Joint Density Functions

One can also describe probabilities when the two variables X and Y are continuous. As a simple example, suppose that one randomly chooses two points X and Y on the interval (0, 2) where $X < Y$. One defines the joint probability density function or joint pdf of X and Y to be the function

$$f(x, y) = \begin{cases} \frac{1}{2}, & 0 < x < y < 2; \\ 0, & \text{elsewhere.} \end{cases}$$

This joint pdf is viewed as a plane of constant height over the set of points (x, y) where $0 < x < y < 2$. This region of points in the plane is shown in Figure 6.1.

In the one variable situation in Chapter 5, a function f is a legitimate density function or pdf if it is nonnegative over the real line and the total area under the curve is equal t to one. Similarly for two variables, any function $f(x, y)$ is considered a pdf if it satisfies two properties:

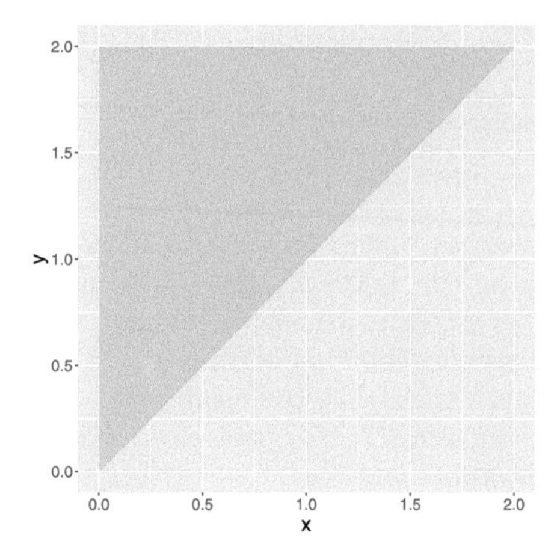

FIGURE 6.1
Region where the joint pdf $f(x, y)$ is positive in the "choose two points" example.

1. Density is nonnegative over the whole plane:

$$f(x, y) \geq 0, \text{ for all } x, y. \tag{6.5}$$

2. The total volume under the density is equal to one:

$$\int\int f(x, y) dx dy = 1. \tag{6.6}$$

One can check that the pdf in our example is indeed a legitimate pdf. It is pretty obvious that the density that was defined is nonnegative, but it is less clear that the integral of the density is equal to one. Since the density is a plane of constant height, one computes this double integral geometrically. Using the familiar "one half base times height" argument, the area of the triangle in the plane is $(1/2)(2)(2) = 2$ and since the pdf has constant height of $1/2$, the volume under the surface is equal to $2(1/2) = 1$.

Probabilities about X and Y are found by finding volumes under the pdf surface. For example, suppose one wants to find the probability that the sum of locations $X + Y > 3$, that is $P(X + Y > 3)$. The region in the (x, y) plane of interest is first identified, and then one finds the volume under the joint pdf over this region. In Figure 6.2, the region where $x + y > 3$ has been shaded. The probability $P(X + Y > 3)$ is the volume under the pdf over this region. Applying a geometric argument, one notes that the area of the shaded region is $1/4$, and so the probability of interest is $(1/4)(1/2) = 1/8$. One also finds

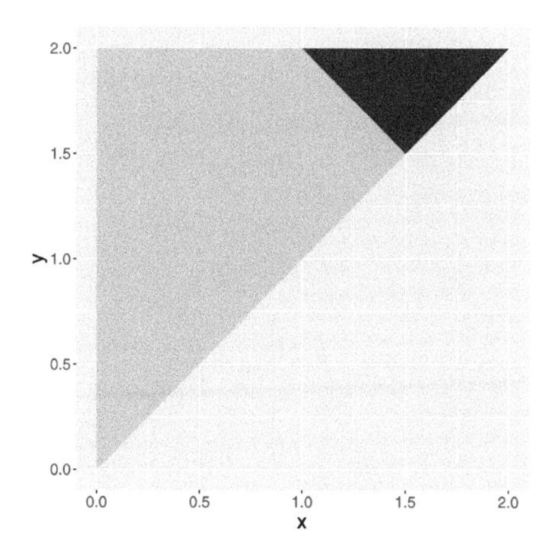

FIGURE 6.2
Shaded region where $x + y > 3$ in the "choose two points" example.

this probability by integrating the joint pdf over the region as follows:

$$\begin{aligned} P(X+Y<3) &= \int_{1.5}^{2}\int_{3-y}^{y} f(x,y)dxdy \\ &= \int_{1.5}^{2}\int_{3-y}^{y} \frac{1}{2}dxdy \\ &= \int_{1.5}^{2} \frac{2y-3}{2}dy \\ &= \frac{y^2-3y}{2}\Big|_{1.5}^{2} \\ &= \frac{1}{8}. \end{aligned}$$

Marginal probability density functions

Given a joint pdf $f(x,y)$ that describes probabilities of two continuous variables X and Y, one summarizes probabilities about each variable individually by the computation of marginal pdfs. The marginal pdf of X, $f_X(x)$, is obtained by integrating out y from the joint pdf.

$$f_X(x) = \int f(x,y)dy. \tag{6.7}$$

In a similar fashion, one defines the marginal pdf of Y by integrating out x from the joint pdf.

$$f_Y(x) = \int f(x,y)dx. \tag{6.8}$$

Let's illustrate the computation of marginal pdfs for our example. One has to be careful about the limits of the integration due to the dependence between x and y in the support of the joint density. Looking back at Figure 6.1, one sees that if the value of x is fixed, then the limits for y go from x to 2. So the marginal density of X is given by

$$\begin{aligned} f_X(x) &= \int f(x,y)dy \\ &= \int_x^2 \frac{1}{2} dy \\ &= \frac{2-x}{2}, \ 0 < x < 2. \end{aligned}$$

By a similar calculation, one can verify that the marginal density of Y is equal to

$$f_Y(y) = \frac{y}{2}, \ 0 < y < 2.$$

Conditional probability density functions

Once a joint pdf $f(x, y)$ has been defined, one can also define conditional pdfs. In our example, suppose one is told that the first random location is equal to $X = 1.5$. What has one learned about the value of the second random variable Y?

To answer this question, one defines the notion of a conditional pdf. The conditional pdf of the random variable Y given the value $X = x$ is defined as the quotient

$$f_{Y|X}(y \mid X = x) = \frac{f(x,y)}{f_X(x)}, \text{ if } f_X(x) > 0. \tag{6.9}$$

In our example one is given that $X = 1.5$. Looking at Figure 6.1, one sees that when $X = 1.5$, the only possible values of Y are between 1.5 and 2. By substituting the values of $f(x, y)$ and $f_X(x)$, one obtains

$$\begin{aligned} f_{Y|X}(y \mid X = 1.5) &= \frac{f(1.5,y)}{f_X(1.5)} \\ &= \frac{1/2}{(2-1.5)/2} \\ &= 2, \ 1.5 < y < 2. \end{aligned}$$

In other words, the conditional density for Y when $X = 1.5$ is uniform from 1.5 to 2.

A conditional pdf is a legitimate density function, so the integral of the pdf over all values y is equal to one. We use this density to compute conditional probabilities. For example, if $X = 1.5$, what is the probability that Y is greater than 1.7? This probability is the conditional probability $P(Y > 1.7 \mid X = 1.5)$ that is equal to an integral over the conditional density $f_{Y|X}(y \mid 1.5)$:

$$\begin{aligned} P(Y > 1.7 \mid X = 1.5) &= \int_{1.7}^{2} f_{Y|X}(y \mid 1.5)dy \\ &= \int_{1.7}^{2} 2dy \\ &= 0.6. \end{aligned}$$

Turn the random variables around

Above, we looked at the pdf of Y conditional on a value of X. One can also consider a pdf of X conditional on a value of Y. Returning to our example, suppose that one learns that Y, the larger random variable on the interval is equal to 0.8. In this case, what would one expect for the random variable X?

This question is answered in two steps – one first finds the conditional pdf of X conditional on $Y = 0.8$. Then once this conditional pdf is found, one finds the mean of this distribution.

The conditional pdf of X given the value $Y = y$ is defined as the quotient

$$f_{X|Y}(x \mid Y = y) = \frac{f(x, y)}{f_Y(y)}, \text{ if } f_Y(y) > 0. \tag{6.10}$$

Looking back at Figure 6.1, one sees that if $Y = 0.8$, the possible values of X are from 0 to 0.8. Over these values the conditional pdf of X is given by

$$\begin{aligned} f_{X|Y}(x \mid 0.8) &= \frac{f(x, 0.8)}{f_Y(0.8)} \\ &= \frac{1/2}{0.8/2} \\ &= 1.25, \ 0 < x < 0.8. \end{aligned}$$

So if one knows that $Y = 0.8$, then the conditional pdf for X is Uniform on (0, 0.8).

To find the "expected" value of X knowing that $Y = 0.8$, one finds the mean of this distribution.

$$\begin{aligned} E(X \mid Y = 0.8) &= \int_0^{0.8} x f_{X|Y}(x \mid 0.8)dx \\ &= \int_0^{0.8} x\, 1.25\, dx \\ &= (0.8)^2/2 \times 1.25 = 0.4. \end{aligned}$$

6.5 Independence and Measuring Association

As a second example, suppose one has two random variables (X, Y) that have the joint density

$$f(x, y) = \begin{cases} x + y, & 0 < x < 1, 0 < y < 1; \\ 0, & \text{elsewhere.} \end{cases}$$

This density is positive over the unit square, but the value of the density increases in X (for fixed y) and also in Y (for fixed x). Figure 6.3 displays a graph of this joint pdf – the density is a section of a plane that reaches its maximum value at the point (1, 1).

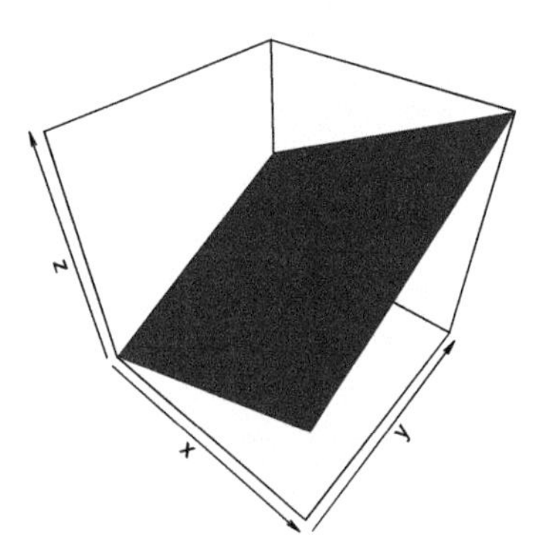

FIGURE 6.3
Three dimensional display of the pdf of $f(x, y) = x + y$ defined over the unit square.

From this density, one computes the marginal pdfs of X and Y. For example, the marginal density of X is given by

$$\begin{aligned} f_X(x) &= \int_0^1 x + y dy \\ &= x + \frac{1}{2}, \quad 0 < x < 1. \end{aligned}$$

Similarly, one can show that the marginal density of Y is given by $f_Y(y) = y + \frac{1}{2}$ for $0 < y < 1$.

Independence

Two random variables X and Y are said to be *independent* if the joint pdf factors into a product of their marginal densities, that is

$$f(x, y) = f_X(x) f_Y(y). \tag{6.11}$$

for all values of X and Y. Are X and Y independent in our example? Since we have computed the marginal densities, we look at the product

$$f_X(x) f_Y(y) = (x + \frac{1}{2})(y + \frac{1}{2})$$

which is clearly not equal to the joint pdf $f(x, y) = x + y$ for values of x and y in the unit square. So X and Y are not independent in this example.

Measuring association by covariance

In the situation like this one where two random variables are not independent, it is desirable to measure the association pattern. A standard measure of association is the covariance defined as the expectation

$$\begin{aligned} Cov(X, Y) &= E\left((X - \mu_X)(Y - \mu_Y)\right) \\ &= \int \int (x - \mu_X)(y - \mu_Y) f(x, y) dx dy. \end{aligned} \tag{6.12}$$

For computational purposes, one writes the covariance as

$$\begin{aligned} Cov(X, Y) &= E(XY) - \mu_X \mu_Y \\ &= \int \int (xy) f(x, y) dx dy - \mu_X \mu_Y. \end{aligned} \tag{6.13}$$

For our example, one computes the expectation $E(XY)$ from the joint density:

$$\begin{aligned} E(XY) &= \int_0^1 \int_0^1 (xy)(x + y) dx dy \\ &= \int \frac{y}{3} + \frac{y^2}{2} dy \\ &= \frac{1}{3}. \end{aligned}$$

One can compute that the means of X and Y are given by $\mu_X = 7/12$ and $\mu_Y = 7/12$, respectively. So then the covariance of X and Y is given by

$$\begin{aligned} Cov(X, Y) &= E(XY) - \mu_X \mu_Y \\ &= \frac{1}{3} - \left(\frac{7}{12}\right)\left(\frac{7}{12}\right) \\ &= -\frac{1}{144}. \end{aligned}$$

It can be difficult to interpret a covariance value since it depends on the scale of the support of the X and Y variables. One standardizes this measure of association by dividing by the standard deviations of X and Y resulting in the correlation measure ρ:

$$\rho = \frac{Cov(X, Y)}{\sigma_X \sigma_Y}. \tag{6.14}$$

In a separate calculation one can find the variances of X and Y to be $\sigma_X^2 = 11/144$ and $\sigma_Y^2 = 11/144$. Then the correlation is given by

$$\begin{aligned}\rho &= \frac{-1/144}{\sqrt{11/144}\sqrt{11/144}} \\ &= -\frac{1}{11}.\end{aligned}$$

It can be shown that the value of the correlation ρ falls in the interval $(-1, 1)$ where a value of $\rho = -1$ or $\rho = 1$ indicates that Y is a linear function of X with probability 1. Here the correlation value is a small negative value indicating weak negative association between X and Y.

6.6 Flipping a Random Coin: The Beta-Binomial Distribution

Suppose one has a box of coins where the coin probabilities vary. If one selects a coin from the box, p, the probability the coin lands heads follows the distribution

$$g(p) = \frac{1}{B(6,6)} p^5 (1-p)^5, \; 0 < p < 1,$$

where $B(6, 6)$ is the beta function, which will be more thoroughly discussed in Chapter 7. This density is plotted in Figure 6.4. A couple of things to notice about this density. First, the density has a significant height over a range of plausible values of the probability – this reflects the idea that we are really unsure about the chance of observing a heads when flipped. Second, the density is symmetric about $p = 0.5$, which means that the coin is equally likely to be biased towards heads or biased towards tails.

One next flips this "random" coin 20 times. Denote the outcome of this experiment by the random variable Y which is equal to the count of heads. If we are given a value of the probability p, then Y has a binomial distribution with $n = 20$ trials and success probability p. This probability function is actually the conditional probability of observing y heads given a value of the probability p:

$$f(y \mid p) = \binom{20}{y} p^y (1-p)^{20-y}, \; y = 0, 1, ..., 20.$$

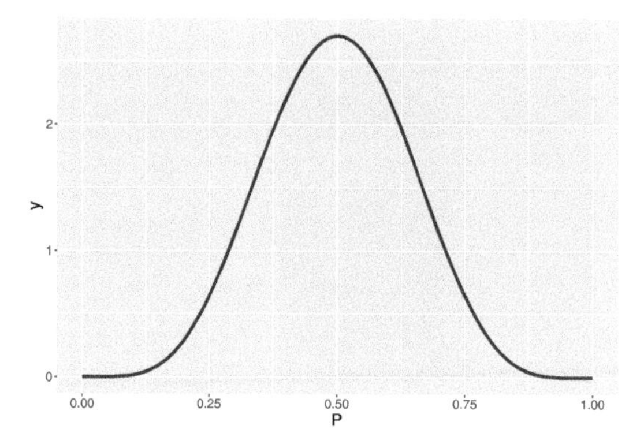

FIGURE 6.4
Beta(6, 6) density representing the distribution of probabilities of heads for a large collection of random coins.

Given the density of p and the conditional density of Y conditional on p, one computes the joint density by the product

$$\begin{aligned} f(y,p) = g(p)f(y \mid p) &= \left[\frac{1}{B(6,6)}p^5(1-p)^5\right]\left[\binom{20}{y}p^y(1-p)^{20-y}\right] \\ &= \frac{1}{B(6,6)}\binom{20}{y}p^{y+5}(1-p)^{25-y}, \ 0 < p < 1, y = 0, 1, ..., 20. \end{aligned}$$

This beta-binomial density is a mixed density in the sense that one variable (p) is continuous and one (Y) is discrete. This will not create any problems in the computation of marginal or conditional distributions, but one should be careful to understand the support of each random variable.

R Simulating from the beta-binomial distribution

Using R it is straightforward to simulate a sample of (p, y) values from the Beta-Binomial distribution. Using the `rbeta()` function, one takes a random sample of 500 draws from the beta(6, 6) distribution. Then for each probability value p, one uses the `rbinom()` function to simulate the number of heads in 20 flips of this "p coin."

```
data.frame(p = rbeta(500, 6, 6)) %>%
  mutate(Y = rbinom(500, size = 20, prob = p)) %>%
  ggplot(aes(p, Y)) + geom_jitter()
```

A scatterplot of the simulated values of p and Y is displayed in Figure 6.5. Note that the variables are positively correlated, which indicates that one

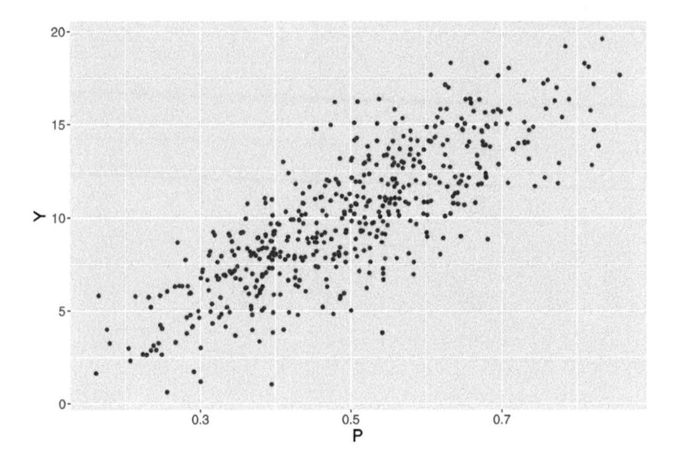

FIGURE 6.5
Scatterplot of 500 simulated draws from the joint density of the probability of heads p and the number of heads Y in 20 flips.

tends to observe a large number of heads with coins with a large probability of heads.

What is the probability that one observes exactly 10 heads in the 20 flips, that is $P(Y = 10)$? One performs this calculation by computing the marginal probability function for Y. This is obtained by integrating out the probability p from the joint density. This density is a special case of the beta-binomial distribution.

$$
\begin{aligned}
f(y) &= \int_0^1 g(p) f(y \mid p) dp \\
&= \int_0^1 \frac{1}{B(6,6)} \binom{20}{y} p^{y+5}(1-p)^{25-y} dp \\
&= \binom{20}{y} \frac{B(y+6, 26-y)}{B(6,6)}, \quad y = 0, 1, 2, ..., 20.
\end{aligned}
$$

Using this formula with the substitution $y = 10$, we use R to find the probability $P(Y = 10)$.

```
choose(20, 10) * beta(10 + 6, 26 - 10) /  beta(6, 6)
[1] 0.1065048
```

6.7 Bivariate Normal Distribution

Suppose one collects multiple body measurements from a group of 30 students. For example, for each of 30 students, one might collect the diameter of the wrist and the diameter of the ankle. If X and Y denote the two body measurements (measured in cm) for a student, then one might think that the density of X and the density of Y are normally distributed. Moreover, the two random variables would be positively correlated – if a student has a large wrist diameter, one would predict her to also have a large forearm length.

A convenient joint density function for two continuous measurements X and Y, each variable measured on the whole real line, is the bivariate normal density with density given by

$$f(x, y) = \frac{1}{2\pi\sigma_X\sigma_Y\sqrt{1-\rho}} \exp\left[-\frac{1}{2(1-\rho^2)}(z_X^2 - 2\rho z_X z_Y + z_Y^2)\right], \quad (6.15)$$

where z_X and z_Y are the standardized scores

$$z_X = \frac{x - \mu_X}{\sigma_X}, \quad z_Y = \frac{y - \mu_Y}{\sigma_Y}, \quad (6.16)$$

and μ_X, μ_Y and σ_X, σ_Y are respectively the means and standard deviations of X and Y. The parameter ρ is the correlation of X and Y and measures the association between the two variables.

Figure 6.6 shows contour plots of four bivariate normal distributions. The bottom right graph corresponds to the values $\mu_X = 17, \mu_Y = 23, \sigma_X = 2, \sigma_Y = 3$ and $\rho = 0.4$ where X and Y represent the wrist diameter and ankle diameter measurements of the student. The correlation value of $\rho = 0.4$ reflects the moderate positive correlation of the two body measurements. The other three graphs use the same means and standard deviations but different values of the ρ parameter. This figure shows that the bivariate normal distribution is able to model a variety of association structures between two continuous measurements.

There are a number of attractive properties of the bivariate normal distribution.

1. **The marginal densities of X and Y are Normal.** So X has a Normal density with parameters μ_X and σ_X and likewise Y is Normal(μ_Y, σ_Y).

2. **X and Y are normal, conditional densities will also be normal.** For example, if one is given that $Y = y$, then the conditional density of X given $Y = y$ is normal where

$$E(X \mid Y = y) = \mu_X + \rho\frac{\sigma_X}{\sigma_Y}(y - \mu_Y), \; Var(X \mid Y = y) = \sigma_X^2(1 - \rho^2). \quad (6.17)$$

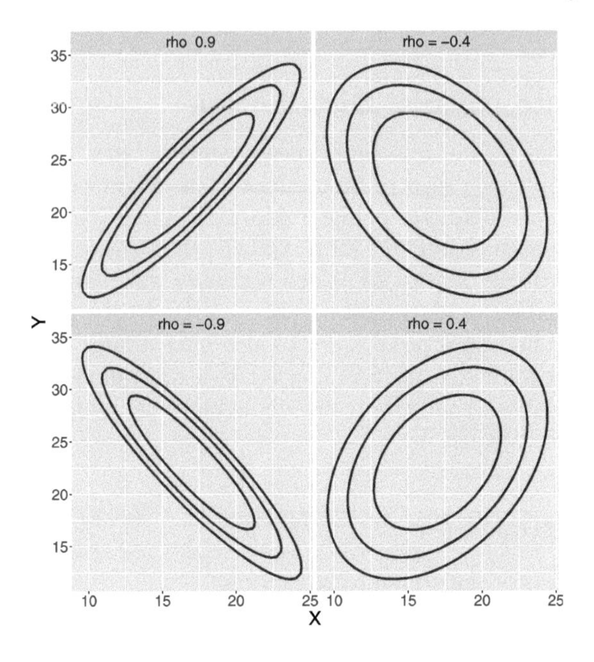

FIGURE 6.6
Contour graphs of four Bivariate Normal distributions with different correlations.

Similarly, if one knows that $X = x$, then the conditional density of Y given $X = x$ is Normal with mean $\mu_Y + \rho\frac{\sigma_Y}{\sigma_X}(x - \mu_X)$ and variance $\sigma_Y^2(1 - \rho^2)$.

3. **For a bivariate normal distribution, X and Y are independent if and only if the correlation $\rho = 0$.** In contrast, as the correlation parameter ρ approaches $+1$ and -1, then all of the probability mass will be concentrated on a line where $Y = aX + b$.

R Bivariate normal calculations

Returning to the body measurements application, different uses of the bivariate normal model can be illustrated. Recall that X denotes the wrist diameter, Y represents the ankle diameter and we are assuming (X, Y) has a bivariate normal distribution with parameters $\mu_X = 17, \mu_Y = 23, \sigma_X = 2, \sigma_Y = 3$ and $\rho = 0.4$

1. **Find the probability a student's wrist diameter exceeds 20 cm.**

   Here one is interested in the probability $P(X > 20)$. From the facts above, the marginal density for X will be normal with mean

$\mu_X = 17$ and standard deviation $\sigma_X = 2$. So this probability is computed using the function `pnorm()`:

```
1 - pnorm(20, 17, 2)
[1] 0.0668072
```

2. **Suppose one is told that the student's ankle diameter is 20 cm – find the conditional probability $P(X > 20 \mid Y = 20)$.**

   By above the distribution of X conditional on the value $Y = y$ is normal with mean $\mu_X + \rho \frac{\sigma_X}{\sigma_Y}(y - \mu_Y)$ and variance $\sigma_X^2(1 - \rho^2)$. Here one is conditioning on the value $Y = 20$ and one computes the mean and standard deviation and apply the `pnorm()` function:

$$
\begin{aligned}
E(X \mid Y = 20) &= \mu_X + \rho\frac{\sigma_X}{\sigma_Y}(y - \mu_Y) \\
&= 17 + 0.4\left(\frac{2}{3}\right)(20 - 23) \\
&= 16.2.
\end{aligned}
$$

$$
\begin{aligned}
SD(X \mid Y = 20) &= \sqrt{\sigma_X^2(1 - \rho^2)} \\
&= \sqrt{2^2(1 - 0.4^2)} \\
&= 1.83.
\end{aligned}
$$

```
1 - pnorm(20, 16.2, 1.83)
[1] 0.01892374
```

3. **Are X and Y independent variables?**

   By the properties above, for a bivariate normal distribution, a necessary and sufficient condition for independence is that the correlation $\rho = 0$. Since the correlation between the two variables is not zero, the random variables X and Y can not be independent.

4. **Find the probability a student's ankle diameter measurement is at 50 percent greater than her wrist diameter measurement, that is $P(Y > 1.5X)$.**

R Simulating Bivariate Normal measurements

The computation of the probability $P(Y > 1.5X)$ is not obvious from the information provided. But simulation provides an attractive method of computing this probability. One simulates a large number, say 1000, draws from the bivariate normal distribution and then finds the fraction of simulated (x, y) pairs where $y > 1.5x$. Figure 6.7 displays a scatterplot of these simulated draws and the line $y = 1.5x$. The probability is estimated by the fraction of

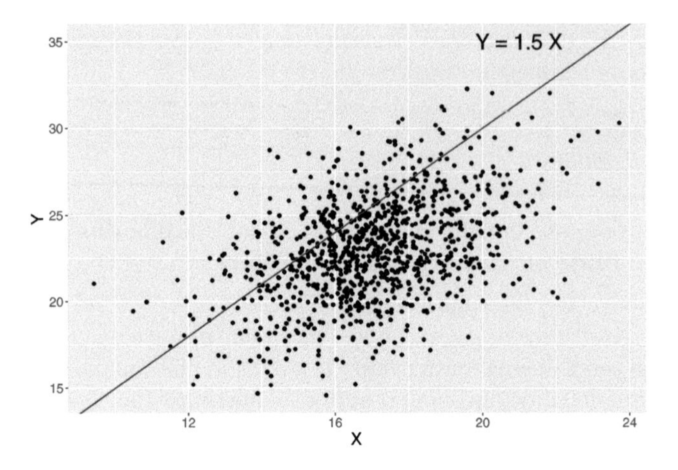

FIGURE 6.7
Scatterplot of simulated draws from the bivariate normal in body measurement example. The probability that $Y > 1.5X$ is approximated by the proportion of simulated points that fall to the left of the line $y = 1.5x$.

points that fall to the left of this line. In the R script below we use a function `sim_binorm()` to simulate 1000 draws from a bivariate normal distribution with inputted parameters $\mu_X, \mu_Y, \sigma_X, \sigma_Y, \phi$. The bivariate normal parameters are set to the values in this example and using the function `sim_binorm()` the probability of interest is approximated by 0.242.

```
sim_binorm <- function(mx, my, sx, sy, r){
  require(ProbBayes)
  v <- matrix(c(sx ^ 2, r * sx * sy,
                r * sx * sy, sy ^ 2),
                2, 2)
  as.data.frame(rmnorm(1000, mean = c(mx, my),
                          varcov = v))}
mx <- 17; my <- 23; sx <- 2; sy <- 3; r <- 0.4
sdata <- sim_binorm(mx, my, sx, sy, r)
names(sdata) <- c("X", "Y")
sdata %>% summarize(mean(Y > 1.5 * X))
  mean(Y > 1.5 * X)
1             0.242
```

6.8 Exercises

1. **Coin Flips**

   Suppose you flip a coin three times with eight equally likely outcomes $HHH, HHT, ..., TTT$. Let X denote the number of heads in the first two flips and Y the number of heads in the last two flips.

   (a) Find the joint probability mass function (pmf) of X and Y and put your answers in the following table.

	y		
x	0	1	2
0			
1			
2			

   (b) Find $P(X > Y)$.

   (c) Find the marginal pmf's of X and Y.

   (d) Find the conditional pmf of X given $Y = 1$.

2. **Selecting Numbers**

   Suppose you select two numbers without replacement from the set $\{1, 2, 3, 4, 5\}$. Let X denote the smaller of the two numbers and Y denote the larger of the two numbers.

   (a) Find the joint probability mass function of X and Y.

   (b) Find the marginal pmf's of X and Y.

   (c) Are X and Y independent? If not, explain why.

   (d) Find $P(Y = 3 \mid X = 2)$.

3. **Die Rolls**

   You roll a die 4 times and record O, the number of ones, and T the number of twos rolled.

   (a) Construct the joint pmf of O and T.

   (b) Find the probability $P(O = T)$.

   (c) Find the conditional pmf of T given $O = 1$.

   (d) Compute $P(T > 0 \mid O = 1)$.

4. **Choosing Balls**

   Suppose you have a box with 3 red and 2 black balls. You first roll a die – if the roll is 1, 2, you sample 3 balls without replacement from the box. If you roll is 3 or higher, you sample 2 balls with replacement from the box. Let X denote the number of balls you sample and Y the number of red balls selected.

(a) Find the joint pmf of X and Y.
(b) Find the probability $P(X = Y)$.
(c) Find the marginal pmf of Y.
(d) Find the conditional pmf of X given $Y = 2$.

5. **Baseball Hitting**

Suppose a player is equally likely to have 4, 5, or 6 at-bats (opportunities) in a baseball game. If N is the number of opportunities, then assume that X, the number of hits, is binomial with probability $p = 0.03$ and sample size N.

(a) Find the joint pmf of N and X.
(b) Find the marginal pmf of X.
(c) Find the conditional pmf of N given $X = 2$.
(d) If the player gets 3 hits, what is the probability he had exactly 5 at-bats?

6. **Multinomial Density**

Suppose a box contains 4 red, 3 black, and 3 green balls. You sample eight balls with replacement from the box and let R denote the number of red and B the number of black balls selected.

(a) Explain why this is a multinomial experiment and given values of the parameters of the multinomial distribution for (R, B).
(b) Compute $P(R = 3, B = 2)$.
(c) Compute the probability that you sample more red balls than black balls.
(d) Find the marginal distribution of B.
(e) If you are given that you sampled $B = 4$ balls, find the probability that you sampled at most 2 red balls.

7. **Joint Density**

Let X and Y have the joint density

$$f(x, y) = ky, \ 0 < x < 2, 0 < y < 2.$$

(a) Find the value of k so that $f()$ is a pdf.
(b) Find the marginal density of X.
(c) Find $P(Y > X)$.
(d) Find the conditional density of Y given $X = x$ for any value $0 < x < 2$.

8. **Joint Density**

Let X and Y have the joint density

$$f(x, y) = x + y, \ 0 < x < 1, 0 < y < 1.$$

(a) Check that f is indeed a valid pdf. If it is not, correct the definition of f so it is valid.
(b) Find the probability $P(X > 0.5, Y < 0.5)$.
(c) Find the marginal density of X.
(d) Are X and Y independent? Answer by a suitable calculation.

9. **Random Division**

Suppose one randomly chooses a values X on the interval (0, 2), and then random choosing a second point Y from 0 to X.

(a) Find the joint density of X and Y.
(b) Are X and Y independent? Explain.
(c) Find the probability $P(Y > 0.5)$.
(d) Find the probability $P(X + Y > 2)$.

10. **Choosing a Random Point in a Circle**

Suppose (X, Y) denotes a random point selected over the unit circle. The joint pdf of (X, Y) is given by

$$f(x, y) = \begin{cases} C, & x^2 + y^2 \leq 1; \\ 0, & \text{elsewhere.} \end{cases}$$

(a) Find the value of the constant C so $f()$ is indeed a joint pdf.
(b) Find the marginal pdf of Y.
(c) Find the probability $P(Y > 0.5)$
(d) Find the conditional pdf of X conditional on $Y = 0.5$.

11. **A Random Meeting**

Suppose John and Jill independently arrive at an airport at a random time between 3 and 4 pm one afternoon. Let X and Y denote respectively the number of minutes past 3 pm that John and Jill arrive.

(a) Find the joint pdf of X and Y.
(b) Find the probability that John arrives later than Jill.
(c) Find the probability that John and Jill meet within 10 minutes of each other.

12. **Defects in Fabric**

Suppose the number of defects per yard in a fabric X is assumed to have a Poisson distribution with mean λ. That is, the conditional density of X given λ has the form

$$f(x \mid \lambda) = \frac{e^{-\lambda}\lambda^x}{x!}, x = 0, 1, 2, ...$$

The parameter λ is assumed to be uniformly distributed over the values 0.5, 1, 1.5, and 2.

(a) Write down the joint pmf of X and λ.
(b) Find the probability that the number of defects X is equal to 0.
(c) Find the conditional pmf of λ if you know that $X = 0$.

13. **Defects in Fabric (continued)**

Again we assume the number of defects per yard in a fabric X given λ has a Poisson distribution with mean λ. But now we assume λ is continuous-valued with the exponential density

$$g(\lambda) = \exp(-\lambda), \ \lambda > 0.$$

(a) Write down the joint density of X and λ.
(b) Find the marginal density of X. [Hint: it may be helpful to use the integral identity

$$\int_0^{\infty} \exp(-a)\lambda^b d\lambda = \frac{b!}{a^b},$$

where b is a nonnegative integer.]
(c) Find the probability that the number of defects X is equal to 0.
(d) Find the conditional density of λ if you know that $X = 0$.

14. **Flipping a Random Coin**

Suppose you plan flipping a coin twice where the probability p of heads has the density function

$$f(p) = 6p(1-p), \ 0 < p < 1.$$

Let Y denote the number of heads of this "random" coin. Y given a value of p is binomial with $n = 2$ and probability of success p.

(a) Write down the joint density of Y and p.
(b) Find $P(Y = 2)$.
(c) If $Y = 2$, then find the probability that p is greater than 0.5.

15. **Passengers on An Airport Limousine**

An airport limousine can accommodate up to four passengers on any one trip. The company will accept a maximum of six reservations for a trip, and a passenger must have a reservation. From previous records, 30% of all those making reservations do not appear for the trip. Answer the following questions, assuming independence whenever appropriate.

(a) If six reservations are made, what is the probability that at least one individual with a reservation cannot be accommodated on the trip?

(b) If six reservations are made, what is the expected number of available places when the limousine departs?

(c) Suppose the probability distribution of the number of reservations made is given in the following table.

Number of observations	3	4	5	6
Probability	0.13	0.18	0.35	0.34

Let X denote the number of passengers on a randomly selected trip. Obtain the probability mass function of X.

x	0	1	2	3	4
$p(x)$					

16. **Heights of Fathers and Sons**

It is well-known that heights of fathers and sons are positively associated. In fact, if X represents the father's height in inches and Y represents the son's height, then the joint distribution of (X, Y) can be approximated by a bivariate normal with means $\mu_X = \mu_Y = 69$, $\sigma_X = \sigma_Y = 3$ and correlation $\rho = 0.4$.

(a) Are X and Y independent? Why or why not?

(b) Find the conditional density of the son's height if you know the father's height is 70 inches.

(c) Using the result in part (b) to find $P(Y > 72 \mid X = 70)$.

(d) By simulating from the bivariate normal distribution, approximate the probability that the son will be more than one inch taller than his father.

17. **Instruction and Students' Scores**

Twenty-two children are given a reading comprehension test before and after receiving a particular instruction method. Assume students' pre-instructional and post-instructional scores follow a Bivariate Normal distribution with: $\mu_{pre} = 47, \mu_{post} = 53, \sigma_{pre} = 13, \sigma_{post} = 15$ and $\rho = 0.7$.

(a) Find the probability that a student's post-instructional score exceeds 60.

(b) Suppose one student's pre-instructional score is 45, find the probability that this student's post-instructional score exceeds 70.

(c) Find the probability that a student has increased the test score by at least 10 points. [Hint: Use R to simulate a large number of draws from the bivariate normal distribution. Refer to the example `sim_binorm()` function in Section 6.7 for simulating Bivariate Normal draws.]

18. **Shooting Free Throws**

Suppose a basketball player will take N free throw shots during a game where N has the following discrete distribution.

N	5	6	7	8	9	10
Probability	0.2	0.2	0.2	0.2	0.1	0.1

If the player takes $N = n$ shots, then the number of makes Y is binomial with sample size n and probability of success $p = 0.7$.

(a) Find the probability the player takes 6 shots and makes 4 of them.

(b) From the joint distribution of (N, Y), find the most likely (n, y) pair.

(c) Find the conditional distribution of the number of shots N if he makes 4 shots.

(d) Find the expectation $E(N \mid Y = 4)$.

19. **Flipping a Random Coin**

Suppose one selects a probability p uniform from the interval (0, 1), and then flips a coin 10 times, where the probability of heads is the probability p. Let X denote the observed number of heads.

(a) Find the joint distribution of p and X.

(b) Use R to simulate a sample of size 1000 from the joint distribution of (p, X).

(c) From inspecting a histogram of the simulated values of X, guess at the marginal distribution of X.

R Exercises

20. **Simulating Multinomial Probabilities**

Revisit Exercise 6.

(a) Write an R function to simulate 10 balls of the special weighted box (4 red, 3 black, and 3 green balls). [Hint: Section 6.3 introduces the `sim_die_rolls()` function for the example of a special weighted die.]

(b) Use the `replicate()` function to simulate the multinomial experiment in Exercise 6 for 5000 iterations, and approximate $P(R = 3, B = 2)$.

(c) Use the 5000 simulated multinomial experiments to approximate the probability that you sample more red balls than black balls, i.e. $P(R > B)$.

(d) Conditional on $B = 4$, approximate the mean number of red balls that will get sampled. Compare the approximated mean value to the exact mean. [Hint: Conditional on $B = 4$, the distribution of R is a binomial distribution.]

21. **Simulating from a Beta-Binomial Distribution**

Consider a box of coins where the coin probabilities vary, and the probability of a selected coin lands heads, p, follows a beta$(2, 8)$ distribution. Jason then continues to flip this "random" coin 10 times, and is interested in the count of heads of the 10 flips, denoted by Y.

(a) Write an R function to simulate 5000 samples of (p, y). [Hint: Use `rbeta()` and `rbinom()` functions accordingly.]

(b) Approximate the probability that Jason observes 3 heads out of 10 flips, using the simulated 5000 samples. Compare the approximated probability to the exact probability. [Hint: Write out $f(y)$ following the work in Section 6.6, and use R to calculate the exact probability.]

22. **Shooting Free Throws (continued)**

Consider the free throws shooting in Exercise 18.

(a) Write an R function to simulate 5000 samples of (n, y).

(b) From the 5000 samples, find the most likely (n, y) pair. Compare your result to Exercise 18 part (b).

(c) Approximate the expectation $E(N \mid Y = 4)$, and compare your result to Exercise 18 part (d).

7

Learning about a Binomial Probability

7.1 Introduction: Thinking Subjectively about a Proportion

In previous chapters, we have seen many examples involving drawing color balls from a box. In those examples, one is given the numbers of balls of various colors in the box, and one considers questions related to calculating probabilities. For example, there are 40 white and 20 red balls in a box. If one draws two balls at random, what is the probability that both balls are white?

Here we consider a new scenario where we do not know the proportions of color balls in the box. In the previous example, one only knows that there are two kinds of color balls in the box, but one doesn't know 40 out of 60 of the balls are white (proportion of white $= 2/3$) and 20 out of the 60 of the balls are red (proportion of red $= 1/3$). How can one learn about the proportions of white and red balls? Since counting 60 balls is tedious, how can one infer those proportions by drawing a sample of balls out of the box and observing the colors of balls in the sample? This becomes an inference question, because one is trying to infer the proportion p of the population, based on a sample from the population.

Let's continue discussing the scenario where one is told that there are 60 balls in total in a box, and the balls are either white or red. One does not know the count of balls of each of the two colors but is given the opportunity to learn about these counts by selecting a random sample of 10 balls. The object of interest is the quantity p, the proportion of red balls in the 60 balls. How can one infer p, the proportion of red balls in the population of 60 balls, based on the numbers of red and white balls observed in the sample of 10 balls?

Proportions are like probabilities. Recall from Chapter 1 the following three views and associated characteristics of probabilities.

1. The classical view: one needs to write down the sample space where each outcome is equally likely.

2. The frequency view: one needs to repeat the random experiment many times under identical conditions.

3. The subjective view: one needs to express one's opinion about the likelihood of a one-time event.

The classical view does not seem to work here, because one only knows there are two kinds of color balls and the total number of balls is 60. Even if one takes a sample of 10 balls, one only observes the proportion of red balls in the sample. There does not seem to be a way for one to write down the sample space where each outcome is equally likely.

The frequency view would work here. One could treat the process of obtaining a sample (i.e. taking a random sample of 10 balls from the box) as an experiment, and obtain a sample proportion $\hat{p}$ from the experiment. One then could repeat the experiment many times under the same condition, get many sample proportions $\hat{p}$, and summarize all the $\hat{p}$. When one repeats the experiment enough times (a large number), one gets a good sense about the proportion p of red balls in the population of 60 balls in the box. This process is doable, but tedious and time-consuming.

The subjective view is one's personal opinion about the location of the unknown proportion p. It does require one to express his or her opinion about the value of p, and he or she could be skeptical or unknown about the opinion. In Chapter 1, a calibration experiment was introduced to help one sharpen an opinion about the likelihood of an event by comparisons with opinion about the likelihood of other events. In this chapter and the chapters to follow, the key ideas will be introduced and the reader will practice thinking subjectively about unknowns and quantifying one's opinions about the values of these unknowns using probability distributions.

As an example, consider plausible values for the proportion p of red balls. As p is a proportion, it takes any possible value between 0 and 1. In the calibration experiment introduced in Chapter 1, we focus on the scenario where only one value of p is of interest. If we think $p = 0.5$, it reflects our belief that the probability of the value $p = 0.5$ is equal to 1. The statement that "the probability that $p = 0.5$ is 1" sounds like a very strong opinion because p is restricted to only one possible value and the probability assigned to it is 1. Since we do not know the exact value of the p proportion, assigning a single possible probability value of 1 appears to be too strong.

Instead suppose that the proportion p takes multiple values between 0 and 1. In particular, consider two scenarios, where in each scenario p takes 10 different values denoted by the set A.

$$A = \{0.1, 0.2, 0.3, 0.4, 0.5, 0.6, 0.7, 0.8, 0.9, 1.0\}$$

Although p takes the same ten values, different probabilities are assigned to the values.

- Scenario 1:

$$f_1(A) = (0.1, 0.1, 0.1, 0.1, 0.1, 0.1, 0.1, 0.1, 0.1, 0.1)$$

- Scenario 2:

$$f_2(A) = (0.05, 0.05, 0.05, 0.175, 0.175, 0.175, 0.175, 0.05, 0.05, 0.05)$$

To visually compare the values of two probability distributions $f_1(A)$ and $f_2(A)$, we plot the distributions using the same scales as in Figure 7.1. Figure

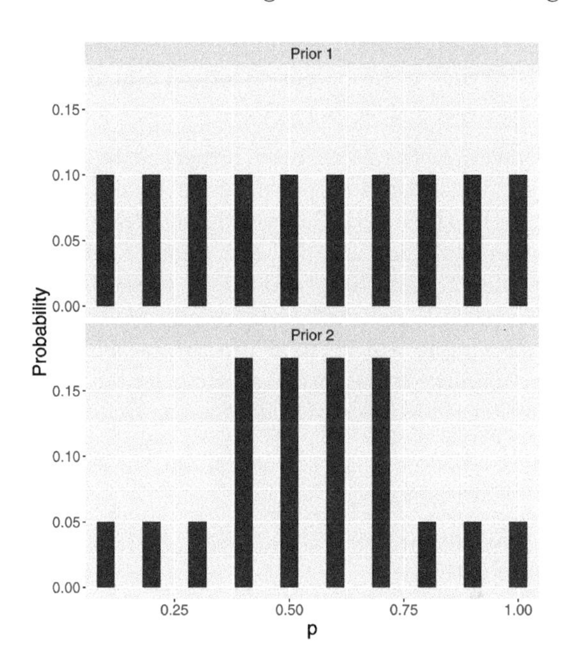

FIGURE 7.1
The same ten possible values of p, but two sets of probabilities.

7.1 labels the x-axis with the values of p (range from 0 to 1) and the y-axis with the probabilities (range from 0 to 1). For both panels, there are ten bars, where the heights represent the associated probabilities of the values of p in the set $A = \{0.1, 0.2, 0.3, 0.4, 0.5, 0.6, 0.7, 0.8, 0.9, 1.0\}$.

The probability assignment in $f_1(A)$ is called a discrete uniform distribution where each possible value of the proportion p is equally likely. Since there are ten possible values of p, each value gets assigned a probability of $1/10 = 0.1$. This assignment expresses the opinion that p can be any value from the set $A = \{0.1, 0.2, 0.3, 0.4, 0.5, 0.6, 0.7, 0.8, 0.9, 1.0\}$, and each value has a probability of 0.1.

The probability assignment in $f_2(A)$ is also discrete, but the pattern of probabilities is not uniform. What one sees is that the probabilities of the first three proportion values (0.1, 0.2, and 0.3) and last three proportion values

(0.8, 0.9, and 1.0) are each 1/3.5 of the probabilities assigned to the middle four values (0.4, 0.5, 0.6, and 0.7). The heights of the bars reflect the opinion that the middle values of p are 3.5 times as likely as the extreme values of p.

Both sets of probabilities follow the three probability axioms in Chapter 1. One sees that within each set,

1. Each probability is nonnegative,
2. The sum of the probabilities is 1,
3. The probability of mutually exclusive values is the sum of probability of each value, e.g. probability of $p = 0.1$ or $p = 0.2$ is $0.1 + 0.1$ in $f_1(A)$, and $0.05 + 0.05$ in $f_2(A)$.

In this introduction, a method has been presented to think about proportions subjectively. This method allows multiple values of p and probability assignments follow the three probability axioms. Each probability distribution expresses a unique opinion about the location of the proportion p.

To answer the inference question "what is the proportion of red balls in the box", a random sample of 10 balls will be sampled and the observed proportion of red balls in that sample will be used to sharpen and update one's belief about p. Bayesian inference is the formal mechanism for updating one's belief given new information. This mode of inference has three general steps.

Step 1: (**Prior**): express an opinion about the location of the proportion p before sampling.

Step 2: (**Likelihood**): take the sample and record the observed proportion of red balls.

Step 3: (**Posterior**): use Bayes' rule to update the previous opinion about p given the information from the sample.

As indicated in the parentheses, the first step "Prior" constructs *prior* opinion about the quantity of interest, and a probability distribution is used (like $f_1(A)$ and $f_2(A)$ earlier) to quantify the prior opinion. The name "prior" indicates that the opinion should be formed before collecting any data.

The second step "Data" is the process of collecting data, where the quantity of interest is observed in the collected data. For example, if our 10-ball sample contains 4 red balls and 6 white balls, the observed proportion of red balls is $4/10 = 0.4$. Informally, how does this information help us revise our opinion about p? Intuitively one would give more probability to $p = 0.4$, but it is unclear how the probabilities would be redistributed among the 10 values in A. Since the sum of all probabilities is 1, is it possible that some of the larger proportion values, such as $p = 0.9$ and $p = 1.0$, will receive probabilities of zero? To address these questions, the third step is needed.

The third step "Posterior" combines one's prior opinion and the collected data, by use of Bayes' rule, to update one's opinion about the quantity of interest. Just like the example of observing 4 red balls in the 10-ball sample, one needs a structured way of updating the opinion from prior to posterior.

Throughout this chapter, the entire inference process will be described for learning about a proportion p. This chapter will discuss how to express prior opinion that matches with one's belief, how to extract information from the likelihood, and how to update our opinion to its posterior.

Section 7.2 introduces inference when a discrete prior distribution is assigned to the proportion p. Section 7.3 introduces the beta class of continuous prior distributions and the inference process with a beta prior is described in detail in Section 7.4. Section 7.5 describes some general Bayesian inference methods for learning about the proportion, namely Bayesian hypothesis testing, Bayesian credible intervals and Bayesian prediction. This chapter will illustrate both the use of exact analytical solutions and approximate simulation calculations (with the help of the R software).

7.2 Bayesian Inference with Discrete Priors

7.2.1 Example: students' dining preference

Let's start our Bayesian inference for proportion p with discrete prior distributions with a students' dining preference example. A popular restaurant in a college town has been in business for about 5 years. Though the business is doing well, the restaurant owner wishes to learn more about his customers. Specifically, he is interested in learning about the dining preferences of the students. The owner plans to conduct a survey by asking students "what is your favorite day for eating out?" In particular, he wants to find out what percentage of students prefer to dine on Friday, so he can plan ahead for ordering supplies and giving promotions.

Let p denote the proportion of all students whose answer is Friday.

7.2.2 Discrete prior distributions for proportion p

Before giving out the survey, let's pause and think about the possible values for the proportion p. Not only does one want to know about possible values, but also the probabilities associated with the values. A probability distribution provides a measure of belief for the proportion and it ultimately will help the restaurant owner improve his business.

One might not know much about students' dining preference, but it is possible to come up with a list of plausible values for the proportion. There are seven days a week. If each day was equally popular, then one would expect 1/7 or approximately 15% of all students to choose Friday. The owner recognizes that Friday is the start of the weekend, therefore there should be a higher chance of being students' preferred day of dining out. So perhaps p starts with 0.3. Then what about the largest plausible value? Letting this largest value be 1 seems unrealistic, as there are six other days in the week. Suppose that one chooses 0.8 to be the largest plausible value, and then comes up with the list of values of p to be the six values going from 0.3 to 0.8 with an increment of 0.1.

$$p = \{0.3, 0.4, 0.5, 0.6, 0.7, 0.8\} \tag{7.1}$$

Next one needs to assign probabilities to the list of plausible values of p. Since one may not know much about the location of the probabilities p, a good place to start is a discrete uniform prior (recall the discrete uniform prior distribution for p, the proportion of red balls, in Section 7.1). A discrete uniform prior distribution expresses the opinion that all plausible values of p are equally likely. In the current students' dining preference example, if one decides on six plausible values of p as in Equation (7.1), each of the six values gets a prior probability of 1/6. One labels this prior as π_l, where l stands for laymen (for all of us who are not in the college town restaurant business).

$$\pi_l(p) = (1/6, 1/6, 1/6, 1/6, 1/6, 1/6) \tag{7.2}$$

With five years of experience of running his restaurant in this college town, the restaurant owner might have different opinions about likely values of p. Suppose he agrees with us that p could take the 6 plausible values from 0.3 to 0.8, but he assigns a different prior distribution for p. In particular, the restaurant owner thinks that values of 0.5 and 0.6 are most likely – each of these values is twice as likely as the other values. His prior is labelled as π_e, where e stands for expert.

$$\pi_e(p) = (0.125, 0.125, 0.250, 0.250, 0.125, 0.125) \tag{7.3}$$

R To obtain $\pi_e(p)$ efficiently, one can use the `ProbBayes` R package. First a data frame is created by providing the list of plausible values of p and corresponding weights assigned to each value using the function `data.frame()`. As one can see here, one does not have to calculate the probability – one only needs to give the weights (e.g. giving $p = 0.3, 0.4, 0.7, 0.8$ weight 1 and giving $p = 0.5, 0.6$ weight 2, to reflect the owner's opinion "0.5 and 0.6 are twice as likely as the other values").

```
bayes_table <- data.frame(p = seq(.3, .8, by=.1),
                          Prior = c(1, 1, 2, 2, 1, 1))

bayes_table
    p Prior
1 0.3     1
2 0.4     1
```

```
3 0.5     2
4 0.6     2
5 0.7     1
6 0.8     1
```

One uses the function `mutate()` to normalize these weights to obtain the prior probabilities in the Prior column.

```
bayes_table %>% mutate(Prior = Prior / sum(Prior)) -> bayes_table
bayes_table
    p Prior
1 0.3 0.125
2 0.4 0.125
3 0.5 0.250
4 0.6 0.250
5 0.7 0.125
6 0.8 0.125
```

One conveniently plots the restaurant owner's prior distribution by use of `ggplot2` functions. This distribution is displayed in Figure 7.2.

```
ggplot(data=bayes_table, aes(x=p, y=Prior)) +
  geom_bar(stat="identity", fill=crcblue, width = 0.06)
```

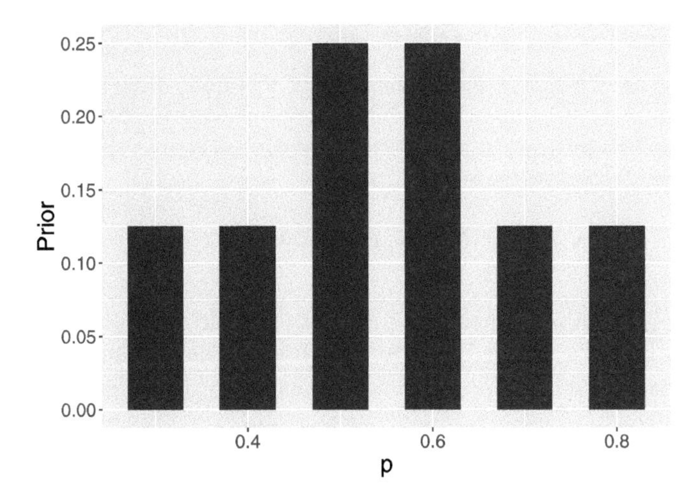

FIGURE 7.2
The restaurant owner's prior distribution for the proportion p.

It is left as an exercise for the reader to compute and plot the laymen's prior $\pi_l(p)$ in Equation (7.2). For the rest of this section, we will work with the expert's prior $\pi_e(p)$.

7.2.3 Likelihood of proportion p

The next step in the inference process is the data collection. The restaurant owner gives a survey to 20 student diners at the restaurant. Out of the 20 student respondents, 12 say that their favorite day for eating out is Friday. Recall the quantity of interest is proportion p of the population of students choosing Friday.

The likelihood is a function of the quantity of interest, which is the proportion p. The owner has conducted an experiment 20 times, where each experiment involves a "yes" or "no" answer from the respondent to the rephrased question "whether Friday is your preferred day to dine out". Then the proportion p is the probability a student answers "yes".

Does this ring a bell of what we have seen before? Indeed, in Chapter 4, one has seen this type of experiment, a binomial experiment, similar to the dining survey. Recall that a binomial experiment needs to satisfy four conditions:

1. One is repeating the same basic task or trial many times – let the number of trials be denoted by n.
2. On each trial, there are two possible outcomes called "success" or "failure".
3. The probability of a success, denoted by p, is the same for each trial.
4. The results of outcomes from different trials are independent.

If one recognizes an experiment as being binomial, then all one needs to know is n and p to determine probabilities for the number of successes Y. The probability of y successes in a binomial experiment is given by

$$Prob(Y = y) = \binom{n}{y} p^y (1-p)^{n-y}, y = 0, \cdots, n. \tag{7.4}$$

Assuming the dining survey is a random sample (thus independent outcomes), this is the result of a binomial experiment. The likelihood is the chance of 12 successes in 20 trials viewed as a function of the probability of success p:

$$Likelihood = L(p) = \binom{20}{12} p^{12} (1-p)^8. \tag{7.5}$$

Generally one uses L to denote a likelihood function — one sees in Equation (7.5), L is a function of p. Note that the value of n, the total number of trials, is known and the number of successes Y is observed to be 12. The proportion

p, is the parameter of the binomial experiment and the likelihood is a function of the proportion p.

The likelihood function $L(p)$ is efficiently computed using the `dbinom()` function in R. In order to use this function, we need to know the sample size n (20 in the dining survey), the number of successes y (12 in the dining survey), and p (the list of 6 plausible values created in Section 7.2.2; $p = \{0.3, 0.4, 0.5, 0.6, 0.7, 0.8\}$). Note that we only need the plausible values of p, not yet the assigned probabilities in the prior distribution. The prior will be used in the third step to update the opinion of p to its posterior.

R Below is the example R code of finding the probability of 12 successes in a sample of 20 for each value of the proportion p. The values are placed in the `Likelihood` column of the `bayes_table` data frame.

```
bayes_table$Likelihood <- dbinom(12, size=20, prob=bayes_table$p)
bayes_table
    p Prior   Likelihood
1 0.3 0.125 0.003859282
2 0.4 0.125 0.035497440
3 0.5 0.250 0.120134354
4 0.6 0.250 0.179705788
5 0.7 0.125 0.114396740
6 0.8 0.125 0.022160877
```

7.2.4 Posterior distribution for proportion p

The posterior probabilities are found as an application of Bayes' rule. This recipe will be illustrated first through a step-by-step calculation process. Next the process is demonstrated with the `bayesian_crank()` function in the `ProbBayes` R package, which implements the Bayes' rule calculation and outputs the posterior probabilities.

Let $\pi(p)$ to be the prior distribution of p, let $L(p)$ denote the likelihood function, and $\pi(p \mid y)$ to be the posterior distribution of p after observing the number of successes y. For discrete parameters, such as the proportion p in our case, one is able to enumerate the list of plausible values and assign prior probabilities to the values. If p_i represents a particular value of p, Bayes' rule for a discrete parameter has the form

$$\pi(p_i \mid y) = \frac{\pi(p_i) \times L(p_i)}{\sum_j \pi(p_j) \times L(p_j)}, \tag{7.6}$$

where $\pi(p_i)$ is the prior probability of $p = p_i$, $L(p_i)$ is the likelihood function evaluated at $p = p_i$, and $\pi(p_i \mid y)$ is the posterior probability of $p = p_i$ given the number of successes y. By the *Law of Total Probability*, the denominator gives the marginal distribution of the observation y.

Bayes' rule can also be expressed as "prior times likelihood":

$$\pi(p_i \mid y) \propto \pi(p_i) \times L(p_i) \tag{7.7}$$

Equation (7.7) ignores the denominator and states that the posterior is proportional to the product of the prior and the likelihood. As one will see soon, the value of the denominator is a constant, meaning that its purpose is to normalize the numerator. It is convenient to work with Bayes' rule as in Equation (7.7) in later chapters. However, it is instructive to show the exact calculation of Equation (7.6), because one has a finite sum in the denominator and it is possible to obtain the analytical solution. In the case where the prior is continuous, it will be more difficult to analytically compute the normalizing constant.

Returning to the students' dining preference example, the list of plausible values of the proportion is $p = \{0.3, 0.4, 0.5, 0.6, 0.7, 0.8\}$ and according to the restaurant owner's expert prior, the assigned probabilities are $\pi_e(p) = (0.125, 0.125, 0.250, 0.250, 0.125, 0.125)$ (recall Figure 7.2). After observing the number of successes, the likelihood values are calculated for the models using `dbinom()` function, as presented in Section 7.2.3.

The denominator is the sum of the products of the prior and the likelihood at each possible p_i, which, given the Law of Total Probability, is equal to the marginal probability of the data $f(y)$. One can think of the above formula as reweighing or normalizing the probability of $\pi(p_i \mid y)$ by all possible values of p. In the case of discrete models like this, the marginal probability of the likelihood is computed through $\sum_j f(p_j) \times L(p_j)$.

In this setup, the computation of the posterior probabilities of different p_i values is straightforward. First, one calculates the denominator and denote the value as D.

$$\begin{aligned} D &= \pi(0.3) \times L(0.3) + \pi(0.4) \times L(0.4) + \cdots + \pi(0.8) \times L(0.8) \\ &= 0.125 \times \binom{20}{12}(0.3)^{12}(1-0.3)^8 + \cdots + 0.125 \times \binom{20}{12}(0.8)^{12}(1-0.8)^8 \\ &\approx 0.0969. \end{aligned}$$

Then the posterior probability of $p = 0.3$ is given by

$$\begin{aligned} \pi(p = 0.3 \mid 12) &= \frac{\pi(0.3) \times L(0.3)}{D} \\ &= \frac{0.125 \times \binom{20}{12}(0.3)^{12}(1-0.3)^8}{D} \\ &\approx 0.005. \end{aligned}$$

In a similar fashion, the posterior probability of $p = 0.5$ is calculated as

$$\begin{aligned}\pi(p = 0.5 \mid 12) &= \frac{\pi(0.5) \times L(0.5)}{D} \\ &= \frac{0.125 \times \binom{20}{12}(0.5)^{12}(1-0.5)^{8}}{D} \\ &\approx 0.310.\end{aligned}$$

One sees that the denominator is the same for the posterior probability calculation of every value of p. This calculation gets tedious for a large number of possible values of p. Relying on statistical software such as `R` helps us simplify the tasks.

R To use the `bayesian_crank()` function, recall that we have already created a data frame with variables `p`, `Prior`, and `Likelihood`. Then the `bayesian_crank()` function is used to compute the posterior probabilities.

```
bayesian_crank(bayes_table) -> bayes_table
bayes_table
    p Prior  Likelihood      Product   Posterior
1 0.3 0.125 0.003859282 0.004824102  0.004975901
2 0.4 0.125 0.035497440 0.0044371799 0.045768032
3 0.5 0.250 0.120134354 0.0300335884 0.309786454
4 0.6 0.250 0.179705788 0.0449264469 0.463401326
5 0.7 0.125 0.114396740 0.0142995925 0.147495530
6 0.8 0.125 0.022160877 0.0027701096 0.028572757
```

As one sees in the `bayes_table` output, the `bayesian_crank()` function computes the product of `Prior` and `Likelihood` and stores the values in the column `Product`, then normalizes each product with the sum of all products to produce the posterior probabilities, stored in the column `Posterior`.

Figure 7.3 compares the prior probabilities in the bottom panel with the posterior probabilities in the top panel. Notice the difference in the two distributions. After observing the survey results (i.e. the likelihood), the owner is more confident that p is equal to 0.5 or 0.6, and it is unlikely for p to be 0.3, 0.4, 0.7, and 0.8. Recall that the data gives an observed proportion $12/20 = 0.6$. Since the posterior is a combination of prior and likelihood, it is not surprising that the likelihood helps the owner to sharpen his belief about proportion p and place a larger posterior probability around 0.6.

7.2.5 Inference: students' dining preference

Let's revisit the posterior distribution table to perform some inference. What is the posterior probability that over half of the students prefer eating out on Friday? One is interested in the probability that $p > 0.5$, in the posterior. Looking at the table, this posterior probability is equal to

$$Prob(p > 0.5) \approx 0.463 + 0.147 + 0.029 = 0.639.$$

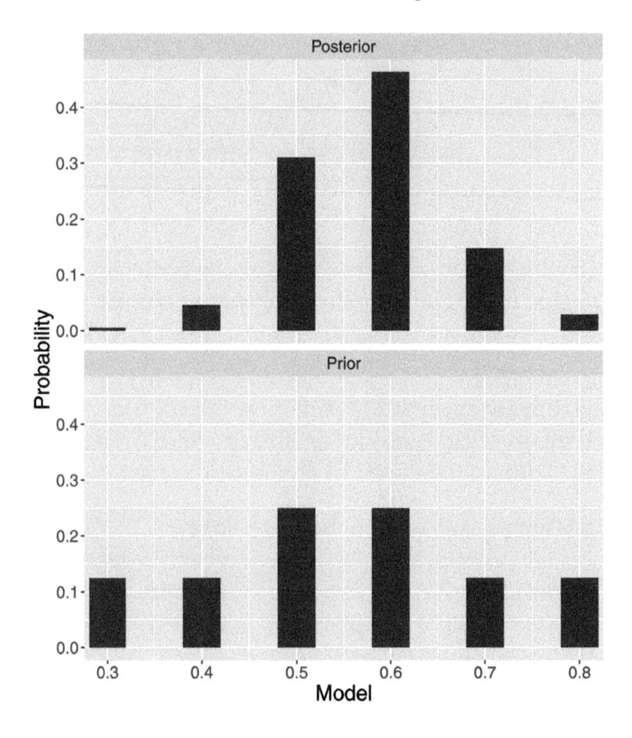

FIGURE 7.3
Prior and posterior distributions on the proportion p.

This means the owner is reasonably confident (with probability 0.639) that over half of the college students prefer to eat out on Friday.

R One easily obtains the probability from the R output, for example.

```
sum(bayes_table$Posterior[bayes_table$p > 0.5])
[1] 0.6394696
```

7.2.6 Discussion: using a discrete prior

Specifying a discrete prior has two steps: (1) specifying a list of plausible values of the parameter of interest, and (2) assigning probabilities to the plausible values. It is important to remember the three probability axioms when specifying a discrete prior.

After the prior specification, the next component is the likelihood, which can also be broken up into two steps. First, one constructs a suitable experiment that works for the particular scenario. Here one has a binomial experiment for a survey to a fixed number of respondents, the answers are classified into "yes" and "no" or "success" and "failure", the outcome of interest is the number of successes and trials are independent. From the binomial distribution, one obtains the likelihood function which is evaluated at each possible value of the parameter of interest. In our example, the `dbinom()` R function was used to calculate the likelihood function.

Last, the posterior probabilities are calculated using Bayes' rule. In particular for the discrete case, follow Equation (7.6). The calculation of the denominator is tedious s, however practice with the Bayes' rule calculation enhances one's understanding of Bayesian inference. R functions such as `bayesian_crank()` are helpful for implementing the Bayes' rule calculations. Bayesian inference follows from a suitable summarization of the posterior probabilities. In our example, inference was illustrated by calculating the probability that over half of the students prefer eating out on Friday.

Let's revisit the list of plausible values of proportion p of students preferring Friday in dining out in the example. Although $p = 1.0$, that is, everyone prefers Friday, is very unlikely, one might not want to eliminate this proportion value from consideration. As one observes in the Bayes' rule calculation process shown in Sections 7.2.3 and 7.2.4, if one does not include $p = 1.0$ as one of the plausible values in the prior distribution in Section 7.2.2, this value will also be given a probability of zero in the posterior.

Alternatively, one could choose the alternative set of values

$$p = \{0.1, 0.2, 0.3, 0.4, 0.5, 0.6, 0.7, 0.8, 0.9, 1.0\},$$

and assign a very small prior probability (e.g. 0.05 or even smaller) for $p = 1.0$ to express the opinion that $p = 1.0$ is very unlikely. One may assign small prior probabilities for other large values of p such as $p = 0.9$.

This comment illustrates a limitation of specifying a discrete prior for a proportion p. If a plausible value is not specified in the prior distribution (e.g. $p = 1.0$ is not in the restaurant owner's prior distribution), it will be assigned a probability of zero in the posterior (e.g. $p = 1.0$ is not in the restaurant owner's posterior distribution).

It generally is more desirable to have p to be any value in [0, 1] including less plausible values such as $p = 1.0$. To make this happen, the proportion p should be allowed to take any value between 0 and 1, which means p will be a continuous variable. In this situation, it is necessary to construct a continuous prior distribution for p. A popular class of continuous prior distributions for proportion is the beta distribution which is the subject of the next section.

7.3 Continuous Priors

Let's continue our students' dining preference example. A restaurant owner is interested in learning about the proportion p of students whose favorite day for eating out is Friday.

The proportion p should be a value between 0 and 1. Previously, we used a discrete prior for p, representing the belief that p only takes the six different values 0.3, 0.4, 0.5, 0.6, 0.7, and 0.8. An obvious limitation of this assumption is, what if the true p is 0.55? If the value 0.55 is not specified in the prior distribution of p (that is, a zero probability is assigned to the value $p = 0.55$), then by the Bayes' rule calculation (either by hand or by the useful `bayesian_crank()` function) there will be zero posterior probability assigned to 0.55. It is therefore preferable to specify a prior that allows p to be any value in the interval [0, 1].

To represent such a prior belief, it is assumed that p is continuous on [0, 1]. Suppose again that one is a layman unfamiliar with the pattern of dining during a week. Then one possible choice of a continuous prior for p is the continuous uniform distribution, which expresses the opinion that p is equally likely to take any value between 0 and 1.

Formally, the probability density function of the continuous uniform on the interval (a, b) is

$$\pi(p) = \begin{cases} \frac{1}{b-a} & \text{for } a \leq p \leq b, \\ 0 & \text{for } p < a \text{ or } p > b. \end{cases} \tag{7.8}$$

In our situation p is a continuous uniform random variable on [0, 1], we have $\pi(p) = 1$ for $p \in [0, 1]$, and $\pi(p) = 0$ everywhere else.

What about other possible continuous prior distributions for p on [0, 1]? Consider a prior distribution for the restaurant owner who has some information about the location (i.e. value) of p. This owner would be interested in a continuous version of the discrete prior distribution where values of p between 0.3 and 0.8 are more likely than the values at the two ends.

The beta family of continuous distributions is useful for representing prior knowledge in this situation. A beta distribution, denoted by Beta(a, b), represents probabilities for a random variable falling between 0 and 1. This distribution has two shape parameters, a and b, with probability density function given by

$$\pi(p) = \frac{1}{B(a,b)} p^{a-1}(1-p)^{b-1}, \; 0 \leq p \leq 1, \tag{7.9}$$

where $B(a, b)$ is the beta function defined by $B(a,b) = \frac{\Gamma(a)\Gamma(b)}{\Gamma(a+b)}$, where Γ is the Gamma function. For future reference, it is useful to know that if $p \sim$ Beta(a, b), its mean $E[p] = \frac{a}{a+b}$ and its variance $V(p) = \frac{ab}{(a+b)^2(a+b+1)}$. The continuous uniform in Equation (7.8) is a special case of the beta distribution: Uniform$(0, 1) =$ Beta$(1, 1)$.

For the remainder of this section, Section 7.3.1 introduces the beta distribution and beta probabilities, and Section 7.3.2 focuses on several ways of choosing a beta prior that reflects one's opinion about the location of a proportion.

7.3.1 The beta distribution and probabilities

The two shape parameters a and b control the shape of the beta density curve. Figure 7.4 shows density curves of beta distributions for several choices of the shape parameters. One observes from this figure that the beta density curve displays vastly different shapes for varying choices of a and b. For example, Beta$(0.5, 0.5)$ represents the prior belief that extreme values of p are likely and $p = 0.5$ is the least probable value. In the students' dining preference example, specifying a Beta$(0.5, 0.5)$ would reflect the owner's belief that the proportion of students dining out on Friday is either very high (near one) or very low (near one) and not likely to be moderate values.

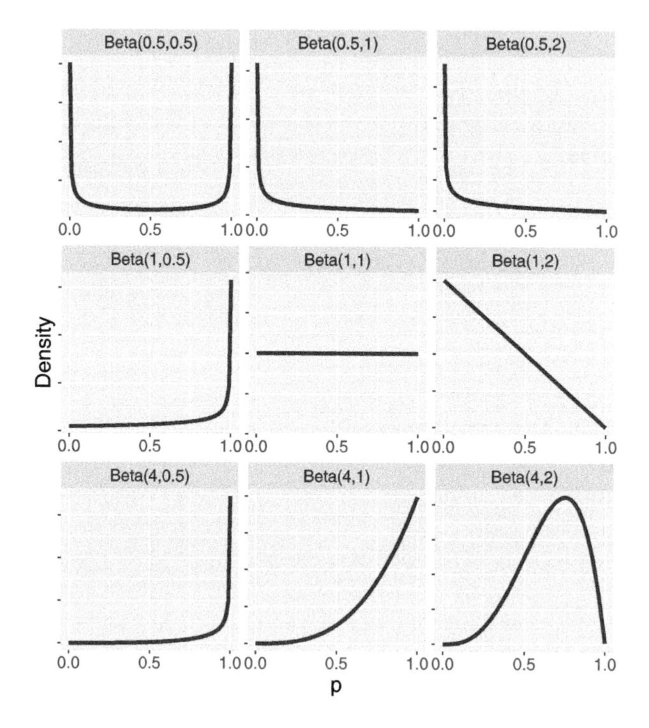

FIGURE 7.4
Illustration of nine beta density curves.

As the beta is a common continuous distribution, R functions are available for beta distribution calculations. We provide a small example of "beta" functions for Beta$(1, 1)$, where the two shape parameters 1 and 1 are the second and third arguments of the functions.

R Recall the following useful results from previous material: (1) a Beta$(1, 1)$ distribution is a uniform density on (0, 1), (2) the density of Uniform$(0, 1)$ is

$\pi(p) = 1$ on [0, 1], and (3) if $p \sim$ Uniform(0, 1), then the cdf $F(x) = Prob(p \leq x) = x$ for $x \in [0, 1]$.

1. `dbeta()`: the probability density function for a Beta(a, b) which takes a value of the random variable as its input and outputs the probability density function at that value.

   For example, we evaluate the density function of Beta(1, 1) at the values $p = 0.5$ and $p = 0.8$, which should be both 1, and 0 at $p = 1.2$ which should be 0 since this value is outside of [0, 1].

   ```
   dbeta(c(0.5, 0.8, 1.2), 1, 1)
   [1] 1 1 0
   ```

2. `pbeta()`: the distribution function of a Beta(a, b) random variable, which takes a value x and gives the value of the random variable at that value, $F(x)$.

   For example, suppose one wishes to evaluate the distribution function of Beta(1, 1) at $p = 0.5$ and $p = 0.8$.

   ```
   pbeta(c(0.5, 0.8), 1, 1)
   [1] 0.5  0.8
   ```

   One calculates the probability of p between 0.5 and 0.8, i.e. $Prob(0.5 \leq p \leq 0.8)$ by taking the difference of the cdf at the two values.

   ```
   pbeta(0.8, 1, 1) - pbeta(0.5, 1, 1)
   [1] 0.3
   ```

3. `qbeta()`: the quantile function of a Beta(a, b), which inputs a probability value p and outputs the value of x such that $F(x) = p$.

   For example, suppose one wishes to calculate the quantile of Beta(1, 1) at $p = 0.5$ and $p = 0.8$.

   ```
   qbeta(c(0.5, 0.8), 1, 1)
   [1] 0.5  0.8
   ```

4. `rbeta()`: the random number generator for Beta(a, b), which inputs the size of a random sample and gives a vector of the simulated random variates.

   For example, suppose one is interested in simulating a sample of size five from Beta(1, 1).

```
rbeta(5, 1, 1)
[1] 0.71242248 0.59102308 0.05953272 0.47189451 0.44856499
```

R There are additional functions in the `ProbBayes` R package that aid in visualizing beta distribution calculations. For example, suppose one has a Beta$(7, 10)$ curve and we want to find the chance that p is between 0.4 and 0.8. Looking at Figure 7.5, this probability corresponds to the area of the shaded region. The special function `beta_area()` will compute and illustrate this probability. Note the use of the vector `c(7, 10)` to input the two shape parameters.

```
beta_area(0.4, 0.8, c(7, 10))
```

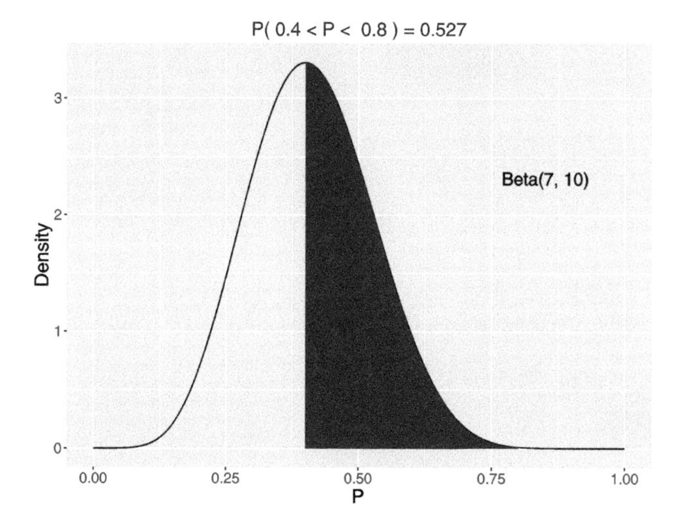

FIGURE 7.5
Area represents the probability that a Beta$(7, 10)$ variable lies between 0.4 and 0.8.

One could also find the chance that p is between 0.4 and 0.8 by subtracting two `pbeta()` functions.

```
pbeta(0.8, 7, 10) - pbeta(0.4, 7, 10)
```

The function `beta_quantile()` works in the same way as `qbeta()`, the quantile function. However, `beta_quantile()` automatically produces a plot with the shaded probability area. Figure 7.6) plots and computes the quantile to be 0.408. The chance that p is smaller than 0.408 is 0.5.

```
beta_quantile(0.5, c(7, 10))
```

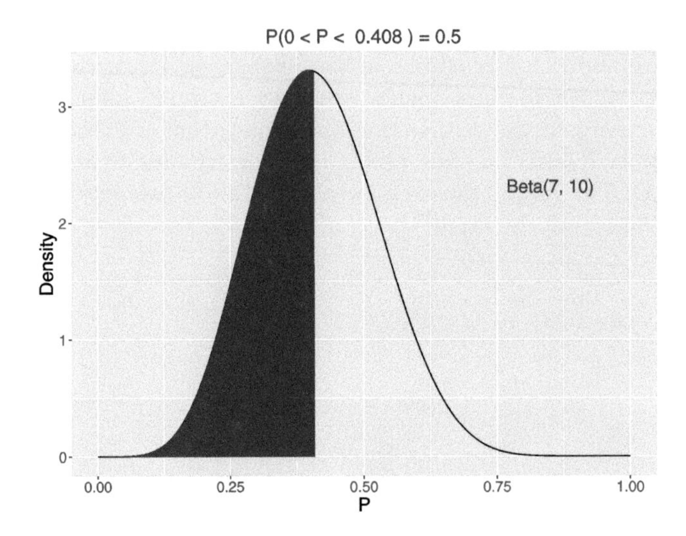

FIGURE 7.6
Illustration of a 0.5 quantile for a Beta(7, 10) variable.

Alternatively, use the `qbeta()` function without returning a plot.

```
qbeta(0.5, 7, 10)
[1] 0.4082265
```

7.3.2 Choosing a beta density to represent prior opinion

One wants to use a Beta(a, b) density curve to represent one's prior opinion about the values of the proportion p and their associated probabilities. It is difficult to guess at values of the shape parameters a and b directly. However, there are indirect ways of guessing their values. We present two general methods here.

The first method is to consider the shape parameter a as the prior count of "successes" and the other shape parameter b as the prior count of "failures". Subsequently, the value $a + b$ represents the *prior sample size* comparable to n, the *data sample size*. Following this setup, one could specify a beta prior with shape parameter a expressing the number of successes in one's prior opinion, and the other shape parameter b expressing the number of failures in one's prior opinion. For example, if one believes that *a priori* there should be about 4 successes and 4 failures, then one could use Beta$(4, 4)$ as the prior distribution for the proportion p.

How can we check if Beta(4, 4) looks like what we believe *a priori*? Recall that `rbeta()` generates a random sample from a beta distribution. The R script below generates a random sample of size 1000 from Beta(4, 4) and we plot a histogram and an overlapping density curve. (See left panel of Figure 7.7.) By an inspection of this graph, one decides if this prior is a reasonable approximation to one's beliefs about the proportion.

```
Beta44samples <- rbeta(1000, 4, 4)
```

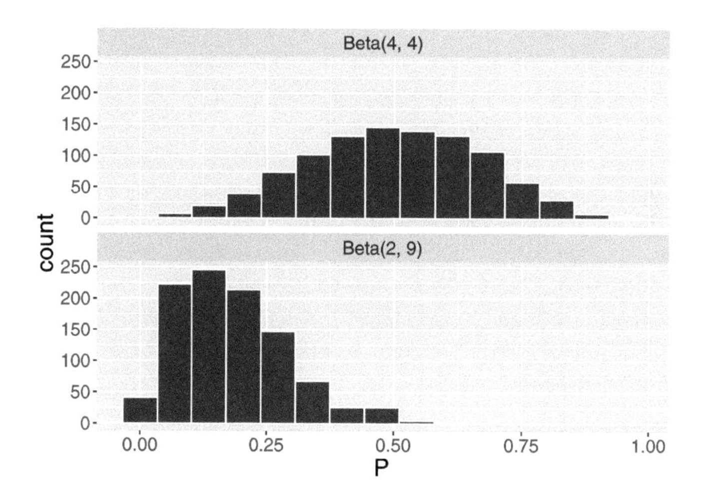

FIGURE 7.7
Histograms of 1000 samples of two beta density curves: Beta(4, 4) and Beta(2, 9).

As a second example, consider a belief that *a priori* there are 2 successes and 9 failures, corresponding to the Beta(2, 9) prior. One can use the `rbeta()` function take a random sample of 1000 from this prior.

```
Beta29samples <- rbeta(1000, 2, 9)
```

Comparing the two distributions, note from Figure 7.7 that Beta(2, 9) favors smaller proportion values than Beta(4, 4).

To further check the quantiles of the prior, one can use the `quantile()` function on the simulated draws from the prior. For example, if one wishes to check the middle 50% range of values of p from the random sample of values from Beta(4, 4), one types

```
quantile(Beta44samples, c(0.25, 0.75))
      25%       75%
0.3890909 0.6254733
```

This tells us that the probability that $p \leq 0.366$ is 0.25 and the probability that $p \geq 0.616$ is also 0.25. These probability statements should be checked against one's prior belief about p. If these quantiles do not seem reasonable, one should make adjustments to the values of the shape parameters a and b .

On the surface the two priors Beta(4, 4) and Beta(40, 40) seem similar in that they both have a mean of 0.5 and represent similar breakdowns of the success and failure counts. However, the aforementioned concept of prior sample size tells us that Beta(4, 4) has a prior sample size of 8 while that of Beta(40, 40) is 80. As we will see in Section 7.4, the prior sample size determines the strength of the prior (i.e. the confidence level in the prior) and so the Beta(40, 40) prior represents a much stronger belief that p is close to the value 0.5.

A second indirect method of determining a beta prior is by specification of quantiles of the distribution. Specifically, one determines the shape parameters a and b by first specifying two quantiles of the beta density curve, and then finding the beta density curve that matches these quantiles. Suppose the restaurant owner uses his knowledge to specify the 0.5 and 0.9 quantiles of the proportion p as follows.

1. First, the restaurant owner thinks of a value p_{50} such that the proportion p is equally likely to be smaller or larger than p_{50}. After some thought, he thinks that $p_{50} = 0.55$.
2. Next, the owner thinks of a value p_{90} that he is pretty sure (with probability 0.90) that the proportion p is smaller than p_{90}. After more thought, he decides $p_{90} = 0.80$.

R One then uses the `beta.select()` function in the `ProbBayes` package to find shape parameters a and b of the beta density curve that match this information. Each quantile is specified by a list with values x and p. From the output, we see Beta(3.06, 2.56) curve represents the owner's prior beliefs.

```
beta.select(list(x = 0.55, p = 0.5),
            list(x = 0.80, p = 0.9))
[1] 3.06 2.56
```

The owner's beta density curve is shown here. To make sure this prior is reasonable, the owner should compute several probabilities and quantiles for his prior distribution and see if these values correspond to his opinion. To illustrate this checking process, Figure 7.8 shows the middle 50% area of the prior distribution. This graph shows that the probability that $p \leq 0.402$ is 0.25 and the probability that $p \geq 0.692$ is also 0.25. If these calculations do not correspond to the owner's opinion, then maybe some change in the prior distribution would be appropriate.

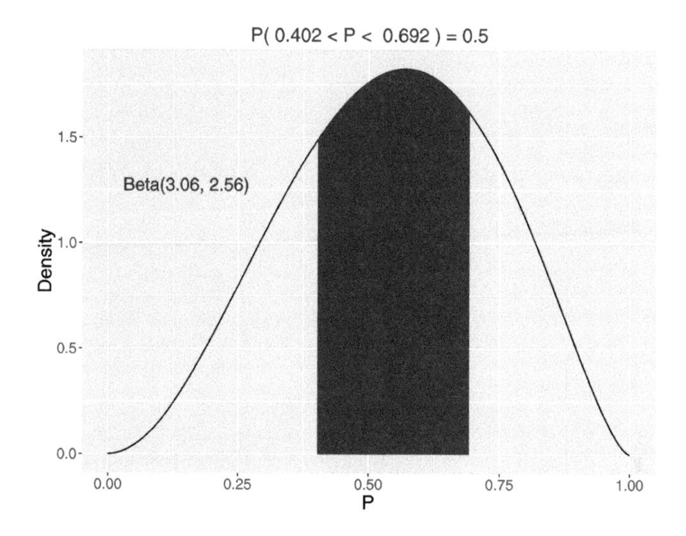

FIGURE 7.8
Illustration of the middle 50% of a Beta(3.06, 2.56) curve.

7.4 Updating the Beta Prior

In the previous section, we have seen that the restaurant owner thinks that a beta curve with shape parameters 3.06 and 2.56 is a reasonable reflection of his prior opinion about the proportion of students p whose favorite day for eating out is Friday. Therefore, we work with Beta$(3.06, 2.56)$ as the prior distribution for p.

Now we have the survey results – the survey was administered to 20 students and 12 say that their favorite day for eating out is Friday. As before in Section 7.2, the likelihood, that is the chance of getting this data if the probability of success is p is given by the binomial formula,

$$Likelihood = L(p) = \binom{20}{12} p^{12}(1-p)^8.$$

In this section, the Bayes' rule calculation of the posterior is presented for the continuous prior case and one discovers an interesting result: if one starts with a beta prior for a proportion p, and the data is binomial, then the posterior will also be a beta distribution. The beta posterior is a natural combination of the information contained in the beta prior and the binomial sampling, as one would expect in typical Bayesian inference. This is an illustration of the use of a conjugate prior where the prior and posterior densities are in the same family of distributions.

7.4.1 Bayes' rule calculation

First we demonstrate the Bayes' rule calculation of the posterior of p through the proportional statement:

$$\pi(p \mid y) \propto \pi(p) \times L(p). \tag{7.10}$$

The prior distribution of p, with density $\pi(p)$, is beta with shape parameters 3.06 and 2.56

$$p \sim \text{Beta}(3.06, 2.56).$$

The symbol "$\sim$" is read "follows", meaning that the random variable before the symbol follows the distribution that is after the symbol.

For the likelihood, we introduce proper notation. Let Y be the random variable of the number of students say that their favorite day for eating out is Friday. We know that the sampling distribution for Y is a binomial distribution with number of trials 20 and success probability p. Using the notation of "$\sim$", we have

$$Y \sim \text{Binomial}(20, p).$$

After the value $Y = y$ is observed, $L(p) = f(y \mid p)$ denotes the likelihood, which is the probability of observing this sample value y viewed as a function of the proportion p. (Note that a small letter y is used to denote the actual data observed, as opposed to the random variable Y.) From the dining survey, we know that $y = 12$.

Now we have the following prior density and the likelihood function.

- The prior distribution:

$$\pi(p) = \frac{1}{B(3.06, 2.56)} p^{3.06-1}(1-p)^{2.56-1}.$$

- The likelihood:

$$f(Y = 12 \mid p) = L(p) = \binom{20}{12} p^{12}(1-p)^8.$$

By Bayes' rule, the posterior density $\pi(p \mid y)$ is proportional to the product of the prior and the likelihood.

$$\pi(p \mid y) \propto \pi(p) \times L(p).$$

Substituting the current prior and likelihood, one can perform the algebra for the posterior density.

$$\begin{aligned}
\pi(p \mid Y = 12) &\propto \pi(p) \times f(Y = 12 \mid p) \\
&= \frac{1}{B(3.06, 2.56)} p^{3.06-1}(1-p)^{2.56-1} \times \\
&\quad \binom{20}{12} p^{12}(1-p)^{8} \\
\texttt{[drop the constants]} &\propto p^{12}(1-p)^{8} p^{3.06-1}(1-p)^{2.56-1} \\
\texttt{[combine the powers]} &= p^{15.06-1}(1-p)^{10.56-1}.
\end{aligned} \tag{7.11}$$

One observes that the posterior density of p given $Y = 12$ is, up to a proportionality constant,

$$\pi(p \mid Y = 12) \propto p^{15.06-1}(1-p)^{10.56-1}.$$

Note that in the posterior derivation, the constants $\binom{20}{12}$ and $\frac{1}{B(3.06,2.56)}$ are dropped due to the proportional sign "$\propto$". That is, the expression of $\pi(p \mid Y = 12)$ is computed up to some constant. In this case, Appendix A demonstrates the calculation of the constant.

Next, one recognizes if the posterior distribution of p is recognizable as a member of a familiar family of distributions. In the computation of the posterior, we have intentionally kept the expression of -1 in the powers of p and $1 - p$ terms, instead of using 14.06 and 9.56 directly. By doing this, one recognizes that the posterior density has the familiar form

$$p^{a-1}(1-p)^{b-1}.$$

As the reader might have guessed, the posterior distribution turns out to be a beta distribution with updated shape parameters. That is, the posterior distribution of p given $Y = 12$ is beta with parameters 15.06 and 10.56.

7.4.2 From beta prior to beta posterior: conjugate priors

The results about a proportion p from the Bayes' rule calculation performed in Section 7.4.1 can be generalized. Suppose one works with the following prior distribution and sampling density:

- The prior distribution:

$$p \sim \text{Beta}(a, b)$$

- The sampling density:

$$Y \sim \text{Binomial}(n, p)$$

One observes the count $Y = y$, the number of successes in the collected data. Then the posterior distribution of p is another beta distribution with shape parameters $a + y$ and $b + n - y$.

- The posterior distribution:

$$p \mid Y = y \sim \text{Beta}(a + y, b + n - y) \tag{7.12}$$

The two shape parameters of the beta posterior distribution, $a + y$ and $b + n - y$, are the sums of the prior and likelihood counts of successes and failures, respectively. We algebraically combine the shape parameters of the beta prior and the binomial likelihood to obtain the shape parameters of the posterior beta distribution.

Table 7.1 demonstrates this process with three rows labelled Prior, Likelihood, and Posterior. The Prior row contains the shape parameters of the beta prior a and b in the Successes and Failures columns, respectively. The Likelihood row contains the number of successes y and the number of failures $n - y$. The shape parameters of the beta posterior are found by adding the prior parameter values and the data values.

TABLE 7.1
Updating the beta prior.

Source	Successes	Failures
Prior	a	b
Likelihood	y	$n - y$
Posterior	$a + y$	$b + n - y$

R In the following R script we update the beta shape parameters. We see that the owner's posterior distribution for p is beta with shape parameters 15.06 and 10.56.

```
ab <- c(3.06, 2.56)
yny <- c(12, 8)
(ab_new <- ab + yny)
[1] 15.06 10.56
```

The function `beta_prior_post()` in the `ProbBayes` R package plots the prior and posterior beta curves together on one graph, see Figure 7.9.

```
beta_prior_post(ab, ab_new)
```

Comparing the two beta curves, several observations can be made.

- One can compare the prior and posterior beta curves using the respective means. The mean of a Beta(a, b) distribution is $\frac{a}{a+b}$. Using this formula, the posterior mean of p is 15.06 / (15.06 + 10.56) = 0.588 which is slightly larger than the prior mean 3.06 / (30.6 + 2.56) = 0.544. Recall that the sample proportion from the survey results is 12/20 = 0.6. The posterior mean lies between the prior mean and sample mean and it is closer to the sample mean.
- Next one compares the spreads of the two curves. One sees a much wider spread of the prior beta curve (dashed line) than that of the posterior beta curve (solid line). Initially the owner was unsure about the proportion of students favoring Friday to dine out. After observing the results of the survey, the solid posterior curve indicates that he is more certain that p is between 0.5 and 0.7. This sheds light on a general feature of Bayesian inference: the data helps sharpen the belief about the parameter of interest, producing a posterior distribution with a smaller spread than the prior distribution.

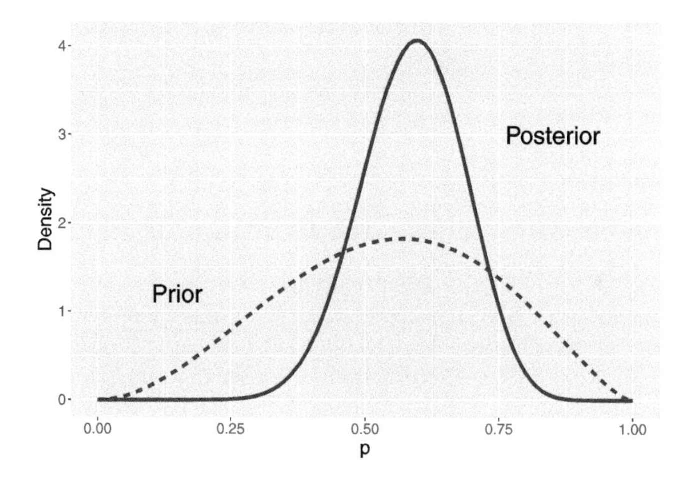

FIGURE 7.9
Prior and posterior curves for the proportion of students who prefer to dine out on Friday.

The attractive combination of a beta prior and a binomial sampling density to obtain a posterior motivates a definition of conjugate priors. If the prior distribution and the posterior distribution come from the same family of distributions, the prior is then called a conjugate prior. Here a beta is a conjugate prior for a success probability p, since the posterior distribution for p is also in the beta family. Conjugate priors are specific to the choice of sampling density. For example, a beta prior is conjugate with binomial sampling, but not to normal sampling which is popular for continuous outcome.

In Chapter 8 we will discover the conjugate prior distribution for a normal sampling distribution.

Conjugate priors are desirable because they simplify the Bayesian inference procedure. In the dining preference example, when a Beta(3.06, 2.56) prior is assigned to p, the posterior is Beta(15.06, 10.56) and inference about p is made in a straightforward way. One can easily plot the prior and posterior beta distributions as in Figure 7.9. One can also make precise comparative statements about the locations of the prior and posterior distribution using quantiles of a beta curve.

Although conjugate priors are convenient and straightforward to use, they may not be appropriate for use in a Bayesian analysis. One should choose a prior that fits one's belief, not one that is convenient to use. In some situations it may be appropriate to choose a prior distribution that does not provide conjugacy. In Chapter 9, we will describe computational methods to facilitate posterior inferences when non-conjugate priors are used. Modern Bayesian posterior computations accommodate a wide variety of choices of prior and sampling distributions. Therefore it is more important to choose a prior that matches one's prior belief than choosing a prior that is computationally convenient.

7.5 Bayesian Inferences with Continuous Priors

We will continue with the dining preference example to illustrate different types of Bayesian inference. The restaurant owner has taken his dining survey and the posterior distribution Beta(15.06, 10.56) reflects his opinion about the proportion p of students whose favorite day for eating out is Friday.

All Bayesian inferences about the proportion p are based on various summaries of this posterior beta distribution. The summary we compute from the posterior will depend on the type of inference. We will focus on three types of inference: (1) testing problems where one is interested in assessing the likelihood of some values of p, (2) interval estimations where one wants to find an interval that is likely to contain p, and (3) Bayesian prediction where one wants to learn about new observation(s) in the future.

Simulation will be incorporated for all three types of Bayesian inference problems. Since one has a conjugate prior distribution, one can derive the exact posterior distribution (a beta) and inferences are performed with the exact posterior beta distribution. In other situations when conjugacy is not available, meaning that no exact representation of the posterior is available, inferences through simulation are much more widely used. It is instructive to present the exact solutions and the approximated simulation-based solutions together, so one learns through practice and prepares for future use of simulation in other settings.

There is nothing magic about simulation. In fact, simulation has been used earlier, when the `rbeta()` function was used to generate simulated samples from Beta(4, 4) and Beta(2, 9) and check the appropriateness of the chosen beta prior (review Section 7.3.2 as needed). Information on simulation and the relevant R code will be introduced in the description of each inferential problem.

7.5.1 Bayesian hypothesis testing

Suppose one of the restaurant workers claims that at least 75% of the students prefer to eat out on Friday. Is this a reasonable claim?

In traditional classical statistics, one might be interested in testing the hypothesis $H : p \geq 0.75$. From a Bayesian viewpoint, it is straightforward to implement this test. Since the hypothesis is an interval of values, one finds the posterior probability that $p \geq 0.75$ and makes a decision based on the value of this probability. If the probability is small, one rejects this claim.

R First the exact solution will be presented. Since the posterior distribution is Beta(15.06, 10.56), the owner's posterior density is graphed and the area under the curve for values of p between 0.75 and 1 is found. The `beta_area()` function is used to display and show the area; see Figure 7.10. Since the probability is only about 4%, one rejects the worker's claim that p is at least 0.75.

```
beta_area(lo = 0.75, hi = 1.0, shape_par = c(15.06, 10.56))
```

This computation can be implemented using simulation. Since the posterior distribution is Beta(15.06, 10.56), one generates a large number of random values from this beta distribution, then summarizes the sample of simulated draws to obtain the probability of $p \geq 0.75$. First a sample of $S = 1000$ from the beta posterior is taken, storing the results in the vector `BetaSamples`.

```
S <- 1000
BetaSamples <- rbeta(S, 15.06, 10.56)
```

The proportion of the 1000 simulated values of p that are at least 0.75 gives an approximation of the probability that $p \geq 0.75$.

```
sum(BetaSamples >= 0.75)/S
[1] 0.037
```

The simulation-based probability estimate is 0.037 which is an accurate approximation to the exact probability 0.04 obtained before.

It would be reasonable to question the choice of the number of simulations $S = 1000$. One can change the simulation sample size to larger or smaller values as one sees fit. In general, the larger the value of S, the more accurate

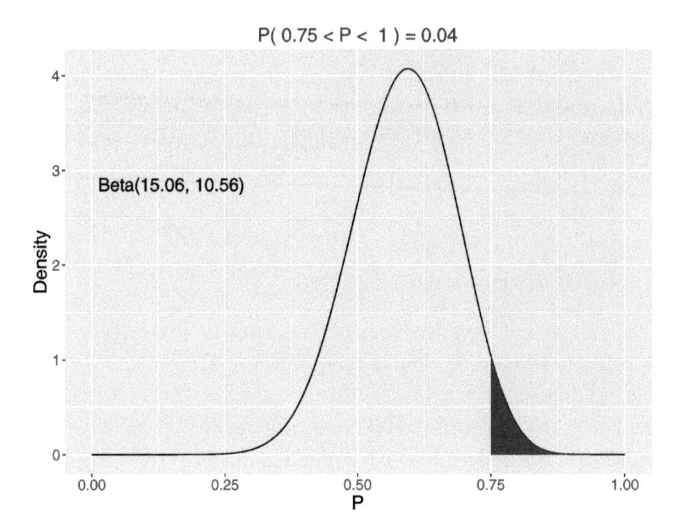

FIGURE 7.10
Probability of the hypothesis from the beta posterior density.

the approximation. Figure 7.11 shows that the shape of a histogram of the simulated values of p approaches the exact posterior density as the value of S changes from 100 to 10,000. The corresponding simulation-based probabilities of $p \geq 0.75$ are $\{0.02, 0.05, 0.033, 0.0422\}$ indicating that the accuracy of the approximation improves for larger simulation sample sizes.

R One will observe variation from one simulation from another (see the two different but similar approximated probabilities 0.037 and 0.033 when $S = 1000$). To replicate one's results one specifies the seed of the random number simulator `set.seed()`. Choose any number that you like to put in – if this `set.seed()` line of code is executed first, then the same sequence of random values will be generated and one replicates the simulation-based computation.

7.5.2 Bayesian credible intervals

Another type of inference is a *Bayesian credible interval*, an interval that one is confident contains p. Such an interval provides an uncertainty estimate for the parameter p. A 90% Bayesian credible interval is an interval that contains 90% of the posterior probability.

R One convenient 90% credible interval is the "equal tails" interval that contains the middle 90% of the probability content. The function `beta_interval()` in `ProbBayes` R package illustrates and computes the equal-tails interval. The shaded area in Figure 7.12 corresponds to 90% of the pos-

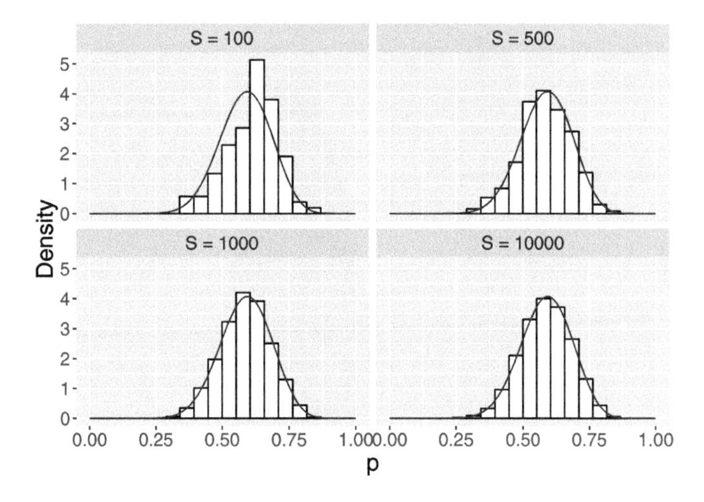

FIGURE 7.11
Histograms of simulated draws from Beta(15.06, 10.56) with exact beta density overlaid for four samples drawn where $S = \{100, 500, 1000, 10,000\}$.

terior probability. The probability p falls between 0.427 and 0.741 is exactly 90%.

```
beta_interval(0.9, c(15.06, 10.56))
```

One obtains this middle 90% credible interval using the `qbeta()` function.

```
qbeta(c(0.05, 0.95), 15.06, 10.56)
[1] 0.4266788 0.7410141
```

This Bayesian credible interval differs from the interpretation of a traditional confidence interval. With a traditional confidence interval, one does not have confidence that one particular interval will contain p. Instead 90% confidence refers to the average coverage of the interval in repeated sampling.

Other types of Bayesian credible intervals can be computed. For example, instead of a credible interval covering the middle 90% of the posterior probability, one could create a credible interval covering the lower 90%, or the upper 90%, or the middle 95%. The `qbeta()` function is helpful in achieving all of these different type of intervals, as long as we know the exact posterior distribution, that is, the two shape parameters of the posterior beta distribution. For example, the following code computes a credible interval that covers the lower 90% of the posterior distribution.

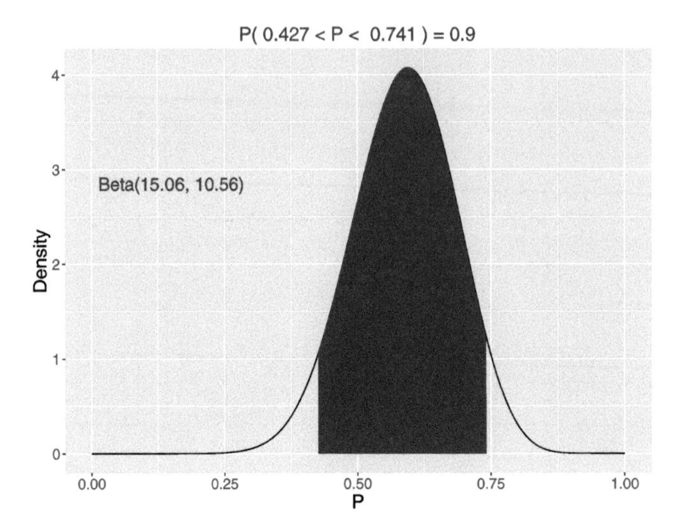

FIGURE 7.12
Display of 90% probability interval for the proportion p.

```
qbeta(c(0.00, 0.90), 15.06, 10.56)
[1] 0.0000000 0.7099912
```

An alternative way of creating credible intervals is by simulation. One first takes a random sample from the Beta(15.06, 10.56) distribution, then summarizes the simulated values by finding the two cutoff points of the middle 90% of the sample. The `quantile()` function is useful for this purpose. As a demonstration, below we simulate $S = 1000$ proportion values and compute the credible interval.

```
S <- 1000
BetaSamples <- rbeta(S, 15.06, 10.56)
quantile(BetaSamples, c(0.05, 0.95))
       5%       95%
0.4266076 0.7333957
```

The approximate middle 90% credible interval is [0.427, 0.733], which is close in value to the exact 90% credible interval [0.427, 0.741] computed using the `qbeta()` and `beta_interval()` functions. In an end-of-chapter exercise the reader is encouraged to practice and experiment with different values of the size of the simulated sample S.

7.5.3 Bayesian prediction

Prediction is a typical task of Bayesian inference and statistical inference in general. Once we are able to make inference about the parameter in our statistical model, we may be interested in predicting future observations.

Denote a new observation by the random variable $\tilde{Y}$. In particular, if the new survey is given to m students, the random variable $\tilde{Y}$ is the number of students preferring Friday to dine out among the m respondents. If again the survey is given to a random sample, the random variable $\tilde{Y}$, conditional on p, follows a binomial distribution with the fixed total number of trails m and success probability p. One's knowledge about the location of p is expressed by the posterior distribution of p.

Mathematically, to make a prediction of a new observation, one is asking for the distribution of $\tilde{Y}$ given the observed data $Y = y$. That is, one is interested in the probability function $f(\tilde{Y} = \tilde{y} \mid Y = y)$ where $\tilde{y}$ is a value of $\tilde{Y}$. But the conditional distribution of $\tilde{Y}$ given a value of the proportion p is binomial(m, p) and the current beliefs about p are described by the posterior density. So one writes the joint density of $\tilde{Y}$ and p as the product

$$f(\tilde{Y} = \tilde{y}, p \mid Y = y) = f(\tilde{Y} = \tilde{y} \mid p)\pi(p \mid Y = y). \tag{7.13}$$

By integrating out p, one obtains the predictive distribution

$$f(\tilde{Y} = \tilde{y} \mid Y = y) = \int f(\tilde{Y} = \tilde{y} \mid p)\pi(p \mid Y = y)dp. \tag{7.14}$$

The density of $\tilde{Y}$ given p is binomial with m trials and success probability p, and the posterior density of p is Beta$(a+y, b+n-y)$. After the substitution of densities and an integration step (see Appendix B for the detail), one finds that the predictive density is given by

$$f(\tilde{Y} = \tilde{y} \mid Y = y) \quad = \quad \binom{m}{\tilde{y}} \frac{B(a + y + \tilde{y}, b + n - y + m - \tilde{y})}{B(a + y, b + n - y)}. \tag{7.15}$$

This is the beta-binomial distribution with parameters m, $a+y$ and $b+n-y$.

$$\tilde{Y} \mid Y = y \sim \text{Beta-Binomial}(m, a + y, b + n - y). \tag{7.16}$$

To summarize, Bayesian prediction of a new observation is a beta-binomial distribution where m is the number of trials in the new sample, a and b are shape parameters from the beta prior, and y and n are quantities from the likelihood.

R Using this beta-binomial distribution in our example, one computes the predictive probability that $\tilde{y}$ students prefer Friday in a new survey of 20 students. We illustrate the use of the `pbetap()` function from the `ProbBayes` package. The inputs to `pbetap()` are the vector of beta posterior shape parameters (a, b), the sample size 20, and the values of $\tilde{y}$ of interest.

```
prob <- pbetap(c(15.06, 10.56), 20, 0:20)
prob_plot(data.frame(Y = 0:20, Probability = prob),
          Color = crcblue, Size = 4) +
  theme(text=element_text(size=18))
```

These predictive probabilities are displayed in Table 7.2 and graphed in Figure 7.13.

TABLE 7.2
Predictive distribution of the number of students preferring Friday in a future sample of 20.

Y	Probability	Y	Probability
0	0	11	0.127
1	0	12	0.134
2	0	13	0.127
3	0.001	14	0.108
4	0.004	15	0.080
5	0.010	16	0.052
6	0.021	17	0.028
7	0.037	18	0.012
8	0.059	19	0.004
9	0.085	20	0.001
10	0.109		

Looking at the table, the most likely number of students preferring Friday is 12. Just as in the inference situation, it is desirable to construct an interval that will contain $\tilde{Y}$ with a high probability. Suppose the desired probability content is 0.90. One constructs this prediction interval by putting in the most likely values of $\tilde{Y}$ until the probability content of the set exceeds 0.90.

R This method is implemented using the following command

```
discint(cbind(0:20, prob), .9)
$prob
[1] 0.9185699

$set
[1]  7  8  9 10 11 12 13 14 15 16
```

One therefore finds that

$$Prob(7 \leq \tilde{Y} \leq 16) = 0.919.$$

This exact predictive distribution is based on the posterior distribution of p, as one uses $\pi(p \mid Y = y)$ in the integration process in Equation (7.14).

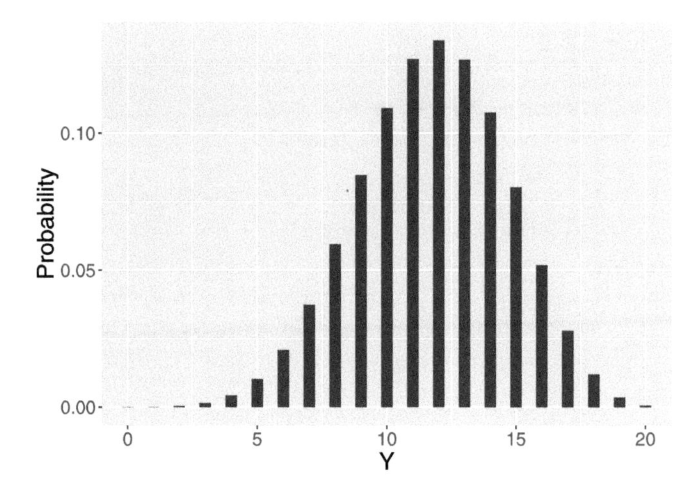

FIGURE 7.13
Display of the exact predictive distribution of the number of students $\tilde{y}$ favoring Friday in a future sample of 20.

For that reason this predictive distribution is called the *posterior predictive distribution*. There also exists a *prior predictive distribution*, a topic we will briefly introduce in Section 7.6.

In situations where it is difficult to derive the exact predictive distribution, one simulates values from this distribution. One implements this predictive simulation by first simulating draws of the parameter (in this case the proportion p) from its posterior distribution, and then simulating values of the future observation (e.g. the new observation $\tilde{Y}$) from the sampling density (here the binomial distribution).

We illustrate this simulation procedure with the generic beta posterior $\text{Beta}(a+y, b+n-y)$. To simulate a single draw from the predictive distribution, one first simulates a single proportion value p from the beta posterior and then simulates a new data point $\tilde{y}$ (the number of successes out of m trials) from a binomial distribution with sample size m and probability of success given by the simulated draw of p.

$$\text{sample } p \sim \text{Beta}(a + y, b + n - y) \quad \rightarrow \quad \text{sample } \tilde{Y} \sim \text{Binomial}(m, p)$$

R This process of simulating a single draw is implemented by the `rbeta()` and `rbinom()` functions. Let $m = n$ (the size of the future sample is the same as the size of the observed sample).

```
a <- 3.06; b <- 2.56
n <- 20; y <- 12
pred_p_sim <- rbeta(1, a + y, b + n - y)
(pred_y_sim <- rbinom(1, n, pred_p_sim))
[1] 14
```

Due to the ability of R to work easily with vectors, the same code is essentially used for simulating $S = 1000$ draws from the predictive distribution. In the following R script, `pred_p_sim` contains 1000 simulated draws from the posterior, and for each element of this posterior sample, the `rbinom()` function is used to simulate a corresponding value of $\tilde{Y}$ from the binomial sampling density.

```
a <- 3.06; b <- 2.56
n <- 20; y <- 12
S = 1000
pred_p_sim <- rbeta(S, a + y, b + n - y)
pred_y_sim <- rbinom(S, n, pred_p_sim)
```

Figure 7.14 displays predictive probabilities for the number of students who prefer Fridays using the exact beta-binomial and simulation methods. One observes good agreement using these two computation methods. For example, using the simulated values of $\tilde{Y}$ one finds that

$$Prob(6 \leq \tilde{Y} \leq 15) = 0.927$$

which is close in value to the range $Prob(7 \leq \tilde{Y} \leq 16) = 0.919$ found using the exact predictive distribution.

```
discint(as.matrix(S1[, 2:3]), .9)
$prob
[1] 0.927

$set
 [1]  6  7  8  9 10 11 12 13 14 15
```

7.6 Predictive Checking

Prior predictive checking

In the previous section, the use of the predictive distribution has been illustrated in learning about future data. This is more precisely described as the

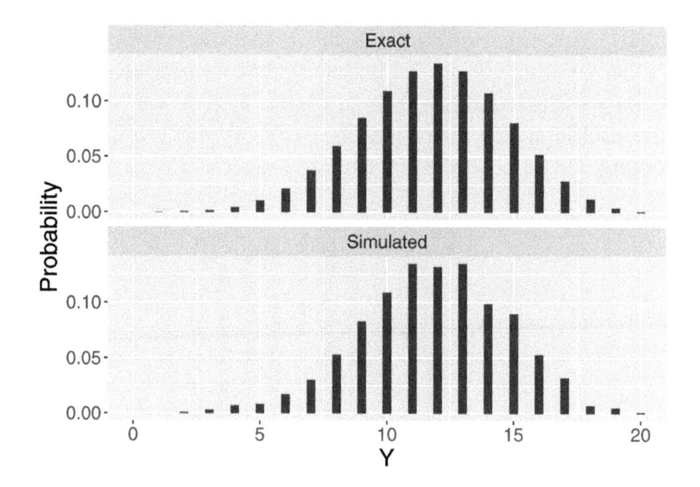

FIGURE 7.14
Display of the exact and simulated predictive probabilities for dining example.

posterior predictive density as one is obtaining this density by integrating the sampling density $f(\tilde{Y} = \tilde{y} \mid p)$ over the posterior density $\pi(p \mid y)$.

The prior predictive density is also useful in model checking. In a Bayesian model where p has a prior $\pi(p)$ and Y has a sampling density $f(Y = y \mid p)$, one writes the joint density of (p, Y) as the product of the sampling density and the prior:

$$f(p, Y = y) = f(Y = y \mid p)\pi(p). \tag{7.17}$$

Suppose one conditions on y instead of p and then one obtains an alternative representation of the joint density:

$$f(p, Y = y) = \pi(p \mid Y = y)f(Y = y). \tag{7.18}$$

The first term in this product, the density $\pi(p \mid Y = y)$, is the posterior density of p given the observation y; this density is useful for performing inference about the proportion. The second term in this product, $f(Y = y)$, is the prior predictive density that represents the density of future data before the observation y is taken. If the actual observation denoted by y_{obs} is not consistent with the prior predictive density $f(Y = y)$, this indicates some problem with the Bayesian model. Basically, this says that the observed data is unlikely to happen if one simulates predictions of data from our model.

To illustrate the use of prior predictive checking, recall that the restaurant owner assigned a Beta(3.06, 2.56) prior to the proportion p of students dining on Friday. A sample of 20 students will be taken. Based on this information, one computes the predictive probability $f(Y = y)$ of y students preferring

Friday dining of the sample of 20. This predictive distribution for all possible values of y is displayed in Figure 7.15. Recall that we actually observed y_{obs} = 12 Friday diners — this value is shown in Figure 7.15 as a large black dot. This value is in the middle of the distribution – the takeaway is that the observed data is consistent with predictions from the owner's Bayesian model.

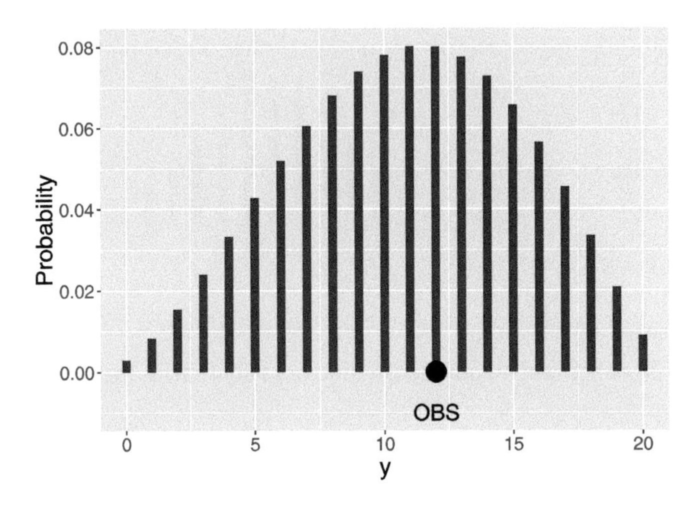

FIGURE 7.15
Prior predictive distribution of y using the owner's beta prior. The observed number of y is indicated with a large black dot. In this case the observed data is consistent with the Bayesian model.

In contrast, suppose another restaurant worker is more pessimistic about the likelihood of students dining on Friday. This worker's prior median of the proportion p is 0.2 and her 90th percentile is 0.4 — this information is matched with a beta prior with shape parameters 2.07 and 7.32. Figure 7.16 displays the predictive density of the number of Friday diners of a sample of 20 using this worker's prior. Here one reaches a different conclusion. The observed number 12 of Friday diners is in the tail of this predictive distribution — this observation is not consistent with predictions from the Bayesian model. In closer examination, one sees conflict between the information in the worker's prior and the data — her prior said that the proportion p was close to 0.20 and the data result (12 out of 20 successes) indicates that the proportion is close to 0.60. Predictive checking is helpful in this case in detecting this prior/data conflict.

Comparing Bayesian models

The prior predictive distribution is also useful in comparing two Bayesian models. To illustrate model comparison, suppose a second worker at the restaurant

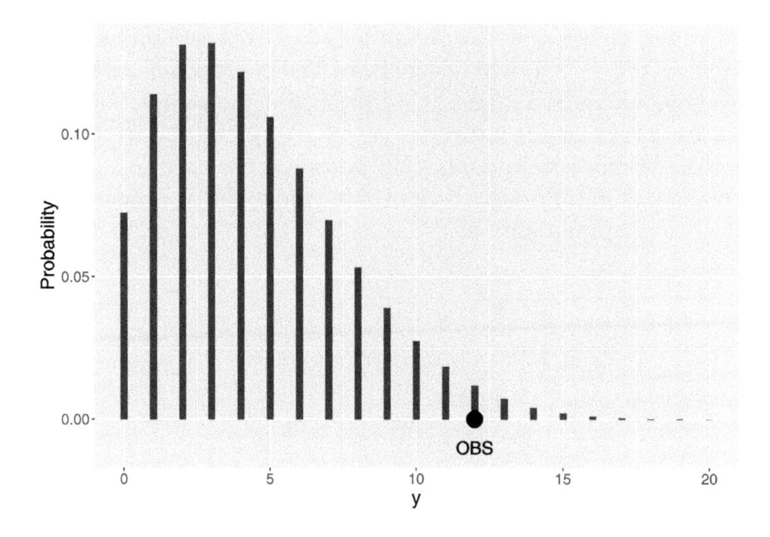

FIGURE 7.16
Prior predictive distribution of y using a worker's beta prior. The observed number of y is indicated with a large black dot. In this case the observed data is not consistent with the Bayesian model.

is also asked about the fraction of students who dine on Friday. He knows that the owner's belief about the proportion p is described by a Beta(3.06, 2.56) density, and the fellow worker's belief about p is represented by a Beta(2.07, 7.32) density. Who should the second worker believe?

Suppose this second worker believes that both the owner's and fellow worker's beliefs about the proportion p are equally plausible. So he places a probability of 0.5 on the Beta(3.06, 2.56) prior and a probability of 0.5 on the Beta(2.07, 7.32) prior. This second worker's prior $\pi(p)$ is written as the mixture

$$\pi(p) = q\pi_1(p) + (1 - q)\pi_2(p), \tag{7.19}$$

where $q = 0.5$ and π_1 and π_2 denote the owner's and worker's beta priors.

Now one observes the survey data – y Fridays in a sample of size n. Using the usual prior times likelihood procedure, the posterior density of p is proportional to the product

$$\pi(p \mid Y = y) \propto \left[q\pi_1(p) + (1 - q)\pi_2(p)\right] \times \binom{n}{y} p^y (1 - p)^{n-y}. \tag{7.20}$$

After some manipulation, one can show that the posterior density for the proportion p has the mixture form

$$\pi(p \mid Y = y) = q(y)\pi_1(p \mid Y = y) + (1 - q(y))\pi_2(p \mid Y = y). \tag{7.21}$$

The posterior densities $\pi_1(p \mid y)$ and $\pi_2(p \mid y)$ are the familiar beta forms. For example, $\pi_1(p \mid Y = y)$ will be the Beta$(3.06 + y, 2.56 + n - y)$ posterior density combining the Beta(3.06, 2.56) prior and the sample data of y successes in a sample of size n. Likewise, $\pi_2(p \mid Y = y)$ will be the beta density combining the worker's Beta(2.07, 7.32) prior and the data.

The quantity $q(y)$ represents the posterior probability of the owner's prior. One expresses this probability as

$$q(y) = \frac{qf_1(Y = y)}{qf_1(Y = y) + (1 - q)f_2(Y = y)} \tag{7.22}$$

where $f_1(Y = y)$ and $f_2(Y = y)$ denote the predictive densities corresponding to the owner's and worker's priors. With a little algebra, one represents the posterior odds of the model probabilities as follows.

$$\frac{P(Prior\,1 \mid Y = y)}{P(Prior\,2 \mid Y = y)} = \frac{q(y)}{1 - q(y)} = \left[\frac{q}{1-q}\right]\left[\frac{f_1(Y = y)}{f_2(Y = y)}\right] \tag{7.23}$$

The posterior odds of the owner's prior $P(Prior\,1 \mid Y = y)/P(Prior\,2 \mid y = y)$ is written as the product of two terms.

- The ratio $q/(1 - q)$ represents the prior odds of the owner's prior.
- The term $f_1(Y = y)/f_2(Y = y)$, the ratio of the predictive densities, is called the Bayes factor. It reflects the relative abilities of the two priors to predict the observation y.

R The function `binomial.beta.mix()` is used to find the Bayes factor for our example. One inputs the prior probabilities of the two models (priors), and the vectors of beta shape parameters that define the owner's prior and the worker's prior. The displayed output is the posterior odds value of 6.77.

```
probs <- c(0.5, 0.5)
beta_par1 <- c(3.06, 2.56)
beta_par2 <- c(2.07, 7.32)
beta_par <- rbind(beta_par1, beta_par2)
output <- binomial.beta.mix(probs, beta_par, c(12, 8))
(posterior_odds <- output$probs[1] / output$probs[2])
 6.777823
```

Since the two priors are given equal probabilities, the prior odds $q/(1 - q)$ is equal to one. In this case the posterior odds is equal to the Bayes factor. The interpretation is that for the given observation (12 successes in 20 trials), there is 6.77 times more support for the owner's prior than for the worker's prior. This conclusion is consistent with the earlier work that showed that the observed value of y was inconsistent with the Bayesian model for the worker's prior.

Posterior predictive checking

Although the prior predictive distribution is useful in model checking, it has some disadvantages. One problem is that the distribution $f(Y = y)$ may not exist in the situation where the prior $\pi()$ is not a proper probability distribution. We will see particular situations in future chapters where a vague or imprecise probability distribution is assigned as our prior and then the prior predictive distribution will not be well-defined. A related issue is that a prior may be assigned that may not accurately reflect one's prior beliefs about a parameter. Small errors in the specification of the prior will result in errors in the prior predictive distribution. So there needs to be some caution in the use of the prior predictive distribution in assessing the goodness of the Bayesian model.

An alternative method of checking the suitability of a Bayesian model is based on the posterior predictive distribution. In this setting, one computes the posterior predictive distribution of a replicated dataset, that is a dataset of the same sample size as our observed sample. One sees if the observed value of y is in the middle of this predictive distribution. If this is true, then this means that the observed sample is consistent with predictions of replicated data. On the other hand, if the observed y is in the tails of the posterior distribution, this indicates some model misspecification which means that there is a possibility of some issue with the specified prior or sampling density.

One attractive aspect of the posterior prediction distribution is that replicated datasets are conveniently simulated. To simulate one replicated dataset, we first simulate a parameter from its posterior distribution, then simulate new data from the data model given the simulated parameter value. In the beta-binomial situation, the posterior of the proportion p is $\text{Beta}(a + y, b + n - y)$. To simulate a new data point $\tilde{Y} = \tilde{y}$, one first simulates a proportion value $p^{(1)}$ from the beta prior and then simulates a new data point $\tilde{y}^{(1)}$ from a binomial distribution with sample size n and probability of success $p^{(1)}$. If we wish to obtain a sample of size S from the posterior predictive distribution, this process is repeated S times as showed in the following diagram.

$$
\begin{array}{lcl}
\text{sample } p^{(1)} \sim \text{Beta}(a + y, b + n - y) & \rightarrow & \text{sample } \tilde{y}^{(1)} \sim \text{Binomial}(n, p^{(1)}) \\
\text{sample } p^{(2)} \sim \text{Beta}(a + y, b + n - y) & \rightarrow & \text{sample } \tilde{y}^{(2)} \sim \text{Binomial}(n, p^{(2)}) \\
 & \vdots & \\
\text{sample } p^{(S)} \sim \text{Beta}(a + y, b + n - y) & \rightarrow & \text{sample } \tilde{y}^{(S)} \sim \text{Binomial}(n, p^{(S)})
\end{array}
$$

The sample $\tilde{y}^{(1)}, ..., \tilde{y}^{(S)}$ is an approximation to the posterior predictive distribution that is used for model checking. In practice, one constructs a histogram of this sample and decides if the observed value of y is in the central portion of this predictive distribution. The reader will be given an opportunity to use this algorithm to see if the observed data is consistent with simulations of replicated data from this predictive distribution.

7.7 Exercises

1. **Laymen's Prior in the Dining Preference Example**

   Revisit Section 7.2.1 for the laymen's prior in Equation (7.2) and the expert's prior in Equation (7.3). Follow the example R code (functions `data.frame()`, `mutate()` and `ggplot()`) to obtain the Bayes table and graph of the laymen's prior distribution. Compare the similarities and differences between the laymen's prior and the expert's prior.

2. **Inference for the Dining Preference (Discrete Priors)**

   Revisit Section 7.2.5 where we show how to find the posterior probability that over half of the students prefer eating out on Friday. Find the following posterior probabilities. (Be careful about the end points.)

   (a) The probability that more than 60% of the students prefer eating out on Friday.
   (b) The probability that less than 40% of the students prefer eating out on Friday.
   (c) The probability that between 20% and 40% of the students prefer eating out on Friday.
   (d) No more than 50% of the students prefer eating out on Friday.

3. **Another Dining Survey (Discrete Priors)**

   Suppose the restaurant owner in the college town gives another survey to a different group of students. This time he gives the survey to 30 students – among these responses 10 of them say that Friday is their preferred day to eat out. Use the owner's prior (restated below) to calculate the following posterior probabilities.

$$\begin{aligned} p &= \{0.3, 0.4, 0.5, 0.6, 0.7, 0.8\} \\ \pi_e(p) &= (0.125, 0.125, 0.250, 0.250, 0.125, 0.125) \end{aligned}$$

   (a) The probability that 30% of the students prefer eating out on Friday.
   (b) The probability that more than half of the students prefer eating out on Friday.
   (c) The probability that between 20% and 40% of the students prefer eating out on Friday.

4. **Interpreting A Beta Curve**

   Revisit Figure 7.4 where nine different beta curves are displayed. In the context of students' dining preference example where p is the proportion of students preferring Friday, interpret the following prior choices in terms of the opinion of p. For example, Beta(0.5, 0.5) represents the prior belief the extreme values $p = 0$ and $p = 1$ are more probable and $p = 0.5$ is the least probable. In the students' dining preference example, specifying a Beta(0.5, 0.5) prior indicates the owner thinks the students' preference of dining out on Friday is either very strong or very weak.

(a) Interpret the Beta(1, 1) curve.
(b) Interpret the Beta(0.5, 1) curve.
(c) Interpret the Beta(4, 2) curve.
(d) Compare the opinion about p expressed by the two beta curves: Beta(4, 1) and Beta(4, 2).

5. **Beta Probabilities**

Use the functions `dbeta()`, `pbeta()`, `qbeta()`, `rbeta()`, `beta_area()`, and `beta_quantile()` to answer the following questions about beta probabilities.

(a) Compute the density of Beta(0.5, 0.5) at the values $p = \{0.1, 0.5, 0.9, 1.5\}$. Check your answers with the Beta(0.5, 0.5) curve in Figure 7.4.
(b) If $p \sim$ Beta(6, 3), compute the probability $Prob(0.2 \leq p \leq 0.6)$.
(c) Compute the quantiles of the Beta(10, 10) distribution at the probability values in the set $\{0.1, 0.5, 0.9, 1.5\}$.
(d) Simulate a sample of 100 random values from Beta(4, 2).

6. **Comparing Beta Distributions**

Consider four Beta curves: (1) Beta(5, 5), (2) Beta(10, 10), (3) Beta(50, 50) and (4) Beta(100, 100). Think of the shape parameters a and b as counts of "successes" and "failures" in a prior sample. Use one of the R beta functions (e.g. `rbeta()`, `beta_area()`, among others) to discuss the similarities and differences between these four beta curves.

7. **Specifying A Continuous Beta Prior**

Consider another dining survey conducted by a restaurant owner in New York. The owner is also interested in knowing about the proportion p of students who prefer eating out on Friday. He believes that its 0.4 quantile is 0.7 and 0.8 quantile is 0.9. Suppose the owner plans on using a beta prior distribution.

(a) Find the values of the beta shape parameters a and b to represent the restaurant owner's belief.

(b) Confirm the choice of beta prior by taking a simulated sample from the prior predictive simulation. [Hint: use the `rbeta()` function to simulate a sample from the selected beta distribution, and then simulate new $\tilde{y}$ values from the binomial data model (function `rbinom()`) with a sample size of 20. Graph and/or calculate a few quantiles of the simulated $\tilde{y}$ sample from the predictive distribution to check the restaurant owner's prior belief.]

8. **Deriving the Beta Posterior**

   Following the derivation process of the dining preference example in Section 7.4.1, derive this more general result. If the proportion has a Beta(a, b) prior and one samples Y from a binomial distribution with parameters n and p, then if one observes $Y = y$, then the posterior density of p is Beta$(a + y, b + n - y)$.

9. **Prior Sample Size and Strength of Priors**

   Another way of specifying a Beta(a, b) prior is to imagine a pre-survey with the same question and represent the beta shape parameters in the form of a successes and b failures in the pre-survey (see Table 7.3). This exercise explores this prior specification method.

TABLE 7.3
Updating the beta prior.

Source	Successes	Failures
Prior	a	b
Likelihood	y	$n - y$
Posterior	$a + y$	$b + n - y$

(a) Recall from Section 7.3 that the mean of the Beta(a, b) distribution is $\frac{a}{a+b}$. Define the prior sample size to be $n_p = a + b$. Consider two beta prior distributions: Beta$(2, 2)$ and Beta$(20, 20)$. Find the prior means and prior sample sizes of these two prior distributions and compare the prior beliefs of these two beta distributions.

(b) Suppose a survey yields four successes out of ten responses. Suppose one wishes to compare the posterior inference obtained by the two different Beta priors Beta$(2, 2)$ and Beta$(20, 20)$. Find and compare the two posterior distributions corresponding to these two priors.

(c) Consider the use of the Beta$(2, 2)$ and Beta$(20, 20)$ prior distributions. Show these two priors have the same prior mean, but different strengths of belief about the location of the proportion. Assuming the survey results in (b), use simulation and graphs to show how different prior sample sizes affect the posterior inference.

(d) Suppose a survey yields 40 successes out of 100 responses. Find the two posterior distributions corresponding to the two prior distributions Beta$(2, 2)$ and Beta$(20, 20)$. Contrast the two posterior distributions and compare with your answer to part (c).

(e) Consider the two prior distributions Beta$(9, 1)$ and Beta$(45, 5)$. Contrast these two beta prior distributions with respect to the mean and strength of belief. Compare the two posterior distributions with data $n = 20, y = 5$, and with the data $n = 200, y = 50$.

10. **Beta Posterior Mean is a Weighted Mean**

If the proportion has a Beta(a, b) prior and one observes Y from a binomial distribution with parameters n and p, then if one observes $Y = y$, then the posterior density of p is Beta$(a + y, b + n - y)$.

Recall that the mean of a Beta(a, b) random variable following is $\frac{a}{a+b}$. Show that the posterior mean of $p \mid Y = y \sim$ Beta$(a + y, b + n - y)$ is a weighted average of the prior mean of $p \sim$ Beta(a, b) and the sample mean $\hat{p} = \frac{y}{n}$. Find the two weights and explain their implication for the posterior being a combination of prior and data.

11. **Sequential Updating**

The restaurant owner's belief about the proportion of students' favorite dining day being Friday is represented by a Beta$(15.06, 10.56)$ distribution. Recall that he obtained this posterior distribution from a Beta$(3.06, 2.56)$ prior and a survey of 12 yes's out of 20 responses. The owner is interested in conducting another dining survey a few months later with the same question and the owner is still interested in p, the proportion of all students who say Friday or Saturday.

(a) The second survey gives a result of 8 yes's out of 20 responses. Use the owner's current beliefs and this information to update the restaurant owner's belief about the proportion p.

(b) Suppose the two surveys are conducted at the same time and the results are 20 yes's $(12 + 8)$ out of 40 responses $(20 + 20)$. Starting with the Beta$(3.06, 2.56)$ prior, update the owner's belief about the proportion of interest.

(c) Are the two posterior distribution the same in parts (a) and (b)? Why or why not?

(d) Suppose the two survey results are reversed. That is, the first survey gives 8 yes's and second survey gives 12 yes's. Do you still observe the same posterior as in part (b)? What does this tell you about the order of different pieces of information shaping the belief about a parameter?

(e) What if the two survey results are slightly different? The first survey gives 15 yes's and second survey gives 5 yes's. What is the posterior distribution in this case?

(f) Should we combine the two survey results together? Describe a scenario in which it would be inappropriate to combine the survey results.

12. **Bayesian Hypothesis Testing**

In the dining preference example, the restaurant owner's posterior distribution of proportion p of students preferring Friday to eat out is Beta$(15.06, 10.56)$. Suppose the owner's wife claims that between 50% and 60% of the students prefer to eat out on Friday. Conduct a Bayesian hypothesis test of this claim.

13. **Simulation Sample Size**

Revisit Section 7.5.2. Use R to simulate random samples of sizes $S = \{10, 100, 500, 1000, 5000\}$ of p from the Beta$(15.06, 10.56)$ distribution. Use the `quantile()` function to find the approximate 90% credible interval of p for each value of S. Comment on the effect of the simulation size S on the accuracy of the simulation results. Recall that the exact middle 90% posterior interval estimate is [0.427, 0.741].

14. **Bayesian Credible Intervals**

In the dining preference example, the restaurant owner's posterior distribution of proportion p of students preferring Friday to eat out is Beta$(15.06, 10.56)$. Find the exact Bayesian credible intervals for the following cases.

(a) The middle 95% credible interval.

(b) The middle 98% credible interval.

(c) The 90% credible interval of the form $(0, B)$.

(d) The 99% credible interval of the form $(A, 1)$

15. **Simulating the Posterior of the Log Odds**

Since one is able to compute exact posterior summaries using the `pbeta()` and `qbeta()` functions, what is the point of using simulation computations? To illustrate the advantage of simulation, suppose one is interested in finding a 90% probability interval about the logit or log odds $\log\left(\frac{p}{1-p}\right)$. One can approximate the posterior of the logit by simulation. First simulate $S = 1000$ values from the beta posterior for p, and then for each simulated value of p, compute a value of the logit. The resulting vector will be a random sample from the posterior distribution of the logit.

(a) If the posterior distribution for p is Beta(12, 20), use R to simulate 1000 draws from the posterior of the logit $\log\left(\frac{p}{1-p}\right)$.

(b) Construct a 90% interval estimate for the logit parameter.

16. **Simulating the Odds**

    Revisit Exercise 15. Instead of the logit or log odds of the proportion p, suppose we are interested in the odds $\frac{p}{1-p}$. If the posterior distribution for p is Beta(12, 20), use R to simulate 1000 values from the posterior distribution of the odds. Construct a histogram of the simulated odds and construct a 90% interval estimate. Experiment with different values of the simulation sample size S and comment on the effect of the value of S on the width of the 90% interval estimates.

17. **Teenagers and Televisions**

    In 1998, the New York Times and CBS News polled 1048 randomly selected $13 - 17$ year olds to ask them if they had a television in their room. Among this group of teenagers, 692 of them said they had a television in their room. Alex and Benedict both want to use the binomial model for this dataset, but they have different prior beliefs about p, the proportion of teenagers having a television in their room.

    (a) Alex asks 10 friends the same question and 8 of them have a television in their room. Alex decides to use this information to construct his prior. Design a continuous beta prior reflecting Alex's belief.

    (b) Benedict thinks the 0.2 quantile is 0.3 and the 0.9 quantile is 0.4. Design a continuous beta prior reflecting Benedict's belief.

    (c) Calculate Alex's posterior and Benedict's posterior distributions. Plot the two priors on one graph, and plot the corresponding posteriors on another graph. In addition, obtain 95% credible intervals for Alex and Benedict.

    (d) Conduct prior predictive checks for Alex and Benedict. For each person, is the prior consistent with the television data? Explain.

18. **Teenagers and Televisions (continued)**

    Revisit Exercise 17. Consider the odds of having a television in the room. Recall that if p is the probability of having a television in room, then the odds is $\frac{p}{1-p}$.

    (a) Find the mean, median and 95% posterior interval of Alex's analysis of the odds of teenagers having a television in their room.

    (b) Find the mean, median and 95% posterior interval of Benedict's analysis of the odds of teenagers having a television in their room.

    (c) Compare the two posterior summaries from parts (a) and (b).

19. **Comparing Two Proportions - Science Majors at Liberal Arts Colleges**

    Many liberal arts colleges and other organizations have been promoting science majors in recent years because of their value on the job market. One wishes to evaluate whether such promotion has any effect on student major preference. A college student, Clara, is interested in this question and collects data from three liberal arts colleges, presented in Table 7.4.

TABLE 7.4
Total numbers of science and non-science majors enrolled in three liberal arts colleges in 2005 and 2015.

Year	Science	Non-Science
2005	264	1496
2015	437	1495

   (a) Let p_{2005} and p_{2015} denote the proportions of science majors in 2005 and 2015, respectively. Assuming that p_{2005} and p_{2015} have independent uniform priors, obtain the joint posterior distribution of p_{2005} and p_{2015}. Recall that the Beta(1, 1) distribution is equivalent to the Uniform(0, 1) distribution.
   (b) Suppose one uses the parameter $\delta = p_{2015} - p_{2005}$ to measure the difference in proportions. Use simulation from the posterior distribution to answer the question "have the proportions of science majors changed from 2005 to 2015?" [Hint: simulate a vector s_{2005} of posterior samples of p_{2005} and another vector s_{2015} of posterior samples of p_{2015} (make sure to use the same number of samples) and subtract s_{2005} from s_{2015} which yields a vector of simulated values from the posterior of δ.]
   (c) Did the proportion of science majors change from 2005 to 2015? Answer this question by a posterior computation.
   (d) Compile a similar dataset for your school type, and answer parts (a) through (c).
   (e) What assumption is made about the proportions p_{2005} and p_{2015} in our assignment of priors? Do you think such assumption is justified? If not, how do you think you can adjust the approach to be more realistic?

20. **Comparing Two Proportions - Number of Depression Cases at a Hospital**

    Data are collected on depression cases at hospitals. For a particular hospital, in the year of 1992, there were 306 diagnosed with depression among 651 patients; in the year of 1993, there were 300 diagnosed with depression among 705 patients. One is interested in learning whether the probability of being diagnosed with depression changed between 1992 and 1993. Conduct a Bayesian analysis of this question. State clearly the inference procedure, the choice of prior distributions, the choice of data model, the posterior distributions and the conclusions.

21. **Prior Predictive Checking - Pizza Popularity**

    Suppose a restaurant is serving pizza of two varieties, cheese and pepperoni, and a manager is interested in the proportion p of customers who prefer pepperoni. After some thought, the manager's prior beliefs about p are represented by a Beta(6, 12) distribution.

    (a) Suppose a random sample of 20 customers is surveyed on their pizza preference and let Y denote the number that prefer pepperoni. Compute and graph the prior predictive density of Y.

    (b) Suppose 20 customers are sampled and 14 prefer pepperoni. Is the value $y = 14$ consistent with the Bayesian model where p has a Beta(6, 12) distribution? Explain why or why not.

22. **Bayes Factor - Pizza Popularity**

    In the restaurant example of Exercise 21, suppose one of the waiters has a different opinion about the popularity of pepperoni pizza. His prior belief about the proportion p preferring pizza is described by a Beta(12, 6) distribution.

    (a) Find and graph the prior predictive density of the number y who prefer pizza in a sample of 20 customers.

    (b) If 14 out of 20 customers prefer pepperoni, is this result consistent with the predictive distribution?

    (c) Compare the two Bayesian models where (1) p is distributed Beta(12, 62) distribution and (2) p is distributed Beta(6, 12) distribution by a Bayes factor. Interpret the value that you compute.

23. **Posterior Predictive Checking - Pizza Popularity**

    Consider the same problem as in Exercise 22 where p is the proportion of customers who prefer pepperoni and the manager's prior beliefs are given by a Beta(6, 12) distribution.

    (a) Suppose 14 out of 20 customers prefer pepperoni. Using the algorithm described in Section 7.6, simulate 1000 values of $\tilde{y}$ (out of 20 customers) from the posterior predictive distribution. Construct a histogram of these values.

    (b) Is the observation (14 preferring pepperoni) consistent with this predictive distribution? Explain.

(c) Repeat parts (a) and (b) using the waiter's Beta(12, 6) distribution.

24. **Learning from a Multinomial Experiment**

In Chapter 6 Section 6.3, we discussed the multinomial distribution, an extension of the binomial distribution where each trial has more than two outcomes. As an application of a multinomial experiment, in an analysis of an election poll, suppose that one wants inferences for three probabilities: θ_A = probability of a vote for a candidate from Party A, θ_B = probability of a vote for a candidate from Party B and θ_C = probability of a vote for a candidate from Party C. One assumes $\theta_A + \theta_B + \theta_C = 1$ as people can vote for only one party. If a random sample of n potential voters is taken, the respective vector counts Y_A, Y_B, Y_C have the probability mass function

$$p(Y_A = y_A, Y_B = y_B, Y_C = y_C) = \frac{n!}{y_A! y_B! y_C!} \theta_A^{y_A} \theta_B^{y_B} \theta_C^{y_C}, \quad (7.24)$$

where $y_A + y_B + y_C = n$ and $y_A, y_B, y_C \geq 0$. This is written Multinomial$(n; \theta_A, \theta_B, \theta_C)$.

(a) A convenient prior distribution for $(\theta_A, \theta_B, \theta_C)$ is the Dirichlet distribution, which has the density function

$$p(\theta_A, \theta_B, \theta_C) = K \theta_A^{\alpha_A - 1} \theta_B^{\alpha_B - 1} \theta_C^{\alpha_C - 1},$$

where $(\alpha_A, \alpha_B, \alpha_C)$ are positive constants, and $K = \frac{\Gamma(\alpha_A + \alpha_B + \alpha_C)}{\Gamma(\alpha_A)\Gamma(\alpha_B)\Gamma(\alpha_C)}$ is a normalizing constant. This is written Dirichlet$(\alpha_A, \alpha_B, \alpha_C)$. Install the `gtools` R package and explore `ddirichlet()` and `rdirichlet()` functions to evaluate the pdf and generate random samples from Dirichlet$(\alpha_A = 2, \alpha_B = 1, \alpha_C = 1)$.

(b) Suppose the prior distribution is Dirichlet$(\alpha_A, \alpha_B, \alpha_C)$ and one collects from n sampled voters, where $(Y_A, Y_B, Y_C) \sim$ Multinomial$(n; \theta_A, \theta_B, \theta_C)$. Find the posterior distribution of $(\theta_A, \theta_B, \theta_C)$ and show that this is a Dirichlet distribution with updated parameters.

(c) Suppose in the sample of $n = 100$ voters, 53 voted for Party A, 18 voted for Party B, and 29 voted for Party C ($y_A = 53, y_B = 18, y_C = 29$). Using the prior distribution Dirichlet$(\alpha_A = 2, \alpha_B = 1, \alpha_C = 1)$ and the generic results from part (b), obtain the posterior distribution for $(\theta_A, \theta_B, \theta_C)$. Plot the prior and the posterior distributions for $(\theta_A, \theta_B, \theta_C)$ and discuss your findings.

(d) Suppose one wants to compute the ratio of odds of voting for Party A to the odds of voting for Party B, $\frac{\theta_A/(1-\theta_A)}{\theta_B/(1-\theta_B)}$. Compute a 95% posterior interval for this odds ratio.

8

Modeling Measurement and Count Data

8.1 Introduction

We first consider the general situation where there is a hypothetical population of individuals of interest and there is a continuous-valued measurement Y associated with each individual. One represents the collection of measurements from all individuals by means of a continuous probability density $f(y)$. As discussed in Chapter 5, one summarizes this probability density with the mean μ:

$$\mu = \int y f(y) dy. \tag{8.1}$$

The value μ gives us a sense of the location of a typical value of the continuous measurement Y.

To learn about the population of measurements, a random sample of individuals $Y_1, ..., Y_n$ will be taken. The general inferential problem is to use these measurements together with any prior beliefs to learn about the population mean μ. In other words, the goal is to use the collected measurements to learn about a typical value of the population of measurements.

8.2 Modeling Measurements

8.2.1 Examples

College applications

How many college applications does a high school senior in the United States complete? Here one imagines a population of all American high school seniors and the measurement is the number of completed college applications. The unknown quantity of interest is the mean number of applications μ completed by these high school seniors. The inferential question may be stated by asking, on average, how many college applications does an American high school senior complete. The answer to this question gives one a sense of the number of completed applications for a typical high school senior. To learn about the average μ, it would be infeasible to collect this measurement from every high school senior in the U.S. Instead, a survey is typically conducted to a sample of high school seniors (ideally a sample representative of all American high school seniors) and based on the measurements from this sample, some inference is performed about the mean number of college applications.

Household spending

How much does a household in San Francisco spend on housing every month? One visualizes the population of households in San Francisco and the continuous measurement is the amount of money spent on housing (either rent for renters and mortgage for homeowners) for a resident. One can ask "on average, how much does a household spend on housing every month in San Francisco?", and the answer to this question gives one a sense of the housing costs for a typical household in San Francisco. To learn about the mean value of housing μ of all San Francisco residents, a sample survey is conducted. The mean value of the housing costs $\bar{y}$ from this sample of surveyed households is informative about the mean housing cost μ for all residents.

Weights of cats

Suppose you have a domestic shorthair cat weighing 14 pounds and you want to find out if she is overweight. One imagines a population of all domestic shorthair cats and the continuous measurement is the weight in pounds. Suppose you were able to compute the mean weight μ of all shorthair cats. Then by comparing 14 pounds (the weight of our cat) to this mean, you would know whether your cat is overweight, or underweight, or close to the mean. If we were able to find the distribution of the weights of all domestic shorthair cats, then one observes the proportion of weights smaller than 14 pounds in the distribution and learns if the cat is severely overweight. To learn if our cat is overweight, you can ask the vet. How does the vet know? Extensive research has been conducted periodically to record weights of a large sample of domestic shorthair cats, and by using these samples of weights, the vet performs an inference about the mean μ of the weights of all domestic shorthair cats.

Common elements of an inference problem

All three examples have common elements:

- One has an underlying population of measurements, where the measurement is an integer, such as the number of college applications, or continuous, such as a housing cost or a cat weight.

- One is interested in learning about the value of the mean μ of the population of measurements.

- It is impossible or impractical to collect all measurements from the population, so one will collect a sample of measurements $Y_1, ..., Y_n$ and use the observed measurements to learn about the unknown population mean μ.

8.2.2 The general approach

Recall the three general steps of Bayesian inference discussed in Chapter 7 in the context of an unknown proportion p.

Step 1 **Prior**: We express an opinion about the location of the proportion p before sampling.

Step 2 **Likelihood**: We take the sample and record the observed proportion.

Step 3 **Posterior**: We use Bayes' rule to sharpen and update the previous opinion about p given the information from the sample.

In this setting, we have a continuous population of measurements that we represent by the random variable Y with density function $f(y)$. It is convenient to assume that this population has a normal shape with mean μ and standard deviation σ. That is, a single measurement Y is assume to come from the density function

$$f(y) = \frac{1}{\sqrt{2\pi}\sigma} \exp\left\{-\frac{(y-\mu)^2}{2\sigma^2}\right\}, -\infty < y < \infty \tag{8.2}$$

displayed in Figure 8.1. To simplify the discussion, it is convenient to assume that the standard deviation σ of the measurement distribution is known. Then the objective is to learn about the single mean measurement μ.

Step 1 in Bayesian inference is to express an opinion about the parameter. In this continuous measurement setting, one constructs a prior for the mean parameter μ that expresses one's opinion about the location of this mean. In this chapter, we discuss different ways to specify a prior distribution for μ. One attractive discrete approach for expressing this prior opinion, similar to the approach in Chapter 7 for a proportion p, has two steps. First one constructs a list of possible values of μ, and then one assigns probabilities to the possible values to reflect one's belief. Alternatively, we will describe the use of a continuous prior to represent one's belief for μ. This is a more realistic approach for constructing a prior since one typically views the mean as a real-valued parameter.

Step 2 of our process is to collect measurements from a random sample to gain more information about the parameter μ. In our first situation, one collects the number of applications from a sample of 100 high school seniors. In the second example, one collects a sample of 2000 housing costs, each from a sampled San Francisco household. The third example collects a sample of 200 different weights of domestic shorthair cats, each from a sampled cat. If these measurements are viewed as independent observations from a normal

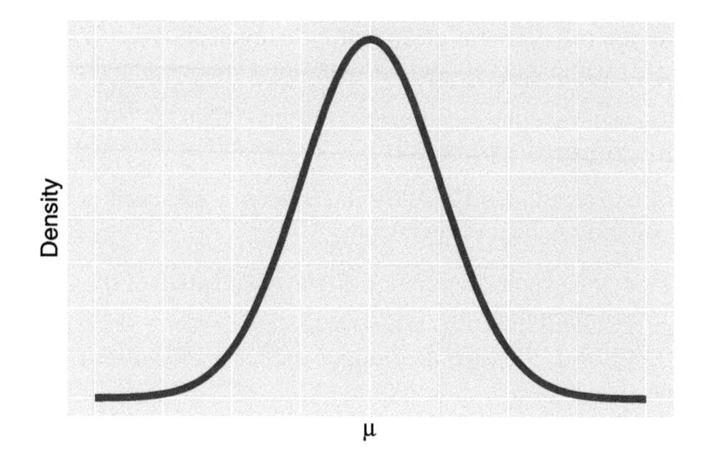

FIGURE 8.1
Normal sampling density with mean μ.

sampling density with mean μ, then one constructs a likelihood function which is the joint density of the sampled measurements viewed as a function of the unknown parameter.

Once the prior is specified and measurements have been collected, one proceeds to Step 3 to use Bayes' rule to update one's prior opinion to obtain a posterior distribution for the mean μ. The algebraic implementation of Bayes' rule is a bit more tedious when dealing with continuous data with a normal sampling density. But we will see there is a simple procedure for computing the posterior mean and standard deviation.

8.2.3 Outline of chapter

Throughout this chapter, the entire inferential process is described for learning about a mean μ assuming a normal sampling density for the measurements. This chapter discusses how to construct a prior distribution that matches one's prior belief, how to extract information from the data by the likelihood function, and how to update one's opinion in the posterior, combining the prior and data information in a natural way.

Section 8.3 introduces inference with a discrete prior distribution for the mean μ and Section 8.4 introduces the continuous family of normal prior distributions for the mean. The inferential process with a normal prior distribution is described in detail in Section 8.5. Section 8.6 describes some general Bayesian inference methods in this normal data and normal prior setting, such as Bayesian hypothesis testing, Bayesian credible intervals and Bayesian prediction. These sections describe the use of both exact analytical solutions and approximation simulation-based calculations. Section 8.7 introduces the use of the posterior predictive distribution as a general tool for checking if the observed data is consistent with predictions from the Bayesian model.

The chapter concludes in Section 8.8 by introducing a popular one-parameter model for counts, the Poisson distribution, and its conjugate gamma distribution for representing prior opinion. Although this section does not deal with the normal mean situation, the exposure to the important gamma-Poisson conjugacy will enhance our understanding and knowledge of the analytical process of combining the prior and likelihood to obtain the posterior distribution.

8.3 Bayesian Inference with Discrete Priors

8.3.1 Example: Roger Federer's time-to-serve

Roger Federer is recognized as one of the greatest players in tennis history. One aspect of his play that people enjoy is his businesslike way of serving to start a point in tennis. Federer appears to be efficient in his preparation to serve and some of his service games are completed very quickly. One measures one's service efficiency by the time-to-serve which is the measured time in seconds between the end of the previous point and the beginning of the current point.

Since Federer is viewed as an efficient server, this raises the question: how long, on average, is Federer's time-to-serve? We know two things about his time-to-serve measurements. First, since they are time measurements, they are continuous variables. Second, due to a number of other variables, the measurements will vary from serve to serve. Suppose one collects a single time-to-serve measurement in seconds. denoted as Y. It seems reasonable to assume Y is normally distributed with unknown mean μ and standard deviation σ. From previous data, we assume that the standard deviation is known and given by $\sigma = 4$ seconds.

Recall the normal probability curve has the general form

$$f(y) = \frac{1}{\sqrt{2\pi}\sigma} \exp\left\{-\frac{(y-\mu)^2}{2\sigma^2}\right\}, -\infty < y < \infty. \tag{8.3}$$

Since $\sigma = 4$ is known, the only parameter in Equation (8.3) is μ. We are interested in learning about the mean time-to-serve μ.

A convenient first method of implementing Bayesian inference is by the use of a discrete prior. One specifies a subjective discrete prior for Federer's mean time-to-serve by specifying a list of plausible values for μ and assigning a probability to each of these values.

In particular suppose one thinks that values of the equally spaced values $\mu = 15, 16, \cdots, 22$ are plausible. In addition, one does not have any good reason to think that any of these values for the mean are more or less likely, so a uniform prior will be assigned where each value of μ is assigned the same probability $\frac{1}{8}$.

$$\pi(\mu) = \frac{1}{8}, \quad \mu = 15, 16, ..., 22. \tag{8.4}$$

Each value of μ corresponds to a particular normal sampling curve for the time-to-serve measurement. Figure 2.1 displays the eight possible normal sampling curves. Our prior says that each of these eight sampling curves has the same prior probability.

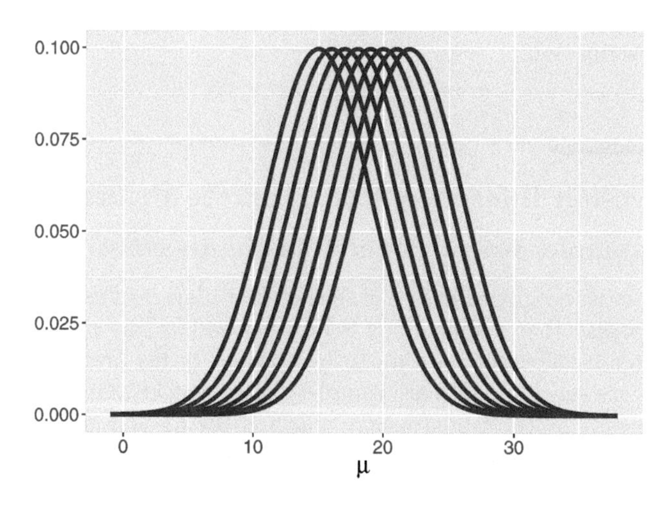

FIGURE 8.2
Eight possible normal sampling curves corresponding to a discrete uniform prior on μ.

To learn more about the mean μ, one collects a single time-to-serve measurement for Federer, and suppose it is 15.1 seconds, that is, one observes $Y = 15.1$. The likelihood function is the normal density of the actual observation y viewed as a function of the mean μ (remember that it was assumed that $\sigma = 4$ was given). By substituting in the observation $y = 15.1$ and the known value of $\sigma = 4$, one writes the likelihood function as

$$L(\mu) = \frac{1}{\sqrt{2\pi}4} \exp\left\{-\frac{1}{2(4)^2}(15.1 - \mu)^2\right\}.$$

For each possible value of μ, we substitute the value into the likelihood expression. For example, the likelihood of $\mu = 15$ is equal to

$$\begin{aligned} L(15) &= \frac{1}{\sqrt{2\pi}(4)} \exp\left(-\frac{1}{2(4)^2}(15.1 - 15)^2\right) \\ &\approx 0.0997. \end{aligned}$$

This calculation is repeated for each of the eight values $\mu = 15, 16, \cdots, 22$, obtaining eight likelihood values.

A discrete prior has been assigned to the list of possible values of μ and one is now able to apply Bayes' rule to obtain the posterior distribution for μ. The posterior probability of the value $\mu = \mu_i$ given the data y for a discrete prior has the form

$$\pi(\mu_i \mid y) = \frac{\pi(\mu_i) \times L(\mu_i)}{\sum_j \pi(\mu_j) \times L(\mu_j)}, \tag{8.5}$$

where $\pi(\mu_i)$ is the prior probability of $\mu = \mu_i$ and $L(\mu_i)$ is the likelihood function evaluated at $\mu = \mu_i$.

If a discrete uniform prior distribution for μ is assigned, one has $\pi(\mu_i) = \frac{1}{8}$ for all $i = 1, \cdots, 8$, and $\pi(\mu_i)$ is canceled out from the numerator and denominator in Equation (8.5). In this case one calculates the likelihood values $L(\mu_i)$ for all $i = 1, \cdots, 8$ and normalizes these values to obtain the posterior probabilities $\pi(\mu_i \mid y)$. Table 8.1 displays the values of μ and the corresponding values of Prior, Likelihood, and Posterior. Readers are encouraged to verify the results shown in the table.

TABLE 8.1
Value, Prior, Likelihood, and Posterior for μ with a single observation.

μ	Prior	Likelihood	Posterior
15	0.125	0.0997	0.1888
16	0.125	0.0972	0.1842
17	0.125	0.0891	0.1688
18	0.125	0.0767	0.1452
19	0.125	0.0620	0.1174
20	0.125	0.0471	0.0892
21	0.125	0.0336	0.0637
22	0.125	0.0225	0.0427

With the single measurement of time-to-serve of $y = 15.1$, one sees from Table 8.1 that the posterior distribution for μ favors values $\mu = 15$, and 16. In fact, the posterior probabilities decrease as a function of μ. The Prior column reminds us that the prior distribution is uniform. Bayesian inference uses the collected data to sharpen one's belief about the unknown parameter from the prior distribution to the posterior distribution. For this single observation, the sample mean is $y = 15.1$ and the μ value closest to the sample mean ($\mu = 15$) is assigned the highest posterior probability.

Typically one collects multiple time-to-serve measurements. Suppose one collects n time-to-serve measurements, denoted as $Y_1, ..., Y_n$, that are normally distributed with mean μ and fixed standard deviation $\sigma = 4$. Each observation follows the same normal density

$$f(y_i) = \frac{1}{\sqrt{2\pi}\sigma} \exp\left\{\frac{-(y_i - \mu)^2}{2\sigma^2}\right\}, -\infty < y_i < \infty. \tag{8.6}$$

Again since $\sigma = 4$ is known, the only parameter in Equation (8.6) is μ and we are interested in learning about this mean parameter μ. Suppose the same discrete uniform prior is used as in Equation (8.4) and graphed in Figure 8.2. The mean μ takes on the values $\{15, 16, \cdots, 22\}$ with each value assigned the same probability of $\frac{1}{8}$.

Suppose one collects a sample of 20 times-to-serve for Federer:

```
15.1 11.8 21.0 22.7 18.6 16.2 11.1 13.2 20.4 19.2
21.2 14.3 18.6 16.8 20.3 19.9 15.0 13.4 19.9 15.3
```

When multiple time-to-serve measurements are taken, the likelihood function is the joint density of the actual observed values $y_1, ..., y_n$ viewed as a function of the mean μ. After some algebra (detailed derivation in Section 8.3.2), one writes the likelihood function as

$$\begin{aligned} L(\mu) &= \prod_{i=1}^{n} \frac{1}{\sqrt{2\pi}\sigma} \exp\left\{-\frac{1}{2\sigma^2}(y_i - \mu)^2\right\} \\ &\propto \exp\left\{-\frac{n}{2\sigma^2}(\bar{y} - \mu)^2\right\} \\ &= \exp\left\{-\frac{20}{2(4)^2}(\bar{y} - \mu)^2\right\}, \end{aligned} \tag{8.7}$$

where we have substituted the known values $n = 20$ and the standard deviation $\sigma = 4$. From our sample, we compute the sample mean $\bar{y} = (15.1 + 11.8 + ... + 15.3)/20 = 17.2$. The value of $\bar{y}$ is substituted into Equation (8.7), and for each possible value of μ, we substitute the value to find the corresponding likelihood. For example, the likelihood of $\mu = 15$ is equal to

$$\begin{aligned} L(15) &= \exp\left\{-\frac{20}{2(4)^2}(17.2 - 15)^2\right\} \\ &\approx 0.022. \end{aligned}$$

This calculation is repeated for each of the eight values $\mu = 15, 16, ..., 22$, obtaining eight likelihood values.

One now applies Bayes' rule to obtain the posterior distribution for μ. The posterior probability of $\mu = \mu_i$ given the sequence of recorded times-to-serve $y_1, \cdots, y_n$ has the form

$$\pi(\mu_i \mid y_1, \cdots, y_n) = \frac{\pi(\mu_i) \times L(\mu_i)}{\sum_j \pi(\mu_j) \times L(\mu_j)}, \tag{8.8}$$

where $\pi(\mu_i)$ is the prior probability of $\mu = \mu_i$ and $L(\mu_i)$ is the likelihood function evaluated at $\mu = \mu_i$. We saw in Equation (8.7) that only the sample mean, $\bar{y}$, is needed in the calculation of the likelihood, so $\bar{y}$ is used in place of $y_1, \cdots, y_n$ in the formula.

With a discrete uniform prior distribution for μ, again one has $\pi(\mu_i) = \frac{1}{8}$ for all $i = 1, \cdots, 8$ and $\pi(\mu_i)$ is canceled out from the numerator and denominator in Equation (8.8). One calculates the posterior probabilities by computing $L(\mu_i)$ for all $i = 1, \cdots, 8$ and normalizing these values. Table 8.2 displays the values of μ and the corresponding values of Prior, Likelihood, and Posterior. Readers are encouraged to verify the results shown in the table.

TABLE 8.2
Value, Prior, Likelihood, and Posterior for μ with n observations.

μ	Prior	Likelihood	Posterior
15	0.125	0.0217	0.0217
16	0.125	0.1813	0.1815
17	0.125	0.4350	0.4353
18	0.125	0.2990	0.2992
19	0.125	0.0589	0.0589
20	0.125	0.0033	0.0033
21	0.125	0.0001	0.0001
22	0.125	0.0000	0.0000

It is helpful to construct a graph (see Figure 8.3) where one contrasts the prior and probability probabilities for the mean time-to-serve μ. While the prior distribution is flat, the posterior distribution for μ favors the values $\mu = 16$, 17, and 18 seconds. Bayesian inference uses the observed data to revise one's belief about the unknown parameter from the prior distribution to the posterior distribution. Recall that the sample mean $\bar{y} = 17.2$ seconds. From Table 8.2 and Figure 8.3 one sees the clear effect of the observed sample mean – μ is likely to be close to the value 17.2.

8.3.2 Simplification of the likelihood

The likelihood function is the joint density of the observations $y_1, ..., y_n$, viewed as a function of the mean μ (since $\sigma = 4$ is given). With n observations being *identically and independently distributed (i.i.d.)* as Normal$(\mu, 4)$, the likelihood function is the product of normal density terms. In the algebra work that will be done shortly, the likelihood, as a function of μ, is found to be normal with mean $\bar{y}$ and standard deviation $\sigma/\sqrt{n}$.

R The calculation of the posterior probabilities is an application of Bayes' rule illustrated in earlier chapters. One creates a data frame of values `mu` and corresponding probabilities `Prior`. One computes the likelihood values in the variable `Likelihood` and the posterior probabilities are found using the `bayesian_crank()` function.

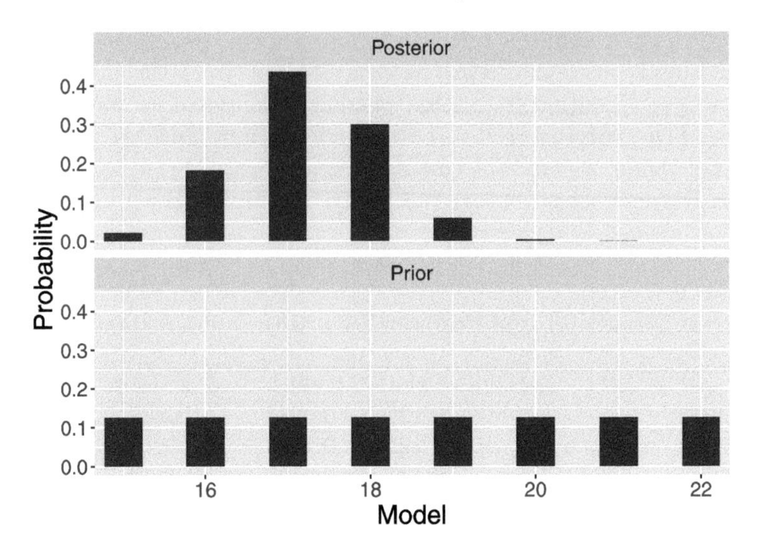

FIGURE 8.3
Prior and posterior probabilities of the normal mean μ with a sample of observations.

```
df <- data.frame(mu = seq(15, 22, 1),
                 Prior = rep(1/8, 8)) %>%
  mutate(Likelihood = dnorm(mu, 17.2, 4 / sqrt(20)))
df <- bayesian_crank(df)
round(df, 4)

  mu Prior Likelihood Product Posterior
1 15 0.125     0.0217  0.0027    0.0217
2 16 0.125     0.1813  0.0227    0.1815
3 17 0.125     0.4350  0.0544    0.4353
4 18 0.125     0.2990  0.0374    0.2992
5 19 0.125     0.0589  0.0074    0.0589
6 20 0.125     0.0033  0.0004    0.0033
7 21 0.125     0.0001  0.0000    0.0001
8 22 0.125     0.0000  0.0000    0.0000
```

Derivation of $L(\mu) \propto \exp\left(-\frac{n}{2\sigma^2}(\bar{y}-\mu)^2\right)$

In the following, we combine the terms in the exponent, expand all of the summation terms, and complete the square to get the result.

$$
\begin{aligned}
L(\mu) &= \prod_{i=1}^{n} \frac{1}{\sqrt{2\pi}\sigma} \exp\left\{-\frac{1}{2\sigma^2}(y_i - \mu)^2\right\} \\
&= \left(\frac{1}{\sqrt{2\pi}\sigma}\right)^n \exp\left\{-\frac{1}{2\sigma^2}\sum_{i=1}^{n}(y_i - \mu)^2\right\} \\
&\propto \exp\left\{-\frac{1}{2\sigma^2}\sum_{i=1}^{n}(y_i^2 - 2\mu y_i + \mu^2)\right\} \\
\text{[expand the } \textstyle\sum \text{ terms]} &= \exp\left\{-\frac{1}{2\sigma^2}\left(\sum_{i=1}^{n} y_i^2 - 2\mu\sum_{i=1}^{n} y_i + n\mu^2\right)\right\} \\
&\propto \exp\left\{-\frac{1}{2\sigma^2}\left(-2\mu\sum_{i=1}^{n} y_i + n\mu^2\right)\right\} \\
\text{[replace } \textstyle\sum \text{ with } n\bar{y}\text{]} &= \exp\left\{-\frac{1}{2\sigma^2}\left(-2n\mu\bar{y} + n\mu^2\right)\right\} \\
\text{[complete the square]} &= \exp\left\{-\frac{n}{2\sigma^2}(\mu^2 - 2\mu\bar{y} + \bar{y}^2) + \frac{n}{2\sigma^2}\bar{y}^2\right\} \\
&\propto \exp\left\{-\frac{n}{2\sigma^2}(\bar{y} - \mu)^2\right\} \qquad (8.9)
\end{aligned}
$$

Sufficient statistic

There are different ways of writing and simplifying the likelihood function. One can choose to keep the product sign and each y_i term, and leave the likelihood function as

$$L(\mu) = \prod_{i=1}^{n} \frac{1}{\sqrt{2\pi}\sigma} \exp\left\{-\frac{1}{2\sigma^2}(y_i - \mu)^2\right\}. \qquad (8.10)$$

Doing so requires one to calculate the individual likelihood from each time-to-serve measurement y_i and multiply these values to obtain the function $L(\mu)$ used to obtain the posterior probability.

If one instead simplifies the likelihood to be

$$L(\mu) \propto \exp\left\{-\frac{n}{2\sigma^2}(\bar{y} - \mu)^2\right\}, \qquad (8.11)$$

all the proportionality constants drop out in the calculation of the posterior probabilities for different values of μ. In the application of Bayes' rule, one only needs to know the number of observations n and the mean time to serve $\bar{y}$ to calculate the posterior. Since the likelihood function depends on the data only through the value $\bar{y}$, the statistic $\bar{y}$ is called a *sufficient statistic* for the mean μ.

8.3.3 Inference: Federer's time-to-serve

What has one learned about Federer's mean time-to-serve from this Bayesian analysis? Our prior said that any of the eight possible values of μ were equally likely with probability 0.125. After observing the sample of 20 measurements, one believes μ is most likely 16, 17, and 18 seconds, with respective probabilities $0.181, 0.425$, and 0.299. In fact, if one adds up the posterior probabilities, one says that μ is in the set $\{16, 17, 18\}$ seconds with probability 0.915.

$$Prob(16 \leq \mu \leq 18) = 0.181 + 0.435 + 0.299 = 0.915$$

This region of values of μ is called a 91.5% posterior probability region for the mean time-to-serve μ.

8.4 Continuous Priors

8.4.1 The normal prior for mean μ

Returning to our example, one is interested in learning about the time-to-serve for the tennis player Roger Federer. His serving times are believed to be normally distributed with unknown mean μ and known standard deviation $\sigma = 4$. The focus is on learning about the mean value μ.

In the prior construction in Section 8.3, we assumed μ was discrete, taking only integer values from 15 to 22. However, the mean time-to-serve μ does not have to be an integer. In fact, it is more realistic to assume μ is continuous-valued. One widely-used approach for representing one's belief about a normal mean is based on a normal prior density with mean μ_0 and standard deviation σ_0, that is

$$\mu \sim \text{Normal}(\mu_0, \sigma_0).$$

There are two parameters for this normal prior: the value μ_0 represents one's best guess at the mean time-to-serve μ and σ_0 indicates how sure one thinks about the guess.

To illustrate the use of different priors for μ, let's consider the opinion of one tennis fan Joe who has strong prior information about the mean. His best guess at Federer's mean time-to-serve is 18 seconds so he lets $\mu_0 = 18$. He is very sure of this guess and so he chooses σ_0 to be the relatively small value of 0.4. In contrast, a second tennis fan Kate also thinks that Federer's mean time-to-serve is 18 seconds, but does not have a strong belief in this guess and chooses the large value 2 of the standard deviation σ_0. Figure 8.4 shows these two normal priors for the mean time-to-serve μ.

Both curves are symmetric and bell-shaped, centered at $\mu_0 = 18$. The main difference is the spread of the two curves: a Normal(8, 0.4) curve is much more

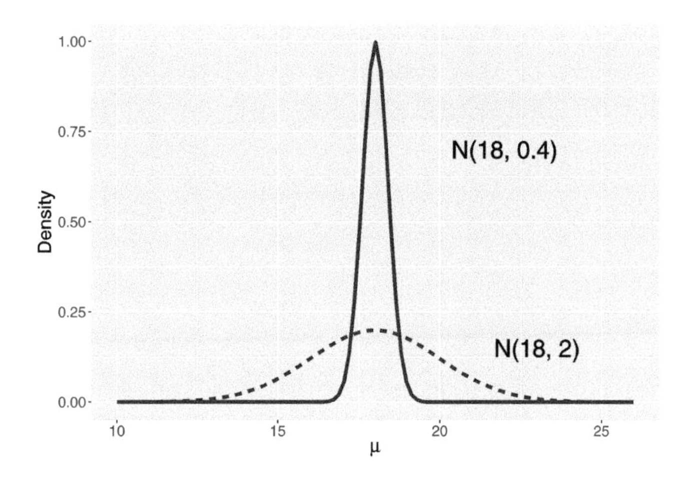

FIGURE 8.4
Two priors for the normal mean μ.

concentrated around the mean $\mu_0 = 18$ compared to the Normal(8, 2) curve. Since the value of the probability density function at a point reflects the probability at that value, the Normal(8, 0.4) prior reflects the belief that the mean time to serve will most likely be around $\mu_0 = 18$ seconds, whereas the Normal(8, 2) prior indicates that the mean μ could be as small as 15 seconds and as large as 20 seconds.

8.4.2 Choosing a normal prior

Informative prior

How does one in practice choose a normal prior for μ that reflects prior beliefs about the location of this parameter? One indirect strategy for selecting values of the prior parameters μ_0 and σ_0 is based on the specification of quantiles. On the basis of one's prior beliefs, one specifies two quantiles of the normal density. Then the normal parameters are found by matching these two quantiles to a particular normal curve.

Recall the definition of a quantile — in this setting it is a value of the mean μ such that the probability of being smaller than that value is a given probability. To construct one's prior for Federer's mean time-to-serve, one thinks first about two quantiles. Suppose one specifies the 0.5 quantile to be 18 seconds — this means that μ is equally likely to be smaller or larger than 18 seconds. Next, one decides that the 0.9 quantile is 20 seconds. This means that one's probability that μ is smaller than 20 seconds is 90%. Given values of these two quantiles, the unique normal curve is found that matches this information.

The matching is performed by the R function `normal.select()`. One inputs two quantiles by `list` statements, and the output is the mean and standard deviation of the normal prior.

```
normal.select(list(p = 0.5, x = 18), list(p = 0.9, x = 20))
$mu
[1] 18

$sigma
[1] 1.560608
```

The normal curve with mean $\mu_0 = 18$ and $\sigma_0 = 1.56$, displayed in Figure 8.5, matches the prior information stated by the two quantiles.

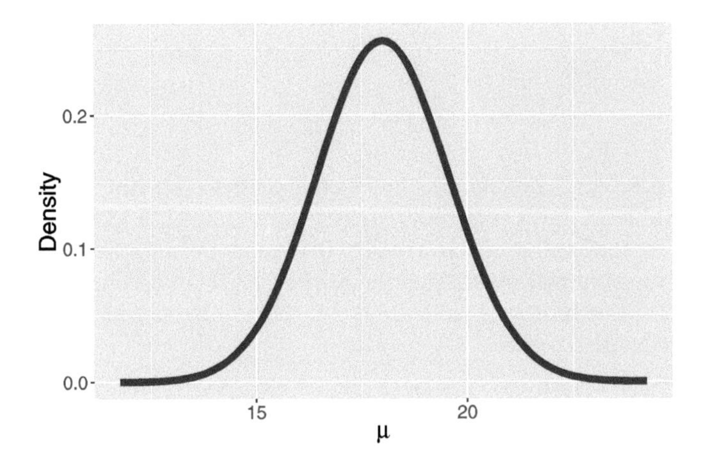

FIGURE 8.5
A person's normal prior for Federer's mean time-to-serve μ.

Since our measurement skills are limited, this prior is just an approximation to our beliefs about μ. We recommend in practice that one perform several checks to see if this normal prior makes sense. Several functions are available to help in this prior checking.

For example, we find the 0.25 quantile of our prior using the `qnorm()` function.

```
qnorm(0.25, 18, 1.56)
[1] 16.9478
```

This prior says that the prior probability that μ is smaller than 16.95 is 25%. If this does not seem reasonable, one would make adjustments in the values of the normal mean and standard deviation until a reasonable normal prior is found.

Weakly informative prior

We have been assuming that we have some information about the mean parameter μ that is represented by a normal prior. What would we do in the situation where little is known about the location on μ? For a normal prior, the standard deviation σ_0 represents the sureness of our belief in our guess μ_0 at the value of the mean. If we are really unsure about any guess at μ, then we can assign the standard deviation σ_0 a large value. Then the choice of the prior mean will not matter, so we suggest using a Normal(0, σ_0) with a large value for σ_0. This prior indicates that μ may plausibly range over a large interval and represents weakly informative prior belief about the parameter.

As will be seen later in this chapter, when a vague prior is chosen, the posterior inference for μ will largely be driven by the data. This behavior is desirable since we know little about the location of μ *a priori* in this situation and we want the data to inform about the location of μ with little influence by the prior.

8.5 Updating the Normal Prior

8.5.1 Introduction

Continuing our discussion on learning about the mean time-to-serve for Roger Federer, the current prior beliefs about Federer's mean time-to-serve μ are represented by a normal curve with mean 18 seconds and standard deviation 1.56 seconds.

Next some data is collected — Federer's time-to-serves are recorded for 20 serves and the sample mean is 17.2 seconds. Recall that we are assuming the population standard deviation $\sigma = 4$ seconds. The likelihood is given by

$$L(\mu) \propto \exp\left\{-\frac{n}{2\sigma^2}(\bar{y}-\mu)^2\right\}, \tag{8.12}$$

and with substitution of the values $\bar{y} = 17.2$, $n = 20$, and $\sigma = 4$, we obtain

$$\begin{aligned} L(\mu) &\propto \exp\left\{-\frac{20}{2(4)^2}(17.2-\mu)^2\right\} \\ &= \exp\left\{-\frac{1}{2(4/\sqrt{20})^2}(\mu-17.2)^2\right\}. \end{aligned} \tag{8.13}$$

Viewing the likelihood as a function of the parameter μ as in Equation (8.13), the likelihood is recognized as a normal density with mean $\bar{y} = 17.2$ and standard deviation $\sigma/\sqrt{n} = 4/\sqrt{20} = 0.89$.

The Bayes' rule calculation is very familiar to the reader — one obtains the posterior density curve by multiplying the normal prior by the likelihood. If one writes down the product of the normal likelihood and the normal prior density and works through some messy algebra, one will discover that the posterior density also has the normal density form.

The normal prior is said to be *conjugate* since the prior and posterior densities come from the same distribution family: normal. To be more specific, suppose the observation has a normal sampling density with unknown mean μ and known standard deviation σ. If one specifies a normal prior for the unknown mean μ with mean μ_0 and standard deviation σ_0, one obtains a normal posterior for μ with updated parameters μ_n and σ_n.

In Section 8.5.2, we provide a quick peak at this posterior updating without worrying about the mathematical derivation and Section 8.5.3 describes the details of the Bayes' rule calculation. Section 8.5.4 looks at the conjugacy more closely and provides some insight on the effects of prior and likelihood on the posterior distribution.

8.5.2 A quick peak at the update procedure

It is convenient to describe the updating procedure by use of a table. In Table 8.3, there are rows corresponding to Prior, Likelihood, and Posterior and columns corresponding to Mean, Precision, and Standard Deviation. The mean and standard deviation of the normal prior are placed in the "Prior" row, and the sample mean and standard error are placed in the "Likelihood" row.

TABLE 8.3
Updating the normal prior: step 1.

Type	Mean	Precision	Stand_Dev
Prior	18.00		1.56
Likelihood	17.20		0.89
Posterior			

We define the *precision*, ϕ, to be the reciprocal of the square of the standard deviation. We compute the precisions of the prior and data from the given standard deviations:

$$\phi_{prior} = \frac{1}{\sigma_0^2} = \frac{1}{1.56^2} = 0.41, \quad \phi_{data} = \frac{1}{\sigma^2/n} = \frac{1}{0.89^2} = 1.21.$$

We enter the precisions in the corresponding rows of Table 8.4.

We will shortly see that the Posterior precision is the sum of the Prior precision and the Likelihood precisions:

$$\phi_{post} = \phi_{prior} + \phi_{data} = 0.41 + 1.26 = 1.67.$$

TABLE 8.4
Updating the normal prior: step 2.

Type	Mean	Precision	Stand_Dev
Prior	18.00	0.41	1.56
Likelihood	17.20	1.26	0.89
Posterior			

Once the posterior precision is computed, the posterior standard deviation is computed as the reciprocal of the square root of the precision.

$$\sigma_n = \frac{1}{\sqrt{\phi_{post}}} = \frac{1}{\sqrt{1.67}} = 0.77.$$

These precisions and standard deviations are entered into Table 8.5.

TABLE 8.5
Updating the normal prior: step 3.

Type	Mean	Precision	Stand_Dev
Prior	18.00	0.41	1.56
Likelihood	17.20	1.26	0.89
Posterior		1.67	0.77

The posterior mean is a weighted average of the Prior and Likelihood means where the weights are given by the corresponding precisions. That is, the formula is given by

$$\mu_n = \frac{\phi_{prior} \times \mu_0 + \phi_{data} \times \bar{y}}{\phi_{prior} + \phi_{data}}. \tag{8.14}$$

By making appropriate substitutions, we obtain the posterior mean:

$$\mu_n = \frac{0.41 \times 18.00 + 1.26 \times 17.20}{0.41 + 1.26} = 17.40.$$

The posterior density is normal with mean 17.40 seconds and standard deviation 0.77 seconds. See Table 8.6 for the final update step.

R The normal updating is performed by the R function `normal_update()`. One inputs two vectors – `prior` is a vector of the prior mean and standard deviation and `data` is a vector of the sample mean and standard error. The output is a vector of the posterior mean and posterior standard deviation.

TABLE 8.6
Updating the normal prior: step 4.

Type	Mean	Precision	Stand_Dev
Prior	18.00	0.41	1.56
Likelihood	17.20	1.26	0.89
Posterior	17.40	1.67	0.77

```
prior <- c(18, 1.56)
data <- c(17.20, 0.89)
normal_update(prior, data)

[1] 17.3964473  0.7730412
```

The prior and posterior densities are displayed in Figure 8.6. As usually the case, the posterior density has a smaller spread since the posterior has more information than the prior about Federer's mean time-to-serve. More information about a parameter indicates less uncertainty and a smaller spread of the posterior density. In the process from prior to posterior, one sees how the data modifies one's initial belief about the parameter μ.

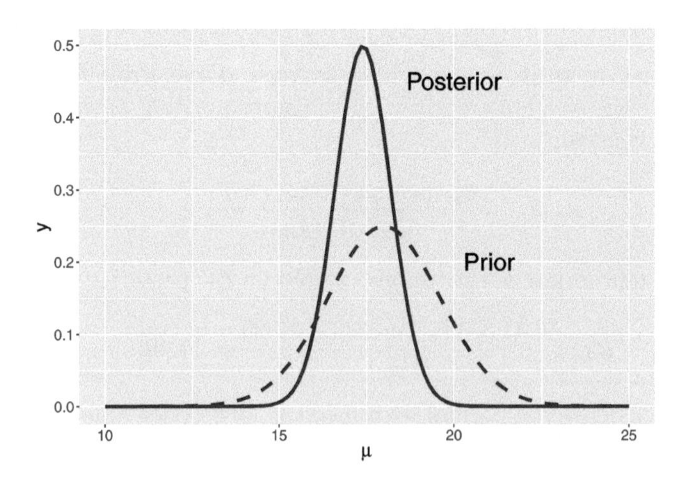

FIGURE 8.6
Prior and posterior curves for Federer's mean time-to-serve μ.

8.5.3 Bayes' rule calculation

Section 8.5.2 gave an overview of the updating procedure for a normal prior and normal sampling. In this section we explain (1) why it is preferable to work with the precisions instead of the standard deviations; (2) why the precisions act as the weights in the calculation of the posterior mean and (3) why the posterior is a normal distribution.

Recall a precision is the reciprocal of the square of the standard deviation. We use $\phi = \frac{1}{\sigma^2}$ to represent the precision of a single observation in the normal likelihood, and $\phi_0 = \frac{1}{\sigma_0^2}$ to represent the precision in the normal prior.

- We write down the likelihood of μ, combining terms, and writing the expression in terms of the precision ϕ.

$$
\begin{aligned}
y_1, \cdots, y_n \mid \mu, \sigma &\overset{i.i.d.}{\sim} \text{Normal}(\mu, \sigma) && (8.15) \\
L(\mu) = f(y_1, \cdots, y_n \mid \mu, \sigma) &= \prod_{i=1}^{n} \frac{1}{\sqrt{2\pi}\sigma} \exp\left\{-\frac{1}{2\sigma^2}(y_i - \mu)^2\right\} \\
&= \prod_{i=1}^{n} \frac{1}{\sqrt{2\pi}} \phi^{\frac{1}{2}} \exp\left\{-\frac{\phi}{2}(y_i - \mu)^2)\right\} \\
&= \left(\frac{1}{\sqrt{2\pi}}\right)^n \phi^{\frac{n}{2}} \exp\left\{-\frac{\phi}{2}\sum_{i=1}^{n}(y_i - \mu)^2)\right\} \\
& && (8.16)
\end{aligned}
$$

Note that σ is assumed known, therefore the likelihood function is only in terms of μ, i.e. $L(\mu)$.

- In similar fashion, we write down the prior density for μ including the prior precision ϕ_0.

$$
\begin{aligned}
\mu &\sim \text{Normal}(\mu_0, \sigma_0) && (8.17) \\
\pi(\mu) &= \frac{1}{\sqrt{2\pi}\sigma_0} \exp\left\{-\frac{1}{2\sigma_0^2}(\mu - \mu_0)^2)\right\} \\
&= \frac{1}{\sqrt{2\pi}} \phi_0^{\frac{1}{2}} \exp\left\{-\frac{\phi_0}{2}(\mu - \mu_0)^2\right\} && (8.18)
\end{aligned}
$$

- Bayes' rule is applied by multiplying the prior by the likelihood to obtain the posterior. In deriving the posterior of μ, the manipulations require careful consideration regarding what is known. The only unknown variable is μ, so any "constants" or known quantities not depending on μ can be added or dropped with the proportionality sign "$\propto$".

$$\begin{aligned}
\pi(\mu \mid y_1, \cdots, y_n, \sigma) &\propto \pi(\mu)L(\mu) \\
&\propto \exp\left\{-\frac{\phi_0}{2}(\mu - \mu_0)^2\right\} \times \exp\left\{-\frac{n\phi}{2}(\mu - \bar{y})^2\right\} \\
&\propto \exp\left\{-\frac{1}{2}(\phi_0 + n\phi)\mu^2 + \frac{1}{2}(2\mu_0\phi_0 + 2n\phi\bar{y})\mu\right\} \\
\texttt{[complete the square]} &\propto \exp\left\{-\frac{1}{2}(\phi_0 + n\phi)(\mu - \frac{\phi_0\mu_0 + n\phi\bar{y}}{\phi_0 + n\phi})^2\right\}
\end{aligned} \tag{8.19}$$

Looking closely at the final expression, one recognizes that the posterior for μ is a normal density with mean and precision parameters. Specifically we recognize $(\phi_0 + n\phi)$ as the posterior precision and $(\frac{\phi_0\mu_0+n\phi\bar{y}}{\phi_0+n\phi})$ as the posterior mean. Summarizing, we have derived the following posterior distribution of μ,

$$\mu \mid y_1, \cdots, y_n, \sigma \sim \text{Normal}\left(\frac{\phi_0\mu_0 + n\phi\bar{y}}{\phi_0 + n\phi}, \sqrt{\frac{1}{\phi_0 + n\phi}}\right). \tag{8.20}$$

In passing, it should be noted that the same result would be attained using the standard deviations, σ and σ_0, instead of the precisions, ϕ and ϕ_0. It is preferable to work with the precisions due to the relative simplicity of the notation. In particular, one sees in Table 8.5 that the posterior precision is the sum of the prior and likelihood precisions, that is, the posterior precision $\phi_n = \phi_0 + n\phi$.

8.5.4 Conjugate normal prior

Let's summarize our calculations in Section 8.5.3. We collect a sequence of continuous observations that are assumed identically and independently distributed as Normal(μ, σ), and a normal prior is assigned to the mean parameter μ.

- The sampling model:

$$Y_1, \cdots, Y_n \mid \mu, \sigma \overset{i.i.d.}{\sim} \text{Normal}(\mu, \sigma) \tag{8.21}$$

  When σ (or ϕ) is known, and mean μ is the only parameter in the likelihood.

- The prior distribution:

$$\mu \sim \text{Normal}(\mu_0, \sigma_0) \tag{8.22}$$

- After $Y_1 = y_1, ..., Y_n = y_n$ are observed, the posterior distribution for the mean μ is another normal distribution with mean $\frac{\phi_0\mu_0 + n\phi\bar{y}}{\phi_0 + n\phi}$ and precision $\phi_0 + n\phi$ (thus standard deviation $\sqrt{\frac{1}{\phi_0+n\phi}}$):

$$\mu \mid y_1, \cdots, y_n, \sigma \sim \text{Normal}\left(\frac{\phi_0\mu_0 + n\phi\bar{y}}{\phi_0 + n\phi}, \sqrt{\frac{1}{\phi_0 + n\phi}}\right). \tag{8.23}$$

In this situation where the sampling standard deviation σ is known, the normal density is a conjugate prior for the mean of a normal distribution, as the posterior distribution for μ is another normal density with updated parameters. Conjugacy is a convenient property as the posterior distribution for μ has a convenient functional form. Conjugacy allows one to conduct Bayesian inference through exact analytical solutions and simulation. Also conjugacy provides insight on how the data and prior are combined in the posterior distribution.

The posterior compromises between the prior and the sample

Recall that Bayesian inference is a general approach where one initializes a prior belief for an unknown quantity, collects data expressed through a likelihood function, and combines prior and likelihood to give an updated belief for the unknown quantity. In Chapter 7, we have seen how the posterior mean of a proportion is a compromise between the prior mean and sample proportion (refer to Section 7.4.2 as needed). In the current normal mean case, the posterior mean is similarly viewed as an estimate that compromises between the prior mean and sample mean. One rewrites the posterior mean in Equation (8.23) as follows:

$$\mu_n = \frac{\phi_0\mu_0 + n\phi\bar{y}}{\phi_0 + n\phi} = \frac{\phi_0}{\phi_0 + n\phi}\mu_0 + \frac{n\phi}{\phi_0 + n\phi}\bar{y}. \tag{8.24}$$

The prior precision is equal to ϕ_0 and the precision in the likelihood for any y_i is ϕ. Since there are n observations, the precision in the joint likelihood is $n\phi$. The posterior mean is a weighted average of the prior mean μ_0 and sample mean $\bar{y}$ where the weights are proportional to the associated precisions.

The posterior accumulates information in the prior and the sample

In addition, the precision of the posterior mean is the sum of the precisions of the prior and likelihood. That is,

$$\phi_n = \phi_0 + n\phi. \tag{8.25}$$

The implication is that the posterior standard deviation will always be smaller than either the prior standard deviation or the sampling standard error:

$$\sigma_n < \sigma_0, \;\; \sigma_n < \frac{\sigma}{\sqrt{n}}.$$

8.6 Bayesian Inferences for Continuous Normal Mean

Continuing with the example about Federer's time-to-serve, our normal prior had mean 18 seconds and standard deviation 1.56 seconds. After collecting 20 time-to-serve measurements with a sample mean of 17.2, the posterior distribution Normal$(17.4, 0.77)$ reflects our opinion about the mean time-to-serve.

Bayesian inferences about the mean μ are based on various summaries of this posterior normal distribution. Because the exact posterior distribution of mean μ is normal, it is convenient to use R functions such as `pnorm()` and `qnorm()` to conduct Bayesian hypothesis testing and construct Bayesian credible intervals. Simulation-based methods utilizing functions such as `rnorm()` are also useful to provide approximations to those inferences. A sequence of examples are given in Section 8.6.1.

Predictions of future data are also of interest. For example, one might want to predict the next time-to-serve measurement based on the posterior distribution of μ being Normal$(17.4, 0.77)$. In Section 8.6.2, details of the prediction procedure and examples are provided.

8.6.1 Bayesian hypothesis testing and credible interval

A testing problem

In a *testing* problem, one is interested in checking the validity of a statement about a population quantity. In our tennis example, suppose someone says that Federer takes on average at least 19 seconds to serve. Is this a reasonable statement?

R The current beliefs about Federer's mean time-to-serve are summarized by a normal distribution with mean 17.4 seconds and standard deviation 0.77 seconds. To assess if the statement "μ is 19 seconds or more" is reasonable, one simply computes its posterior probability, $Prob(\mu \geq 19 \mid \mu_n = 17.4, \sigma_n = 0.77)$.

```
1 - pnorm(19, 17.4, 0.77)
[1] 0.01885827
```

This probability is about 0.019, a small value, so one would conclude that this person's statement is unlikely to be true.

This is the exact solution using the `pnorm()` function with mean 17.4 and standard deviation 0.77. As seen in Chapter 7, simulation provides an alternative approach to obtaining the probability $Prob(\mu \geq 19 \mid \mu_n = 17.4, \sigma_n = 0.77)$. To implement the simulation approach, recall that one generates a large number of values from the posterior distribution and summarizes this simulated sample. In particular, using the following R script, one generates 1000 values from the Normal(17.4, 0.77) distribution and approximates the probability of "μ is 19 seconds or more" by computing the percentage of values that falls above 19.

```
S <- 1000
NormalSamples <- rnorm(S, 17.4, 0.77)
sum(NormalSamples >= 19) / S
[1] 0.024
```

The reader might notice that the approximated value of 0.024 differs from the exact answer of 0.019 using the `pnorm()` function. One way to improve the accuracy of the approximation is by increasing the number of simulated values. For example, increasing S from 1000 to 10,000 provides a better approximation to the exact probability 0.019.

```
S <- 10000
NormalSamples <- rnorm(S, 17.4, 0.77)
sum(NormalSamples >= 19) / S
[1] 0.0175
```

A Bayesian interval estimate

Bayesian credible intervals for the mean parameter μ can be achieved both by exact calculation and simulation. Recall that a Bayesian credible interval is an interval that contains the unknown parameter with a certain probability content. For example, a 90% Bayesian credible interval for the parameter μ is an interval containing μ with a probability of 0.90.

R The exact interval is obtained by using the R function `qnorm()`. For example, with the posterior distribution for μ being Normal(17.4, 0.77), the following R script shows that a 90% central Bayesian credible interval is (16.133, 18.667). That is, the posterior probability of μ falls between 16.133 and 18.667 is exactly 90%.

```
qnorm(c(0.05, 0.95), 17.4, 0.77)
[1] 16.13346 18.66654
```

For simulation-based inference, one generates a large number of values from its posterior distribution, then finds the 5th and 95th sample quantiles to obtain the middle 90% of the generated values. Below one sees that a 90% credible interval for posterior of μ is approximately (16.151, 18.691).

```
S <- 1000
NormalSamples <- rnorm(S, 17.4, 0.77)
quantile(NormalSamples, c(0.05, 0.95))
      5%      95%
16.15061 18.69062
```

The Bayesian credible intervals can also be used for testing hypothesis. Suppose one again wants to evaluate the statement " Federer takes on average at least 19 seconds to serve." One answers this question by computing the 90% credible interval. One notes that the values of μ "at least 19" are not included in the exact 90% credible interval (16.15, 18.69). The interpretation is that the probability is at least 0.90 that Federer's average time-to-service is smaller than 19 seconds. One could obtain a wider credible interval, say by computing a central 95% credible interval (see the R output below), and observe that 19 is out of the interval. This indicates we are 95% confident that 19 seconds is not the value of Federer's average time-to-serve.

```
qnorm(c(0.025, 0.975), 17.4, 0.77)
[1] 15.89083 18.90917
```

On the basis of this credible interval calculation, one concludes that the statement about Federer's time-to-serve is unlikely to be true. This conclusion is consistent with the typical Bayesian hypothesis testing procedure given at the beginning of this section.

8.6.2 Bayesian prediction

Suppose one is interested in predicting Federer's future time-to-serve. Since one has already updated the belief about the parameter, the mean μ, the prediction is made based on its posterior predictive distribution.

How to make one future prediction of Federer's time-to-serve? In Chapter 7, we have seen two different approaches for predicting of a new survey outcome of students' dining preferences. One approach in Chapter 7 is based on the derivation of the exact posterior predictive distribution $f(\tilde{Y} = \tilde{y} \mid Y = y)$ which was shown to be a beta-binomial distribution. The second approach is a simulation-based approach, which involves two steps: first, sample a value of the parameter from its posterior distribution (a beta distribution), and second, sample a prediction from the data model based on the sampled parameter draw (a binomial distribution). When the sample size in the simulation-based approach is sufficiently large, a prediction interval from the simulation-based approach is an accurate approximation to the exact prediction interval.

Exact predictive distribution

We first describe the exact posterior predictive distribution. Consider making a prediction of a single Federer's time-to-serve $\tilde{Y}$. In general, suppose the sampling density of $\tilde{Y}$ given μ and σ is $f(\tilde{Y} = \tilde{y} \mid \mu)$ and suppose the current beliefs about μ are represented by the density $\pi(\mu)$. The joint density of $(\tilde{y}, \mu)$ is given by the product

$$f(\tilde{Y} = \tilde{y}, \mu) = f(\tilde{Y} = \tilde{y} \mid \mu)\pi(\mu), \tag{8.26}$$

and by integrating out μ, the predictive density of $\tilde{Y}$ is given by

$$f(\tilde{Y} = \tilde{y}) = \int f(\tilde{Y} = \tilde{y} \mid \mu)\pi(\mu)d\mu. \tag{8.27}$$

The computation of the predictive density is possible for this normal sampling model with a normal prior. It is assumed that $f(\tilde{Y} = \tilde{y} \mid \mu)$ is normal with mean μ and standard deviation σ and that the current beliefs about μ are described by a normal density with mean μ_0 and standard deviation σ_0. Then it is possible to integrate out μ from the joint density of $(\tilde{y}, \mu)$ and one finds that the predictive density for $\tilde{Y}$ is normal with mean and standard deviation given by

$$E(\tilde{Y}) = \mu_0, \; SD(\tilde{Y}) = \sqrt{\sigma^2 + \sigma_0^2}. \tag{8.28}$$

This result can be used to derive the posterior predictive distribution of $f(\tilde{Y} = \tilde{y} \mid Y_1, \cdots, Y_n)$, where $\tilde{Y}$ is a future observation and $Y_1, \cdots, Y_n$ are n *i.i.d.* observations from a normal sampling density with unknown mean μ and known standard deviation σ. After observing the sample values $y_1, \cdots, y_n$, the current beliefs about the mean μ are represented by a Normal(μ_n, σ_n) density, where the mean and standard deviation are given by

$$\mu_n = \frac{\phi_0\mu_0 + n\phi\bar{y}}{\phi_0 + n\phi}, \sigma_n = \sqrt{\frac{1}{\phi_0 + n\phi}}. \tag{8.29}$$

Then by applying our general result in Equation (8.28), the posterior predictive density of the single future observation $\tilde{Y}$ is normal with mean μ_n and standard deviation $\sqrt{\sigma^2 + \sigma_n^2}$. That is,

$$\tilde{Y} = \tilde{y} \mid y_1, \cdots, y_n, \sigma \sim \text{Normal}(\mu_n, \sqrt{\sigma^2 + \sigma_n^2}). \tag{8.30}$$

An important aspect of the predictive distribution for $\tilde{Y}$ is on the variance term $\sigma^2 + \sigma_n^2$. The variability of a future prediction comes from two sources: (1) the data model variance σ^2, and (2) the posterior variance σ_n^2. Recall that the posterior variance $\sigma_n^2 = \frac{1}{\phi_0 + n\phi}$. If one fixes values of ϕ_0 and ϕ and allows the sample size n to grow, the posterior variance will approach zero. In this "large n" case, the uncertainty in inference about the population mean μ will decrease – essentially we are certain about the location of μ. However the uncertainty in prediction will not decrease towards zero. In contrast, in this large sample case, the variance of $\tilde{Y}$ will decrease and approach the sampling variance σ^2.

Predictions by simulation

The alternative method of computing the predictive distribution is by simulation. In this setting, there are two unknowns – the mean parameter μ and the future observation $\tilde{Y}$. One simulates a value from the predictive distribution in two steps: first, one simulates a value of the parameter μ from its posterior distribution; second, use this simulated parameter draw to simulate a future observation $\tilde{Y}$ from the data model. In particular, the following algorithm is used to simulate a single value from the posterior predictive distribution.

1. Sample a value of μ from its posterior distribution

$$\mu \sim \text{Normal}\left(\frac{\phi_0\mu_0 + n\phi\bar{y}}{\phi_0 + n\phi}, \sqrt{\frac{1}{\phi_0 + n\phi}}\right), \tag{8.31}$$

2. Sample a new observation $\tilde{Y}$ from the data model (i.e. a prediction)

$$\tilde{Y} \sim \text{Normal}(\mu, \sigma). \tag{8.32}$$

R This two-step procedure is implemented for our time-to-serve example using the following R script.

```
sigma <- 4
mu_n <- 17.4
sigma_n <- 0.77
pred_mu_sim <- rnorm(1, mu_n, sigma_n)
(pred_y_sim <- rnorm(1, pred_mu_sim, sigma))
[1] 16.04772
```

The script can easily be updated to create $S = 1000$ predictions, which is helpful to make summary about predictions.

```
S <- 1000
pred_mu_sim <- rnorm(S, mu_n, sigma_n)
pred_y_sim <- rnorm(S, pred_mu_sim, sigma)
```

The vector `pred_y_sim` contains 1000 predictions of Federer's time-to-serve.

To evaluate the accuracy of the simulation-based predictions, Figure 8.7 displays the exact and a density estimate of the simulation-based predictive densities for a single time-to-serve measurement. One observes pretty good agreement using these two computation methods in this example.

8.7 Posterior Predictive Checking

In Section 8.6, the use of the posterior predictive distribution for predicting a future time-to-serve measurement was described. As discussed in Chapter 7, this distribution is also helpful for assessing the suitability of the Bayesian model.

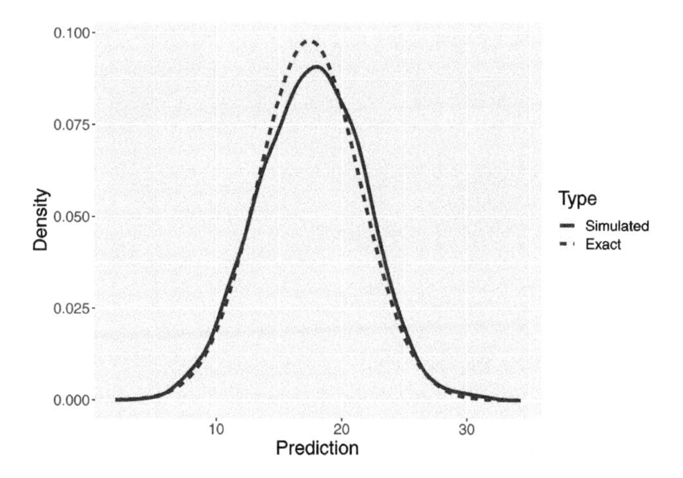

FIGURE 8.7
Display of the exact and simulated predictive time-to-serve for Federer's example.

In our example, we observed 20 times-to-serve for Federer. The question is whether these observed times are consistent with replicated data from the posterior predictive distribution. In this setting, replicated refers to the same sample size as our original sample. In other words, if one takes samples of 20 from the posterior predictive distribution, do these replicated datasets resemble the observed sample?

Since the population standard deviation is known as $\sigma = 4$ seconds, the sampling distribution of Y is normal with mean μ and standard deviation σ. One simulates replicated data $\tilde{Y}_1, ..., \tilde{Y}_{20}$ from the posterior predictive distribution in two steps:

1. Sample a value of μ from its posterior distribution

$$\mu \sim \text{Normal}\left(\frac{\phi_0\mu_0 + n\phi\bar{y}}{\phi_0 + n\phi}, \sqrt{\frac{1}{\phi_0 + n\phi}}\right). \tag{8.33}$$

2. Sample $\tilde{Y}_1, ..., \tilde{Y}_{20}$ from the data model

$$\tilde{Y} \sim \text{Normal}(\mu, \sigma). \tag{8.34}$$

R This method is implemented in the following R script to simulate 1000 replicated samples from the posterior predictive distribution. The vector `pred_mu_sim` contains draws from the posterior distribution and the matrix `ytilde` contains the simulated predictions where each row of the matrix is a simulated sample of 20 future times.

```
sigma <- 4
mu_n <- 17.4
sigma_n <- 0.77
S <- 1000
pred_mu_sim <- rnorm(S, mu_n, sigma_n)
sim_ytilde <- function(j){
  rnorm(20, pred_mu_sim[j], sigma)
}
ytilde <- t(sapply(1:S, sim_ytilde))
```

To judge goodness of fit, we wish to compare these simulated replicated datasets from the posterior predictive distribution with the observed data. One convenient way to implement this comparison is to compute some "testing function", $T(\tilde{y})$, on each replicated dataset. If we have 1000 replicated datasets, one has 1000 values of the testing function. One constructs a graph of these values and overlays the value of the testing function on the observed data $T(y)$. If the observed value is in the tail of the posterior predictive distribution of $T(\tilde{y})$, this indicates some misfit of the observed data with the Bayesian model.

To implement this procedure, one needs to choose a testing function $T(\tilde{y})$. Suppose, for example, one decides to use the sample mean $T(\tilde{y}) = \sum \tilde{y}_j / 20$. In the R script, we compute the sample mean on each row of the simulated prediction matrix.

```
pred_ybar_sim <- apply(ytilde, 1, mean)
```

Figure 8.8 displays a density estimate of the simulated values from the posterior predictive distribution of $\bar{Y}$ and the observed value of the sample mean $\bar{Y} = 17.20$ is displayed as a vertical line. Since this observed mean is in the middle of this distribution, one concludes that this observation is consistent with samples predicted from the Bayesian model. It should be noted that this conclusion about model fit is sensitive to the choice of checking function $T()$. In the end-of-chapter exercises, the reader will explore the suitability of this model using alternative choices for the checking function.

8.8 Modeling Count Data

To further illustrate the Bayesian approach to inference for measurements, consider Poisson sampling, a popular model for count data. One assumes that one observes a random sample from a Poisson distribution with an unknown rate parameter λ. The conjugate prior for the Poisson mean is the gamma distribution. This scenario provides further practice in various

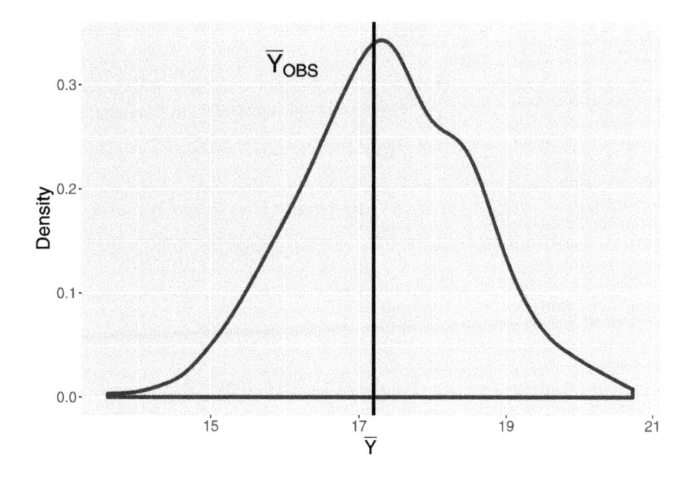

FIGURE 8.8
Display of the posterior predictive mean time-to-serve for twenty observations. The observed mean time-to-serve value is displayed by a vertical line.

Bayesian computations, such as computing the likelihood function and posterior distribution, and obtaining the predictive distribution to learn about future data. In this section, we focus on the main results and the detailed derivations are left as end-of-chapter exercises.

8.8.1 Examples

Counts of patients in an emergency room

A hospital wants to determine how many doctors and nurses to assign on its emergency room (ER) team between 10 pm and 11 pm during the week. An important piece of information is the count of patients arriving in the ER in this one-hour period.

For a count measurement variable such as the count of patients, a popular sampling model is the Poisson distribution. This distribution is used to model the number of times an event occurs in an interval of time or space. In the current example, the event is a patient's arrival to the ER, and the time interval is the period between 10 pm and 11 pm. The hospital wishes to learn about the average count of patients arriving to the ER each hour. Perhaps more importantly, the hospital wants to predict the patient count since that will directly address the scheduling of doctors and nurses question.

Counts of visitors to a website

As a second example, suppose one is interested in monitoring the popularity of a particular blog focusing on baseball analytics. Table 8.7 displays the number of visitors viewing this blog for 28 days during June of 2019. In this setting, the event of interest is a visit to the blog website and the time interval is a single day. The blog author is particularly interested in learning about the average number of visitors during the days Monday through Friday and predicting the number of visits for a future day in the summer of 2019.

TABLE 8.7
Count of visitors to blog during 28 days in June 2019.

	Fri	Sat	Sun	Mon	Tue	Wed	Thu
Week 1	95	81	85	100	111	130	113
Week 2	92	65	78	96	118	120	104
Week 3	91	91	79	106	91	114	110
Week 4	98	61	84	96	126	119	90

8.8.2 The Poisson distribution

Let the random variable Y denote the number of occurrences of an event in an interval with sample space $\{0, 1, 2, \cdots\}$. In contrast to the normally distributed continuous measurement, note that Y only takes integer values from 0 to infinity. The variable Y follows a Poisson distribution with rate parameter λ when the probability mass function (pmf) of observing y events in an interval is given by

$$f(Y = y \mid \lambda) = e^{-\lambda}\frac{\lambda^y}{y!}, \quad y = 0, 1, 2, ... \tag{8.35}$$

where λ is the average number of events per interval, $e = 2.71828...$ is Euler's number, and $y!$ is the factorial of y.

The Poisson sampling model is based on several assumptions about the sampling process. One assumes that the time interval is fixed, counts of arrivals occurring during different time intervals are independent, and the rate λ at which the arrivals occur is constant over time. To check the suitability of the Poisson distribution for the examples, one needs to check the conditions individually.

1. The time interval is fixed in the ER example as we observe patient arrivals during a one hour period between 10 pm and 11 pm. For the blog visits example, the fixed time period is one day.

2. In both examples, one assumes that events occur independently during different time intervals. In the ER example it is reasonable to assume that the time of one patient's arrival does not influence the time of another patient's arrival. For the website visits example, if different people are visiting the website on different days, then the number of visits in a single day would be independent of the number of visits on another day.

3. Is it reasonable to assume the rate λ at which events occur is constant through the time interval? In the ER example, one might not think that the rate of patient arrivals would change much through one hour during the evening, so it seems reasonable to assume that the average number of events is constant in the fixed interval. Similarly, if one focuses on weekdays, then for the website visits example, it is reasonable to assume that the average number of visits remains constant across days.

In some situations, the second and third conditions will be violated. In our ER example, the occurrence of serious accidents may bring multiple groups of patients to the ER at certain time intervals. In this case, arrival times of patients may not be independent and the arrival rate λ in one subinterval will be higher than the arrival rate of another subinterval. When such situations occur, one needs to decide about the severity of the violation of the conditions and possibly use an alternative sampling model instead of the Poisson.

As evident in Equation (8.35), the Poisson distribution has only one parameter, the rate parameter λ, so the Poisson sampling model belongs to the family of one-parameter sampling models. The binomial data model with success probability p and the normal data model with mean parameter μ (with known standard deviation) are two other examples of one-parameter models. One distinguishes these models by the type of possible sample values, discrete or continuous. The binomial random variable is the number of successes and the Poisson random variable is a count of arrivals, so they both are discrete one-parameter models. In contrast, the normal sampling data model is a continuous one-parameter model.

8.8.3 Bayesian inferences

The reader should be familiar with the typical procedure of Bayesian inference and prediction for one-parameter models. We rewrite this procedure in the context of the Poisson sampling model.

Step 1 One constructs a prior expressing an opinion about the location of the rate λ before any data is collected.

Step 2 One takes the sample of intervals and records the number of arrivals in each interval. From this data, one forms the likelihood, the probability of these observations expressed as a function of λ.

Step 3 One uses Bayes' rule to compute the posterior – this distribution updates the prior opinion about λ given the information from the data.

In addition, one computes the predictive distribution to learn about the number of arrivals in future intervals. The posterior predictive distribution is also useful in checking the appropriateness of our model.

Gamma prior distribution

One begins by constructing a prior density to express one's opinion about the rate parameter λ. Since the rate is a positive continuous parameter, one needs to construct a prior density that places its support only on positive values. The convenient choice of prior distributions for Poisson sampling is the gamma distribution which has a density function given by

$$\pi(\lambda \mid \alpha, \beta) = \frac{\beta^\alpha}{\Gamma(\alpha)}\lambda^{\alpha-1}e^{-\beta\lambda}, \text{ for } \lambda > 0, \text{ and } \alpha, \beta > 0, \tag{8.36}$$

where $\Gamma(\alpha)$ is the gamma function evaluated at α. The gamma density is a continuous density where the support is on positive values. It depends on two parameters, a positive shape parameter α and a positive rate parameter β.

The gamma density is a flexible family of distributions that can reflect many different types of prior beliefs about the location of the parameter λ. One chooses values of the shape α and the rate β so that the gamma density matches one's prior information about the location of λ. In R, the function `dgamma()` gives the density, `pgamma()` gives the distribution function and `qgamma()` gives the quantile function for the gamma distribution. These functions are helpful in graphing the prior and choosing values of the shape and rate parameters that match prior statements about gamma percentiles and probabilities. We provide an illustration of choosing a subjective gamma prior in the example.

Sampling and the likelihood

Suppose that $Y_1, ..., Y_n$ represent the observed counts in n time intervals where the counts are independent and each Y_i follows a Poisson distribution with rate λ. The joint mass function of $Y_1, ..., Y_n$ is obtained by multiplying the Poisson densities.

$$\begin{aligned} f(Y_1 = y_1, ..., Y_n = y_n \mid \lambda) &= \prod_{i=1}^{n} f(y_i \mid \lambda) \\ &\propto \lambda^{\sum_{i=1}^{n} y_i} e^{-n\lambda}. \end{aligned} \tag{8.37}$$

Once the counts $y_1, ..., y_n$ are observed, the likelihood of λ is the joint probability of observing this data, viewed as a function of the rate parameter λ.

$$L(\lambda) = \lambda^{\sum_{i=1}^{n} y_i} e^{-n\lambda}. \tag{8.38}$$

The Gamma posterior

If the rate parameter λ in the Poisson sampling model follows a gamma prior distribution, then it turns out that the posterior distribution for λ will also have a gamma density with updated parameters. This demonstrates that the gamma density is the conjugate distribution for Poisson sampling as the prior and posterior densities both come from the same family of distribution: gamma.

We begin by assuming that the Poisson parameter λ has a gamma distribution with shape and rate parameters α and β, that is, $\lambda \sim \textrm{Gamma}(\alpha, \beta)$. If one multiplies the gamma prior by the likelihood function $L(\lambda)$, then in an end-of-chapter exercise you will show that the posterior density of λ is $\textrm{Gamma}(\alpha_n, \beta_n)$, where the updated parameters α_n and β_n are given by

$$\alpha_n = \alpha + \sum_{i=1}^{n} y_i, \;\; \beta_n = \beta + n. \tag{8.39}$$

Inference about λ

Once the posterior distribution has been derived, then all inferences about the Poisson parameter λ are performed by computing particular summaries of the gamma posterior distribution. In particular, one may be interested in testing if λ falls in a particular region by computing a posterior probability. All of these computations are facilitated using the `pgamma()`, `qgamma()`, and `rgamma()` functions. Or one may be interested in constructing an interval estimate for λ. In the end-of-chapter exercises, there are opportunities to perform these inferences using a dataset containing a sample of ER arrival counts.

Prediction of future data

One advantage of using a conjugate prior is that the predictive density for a future observation $\tilde{Y}$ is available in closed form. Suppose λ is assigned a $\textrm{Gamma}(\alpha, \beta)$ prior. Then the prior predictive density of $\tilde{Y}$ is given by

$$\begin{aligned} f(\tilde{Y} = \tilde{y}) &= \int f(\tilde{Y} = \tilde{y} \mid \lambda)\pi(\lambda)\lambda \\ &= \int \frac{e^{-\lambda}\lambda^{\tilde{y}}}{\tilde{y}!} \frac{\beta^{\alpha}}{\Gamma(\alpha)} \lambda^{\alpha-1} e^{-\beta\lambda} d\lambda \\ &= \frac{\Gamma(\alpha + \tilde{y})}{\Gamma(\alpha)} \frac{\beta^{\alpha}}{(\beta+1)^{\tilde{y}+\alpha}}. \end{aligned} \tag{8.40}$$

In addition, the posterior distribution of λ also has the gamma form with updated parameters α_n and β_n. So Equation (8.40) also provides the posterior predictive distribution for a future count $\tilde{Y}$ using the updated parameter values.

For prediction purposes, there are several ways of summarizing the predictive distribution. One can use the formula in Equation (8.40) to directly compute $f(\tilde{Y})$ for a list of values of $\tilde{Y}$ and then one uses the computed probabilities to form a prediction interval for $\tilde{Y}$. Alternately, one simulates values of $\tilde{Y}$ in a two-step process. For example, if one wants to simulate a draw from the posterior predictive distribution, one would first simulate a value λ from its posterior distribution, and given that simulated draw λ^*, simulate $\tilde{Y}$ from a Poisson distribution with mean λ^*. Repeating this process for a large number of iterations provides a sample from the posterior prediction distribution that one uses to construct a prediction interval.

8.8.4 Case study: Learning about website counts

Let's return to the website example where one is interested in learning about the average weekday visits to a baseball analytics blog site. One observes the counts $y_1, ..., y_{20}$ displayed in the "Mon", "Tue", "Wed", "Thu", "Fri" columns of Table 8.7. We assume the $\{y_i\}$ represent a random sample from a Poisson distribution with mean parameter λ.

Suppose one's prior guess at the value of λ is 80 and one wishes to match this information with a Gamma(α, β) prior. Two helpful facts about the gamma distribution are that the mean and variance are equal to $\mu = \alpha/\beta$ and $\sigma^2 = \alpha/\beta^2 = \mu/\beta$, respectively. Figure 8.9 displays three gamma curves for values (α, β) = (80, 1), (40, 0.5), and (20, 0.25). Each of these gamma curves has a mean of 80 and the curves become more diffuse as the parameter β moves from 1 to 0.25. After some thought, the user believes that the Gamma(80, 1) matches her prior beliefs. To check, she computes a prior probability interval. Using the `qgamma()` function, she finds that her 90% prior probability interval is $Prob(65.9 < \lambda < 95.3) = 0.90$ and this appears to be a reasonable approximation to her prior beliefs.

From the data, we compute $\sum_{i=1}^{20} y_i = 2120$ and the sample size is $n = 20$. The posterior distribution is Gamma(α_n, β_n) where the updated parameters are

$$\alpha_n = 80 + 2120 = 2200, \;\; \beta_n = 1 + 20 = 21.$$

Figure 8.10 displays the gamma posterior curve for λ. This figure displays a 90% probability interval which is found using the `qgamma()` function to be (101.1, 108.5). The interpretation is that the average number of visits lies between 101.1 and 108.5 with probability 0.90.

Suppose the user is interested in predicting the number of blog visits $\tilde{Y}$ at a future summer weekday. One simulates the posterior predictive distribution by first simulating 1000 values from the gamma posterior, and then simulating values of $\tilde{Y}$ from Poisson distributions where the Poisson means come from the posterior. Figure 8.11 displays a histogram of the simulated values from the predictive distribution. The 5th and 95th quantiles of this distribution are computed to be 88 and 123 – there is a 90% probability that that the number of visitors in a future weekday will fall in the interval (88, 123).

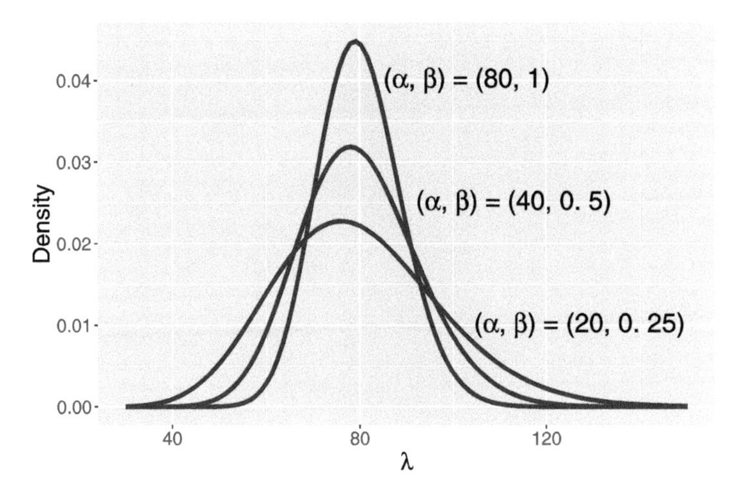

FIGURE 8.9
Three Gamma(α, β) plausible prior distributions for the average number of weekday visits to the website.

8.9 Exercises

1. **Another Set of Federer's Time-to-Serve Measurements (Discrete Priors)**

   Suppose another set of thirty Federer's time-to-serve measurements are collected with an observed mean of 19 seconds. Assume the same discrete uniform prior on the values μ = 15, 16, ..., 22. The prior and the likelihood function are displayed below.

$$\pi(\mu) = \frac{1}{8}, \quad \mu = 15, 16, ..., 22,$$

$$L(\mu) \propto \exp\left(-\frac{n}{2\sigma^2}(\bar{y} - \mu)^2\right).$$

   (a) Assuming $\sigma = 4$, perform the Bayes' rule calculation to find the posterior distribution for μ.
   (b) Using the posterior, find a "best" estimate at μ and an interval of values that contains μ with probability 0.5.

2. **Temperature in Bismarck**

   Suppose one is interested in learning about the average January daily temperature (in degrees Fahrenheit) in Bismarck, North

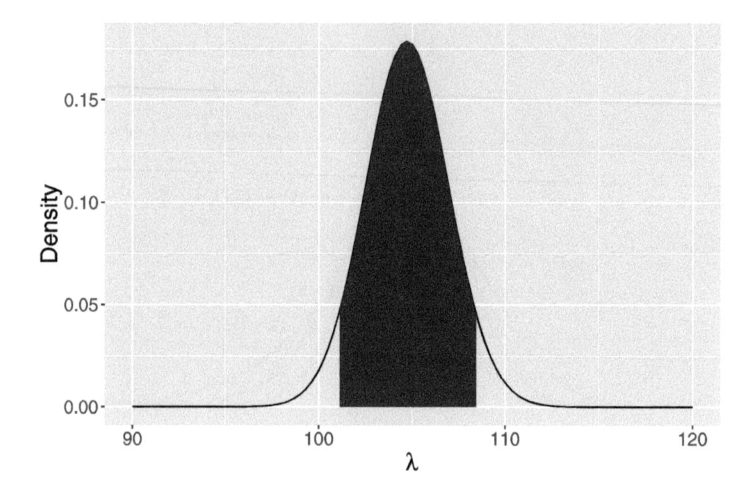

FIGURE 8.10
Posterior curve for the mean number of visits λ to the website. The shaded region shows the limits of a 90% interval estimate.

Dakota. One assumes that the daily temperature Y is normally distributed with mean μ and known standard deviation $\sigma = 10$. Suppose that one's prior is uniformly distributed over the values $\mu = 5, 10, 15, 20, 25, 30, 25$. Suppose one observes the temperature for one January day to be 28 degrees. Find the posterior distribution of μ and compute the posterior probability the mean is at least as large as 30 degrees.

3. **Choosing A Normal Prior**

   (a) Suppose Sam believes that the 0.25 quantile of the mean of Federer's time-to-serve μ is 14 seconds and the 0.8 quantile is 21 seconds. Using the `normal.select()` function, construct a normal prior distribution to match this belief.

   (b) Suppose Sam also believes that the 0.10 quantile of his prior is equal to 10.5 seconds. Is this statement consistent with the normal prior chosen in part (a)? If not, how could you adjust the prior to reconcile this statement about the 0.10 quantile?

4. **Choosing A Normal Prior**

   Another way of choosing a normal prior for Federer's mean time-to-serve μ is to specify statements about the prior predictive distribution for a future time-to-serve measurement $\tilde{Y}$. Using results from Section 8.5.2, if μ has a normal prior with mean μ_0 and σ_0, then the predictive density of $\tilde{Y}$ is normal with mean μ_0 and

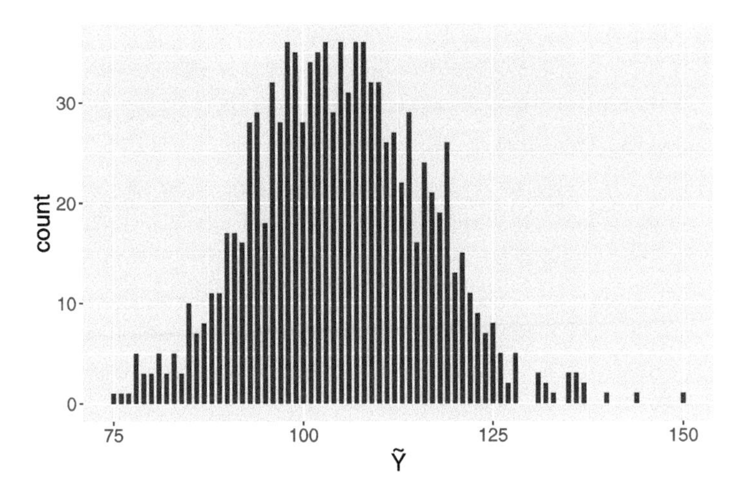

FIGURE 8.11
Histogram of a simulated sample from the posterior predictive distribution of the number of visitors to the website on a future day.

standard deviation $\sqrt{\sigma^2 + \sigma_0^2}$, where we are assuming that the sampling standard deviation $\sigma = 4$ seconds.

(a) Suppose your best guess at $\tilde{Y}$ is 15 seconds, and you are 90 percent confident that $\tilde{Y}$ is smaller than 25 seconds. Find the normal prior for μ that matches this prior information about the future time-to-serve.

(b) Suppose instead that you are 90% confident that the future time-to-serve is between 18 and 24 seconds. Find the normal prior for μ that matches this prior belief.

5. **Bayesian Hypothesis Testing**

   The posterior distribution for the mean time-to-serve μ for Federer is normal with mean 17.4 seconds and standard deviation 0.77 seconds.

   (a) Using this posterior, evaluate the plausibility of the statement "Federer's mean time-to-serve is at least 16.5 seconds."

   (b) Is it reasonable to say that Federer's mean time-to-serve falls between 17 and 18 seconds? Explain.

6. **Bayesian Credible Interval**

   The posterior distribution for the mean time-to-serve μ for Federer is normal with mean 17.4 seconds and standard deviation 0.77 seconds.

(a) Construct a central 98% credible interval for μ.

(b) Can you use the credible interval to test the hypothesis "Federer's mean time-to-serve is 16.5 seconds"? Explain.

7. **Posterior Predictive Distribution**

   Write an R script to generate $S = 1000$ predictions of a single time-to-serve of Federer based on the posterior predictive distribution using the results given in Equation (8.31) and Equation (8.32).

   (a) Compare the exact posterior predictive distribution (Equation (8.30)) with the density estimate of the simulated predictions.

   (b) Construct a 90% prediction interval for the future time-to-serve.

8. **Posterior Predictive Checking**

   The posterior predictive distribution can be used to check the suitability of the normal sampling and normal prior model for Federer's time-to-serve data. The function `post_pred_check()` simulates samples of $n = 20$ from the posterior predictive function, and for each sample, computes a value of the checking function $T(\tilde{y})$.

```
post_pred_check <- function(test_function){
  mu_n <- 17.4
  sigma_n <- 0.77
  sigma <- 4
  n <- 20
  one_sim <- function(){
    mu <- rnorm(1, mu_1, sigma_1)
    test(rnorm(n, mu, sigma))
  }
  replicate(1000, one_sim())
}
```

   The output of the function is 1000 draws from the posterior predictive distribution of T. If the checking function is $\max(y)$, then one would obtain 1000 draws from the posterior predictive distribution by typing

```
post_pred_check(max)
```

   If the value of the checking function on the observed time-to-serves $T(y)$ is unusual relative to this posterior predictive distribution of T, this would cast doubt on the model. The observed times-to-serve for Federer are displayed in Section 8.3.1. and repeated below.

```
15.1 11.8 21.0 22.7 18.6 16.2 11.1 13.2 20.4 19.2
```

```
21.2 14.3 18.6 16.8 20.3 19.9 15.0 13.4 19.9 15.3
```

(a) Use the function `post_pred_check()` with the checking function $T(y) = \max(y)$ to check the suitability of the Bayesian model.

(b) Use the function `post_pred_check()` with the checking function $T(y) = \text{sd}(y)$ to check the suitability of the Bayesian model.

9. **Taxi Cab Fares**

   Suppose a city manager is interested in learning about the mean fare μ for taxi cabs in New York City.

   (a) Suppose the manager believes that μ is smaller than \$8 with probability 0.25, and that μ is smaller than \$12 with probability 0.75. Find a normal prior that matches this prior information.

   (b) The manager reviews 20 fares and observes the values (in dollars): 7.5, 8.5, 9.5, 6.5, 7.0, 6.0, 7.0, 16.0, 8.0, 8.5, 9.5, 13.5, 4.5, 8.5, 7.5, 13.0, 6.5, 9.5, 21.0, 6.5. Assuming these fares are normally distributed with mean μ and standard deviation $\sigma = 4$, find the posterior distribution for the mean μ.

   (c) Construct a 90% interval estimate for the mean fare μ.

10. **Taxi Cab Fares (continued)**

    Suppose that a visitor to New York City has little knowledge about the mean taxi cab fare.

    (a) Construct a weakly informative prior for μ.

    (b) Use the data from Exercise 9 to compute the posterior distribution for the mean fare.

    (c) Construct a 90% interval estimate for the mean fare and compare your interval with the interval computed in Exercise 9 using an informative prior.

11. **Taxi Cab Fares (continued)**

    (a) In Exercise 9, one finds the posterior distribution for the mean fare μ. Write an R function to simulate a sample of twenty fares from the posterior predictive distribution.

    (b) Looking at the observed data, one sees an unusually large fare of \$21. To see if this fare is unusual for our model, first revise your function in part (a) to simulate the maximum fare of a sample of twenty fares from the posterior predictive distribution. Then repeat this process 1000 times, collecting the maximum fares for 1000 predictive samples.

    (c) Construct a graph of the maximum fares. Is the fare of \$21 large relative to the prediction distribution of maximum fares?

(d) Based on the answer to part (c), what does that say about the suitability of our model?

12. **Student Sleeping Times**

How many hours do college students sleep, on the average? Recently, some introductory students were asked when they went to bed and when they woke the following morning. A following random sample of 14 sleeping times (in hours) were recorded: 9.0, 7.5, 7.0, 8.0, 5.0, 6.5, 8.5, 7.0, 9.0, 7.0, 5.5, 6.0, 8.5, 7.5. Assume that these measurements follow a normal sampling distribution with mean μ and standard deviation σ, where we are given that $\sigma = 1.5$.

(a) Suppose that John believes a priori that the mean amount of sleep μ is normal with mean 8 hours and standard deviation 1 hour. Find the posterior distribution of μ.

(b) Construct a 90% interval estimate for the mean μ.

(c) Let y^* denote the sleeping time for a randomly selected student. Find the predictive distribution for y^* and use this to construct a 90% prediction interval.

13. **Student Sleeping Times (continued)**

Suppose two other people are interested in learning about the mean sleeping times of college students. Mary's prior is normal with mean 8 hours and standard deviation 0.1 – she is pretty confident that the mean sleeping time is close to 8 hours. In contrast, Larry is very uncertain about the location of μ and assigns a normal prior with mean 8 hours and standard deviation 3 hours.

(a) Find the posterior distributions of μ using Mary's prior and using Larry's prior.

(b) Construct 90% interval estimates for μ using Mary's and Larry's priors.

(c) Compare the interval estimates with the interval estimates constructed in Exercise 12(b) using Mary's prior. Is the location of the interval estimate sensitive to the choice of prior? If so, explain the differences.

14. **Comparing Two Means - IQ Tests on School Children**

Do teachers' expectations impact academic development of children? To find out, researchers gave an IQ test to a group of 12 elementary school children. They randomly picked six children and told teachers that the test predicts them to have high potential for accelerated growth (accelerated group); for the other six students in the group, the researchers told teachers that the test predicts them to have no potential for growth (no growth group). At the end of school year, they gave IQ tests again to all 12 students, and the change in IQ scores of each student is recorded. Table 8.8 shows the IQ score change of students in the accelerated group and the no growth group.

TABLE 8.8
Data from IQ score change of 12 students; 6 are in the accelerated group, and 6 are in the no growth group.

Group	IQ score change
Accelerated	20, 10, 19, 15, 9, 18
No growth	3, 2, 6, 10, 11, 5

The sample means of the accelerated group and the no growth group are respectively $\bar{y}_A = 15.2$ and $\bar{y}_N = 6.2$. Consider independent sampling models, where the IQ scores for the accelerated group (no growth group) are assumed normal with mean μ_A (μ_N) with known standard deviation $\sigma = 4$.

$$Y_{A,i} \overset{i.i.d.}{\sim} \text{Normal}(\mu_A, 4), \text{ for } i = 1, \cdots n_A,$$
$$Y_{N,i} \overset{i.i.d.}{\sim} \text{Normal}(\mu_N, 4), \text{ for } i = 1, \cdots n_N,$$

where $n_A = n_N = 6$.

(a) Assuming independent sampling, write down the likelihood function of the means (μ_A, μ_N).

(b) Assume that one's prior beliefs about μ_A and μ_N are independent, where $\mu_A \sim \text{Normal}(\gamma_A, \tau_A)$ and $\mu_N \sim \text{Normal}(\gamma_N, \tau_N)$. Show that the posterior distributions for μ_A and μ_N are independent normal and find the mean and standard deviation parameters for each distribution.

15. **Comparing Two Means - IQ Tests on School Children (continued)**

    In Exercise 14, you should have established that the mean IQ score changes μ_A and μ_N have independent normal posterior distributions. Assume that one has vague prior beliefs and $\mu_A \sim$ Normal(0, 20) and $\mu_N \sim$ Normal(0, 20).

    (a) Is the average improvement for the accelerated group larger than that for the no growth group? Consider the parameter $\delta = \mu_A - \mu_N$ to measure the difference in means. The question now becomes finding the posterior probability of $\delta > 0$, i.e. $p(\mu_A - \mu_N > 0 \mid \mathbf{y}_A, \mathbf{y}_N)$, where $\mathbf{y}_A$ and $\mathbf{y}_N$ are the vectors of recorded IQ score change. [Hint: simulate a vector s_A of posterior samples of μ_A and another vector s_N of posterior samples of μ_N (make sure to use the same number of samples) and subtract s_N from s_A, which gives us a vector of posterior differences between s_N and s_A. This vector of posterior differences serves as an approximation to the posterior distribution of δ.]

(b) What is the probability that a randomly selected child assigned to the accelerated group will have larger improvement than a randomly selected child assigned to the no growth group? Consider $\tilde{Y}_A$ and $\tilde{Y}_N$ to be random variables for predicted IQ score change for the accelerated group and the no growth group, respectively. The question now becomes finding the posterior predictive probability of $\tilde{Y}_A > \tilde{Y}_N$, i.e. $p(\tilde{Y}_A > \tilde{Y}_N \mid \mathbf{y}_A, \mathbf{y}_N)$, where $\mathbf{y}_A$ and $\mathbf{y}_N$ are the vectors of recorded IQ score change, each of length 6. [Hint: Show that the posterior predictive distributions of $\tilde{Y}_A$ and $\tilde{Y}_N$ are independent. Simulate predicted IQ score changes from the posterior predictive distributions for the two groups, then simulate the posterior predictive distribution of $\tilde{Y}_A - \tilde{Y}_N$ by taking the difference of simulated draws.]

16. **Comparing Two Means - Prices of Diamonds**

Weights of diamonds are measured in carats. The difference between the size of a 0.99 carat diamond and a 1 carat diamond is most likely undetectable to the naked human eye, but the price of a 1 carat diamond tends to be much higher than the price of a 0.99 carat diamond. To find out if it is truly the case, data on point prices (the prices of 0.99 carat diamonds divided by 99, and the prices of 1 carat diamonds divided by 100) of $n_{99} = 23$ of 0.99 carat diamonds and $n_{100} = 25$ of 1 carat diamonds were collected and stored in the files `pt99price.csv` and `pt100price.csv`.

(a) Explore the two datasets by making plots and computing summary statistics. What are the findings?

(b) Consider independent normal sampling models for these datasets with a fixed and known value of the standard deviation. From your exploratory work, choose a value for the standard deviation.

(c) Choose appropriate weakly informative prior distributions, and use posterior simulation to answer whether the average point price of the 1 carat diamonds is higher than that of the 0.99 diamonds.

(d) Perform posterior predictive checks of the Bayesian inferences obtained in part (c).

17. **Gamma-Poisson Conjugacy Derivation**

Section 8.8.3 presents the Bayesian update results for Poisson sampling with the use of the gamma conjugate prior.

(a) Verify the equation for the likelihood in Equation (8.37). [Hint:

$$
\begin{aligned}
f(Y_1 = y_1, ..., Y_n = y_n \mid \lambda) &= \prod_{i=1}^{n} f(y_i \mid \lambda) \\
&= \prod_{i=1}^{n} \frac{1}{y_i!} \lambda^{y_i} e^{-\lambda},
\end{aligned}
$$

the joint sampling density of n *i.i.d.* Poisson distributed random variables.]

(b) Assuming that the Poisson parameter λ has a gamma prior with shape α and rate β, show that the posterior distribution of λ has a gamma functional form and find the parameters of this gamma distribution.

18. **The Number of ER Visits: the Prior**

Suppose two people, Pedro and Mia, have different prior beliefs about the average number of ER visits during the 10 pm - 11 pm time period. Pedro's prior information is matched to a gamma distribution with parameters $\alpha = 70$ and $\beta = 10$, and Mia's beliefs are matched to a gamma distribution with $\alpha = 33.3$ and $\beta = 3.3$. The two gamma priors are displayed in Figure 8.12.

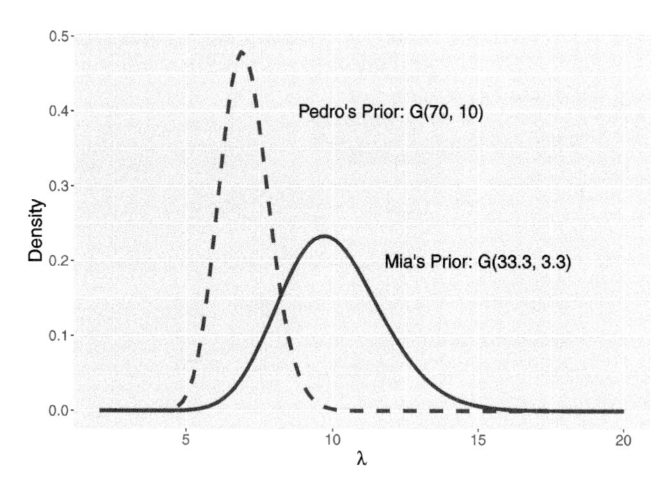

FIGURE 8.12
Two gamma priors for the average number of visits to ER during a particular hour in the evening.

(a) Compare the priors of Pedro and Mia with respect to average value and spread. Which person believes that there will be more

ER visits, on average? Which person is more confident of his or her best guess at the average number of ER visits?

(b) Using the `qgamma()` function, construct 90% interval estimates for λ using Pedro's prior and Mia's prior.

(c) After some thought, Pedro believes that his best prior guess at λ is correct, but he is less confident in this new guess. Explain how Pedro can adjust the parameters of his gamma prior to reflect this new prior belief.

(d) Mia also revisits her prior. Her best guess at the average number of ER visits is now 3 larger than her previous best guess, but the degree of confidence in this guess hasn't changed. Explain how Mia can adjust the parameters of her gamma prior to reflect this new prior belief.

19. **The Number of ER Visits**

A hospital collects the number of patients in the emergency room (ER) admitted between 10 pm and 11 pm for each day of a week. Table 8.9 records the day and the number of ER visits for the given day.

TABLE 8.9
Data for ER visits in a given week.

Day	Number of ER visits
Sunday	8
Monday	6
Tuesday	6
Wednesday	9
Thursday	8
Friday	9
Saturday	7

Suppose one assumes Poisson sampling for the counts, and a conjugate gamma prior with parameters $\alpha = 70$ and $\beta = 10$ for the Poisson rate parameter λ.

(a) Given the sample shown in Table 8.9, obtain the posterior distribution for λ through the gamma-Poisson conjugacy. Obtain a 95% posterior credible interval for λ.

(b) Suppose a hospital administrator states that the average number of ER visits during any evening hour does not exceed 6. By computing a posterior probability, evaluate the validity of the administrator's statement.

(c) The hospital is interested in predicting the number of ER visits between 10 pm and 11 pm for another week. Use simulations to generate posterior predictions of the number of ER visits for another week (seven days).

20. **Times Between Traffic Accidents**

The exponential distribution is often used as a model to describe the time between events, such as traffic accidents. A random variable Y has an Exponential distribution if its pdf is as follows.

$$f(y \mid \lambda) = \begin{cases} \lambda \exp(-\lambda y), & \text{if } y \geq 0. \\ 0, & \text{if } y < 0. \end{cases} \tag{8.41}$$

Here, the parameter $\lambda > 0$, considered as the rate of event occurrences. This is a one-parameter model.

(a) The gamma distribution is a conjugate prior distribution for the rate parameter λ in the Exponential data model. Use the prior distribution $\lambda \sim \text{Gamma}(a, b)$, and find its posterior distribution $\pi(\lambda \mid y_1, \cdots, y_n)$, where $y_i \overset{i.i.d.}{\sim} \text{Exponential}(\lambda)$ for $i = 1, \cdots, n$.

(b) Suppose 10 times between traffic accidents are collected: 1.5, 15, 60.3, 30.5, 2.8, 56.4, 27, 6.4, 110.7, 25.4 (in minutes). With the posterior distribution derived in part (a), use Monte Carlo approximation to calculate the posterior mean, median, and a middle 95% credible interval for the rate λ. [Hint: choose the appropriate R functions from `dgamma()`, `pgamma()`, `qgamma()`, and `rgamma()`.]

(c) Use Monte Carlo approximation to generate another set of 10 predicted times between events. [Hint: `rexp()` generates random draws from an Exponential distribution.]

21. **Modeling Survival Times**

The Weibull distribution is often used as a model for survival times in biomedical, demographic, and engineering analyses. A random variable Y has a Weibull distribution if its pdf is as follows.

$$f(y \mid \alpha, \lambda) = \lambda \alpha y^{\alpha - 1} \exp(-\lambda y^{\alpha}) \quad \text{for } y > 0. \tag{8.42}$$

Here, $\alpha > 0$ and $\lambda > 0$ are parameters of the distribution. For this problem, assume that $\alpha = \alpha_0$ is known, but λ is not known, i.e. a simplified case of a one-parameter model. Also assume that software routines for simulating from Weibull distributions are available (e.g. `rweibull()`)

(a) Assuming a prior distribution $\pi(\lambda \mid \alpha = \alpha_0) \propto 1$, find its posterior $\pi(\lambda \mid y_1, \ldots, y_n, \alpha = \alpha_0)$, where $y_i \overset{i.i.d.}{\sim} \text{Weibull}(\lambda, \alpha = \alpha_0)$ for $i = 1, \cdots, n$. Write the name of the distribution and expressions for its parameter values.

(b) Using the posterior distribution derived in part (a), explain step-by-step how you would use Monte Carlo simulation to approximate the posterior median survival time, assuming that $\alpha = \alpha_0$.

(c) What family of distributions represents the conjugate prior distributions for λ, assuming that $\alpha = \alpha_0$.

9

Simulation by Markov Chain Monte Carlo

9.1 Introduction

9.1.1 The Bayesian computation problem

The Bayesian models in Chapters 7 and 8 describe the application of conjugate priors where the prior and posterior belong to the same family of distributions. In these cases, the posterior distribution has a convenient functional form such as a beta density or normal density, and the posterior distributions are easy to summarize. For example, if the posterior density has a normal form, one uses the R functions `pnorm()` and `qnorm()` to compute posterior probabilities and quantiles.

In a general Bayesian problem, the data Y comes from a sampling density $f(y \mid \theta)$ and the parameter θ is assigned a prior density $\pi(\theta)$. After $Y = y$ has been observed, the likelihood function is equal to $L(\theta) = f(y \mid \theta)$ and the posterior density is written as

$$\pi(\theta \mid y) = \frac{\pi(\theta)L(\theta)}{\int \pi(\theta)L(\theta)d\theta}. \tag{9.1}$$

If the prior and likelihood function do not combine in a helpful way, the normalizing constant $\int \pi(\theta)L(\theta)d\theta$ can not be evaluated analytically. In addition, summaries of the posterior distribution are expressed as ratios of integrals. For example, the posterior mean of θ is given by

$$E(\theta \mid y) = \frac{\int \theta\pi(\theta)L(\theta)d\theta}{\int \pi(\theta)L(\theta)d\theta}. \tag{9.2}$$

Computation of the posterior mean requires the evaluation of two integrals, each not expressible in closed-form.

The following sections illustrate this general problem where integrals of the product of the likelihood and prior can not be evaluated analytically and so there are challenges in summarizing the posterior distribution.

9.1.2 Choosing a prior

Suppose you are planning to move to Buffalo, New York. You currently live on the west coast of the United States where the weather is warm and you are wondering about the snowfall you will encounter in Buffalo in the following winter season.

Suppose you focus on the quantity μ, the average snowfall during the month of January. After some reflection, you are 50 percent confident that μ falls between 8 and 12 inches. That is, the 25th percentile of your prior for μ is 8 inches and the 75th percentile is 12 inches.

A normal prior

Once you have figured out your prior information, you construct a prior density for μ that matches this information. In one of the end-of-chapter exercises, you can confirm that one possible density matching this information is a normal density with mean 10 and standard deviation 3.

We collect data for the last 20 seasons in January. Assume that these observations of January snowfall are normally distributed with mean μ and standard deviation σ. For simplicity we assume that the sampling standard deviation σ is equal to the observed standard deviation s. The observed sample mean $\bar{y}$ and corresponding standard error are given by $\bar{y} = 26.785$ and $se = s/\sqrt{n} = 3.236$.

R With this normal prior and normal sampling, results from Chapter 8 are applied to find the posterior distribution of μ. The `normal_update()` function is used to find the mean and standard deviation of the normal posterior distribution.

```
(post1 <- normal_update(c(10, 3), c(ybar, se)))
[1] 17.75676  2.20020
```

In Figure 9.1 the prior, likelihood, and posterior are displayed on the same graph. Initially you believed that μ was close to 10 inches, the data says that the mean is in the neighborhood of 26.75 inches, and the posterior is a compromise, where μ is in an interval about 17.75 inches.

An alternative prior

Looking at Figure 9.1, there is some concern about this particular Bayesian analysis. Since the the main probability contents of the prior and likelihood functions have little overlap, there is serious conflict between the information in your prior and the information from the data.

Since we have a prior-data conflict, it would make sense to revisit our choice for a prior density on μ. Remember you specified the quartiles for μ to be 8 and 12 inches. Another symmetric density that matches this information is a Cauchy density with location 10 inches and scale parameter 2 inches. The reader can confirm that the quantiles of a Cauchy(10, 2) do match your prior information. [Hint: use the `qcauchy()` R command.]

In Figure 9.2 we compare the normal and Cauchy priors graphically. Remember these two densities have the same quartiles at 8 and 12 inches.

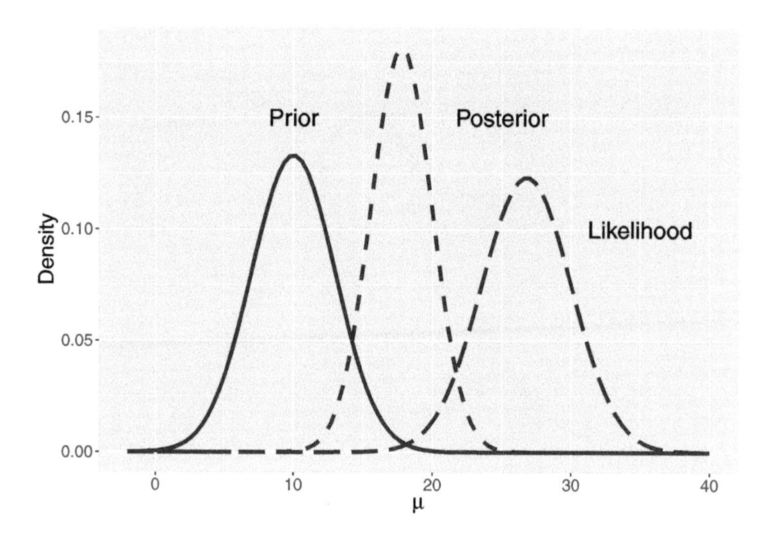

FIGURE 9.1
Prior, likelihood, and posterior of a normal mean with a normal prior.

But the two priors have different shapes – the Cauchy prior is more peaked near the median value 10 and has tails that decrease to zero at a slower rate than the normal. In other words, the Cauchy curve has flatter tails than the normal curve.

With the use of a Cauchy(10, 2) prior and the same normal likelihood, the posterior density of μ is

$$\pi(\mu \mid y) \propto \pi(\mu)L(\mu) \propto \frac{1}{1 + \left(\frac{\mu - 10}{2}\right)^2} \times \exp\left\{-\frac{n}{2\sigma^2}(\bar{y} - \mu)^2\right\}. \tag{9.3}$$

In contrast with a normal prior, one can not algebraically simplify this likelihood times prior product to obtain a "nice" functional expression for the posterior density in terms of the mean μ. That raises the question – how does one implement a Bayesian analysis when one can not easily express the posterior density in a convenient functional form?

9.1.3 The two-parameter normal problem

In the problem in learning about a normal mean μ in Chapter 8, it was assumed that the sampling standard deviation σ was known. This is unrealistic – in most settings, if one is uncertain about the mean of the population, then likely the population standard deviation will also be unknown. From a Bayesian perspective, since we have two unknown parameters μ and σ, this

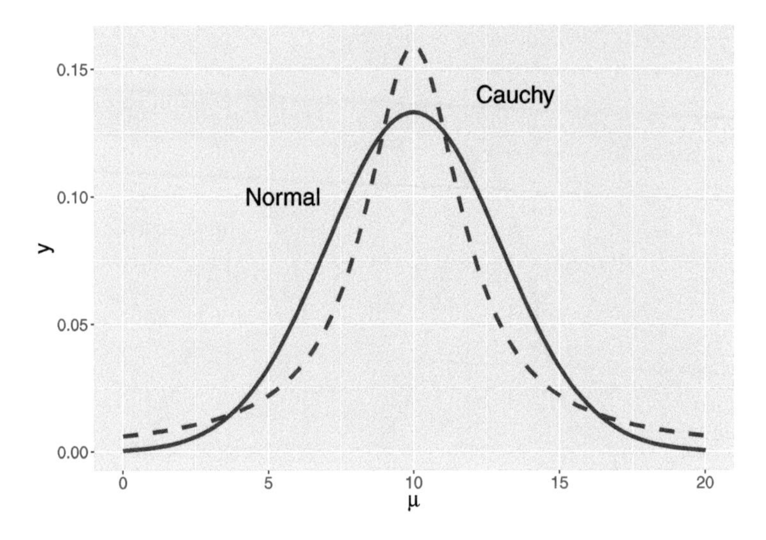

FIGURE 9.2
Two priors for representing prior opinion about a normal mean.

situation presents new challenges. One needs to construct a joint prior $\pi(\mu, \sigma)$ for the two parameters – up to this point, we have only discussed constructing a prior distribution for a single parameter. Also, although one can compute the posterior density by the usual "prior times likelihood" recipe, it may be difficult to get nice analytic answers with this posterior to obtain particular inferences of interest.

9.1.4 Overview of the chapter

In Chapters 7 and 8, we illustrated the use of simulation to summarize posterior distributions of a specific functional form such as the beta and normal. In this chapter, we introduce a general class of algorithms, collectively called Markov chain Monte Carlo (MCMC), that can be used to simulate the posterior from general Bayesian models. These algorithms are based on a general probability model called a Markov chain and Section 9.2 describes this probability model for situations where the possible models are finite. Section 9.3 introduces the Metropolis sampler, a general algorithm for simulating from an arbitrary posterior distribution. Section 9.4 describes the implementation of this simulation algorithm for the normal sampling problem with a Cauchy prior. Section 9.5 introduces another MCMC simulation algorithm, Gibbs sampling, that is well-suited for simulation from posterior distributions of many parameters. One issue in the implementation of these MCMC algorithms is that the simulation draws represent an approximate sample from the posterior distribution. Section 9.6 describes some common diagnostic methods for seeing if the simulated sample is a suitable exploration of the posterior distribution. Finally in Section 9.7, we describe the use of a general-purpose software program Just Another Gibbs Sampler (JAGS) and R interface for implementing these MCMC algorithms.

9.2 Markov Chains

9.2.1 Definition

Since our simulation algorithms are based on Markov chains, we begin by defining this class of probability models in the situation where the possible outcomes are finite. Suppose a person takes a random walk on a number line on the values 1, 2, 3, 4, 5, 6. If the person is currently at an interior value (2, 3, 4, or 5), in the next second she is equally likely to remain at that number or move to an adjacent number. If she does move, she is equally likely to move left or right. If the person is currently at one of the end values (1 or 6), in the next second she is equally likely to stay still or move to the adjacent location.

This is a simple example of a discrete Markov chain. A Markov chain describes probabilistic movement between a number of states. Here there are six possible states, 1 through 6, corresponding to the possible locations of the walker. Given that the person is at a current location, she moves to other locations with specified probabilities. The probability that she moves to another location depends only on her current location and not on previous locations visited. We describe movement between states in terms of transition probabilities – they describe the likelihoods of moving between all possible states in a single step in a Markov chain. We summarize the transition probabilities by means of a transition matrix P:

$$P = \begin{bmatrix} .50 & .50 & 0 & 0 & 0 & 0 \\ .25 & .50 & .25 & 0 & 0 & 0 \\ 0 & .25 & .50 & .25 & 0 & 0 \\ 0 & 0 & .25 & .50 & .25 & 0 \\ 0 & 0 & 0 & .25 & .50 & .25 \\ 0 & 0 & 0 & 0 & .50 & .50 \end{bmatrix}$$

The first row in P gives the probabilities of moving to all states 1 through 6 in a single step from location 1, the second row gives the transition probabilities in a single step from location 2, and so on.

There are several important properties of this particular Markov chain. It is possible to go from every state to every state in one or more steps – a Markov chain with this property is said to be *irreducible*. Given that the person is in a particular state, if the person can only return to this state at regular intervals, then the Markov chain is said to be *periodic*. This example is aperiodic since the walker cannot return to the current state at regular intervals.

9.2.2 Some properties

We represent the person's current location as a probability row vector of the form

$$p = (p_1, p_2, p_3, p_4, p_5, p_6),$$

where p_i represents the probability that the person is currently in state i. If $p^{(j)}$ represents the location of the person at step j, then the location of the person at the $j+1$ step is given by the matrix product

$$p^{(j+1)} = p^{(j)} P.$$

Moreover, if $p^{(j)}$ represents the location at step j, then the location of the traveler after m additional steps, $p^{(j+m)}$, is given by the matrix product

$$p^{(j+m)} = p^{(j)} P^m,$$

where P^m indicates the matrix multiplication $P \times P \times ... \times P$ (the matrix P multiplied by itself m times).

R To illustrate for our example using R, suppose that the person begins at state 3 that is represented in R by the vector p with a 1 in the third entry:

```
p <- c(0, 0, 1, 0, 0, 0)
```

We also define the transition matrix by use of the `matrix()` function.

```
P <- matrix(c(.5, .5, 0, 0, 0, 0,
              .25, .5, .25, 0, 0, 0,
              0, .25, .5, .25, 0, 0,
              0, 0, .25, .5, .25, 0,
              0, 0, 0, .25, .5, .25,
              0, 0, 0, 0, .5, .5),
              nrow=6, ncol=6, byrow=TRUE)
```

If one multiplies this vector by the matrix P, one obtains the probabilities of being in all six states after one move.

```
print(p %*% P, digits = 5)
     [,1] [,2] [,3] [,4] [,5] [,6]
[1,]    0 0.25  0.5 0.25    0    0
```

After one move (starting at state 3), our walker will be at states 2, 3, and 4 with respective probabilities 0.25, 0.5, and 0.25. If one multiplies p by the matrix P four times, one obtains the probabilities that the walker will be in the different states after four moves.

```
print(p %*% P %*% P %*% P %*% P, digits = 5)
        [,1] [,2]    [,3]    [,4]    [,5]    [,6]
[1,] 0.10938 0.25 0.27734 0.21875 0.11328 0.03125
```

Starting from state 3, this person will most likely be in states 2, 3, and 4 after four moves.

For an irreducible, aperiodic Markov chain, there is a limiting behavior of the matrix power P^m as m approaches infinity. Specifically, this limit is equal to

$$W = \lim_{m \to \infty} P^m, \tag{9.4}$$

where W has common rows equal to w. The implication of this result is that, as one takes an infinite number of moves, the probability of landing at a particular state does not depend on the initial starting state.

One can demonstrate this result empirically for our example. Using a loop, we take the transition matrix P to the 100th power by repeatedly multiplying the transition matrix by itself. From this calculation below, note that the rows of the matrix `Pm` appear to be approaching a constant vector. Specifically, it appears the constant vector w is equal to (0.1, 0.2, 0.2, 0.2, 0.2, 0.1).

```
Pm <- diag(rep(1, 6))
for(j in 1:100){
  Pm <- Pm %*% P
}
print(Pm, digits = 5)
         [,1]    [,2]    [,3]    [,4]    [,5]     [,6]
[1,] 0.100009 0.20001 0.20001 0.19999 0.19999 0.099991
[2,] 0.100007 0.20001 0.20000 0.20000 0.19999 0.099993
[3,] 0.100003 0.20000 0.20000 0.20000 0.20000 0.099997
[4,] 0.099997 0.20000 0.20000 0.20000 0.20000 0.100003
[5,] 0.099993 0.19999 0.20000 0.20000 0.20001 0.100007
[6,] 0.099991 0.19999 0.19999 0.20001 0.20001 0.100009
```

From this result about the limiting behavior of the matrix power P^m, one can derive a rule for determining this constant vector. Suppose we can find a probability vector w such that $wP = w$. This vector w is said to be the *stationary distribution*. If a Markov chain is irreducible and aperiodic, then it has a unique stationary distribution. Moreover, as illustrated above, the limiting distribution of this Markov chain, as the number of steps approaches infinity, will be equal to this stationary distribution.

9.2.3 Simulating a Markov chain

Another method for demonstrating the existence of the stationary distribution of our Markov chain is by running a simulation experiment. We start our random walk at a particular state, say location 3, and then simulate many steps of the Markov chain using the transition matrix P. The relative frequencies of our traveler in the six locations after many steps will eventually approach the stationary distribution w.

R In R we have already defined the transition matrix `P`. To begin the simulation exercise, we set up a storage vector `s` for the locations of our traveler in the random walk. We indicate that the starting location for our traveler is state 3 and perform a loop to simulate 10,000 draws from the Markov chain. We use the `sample()` function to simulate one step – the arguments to this function indicate that we are sampling a single value from the set {1, 2, 3, 4, 5, 6} with probabilities given by the s_j^1 row of the transition matrix P, where s_j^1 is the current location of our traveler.

```
s <- vector("numeric", 10000)
s[1] <- 3
for (j in 2:10000)
s[j] <- sample(1:6, size=1, prob=P[s[j - 1], ])
```

Suppose that we record the relative frequencies of each of the outcomes 1, 2, ..., 6 after each iteration of the simulation. Figure 9.3 graphs the relative frequencies of each of the outcomes as a function of the iteration number. It appears from Figure 9.3 that the relative frequencies of the states are converging to the stationary distribution $w = (0.1, 0.2, 0.2, 0.2, 0.2, 0.1)$. We confirm that this specific vector w is indeed the stationary distribution of this chain by multiplying w by the transition matrix P and noticing that the product is equal to w.

```
w <- matrix(c(.1,.2,.2,.2,.2,.1), nrow=1, ncol=6)
 w %*% P
[,1] [,2] [,3] [,4] [,5] [,6]
[1,] 0.1 0.2 0.2 0.2 0.2 0.1
```

9.3 The Metropolis Algorithm

9.3.1 Example: Walking on a number line

Markov chains can be used to sample from an arbitrary probability distribution. To introduce a general Markov chain sampling algorithm, we illustrate sampling from a discrete distribution. Suppose one defines a discrete probability distribution on the integers 1, ..., K.

R As an example, we write a short function `pd()` in R taking on the values 1, ..., 8 with probabilities proportional to the values 5, 10, 4, 4, 20, 20, 12,

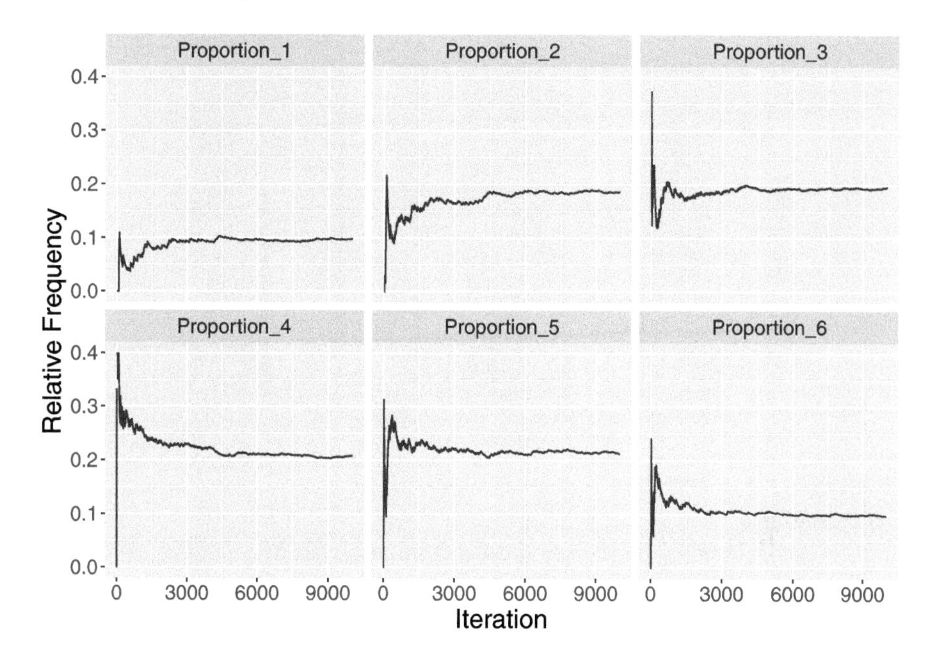

FIGURE 9.3
Relative frequencies of the states 1 through 6 as a function of the number of iterations for Markov chain simulation. As the number of iterations increases, the relative frequencies appear to approach the probabilities in the stationary distribution $w = (0.1, 0.2, 0.2, 0.2, 0.2, 0.1)$.

and 5. Note that these probabilities don't sum to one, but we will shortly see that only the relative sizes of these values are relevant in this algorithm. A line graph of this probability distribution is displayed in Figure 9.4.

```
pd <- function(x){
  values <- c(5, 10, 4, 4, 20, 20, 12, 5)
  ifelse(x %in% 1:length(values), values[x], 0)
}
prob_dist <- data.frame(x = 1:8,
                        prob = pd(1:8))
```

To simulate from this probability distribution, we will take a simple random walk described as follows.

1. We start at any possible location of our random variable from 1 to $K = 8$.

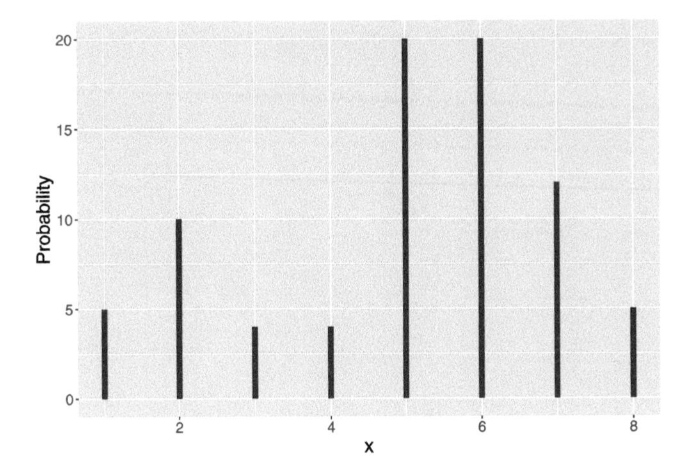

FIGURE 9.4
A discrete probability distribution on the values 1, ..., 8.

2. To decide where to visit next, a fair coin is flipped. If the coin lands heads, we think about visiting the location one value to the left, and if coin lands tails, we consider visiting the location one value to right. We call this the "candidate" location.

3. We compute

$$R = \frac{pd(candidate)}{pd(current)}, \tag{9.5}$$

   the ratio of the probabilities at the candidate and current locations.

4. We spin a continuous spinner that lands anywhere from 0 to 1 – call the random spin X. If X is smaller than R, we move to the candidate location, and otherwise we remain at the current location.

Steps 1 through 4 define an irreducible, aperiodic Markov chain on the state values {1, 2, ..., 8} where Step 1 gives the starting location and the transition matrix P is defined by Steps 2 through 4. One way of "discovering" the discrete probability distribution *pd* is by starting at any location and walking through the distribution many times repeating Steps 2, 3, and 4 (propose a candidate location, compute the ratio, and decide whether to visit the candidate location). If this process is repeated for a large number of steps, the distribution of our actual visits should approximate the probability distribution *pd*.

R A R function `random_walk()` is written implementing this random walk algorithm. There are three inputs to this function, the probability distribution `pd`, the starting location `start` and the number of steps of the algorithm `s`.

```
random_walk <- function(pd, start, num_steps){
  y <- rep(0, num_steps)
  current <- start
  for (j in 1:num_steps){
    candidate <- current + sample(c(-1, 1), 1)
    prob <- pd(candidate) / pd(current)
    if (runif(1) < prob) current <- candidate
    y[j] <- current
  }
  return(y)
}
```

We have already defined the probability distribution by use of the function `pd()`. Below, we implement the random walk algorithm by inputting this probability function, starting at the value $X = 4$ and running the algorithm for $s = 10{,}000$ iterations.

```
out <- random_walk(pd, 4, 10000)
data.frame(out) %>% group_by(out) %>%
  summarize(N = n(), Prob = N / 10000) -> S
```

In Figure 9.5 a histogram of the simulated values from the random walk is compared with the actual probability distribution. Note that the collection of simulated draws appears to be a close match to the true probabilities.

9.3.2 The general algorithm

A popular way of simulating from a general continuous posterior distribution is by using a generalization of the discrete Markov chain setup described in the random walk example in the previous section. The Markov chain Monte Carlo sampling strategy sets up an irreducible, aperiodic Markov chain for which the stationary distribution equals the posterior distribution of interest. This method, called the Metropolis algorithm, is applicable to a wide range of Bayesian inference problems.

Here the Metropolis algorithm is presented and illustrated. This algorithm is a special case of the Metropolis-Hastings algorithm, where the proposal distribution is symmetric (e.g. uniform or normal).

Suppose the posterior density is written as

$$\pi_n(\theta) \propto \pi(\theta)L(\theta),$$

where $\pi(\theta)$ is the prior and $L(\theta)$ is the likelihood function. In this algorithm, it is not necessary to compute the normalizing constant – only the product of likelihood and prior is needed.

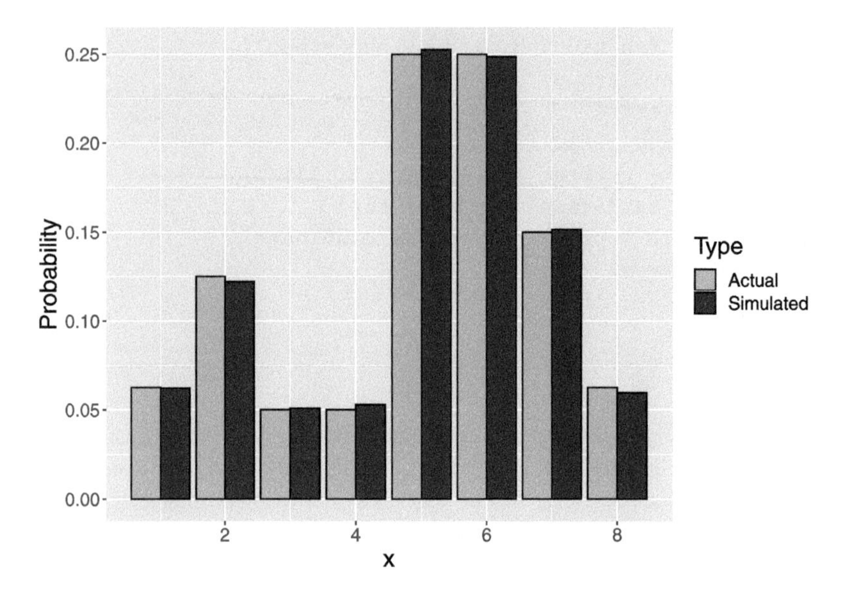

FIGURE 9.5
Histogram of simulated draws from the random walk compared with the actual probabilities of the distribution.

1. (START) As in the random walk algorithm, we begin by selecting any θ value where the posterior density is positive – the value we select $\theta^{(0)}$ is the starting value.
2. (PROPOSE) Given the current simulated value $\theta^{(j)}$ we propose a new value θ^P which is selected at random in the interval $(\theta^{(j)} - C, \theta^{(j)} + C)$ where C is a preselected constant.
3. (ACCEPTANCE PROBABILITY) One computes the ratio R of the posterior density at the proposed value and the current value:
$$R = \frac{\pi_n(\theta^P)}{\pi_n(\theta^{(j)})}. \tag{9.6}$$
The acceptance probability is the minimum of R and 1:
$$PROB = \min\{R, 1\}. \tag{9.7}$$
4. (MOVE OR STAY?) One simulates a uniform random variable U. If U is smaller than the acceptance probability $PROB$, one moves to the proposed value θ^P; otherwise one stays at the current value $\theta^{(j)}$. In other words, the next simulated draw $\theta^{(j+1)}$
$$\theta^{(j+1)} = \begin{cases} \theta^p & \text{if } U < PROB, \\ \theta^{(j)} & \text{elsewhere.} \end{cases} \tag{9.8}$$

5. (CONTINUE) One continues by returning to Step 2 – propose a new simulated value, compute an acceptance probability, decide to move to the proposed value or stay, and so on.

Figure 9.6 illustrates how the Metropolis algorithm works. The bell-shaped curve is the posterior density of interest. In the top-left panel, the solid dot represents the current simulated draw and the black bar represents the proposal region. One simulates the proposed value represented by the "P" symbol. One computes the probability of accepting this proposed value – in this case, this probability is 0.02. By simulating a uniform draw, one decides not to accept this proposal and the new simulated draw is the current value shown in the top-right panel. A different scenario is shown in the bottom panels. One proposes a value corresponding to a higher posterior density value. The probability of accepting this proposal is 1 and the bottom left graph shows that the new simulated draw is the proposed value.

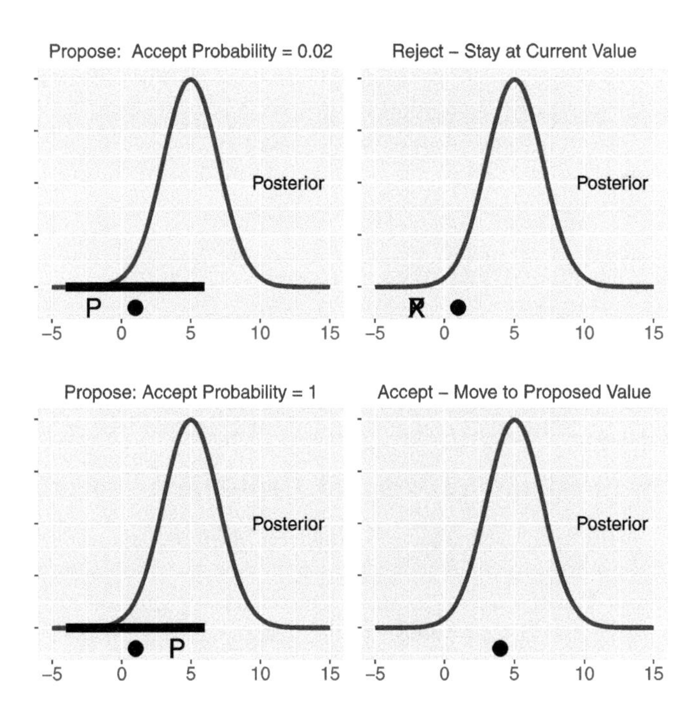

FIGURE 9.6
Illustration of the Metropolis algorithm. The left graphs show the proposal region and two possible proposal values and the right graphs show the result of either accepting or rejecting the proposal.

9.3.3 A general function for the Metropolis algorithm

Since the Metropolis is a relatively simple algorithm, one writes a short function in R to implement this sampling for an arbitrary probability distribution.

R The function `metropolis()` has five inputs: `logpost` is a function defining the logarithm of the density, `current` is the starting value, `C` defines the neighborhood where one looks for a proposal value, `iter` is the number of iterations of the algorithm, and `...` denotes any data or parameters needed in the function `logpost()`.

```
metropolis <- function(logpost, current, C, iter, ...){
  S <- rep(0, iter)
  n_accept <- 0
  for(j in 1:iter){
  candidate <- runif(1, min=current - C,
                     max=current + C)
  prob <- exp(logpost(candidate, ...) -
             logpost(current, ...))
  accept <- ifelse(runif(1) < prob, "yes", "no")
  current <- ifelse(accept == "yes",
                    candidate, current)
  S[j] <- current
  n_accept <- n_accept + (accept == "yes")
  }
  list(S=S, accept_rate=n_accept / iter)
}
```

9.4 Example: Cauchy-Normal Problem

To illustrate using the `metropolis()` function, suppose we wish to simulate 1000 values from the posterior distribution in our Buffalo snowfall problem where we use a Cauchy prior to model our prior opinion about the mean snowfall amount. Recall that the posterior density of μ is proportional to

$$\pi(\mu \mid y) \propto \frac{1}{1 + \left(\frac{\mu - 10}{2}\right)^2} \times \exp\left\{-\frac{n}{2\sigma^2}(\bar{y} - \mu)^2\right\}. \tag{9.9}$$

There are four inputs to this posterior – the mean $\bar{y}$ and corresponding standard error $\sigma/\sqrt{n}$, and the location parameter 10 and the scale parameter 2 for the Cauchy prior. Recall that for the Buffalo snowfall, we observed $\bar{y} = 26.785$ and $\sigma/\sqrt{n} = 3.236$.

R First we need to define a short function defining the logarithm of the posterior density function. Ignoring constants, the logarithm of this density is given by

$$\log \pi(\mu \mid y) = -\log \left\{ 1 + \left(\frac{\mu - 10}{2} \right)^2 \right\} - \frac{n}{2\sigma^2}(\bar{y} - \mu)^2. \tag{9.10}$$

The function `lpost()` returns the value of the logarithm of the posterior where `s` is a list containing the four inputs `ybar`, `se`, `loc`, and `scale`.

```
lpost <- function(theta, s){
    dcauchy(theta, s$loc, s$scale, log = TRUE) +
    dnorm(s$ybar, theta, s$se, log = TRUE)
}
```

A list named `s` is defined that contains these inputs for this particular problem.

```
s <- list(loc = 10, scale = 2,
          ybar = mean(data$JAN),
          se = sd(data$JAN) / sqrt(20))
```

Now we are ready to apply the Metropolis algorithm as coded in the function `metropolis()`. The inputs to this function are the log posterior function `lpost`, the starting value $\mu = 5$, the width of the proposal density $C = 20$, the number of iterations 10,000, and the list `s` that contains the inputs to the log posterior function.

```
out <- metropolis(lpost, 5, 20, 10000, s)
```

The output variable `out` has two components – `S` is a vector of the simulated draws and `accept_rate` gives the acceptance rate of the algorithm.

9.4.1 Choice of starting value and proposal region

In implementing this Metropolis algorithm, the user has to make two choices. He or she needs to select a starting value for the algorithm and select a value of C which determines the width of the proposal region.

Assuming that the starting value is a place where the density is positive, then this particular choice in usual practice is not critical. In the event where the probability density at the starting value is small, the algorithm will move towards the region where the density is more probable.

The choice of the constant C is more critical. If one chooses a very small value of C, then the simulated values from the algorithm tend to be strongly correlated and it takes a relatively long time to explore the entire probability distribution. In contrast, if C is chosen too large, then it is more likely that proposal values will not be accepted and the simulated values tend to get stuck at the current values. One monitors the choice of C by computing the acceptance rate, the proportion of proposal values that are accepted. If the acceptance rate is large, that indicates that the simulated values are highly correlated and the algorithm is not efficiently exploring the distribution. If the acceptance rate is low, then few candidate values are accepted and the algorithm tends to be "sticky" or stuck at current draws.

We illustrate different choices of C for the mean amount of Buffalo snowfall problem. In each case, we start with the value $\mu = 20$ and try the C values 0.3, 3, 30, and 200. In each case, we simulate 5000 values of the MCMC chain. Figure 9.7 shows in each case a line graph of the simulated draws against the iteration number and the acceptance rate of the algorithm is displayed.

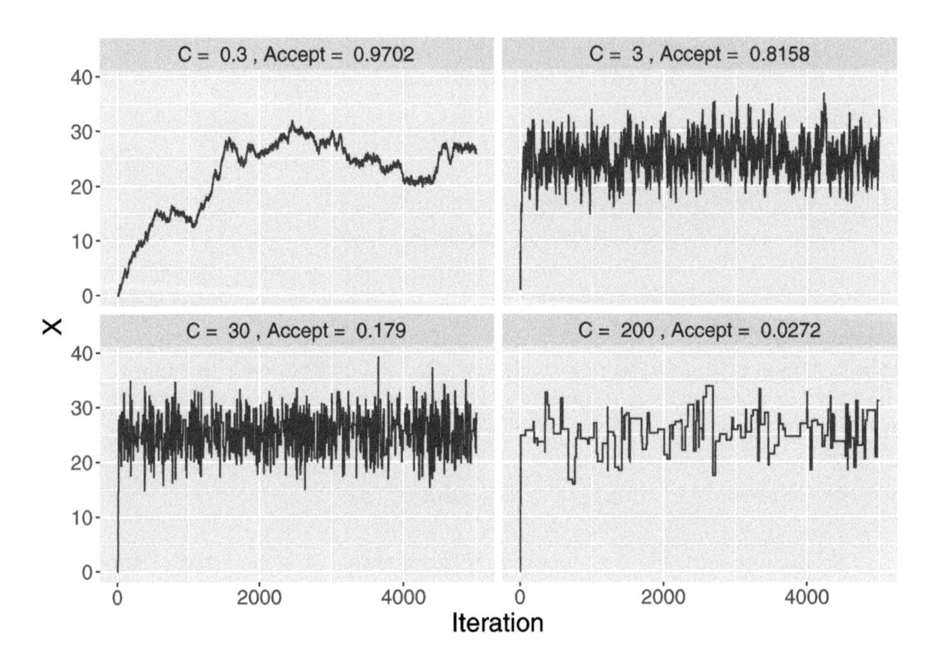

FIGURE 9.7
Trace plots of simulated draws using different choices of the constant C.

When one chooses a small value $C = 0.3$ (top-left panel in Figure 9.7), note that the graph of simulated draws has a snake-like appearance. Due to the strong autocorrelation of the simulated draws, the sampler does a relatively poor job of exploring the posterior distribution. One measure that this sampler is not working well is the large acceptance rate of 0.9702. On the other hand, if one uses a large value $C = 200$ (bottom-right panel in Figure 9.7), the flat-portions in the graph indicates there are many occurrences where the chain will not move from the current value. The low acceptance rate of 0.0272 indicates this problem. The more moderate values of $C = 3$ and $C = 30$ (top-right and bottom-left panels in Figure 9.7) produce more acceptable streams of simulated values, although the respective acceptance rates (0.8158 and 0.179) are very different.

In practice, it is recommended that the Metropolis algorithm has an acceptance rate between 20% and 40%. For this example, this would suggest trying an alternative choice of C between 2 and 20.

9.4.2 Collecting the simulated draws

Using MCMC diagnostic methods that will be described in Section 9.6, one sees that the simulated draws are a reasonable approximation to the posterior density of μ. One displays the posterior density by computing a density estimate of the simulated sample. In Figure 9.8, we plot the prior, likelihood, and posterior density for the mean amount of Buffalo snowfall μ using the Cauchy prior. Recall that we have prior-data conflict, the prior says that the mean snowfall is about 10 inches and the likelihood indicates that the mean snowfall was around 27 inches. When a normal prior was applied, we found that the posterior mean was 17.75 inches – actually the posterior density has little overlap with the prior or the likelihood in Figure 9.1. In contrast, it is seen from Figure 9.8 that the posterior density using the Cauchy density resembles the likelihood. Essentially this posterior analysis says that our prior information was off the mark and the posterior is most influenced by the data.

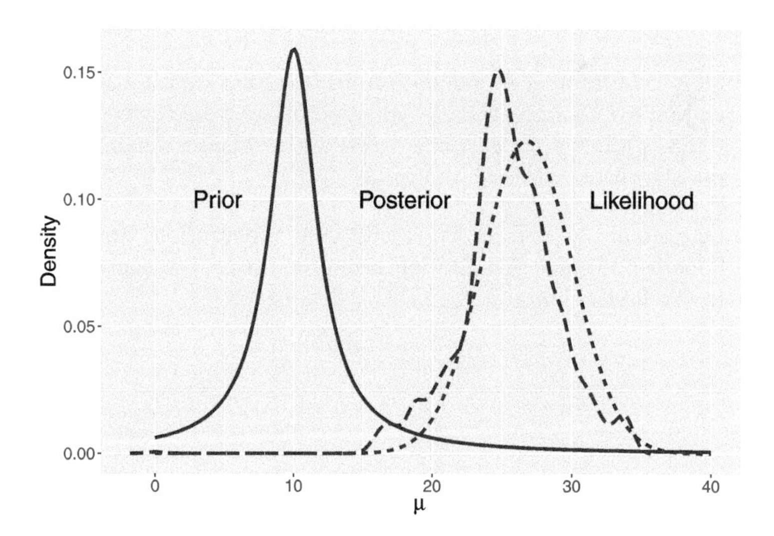

FIGURE 9.8
Prior, likelihood, and posterior of a normal mean with a Cauchy prior.

9.5 Gibbs Sampling

In our examples, we have focused on the use of the Metropolis sampler in simulating from a probability distribution of a single variable. Here we introduce an MCMC algorithm for simulating from a probability distribution of several variables based on conditional distributions: the Gibbs sampling algorithm. As we will see, it facilitates parameter estimation in Bayesian models with more than one parameter, providing data analysts much flexibility in specifying Bayesian models.

9.5.1 Bivariate discrete distribution

To introduce the Gibbs sampling method, suppose that the random variables X and Y each take on the values 1, 2, 3, 4, and the joint probability distribution is given in the following table.

	X			
Y	1	2	3	4
1	0.100	0.075	0.050	0.025
2	0.075	0.100	0.075	0.050
3	0.050	0.075	0.100	0.075
4	0.025	0.050	0.075	0.100

Suppose it is of interest to simulate from this joint distribution of (X, Y). We set up a Markov chain by taking simulated draws from the conditional distributions $f(x \mid y)$ and $f(y \mid x)$. Let's describe this Markov chain by example. Suppose the algorithm starts at the value $X = 1$.

Step 1 One simulates Y from the conditional distribution $f(y \mid X = 1)$. This conditional distribution is represented by the probabilities in the first column of the probability matrix.

Y	Probability
1	0.100
2	0.075
3	0.050
4	0.025

(Actually these values are proportional to the distribution $f(y \mid X = 1)$.) Suppose we perform this simulation and obtain $Y = 2$.

Step 2 Next one simulates X from the conditional distribution of $f(x \mid Y = 2)$. This distribution is found by looking at the probabilities in the second row of the probability matrix.

X	1	2	3	4
Probability	0.075	0.100	0.075	0.050

Suppose the simulated draw from this distribution is $X = 3$.

By implementing Steps 1 and 2, we have one iteration of Gibbs sampling, obtaining the simulated pair $(X, Y) = (3, 2)$. To continue this algorithm, we repeat Steps 1 and 2 many times where we condition in each case on the most recently simulated values of X or Y.

By simulating successively from the distributions $f(y \mid x)$ and $f(x \mid y)$, one defines a Markov chain that moves from one simulated pair $(X^{(j)}, Y^{(j)})$ to the next simulated pair $(X^{(j+1)}, Y^{(j+1)})$. In theory, after simulating from these two conditional distributions a large number of times, the distribution will converge to the joint probability distribution of (X, Y).

R We write a short R function `gibbs_discrete()` to implement Gibbs sampling for a two-parameter discrete distribution where the probabilities are represented in a matrix. One inputs the matrix `p` and the output is a matrix of simulated draws of X and Y where each row corresponds to a simulated pair. By default, the sampler starts at the value $X = 1$ and 1000 iterations of the algorithm will be taken.

```
gibbs_discrete <- function(p, i = 1, iter = 1000){
  x <- matrix(0, iter, 2)
  nX <- dim(p)[1]
  nY <- dim(p)[2]
  for(k in 1:iter){
    j <- sample(1:nY, 1, prob = p[i, ])
    i <- sample(1:nX, 1, prob = p[, j])
    x[k, ] <- c(i, j)
  }
  x
}
```

The function `gibbs_discrete()` is run using the probability matrix for our example. The output is converted to a data frame and we tally the counts for each possible pair of values of (X, Y), and then divide the counts by the simulation sample size of 1000. One can check that the relative frequencies of these pairs are good approximations to the joint probabilities.

```
sp <- data.frame(gibbs_discrete(p))
names(sp) <- c("X", "Y")
table(sp) / 1000
    Y
X        1     2     3     4
  1 0.086 0.058 0.050 0.020
```

```
2 0.061 0.081 0.079 0.048
3 0.046 0.070 0.090 0.079
4 0.017 0.036 0.068 0.111
```

9.5.2 Beta-binomial sampling

The previous example demonstrated Gibbs sampling for a two-parameter discrete distribution. In fact, the Gibbs sampling algorithm works for any two-parameter distribution. To illustrate, consider a familiar Bayesian model discussed in Chapter 7. Suppose we flip a coin n times and observe y heads where the probability of heads is p, and our prior for the heads probability is described by a beta curve with shape parameters a and b. It is convenient to write $X \mid Y = y$ as the conditional distribution of X given $Y = y$. Using this notation we have

$$
\begin{aligned}
Y \mid p &\sim \text{Binomial}(n, p), && (9.11)\\
p &\sim \text{Beta}(a, b). && (9.12)
\end{aligned}
$$

To implement Gibbs sampling for this situation, one needs to identify the two conditional distributions $Y \mid p$ and $p \mid Y$. First write down the joint density of (Y, p) which is found by multiplying the marginal density $\pi(p)$ with the conditional density $f(y \mid p)$.

$$
\begin{aligned}
f(Y = y, p) &= \pi(p) f(Y = y \mid p)\\
&= \left[\frac{1}{B(a,b)} p^{a-1}(1-p)^{b-1}\right]\left[\binom{n}{y} p^{y}(1-p)^{n-y}\right].
\end{aligned}
\tag{9.13}
$$

1. The conditional density $f(Y = y \mid p)$ is found by fixing a value of the proportion p and then the only random variable is Y. This distribution is Binomial(n, p) which actually was given in the statement of the problem.
2. Turning things around, the conditional density $\pi(p \mid y)$ takes the number of successes y and views the joint density as a function only of the random variable p. Ignoring constants, we see this conditional density is proportional to

$$
p^{y+a-1}(1-p)^{n-y+b-1}, \tag{9.14}
$$

which we recognize as a beta distribution with shape parameters $y + a$ and $n - y + b$. Using our notation, we have $p \mid y \sim \text{Beta}(y + a, n - y + b)$.

R Once these conditional distributions are identified, it is straightforward to write an algorithm to implement Gibbs sampling. For example, suppose $n = 20$ and the prior density for p is Beta(5, 5). Suppose that the current simulated value of p is $p^{(j)}$.

1. Simulate $Y^{(j)}$ from a Binomial$(20, p^{(j)})$ distribution.

   ```
   y <- rbinom(1, size = 20, prob = p)
   ```

2. Given the current simulated value $y^{(j)}$, simulate $p^{(j+1)}$ from a beta distribution with shape parameters $y^{(j)} + 5$ and $20 - y^{(j)} + 5$.

   ```
   p <- rbeta(1, y + a, n - y + b)
   ```

The R function `gibbs_betabin()` will implement Gibbs sampling for this problem. One inputs the sample size n and the shape parameters a and b. By default, one starts the algorithm at the proportion value $p = 0.5$ and one takes 1000 iterations of the algorithm.

```
gibbs_betabin <- function(n, a, b, p = 0.5, iter = 1000){
  x <- matrix(0, iter, 2)
  for(k in 1:iter){
    y <- rbinom(1, size = n, prob = p)
    p <- rbeta(1, y + a, n - y + b )
    x[k, ] <- c(y, p)
  }
  x
}
```

Below we run Gibbs sampling for this beta-binomial model with $n = 20$, $a = 5$, and $b = 5$. After performing 1000 iterations, one regards the matrix `sp` as an approximate simulated sample from the joint distribution of Y and p. A histogram is constructed of the simulated draws of Y in Figure 9.9. This graph represents an approximate sample from the marginal distribution $f(y)$ of Y.

```
sp <- data.frame(gibbs_betabin(20, 5, 5))
```

9.5.3 Normal sampling – both parameters unknown

In Chapter 8, we considered the situation of sampling from a normal distribution with mean μ and standard deviation σ. To simplify this to a one-parameter model, we assumed that the value of σ was known and focused on the problem of learning about the mean μ. Since Gibbs sampling allows

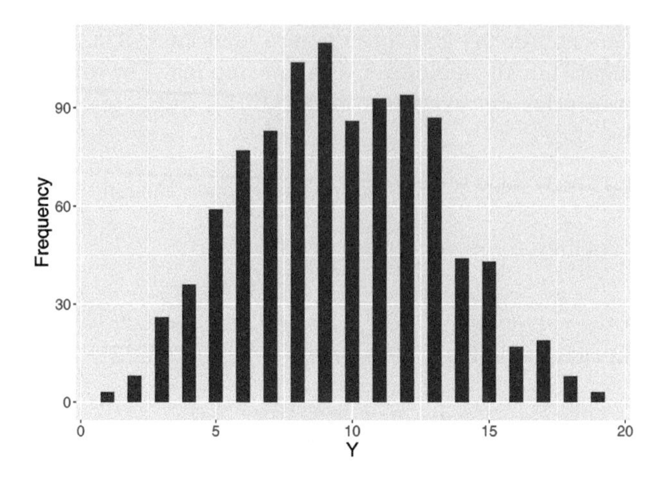

FIGURE 9.9
Histogram of simulated draws of Y from Gibbs sampling for the beta-binomial model with $n = 20$, $a = 5$, and $b = 5$.

us to simulate from posterior distributions of more than one parameter, we can generalize to the more realistic situation where both the mean and the standard deviation are unknown.

Suppose we take a sample of n observations $Y_1, .., Y_n$ from a normal distribution with mean μ and variance σ^2. Recall the sampling density of Y_i has the form

$$f(y_i \mid \mu, \sigma) = \frac{1}{\sqrt{2\pi}\sigma} \exp\left\{-\frac{1}{2\sigma^2}(y_i - \mu)^2\right\}. \tag{9.15}$$

It will be convenient to reexpress the variance σ by the *precision* ϕ where

$$\phi = \frac{1}{\sigma^2}. \tag{9.16}$$

The precision ϕ reflects the strength in knowledge about the location of the observation Y_i. If Y_i is likely to be close to the mean μ, then the variance σ^2 would be small and so the precision ϕ would be large. So we restate the sampling model as follows. The observations $Y_1, .., Y_n$ are a random sample from a normal density with mean μ and precision ϕ, where the sampling density of Y_i is given by

$$f(y_i \mid \mu, \phi) = \frac{\sqrt{\phi}}{\sqrt{2\pi}} \exp\left\{-\frac{\phi}{2}(y_i - \mu)^2\right\}. \tag{9.17}$$

The next step is to construct a prior density on the parameter vector (μ, ϕ). A convenient choice for this prior is to assume that one's opinion about the location of the mean μ is independent of one's belief about the location of the precision ϕ. So we assume that μ and ϕ are independent, so one writes the joint prior density as

$$\pi(\mu, \phi) = \pi_\mu(\mu)\pi_\phi(\phi), \tag{9.18}$$

where $\pi_\mu()$ and $\pi_\phi()$ are marginal densities. For convenience, each of these marginal priors are assigned conjugate forms: we assume that μ is normal with mean μ_0 and precision ϕ_0:

$$\pi_\mu(\mu) = \frac{\sqrt{\phi_0}}{\sqrt{2\pi}} \exp\left\{-\frac{\phi_0}{2}(\mu - \mu_0)^2\right\}. \tag{9.19}$$

The prior for the precision parameter ϕ is assumed gamma with parameters a and b:

$$\pi_\phi(\phi) = \frac{b^a}{\Gamma(a)}\phi^{a-1} \exp(-b\phi), \ \phi > 0. \tag{9.20}$$

Once values of $y_1, ..., y_n$ are observed, the likelihood is the density of these normal observations viewed as a function of the mean μ and the precision parameter ϕ. Simplifying the expression and removing constants, one obtains:

$$\begin{aligned} L(\mu, \phi) &= \prod_{i=1}^{n} \frac{\sqrt{\phi}}{\sqrt{2\pi}} \exp\left\{-\frac{\phi}{2}(y_i - \mu)^2\right\} \\ &\propto \phi^{n/2} \exp\left\{-\frac{\phi}{2}\sum_{i=1}^{n}(y_i - \mu)^2\right\}. \end{aligned} \tag{9.21}$$

To implement Gibbs sampling, one first writes down the expression for the posterior density as the product of the likelihood and prior where any constants not involving the parameters are removed.

$$\begin{aligned} \pi(\mu, \phi \mid y_1, \cdots, y_n) \quad &\propto \quad \phi^{n/2} \exp\left\{-\frac{\phi}{2}\sum_{i=1}^{n}(y_i - \mu)^2\right\} \\ &\times \quad \exp\left\{-\frac{\phi_0}{2}(\mu - \mu_0)^2\right\} \phi^{a-1} \exp(-b\phi). \end{aligned} \tag{9.22}$$

Next, the two conditional posterior distributions $\pi(\mu \mid \phi, y_1, \cdots, y_n)$ and $\pi(\phi \mid \mu, y_1, \cdots, y_n)$ are identified.

1. The first conditional density $\pi(\mu \mid \phi, y_1, \cdots, y_n)$ follows from the work in Chapter 8 on Bayesian inference about a mean with a conjugate prior when the sampling standard deviation was assumed known. One obtains that this conditional distribution $\pi(\mu \mid \phi, y_1, \cdots, y_n)$ is normal with mean

$$\mu_n = \frac{\phi_0\mu_0 + n\phi\bar{y}}{\phi_0 + n\phi}. \tag{9.23}$$

and standard deviation

$$\sigma_n = \sqrt{\frac{1}{\phi_0 + n\phi}}. \tag{9.24}$$

2. Collecting terms, the second conditional density $\pi(\phi \mid \mu, y_1, \cdots, y_n)$ is proportional to

$$\pi(\phi \mid \mu, y_1, \cdots y_n) \propto \phi^{n/2+a-1} \exp\left\{-\phi\left[\frac{1}{2}\sum_{i=1}^{n}(y_i - \mu)^2 + b\right]\right\}. \tag{9.25}$$

The second conditional distribution $\pi(\phi \mid \mu, y_1, \cdots, y_n)$ is seen to be a gamma density with parameters

$$a_n = \frac{n}{2} + a, \tag{9.26}$$

$$b_n = \frac{1}{2}\sum_{i=1}^{n}(y_i - \mu)^2 + b. \tag{9.27}$$

R An R function `gibbs_normal()` is written to implement this Gibbs sampling simulation. The inputs to this function are a list `s` containing the vector of observations `y` and the prior parameters `mu0`, `phi0`, `a`, and `b`, the starting value of the precision parameter ϕ, `phi`, and the number of Gibbs sampling iterations `S`. This function is similar in structure to the `gibbs_betabin()` function – the two simulations in the Gibbs sampling are accomplished by use of the `rnorm()` and `rgamma()` functions.

```
gibbs_normal <- function(s, phi = 0.002, iter = 1000){
  ybar <- mean(s$y)
  n <- length(s$y)
  mu0 <- s$mu0
  phi0 <- s$phi0
  a <- s$a
  b <- s$b
  x <- matrix(0, iter, 2)
  for(k in 1:iter){
   mun <- (phi0 * mu0 + n * phi * ybar) /
      (phi0 + n * phi)
    sigman <- sqrt(1 / (phi0 + n * phi))
    mu <- rnorm(1, mean = mun, sd = sigman)
    an <- n / 2 + a
    bn <- sum((s$y - mu) ^ 2) / 2 + b
    phi <- rgamma(1, shape = an, rate = bn)
    x[k, ] <- c(mu, phi)
```

```
  }
  x
}
```

We run this function for our Buffalo snowfall example where now the sampling model is normal with both the mean μ and standard deviation σ unknown. The prior distribution assumes that μ and the precision ϕ are independent, where μ is normal with mean 10 and standard deviation 3 (i.e. precision $1/3^2$), and ϕ is gamma with $a = b = 1$. The output of this function is a matrix `out` where the two columns of the matrix correspond to random draws of μ and ϕ from the posterior distribution.

```
s <- list(y = data$JAN, mu0 = 10, phi0 = 1/3^2, a = 1, b = 1)
out <- gibbs_normal(s, iter=10000)
```

By performing the transformation $\sigma = \sqrt{1/\phi}$, one obtains a sample of the simulated draws of the standard deviation σ. Figure 9.10 displays a scatterplot of the posterior draws of μ and σ.

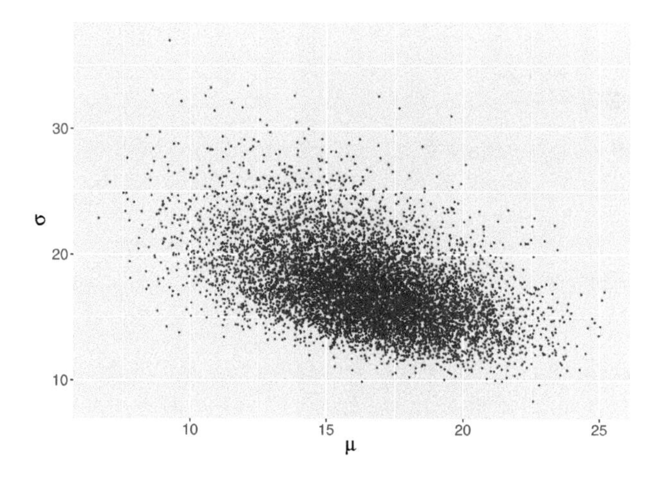

FIGURE 9.10
Scatterplot of simulated draws of the posterior distribution of μ and σ from Gibbs sampling for the normal sampling model with independent priors on μ and the precision ϕ.

9.6 MCMC Inputs and Diagnostics

9.6.1 Burn-in, starting values, and multiple chains

In theory, the Metropolis and Gibbs sampling algorithms will produce simulated draws that converge to the posterior distribution of interest. But in typical practice, it may take a number of iterations before the simulation values are close to the posterior distribution. So in general it is recommended that one run the algorithm for a number of "burn-in" iterations before one collects iterations for inference. The JAGS software that is introduced in Section 9.7 will allow the user to specify the number of burn-in iterations.

In the examples, we have illustrated running a single "chain" where one has a single starting value and one collects simulated draws from many iterations. It is possible that the MCMC sample will depend on the choice of starting value. So a general recommendation is to run the MCMC algorithm several times using different starting values. In this case, one will have multiple MCMC chains. By comparing the inferential summaries from the different chains one explores the sensitivity of the inference to the choice of starting value. Although we will focus on the use of a single chain, we will explore the use of different starting values and multiple chains in an example in this chapter. The JAGS software and other programs to implement MCMC will allow for different starting values and several chains.

9.6.2 Diagnostics

The output of a single chain from the Metropolis and Gibbs algorithms is a vector or matrix of simulated draws. Before one believes that a collection of simulated draws is a close approximation to the posterior distribution, some special diagnostic methods should be initially performed.

Trace plot

It is helpful to construct a trace plot which is a line plot of the simulated draws of the parameter of interest graphed against the iteration number. Figure 9.11 displays a trace plot of the simulated draws of μ from the Metropolis algorithm for our Buffalo snowfall example for normal sampling (known standard deviation) with a Cauchy prior. Section 9.4.1 shows some sample trace plots for Metropolis sampler. As discussed in that section, it is undesirable to have a snake-like appearance in the trace plot indicating a high acceptance rate. Also, Section 9.4.1 displays a trace plot with many flat portions that indicates a sampler with a low acceptance rate. From the authors' experience, the trace plot in Figure 9.11 indicates that the sampler is using a good value of the constant C and efficiently sampling from the posterior distribution.

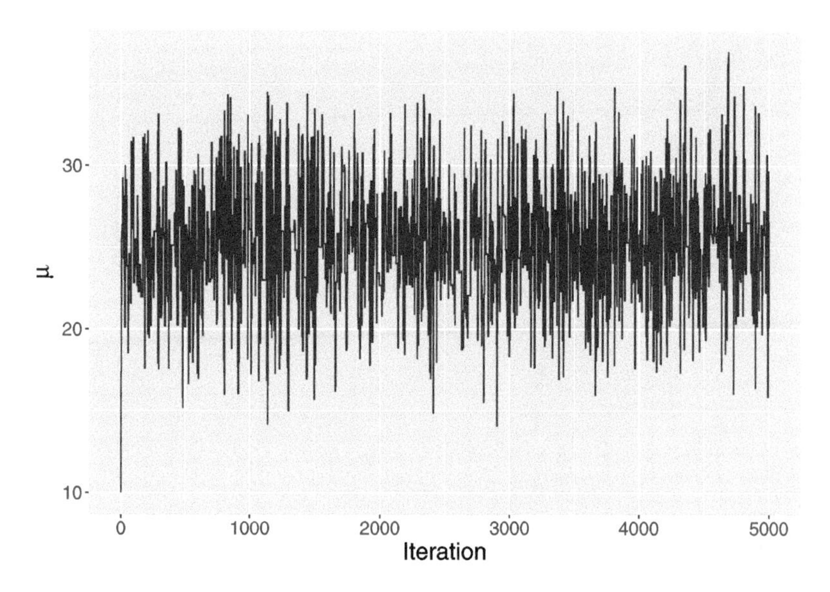

FIGURE 9.11
Trace plot of simulated draws of μ using the Metropolis algorithm with $C = 20$.

Autocorrelation plot

Since one is simulating a dependent sequence of values of the parameter, one is concerned about the possible strong correlation between successive draws of the sampler. One visualizes this dependence by computing the correlation of the pairs $\{\theta^{(j)}, \theta^{(j+l)}\}$ and plotting this "lag-correlation" as a function of the lag value l. This autocorrelation plot of the simulated draws from our example is displayed in Figure 9.12. If there is a strong degree of autocorrelation in the sequence, then there will be a large correlation of these pairs even for large values of the lag value. Figure 9.12 is an example of a suitable autocorrelation graph where the lag correlation values quickly drop to zero as a function of the lag value. This autocorrelation graph is another indication that the Metropolis algorithm is providing an efficient sampler of the posterior.

9.6.3 Graphs and summaries

If the trace plot or autocorrelation plot indicate issues with the Metropolis sampler, then the width of the proposal C should be adjusted and the algorithm run again. Since we believe that the Metropolis simulation stream is reasonable with the use of the value $C = 20$, then we use a histogram of simulated draws, as displayed in Figure 9.13 to represent the posterior distribution. Alternatively, a density estimate of the simulated draws can be used to show a

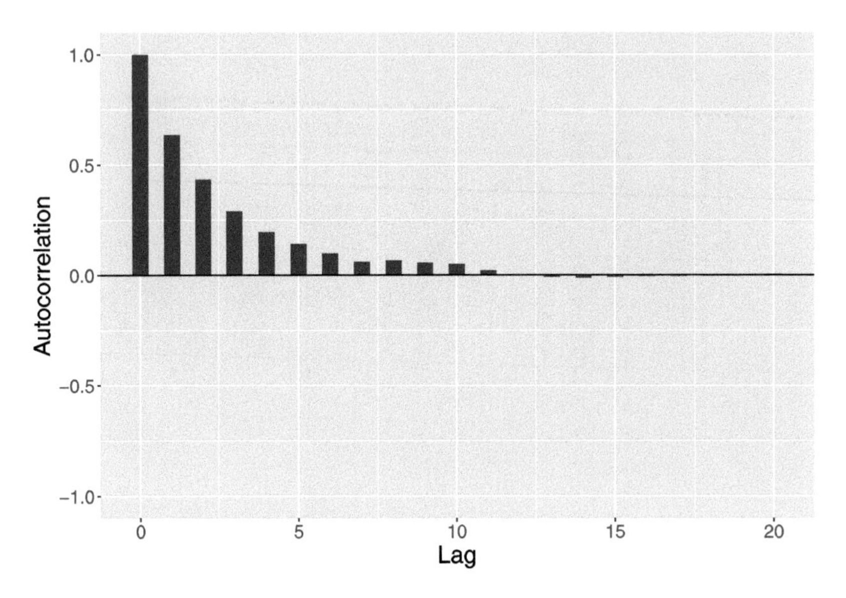

FIGURE 9.12
Autocorrelation plot of simulated draws of μ using the Metropolis algorithm with $C = 20$.

smoothed representation of the posterior density. Figure 9.13 places a density estimate on top of the histogram of the simulated values of the parameter μ.

One estimates different summaries of the posterior distribution by computing different summaries of the simulated sample. In our Cauchy-normal model, one estimates, for example, the posterior mean of μ by computing the mean of the simulated posterior draws:

$$E(\mu \mid y) \approx \frac{\sum_{j=1}^{S} \mu^{(j)}}{S}. \tag{9.28}$$

One typically wants to estimate the simulation standard error of this MCMC estimate. If the draws from the posterior were independent, then the Monte Carlo standard error of this posterior mean estimate would be given by the standard deviation of the draws divided by the square root of the simulation sample size:

$$se = \frac{sd(\{\mu^{(j)}\})}{\sqrt{S}}. \tag{9.29}$$

However, this estimate of the standard error is not correct since the MCMC sample is not independent (the simulated value $\mu^{(j)}$ depends on the value of the previous simulated value $\mu^{(j-1)}$). One obtains a more accurate estimate of Monte Carlo standard error by using time-series methods. As we will see in the examples of Section 9.7, this standard error estimate will be larger than

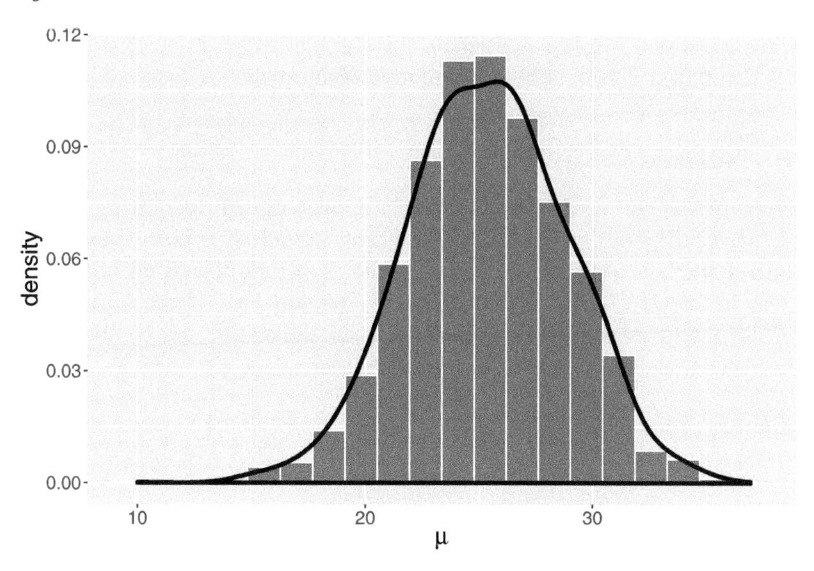

FIGURE 9.13
Histogram of simulated draws of μ using the Metropolis algorithm with $C = 20$. The solid curve is a density estimate of the simulated values.

the "naive" standard error estimate that assumes the MCMC sample values are independent.

9.7 Using JAGS

Sections 9.3 and 9.5 have illustrated general strategies for simulating from a posterior distribution of one or more parameters. Over the years, there has been an effort to develop general-purpose Bayesian computing software that would take a Bayesian model (i.e. the specification of a prior and sampling density as input), and use an MCMC algorithm to output a matrix of simulated draws from the posterior. One of the earliest Bayesian simulation-based computing software was BUGS (for Bayesian inference Using Gibbs Sampling) and we illustrate in this text applications of a similar package JAGS (for Just Another Gibbs Sampler).

The use of JAGS has several attractive features. One defines a Bayesian model for a particular problem by writing a short script. One then inputs this script together with data and prior parameter values in a single R function from the `runjags` package that decides on the appropriate MCMC sampling algorithm for the particular Bayesian model. In addition, this function simulates from the MCMC algorithm for a specified number of samples and collects simulated draws of the parameters of interest.

9.7.1 Normal sampling model

To illustrate the use of JAGS, consider the problem of estimating the mean Buffalo snowfall assuming a normal sampling model with both the mean and standard deviation unknown, and independent priors placed on both parameters. As in Section 9.5.3 one expresses the parameters of the normal distribution as μ and ϕ, where the precision ϕ is the reciprocal of the variance $\phi = 1/\sigma^2$. One then writes this Bayesian model as

- Sampling, for $i = 1, \cdots, n$:

$$Y_i \overset{i.i.d.}{\sim} \text{Normal}(\mu, \sqrt{1/\phi}). \tag{9.30}$$

- Independent priors for μ and ϕ:

$$\mu \sim \text{Normal}(\mu_0, \sqrt{1/\phi_0}), \tag{9.31}$$
$$\phi \sim \text{Gamma}(a, b). \tag{9.32}$$

The JAGS program parameterizes a normal density in terms of the precision, so the prior precision is equal to $\phi_0 = 1/\sigma_0^2$. As in Section 9.5.3, the parameters of the normal and gamma priors are set at $\mu_0 = 10, \phi_0 = 1/3^2, a = 1, b = 1$.

Describe the model by a script

R To begin, one writes the following script defining this model. The model is saved in the character string `modelString`.

```
modelString = "
model{
## sampling
for (i in 1:N) {
   y[i] ~ dnorm(mu, phi)
}
## priors
mu ~ dnorm(mu0, phi0)
phi ~ dgamma(a, b)
sigma <- sqrt(pow(phi, -1))
}
"
```

Note that this script closely resembles the statement of the model. In the sampling part of the script, the loop structure starting with `for (i in`

1:N) is used to assign the distribution of each value in the data vector y the same normal distribution, represented by dnorm. The ~ operator is read as "is distributed as".

In the priors part of the script, in addition to setting the normal prior and gamma prior for mu and phi respectively, sigma <- sqrt(pow(phi, -1)) is added to help track sigma directly.

Define the data and prior parameters

The next step is to define the data and provide values for parameters of the prior. In the script below, a list the_data is used to collect the vector of observations y, the number of observations N, and values of the normal prior parameters mu0, phi0, and of the gamma prior parameters a and b.

```
buffalo <- read.csv("../data/buffalo_snowfall.csv")
data <- buffalo[59:78, c("SEASON", "JAN")]
y <- data$JAN
N <- length(y)
the_data <- list("y" = y, "N" = N,
                 "mu0"=10, "phi0"=1/3^2,
                 "a"=1,"b"=1)
```

Define initial values

One needs to supply initial values in the MCMC simulation for all of the parameters in the model. To obtain reproducible results, one can use the initsfunction() function shown below to set the seed for the sequence of simulated parameter values in the MCMC.

```
initsfunction <- function(chain){
  .RNG.seed <- c(1,2)[chain]
  .RNG.name <- c("base::Super-Duper",
                 "base::Wichmann-Hill")[chain]
  return(list(.RNG.seed=.RNG.seed,
              .RNG.name=.RNG.name))
}
```

Alternatively, one can specify the initial values by means of a function – this will be implemented when multiple chains are discussed. If no initial values are specified, then JAGS will select initial values – these are usually a "typical" value such as a mean or median from the prior distribution.

Generate samples from the posterior distribution

Now that the model definition and data have been defined, one is ready to draw samples from the posterior distribution. The runjags provides the R interface to the use of the JAGS software. The run.jags() function sets up the

Bayesian model defined in `modelString`. The input `n.chains = 1` indicates that one stream of simulated values will be generated. `adapt = 1000` says that 1000 simulated iterations are used in "adapt" period to prepare for MCMC, `burnin = 1000` indicates 5000 simulated iterations are used in a "burn-in" period where the iterations are approaching the main probability region of the posterior distribution. The `sample = 5000` arguments indicates that 5000 additional iterations of the MCMC algorithm will be collected. The `monitor` arguments says that we are collecting simulated values of the mean `mu` and the standard deviation `sigma`. The output variable `posterior` includes a matrix of the simulated draws. The `inits = initsfunction` argument indicates that initial parameter values are chosen by the `initsfunction()` function.

```
posterior <- run.jags(modelString,
                      n.chains = 1,
                      data = the_data,
                      monitor = c("mu", "sigma"),
                      adapt = 1000,
                      burnin = 5000,
                      sample = 5000,
                      inits = initsfunction)
```

MCMC diagnostics and summarization

Before summarizing the simulated sample, some graphical diagnostics methods should be implemented to judge if the sample appears to "mix" or move well across the space of likely values of the parameters. The `plot()` function in the `runjags` package constructs a collection of four graphs for a parameter of interest. By running `plot()` for `mu` and `sigma`, we obtain the graphs displayed in Figures 9.14 and 9.15.

```
plot(posterior, vars = "mu")
plot(posterior, vars = "sigma")
```

The trace and autocorrelation plots in the top left and bottom right sections of the display are helpful for seeing how the sampler moves across the posterior distribution. In Figures 9.14 and 9.15, the trace plots show little autocorrelation in the streams of simulated draws and both simulated samples of μ and σ appear to mix well. In the autocorrelation plots, the value of the autocorrelation drops sharply to zero as a function of the lag which confirms that we have modest autocorrelation in these samples. In each display, the bottom left graph is a histogram of the simulated draws and the top right graph is an estimate at the cumulative distribution function of the variable.

Since we are encouraged by these diagnostic graphs, we go ahead and obtain summaries of the simulated samples of μ and σ by the `print()` function on our MCMC object. The posterior mean of μ is 16.5. The standard error of this simulation estimate is the "MCerr" value of 0.0486 – this standard

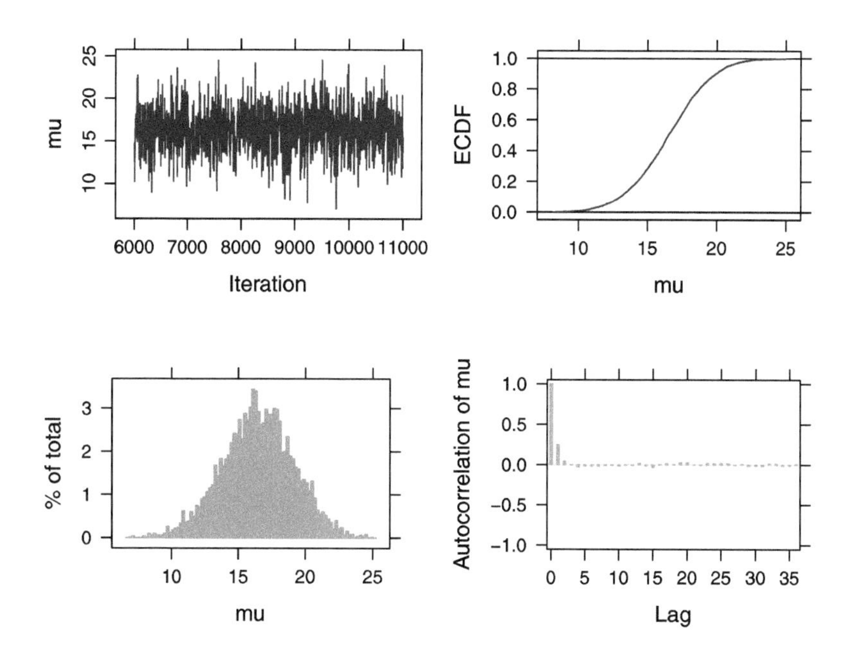

FIGURE 9.14
Diagnostic plots of simulated draws of μ using the JAGS software with the `runjags` package.

error takes in account the correlated nature of these simulated draws. A 90% probability interval for the mean μ is found from the output to be (10.8, 21.4). For σ, it has a posterior mean of 17.4, and a 90% probability interval (11.8, 24).

```
print(posterior, digits = 3)
      Lower95 Median Upper95 Mean   SD Mode  MCerr
mu       10.8   16.5    21.4 16.5 2.68   -- 0.0486
sigma    11.8   17.1      24 17.4 3.18   -- 0.0576
```

9.7.2 Multiple chains

In Section 9.6.1, we explained the benefit of trying different starting values and running several MCMC chains. This is facilitated by arguments in the `run.jags()` function. Suppose one considers the very different pairs of starting values, $(\mu, \phi) = (2, 1/4)$ and $(\mu, \phi) = (30, 1/900)$. Note that both pairs of parameter values are far outside of the region where the posterior density is

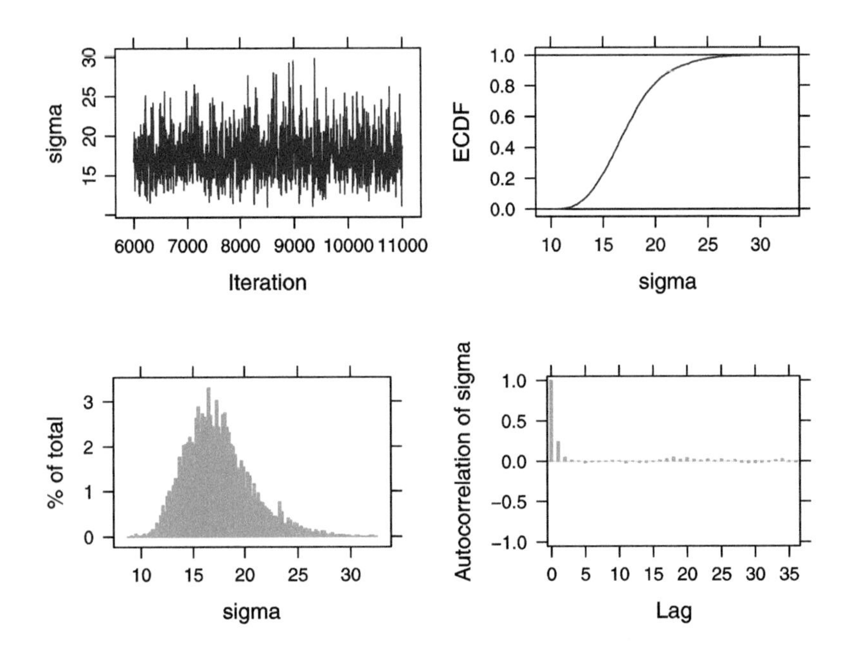

FIGURE 9.15
Diagnostic plots of simulated draws of σ using the JAGS software with the `runjags` package.

concentrated. One defines `InitialValues` as containing two lists, each containing a starting value.

```
InitialValues <- list(
  list(mu = 2, phi = 1 / 4),
  list(mu = 30, phi = 1 / 900)
)
```

The `run.jags()` function is run with two modifications – one chooses `n.chains = 2` and the initial values are input through the `inits = InitialValues` option.

```
posterior <- run.jags(modelString,
                      n.chains = 2,
                      data = the_data,
                      monitor = c("mu", "sigma"),
                      adapt = 1000,
                      burnin = 5000,
```

```
                  sample = 5000,
                  inits = InitialValues)
```

The output variable `posterior` contains a component `mcmc` which is a list of two components where `posterior$mcmc[[1]]` contains the simulated draws from the first chain and `posterior$mcmc[[2]]` contains the simulated draws from the second chain. To see if the MCMC run is sensitive to the choice of starting value, one compares posterior summaries from the two chains. Below, we display posterior quantiles for the parameters μ and σ for each chain. Note that these quantiles are very close in value indicating that the MCMC run is insensitive to the choice of starting value.

```
summary(posterior$mcmc[[1]], digits = 3)
2. Quantiles for each variable:

       2.5%   25%   50%   75% 97.5%
mu    10.99 14.64 16.49 18.35 21.62
sigma 12.26 15.15 17.03 19.31 25.07

summary(posterior$mcmc[[2]], digits = 3)
2. Quantiles for each variable:

       2.5%   25%   50%   75% 97.5%
mu    10.97 14.59 16.55 18.33 21.54
sigma 12.21 15.08 16.96 19.18 24.99
```

9.7.3 Posterior predictive checking

In Chapter 8 Section 8.7, we illustrated the usefulness of the posterior predictive checking in model checking. The basic idea is to simulate a number of replicated datasets from the posterior predictive distribution and see how the observed sample compares to the replications. If the observed data does resemble the replications, one says that the observed data is consistent with predicted data from the Bayesian model.

For our Buffalo snowfall example, suppose we wish to simulate a replicated sample from the posterior predictive distribution. Since our original sample size was $n = 20$, the intent is to simulate a sample of values $\tilde{y}_1, ..., \tilde{y}_{20}$ from the posterior predictive distribution. A single replicated sample is simulated in the following two steps.

1. We draw a set of parameter values, say μ^*, σ^* from the posterior distribution of (μ, σ).

2. Given these parameter values, we simulate $\tilde{y}_1, ..., \tilde{y}_{20}$ from the normal sampling density with mean μ^* and standard deviation σ^*.

R Recall that the simulated posterior values are stored in the matrix post. We write a function postpred_sim() to simulate one sample from the predictive distribution.

```
post <- data.frame(posterior$mcmc[[1]])
postpred_sim <- function(j){
  rnorm(20, mean = post[j, "mu"],
        sd = post[j, "sigma"])
}
print(postpred_sim(1), digits = 3)
 [1]   5.37  10.91  40.87  15.94  16.93  43.49  22.48
 [8]  -6.43   3.26   7.30  35.27  20.79  21.47  16.62
[15]   5.45  44.69  23.10 -18.18  26.51   6.84
```

If this process is repeated for each of the 5000 draws from the posterior distribution, then one obtains 5000 samples of size 20 drawn from the predictive distribution. In R, the function sapply() is used together with postpred_sim() to simulate 5000 samples that are stored in the matrix ypred.

```
ypred <- t(sapply(1:5000, postpred_sim))
```

Figure 9.16 displays histograms of the predicted snowfalls from eight of these simulated samples and the observed snowfall measurements are displayed in the lower right panel. Generally, the center and spread of the observed snowfalls appear to be similar in appearance to the eight predicted snowfall samples from the fitted model. Can we detect any differences between the distribution of observed snowfalls and the distributions of predicted snowfalls? One concern is that some of the predictive samples contain negative snowfall values. Another concern from this inspection is that we observed a snowfall of 65.1 inches in our sample and none of our eight samples had a snowfall this large. Perhaps there is an outlier in our sample that is not consistent with predictions from our model.

When one notices a possible discrepancy between the observed sample and simulated prediction samples, one thinks of a checking function $T()$ that will distinguish the two types of samples. In this situation since we noticed the extreme snowfall of 65.1 inches, that suggests that we use $T(y) = \max y$ as a checking function.

Once one decides on a checking function $T()$, then one simulates the posterior predictive distribution of $T(\tilde{y})$. This is conveniently done by evaluating the function $T()$ on each simulated sample from the predictive distribution. In R, this is conveniently done using the apply() function and the values of $T(\tilde{y})$ are stored in the vector postpred_max.

```
postpred_max <- apply(ypred, 1, max)
```

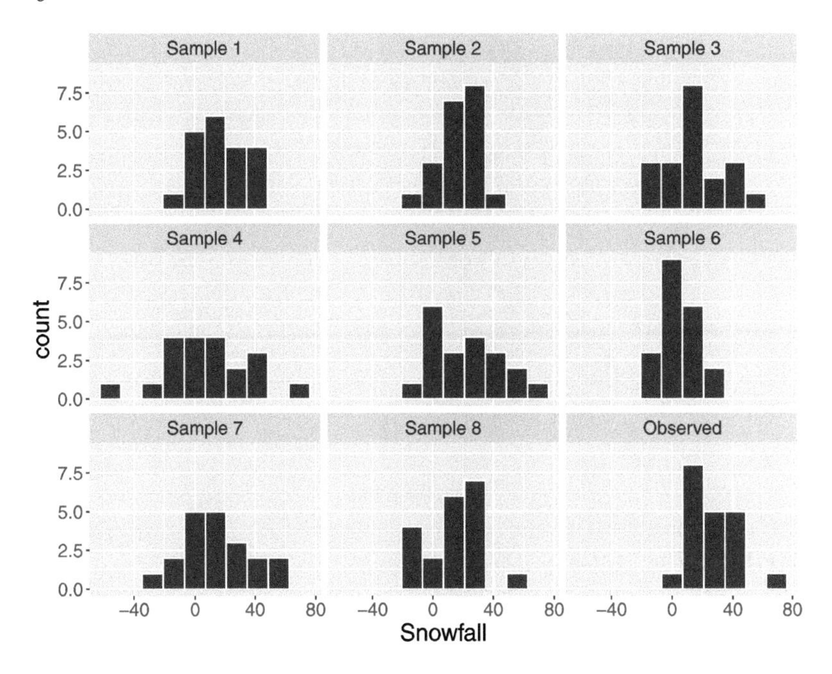

FIGURE 9.16
Histograms of eight simulated predictive samples and the observed sample for the snowfall example.

If the checking function evaluated at the observed sample $T(y)$ is not consistent with the distribution of $T(\tilde{y})$, then predictions from the model are not similar to the observed data and there is some issue with the model assumptions. Figure 9.17 displays a histogram of the predictive distribution of $T(y)$ in our example where $T()$ is the maximum function, and the observed maximum snowfall is shown by a vertical line. Here the observed maximum is in the right tail of the posterior predictive distribution – the interpretation is that this largest snowfall of 65.1 inches is not predicted from the model. In this case, one might want to think about revising the sampling model, say, by assuming that the data follow a distribution with flatter tails than the normal.

9.7.4 Comparing two proportions

To illustrate the usefulness of the JAGS software, we consider a problem comparing two proportions from independent samples. The model is defined in a JAGS script, the data and values of prior parameters are entered through a list, and the `run.jags()` function is used to simulate from the posterior of the parameters by an MCMC algorithm.

To better understand the behavior of Facebook users, a survey was administered in 2011 to 244 students. Each student was asked his or her gender and

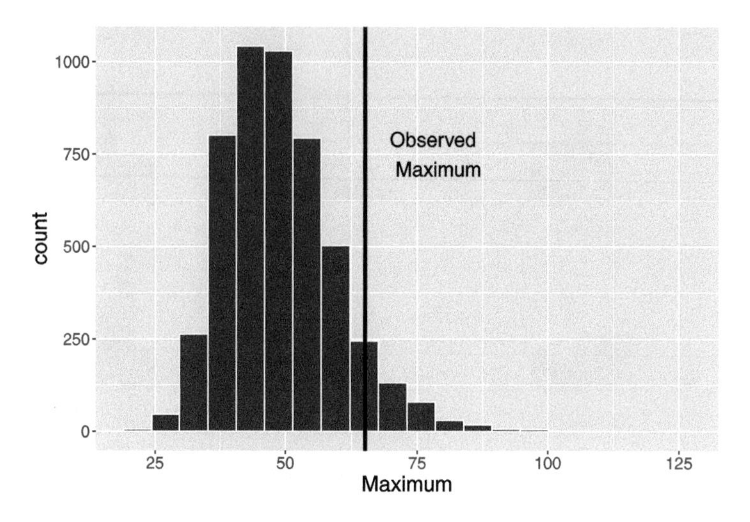

FIGURE 9.17
Histogram of the posterior predictive distribution of $T(\tilde{y})$ where $T()$ is the maximum function. The vertical line shows the location of the observed value $T(y)$.

the average number of Facebook visits in a day. We say that the number of daily visits is "high" if the number of visits is 5 or more; otherwise it is "low". If we classify the sample by gender and daily visits, we obtain the two by two table of counts as shown in Table 9.1.

TABLE 9.1
Two-way table of counts of students by gender and Facebook visits.

	Visits to Facebook	
Gender	High	Low
Male	y_M	$n_M - y_M$
Female	y_F	$n_F - y_F$

In Table 9.1, the random variable Y_M represents the number of males who have a high number of Facebook visits in a sample of n_M, and Y_F and n_M are the analogous count and sample size for women. Assuming that the sample survey represents a random sample from all students using Facebook, then it is reasonable to assume that Y_M and Y_F are independent with Y_M distributed binomial with parameters n_M and p_M, and Y_F is binomial with parameters n_F and p_F.

TABLE 9.2
Probability structure in two-way table.

	Visits to Facebook	
Gender	High	Low
Male	p_M	$1 - p_M$
Female	p_F	$1 - p_F$

The probabilities p_M and p_F are displayed in Table 9.2. In this type of data structure, one is interested in the association between gender and Facebook visits. Define the odds as the ratio of the probability of high to the probability of low. The odds of high for the men and odds of high for the women are defined by

$$\frac{p_M}{1 - p_M}, \tag{9.33}$$

and

$$\frac{p_F}{1 - p_F}, \tag{9.34}$$

respectively. The odds ratio

$$\alpha = \frac{p_M/(1 - p_M)}{p_F/(1 - p_F)}, \tag{9.35}$$

is a measure of association in this two-way table. If $\alpha = 1$, this means that $p_M = p_L$ – this says that tendency to have high numbers of visits to Facebook does not depend on gender. If $\alpha > 1$, this indicates that men are more likely to have high numbers of visits to Facebook, and a value $\alpha < 1$ indicates that women are more likely to have high numbers of visits. Sometimes association is expressed on a log scale – the log odds ratio λ is written as

$$\lambda = \log \alpha = \log\left(\frac{p_M}{1 - p_M}\right) - \log\left(\frac{p_F}{1 - p_F}\right). \tag{9.36}$$

That is, the log odds ratio is expressed as the difference in the logits of the men and women probabilities, where the logit of a probability p is equal to $\text{logit}(p) = \log(p) - \log(1 - p)$. If gender is independent of Facebook visits, then $\lambda = 0$.

One's prior beliefs about association in the two-way table is expressed in terms of logits and the log odds ratio. If one believes that gender and Facebook visits are independent, then the log odds ratio is assigned a normal prior with mean 0 and standard deviation σ. The mean of 0 reflects the prior guess of independence and σ indicates the strength of the belief in independence. If one believed strongly in independence, then one would assign σ a small value.

In addition, let

$$\theta = \frac{\text{logit}(p_M) + \text{logit}(\text{p}_\text{F})}{2} \tag{9.37}$$

be the mean of the logits, and assume that θ has a normal prior with mean θ_0 and standard deviation σ_0 (precision ϕ_0). The prior on θ reflects beliefs about the general size of the proportions on the logit scale.

R To fit this model using JAGS, the following script, saved in `modelString`, is written defining the model.

```
modelString = "
model{
## sampling
yF ~ dbin(pF, nF)
yM ~ dbin(pM, nM)
logit(pF) <- theta - lambda / 2
logit(pM) <- theta + lambda / 2
## priors
theta ~ dnorm(mu0, phi0)
lambda ~ dnorm(0, phi)
}
"
```

In the sampling part of the script, the two first lines define the binomial sampling models, and the logits of the probabilities are defined in terms of the log odds ratio `lambda` and the mean of the logits `theta`. In the priors part of the script, note that `theta` is assigned a normal prior with mean `mu0` and precision `phi0`, and `lambda` is assigned a normal prior with mean 0 and precision `phi`.

When the sample survey is conducted, one observes that 75 of the 151 female students say that they are frequent visitors of Facebook, and 39 of the 93 male students are frequent visitors. This data and the values of the prior parameters are entered into R by use of a list. Note that `phi = 2` indicating some belief that gender is independent of Facebook visits, and `mu0 = 0` and `phi0 = 0.001` reflecting little knowledge about the location of the logit proportions. Using the `run.jags()` function, we take an adapt period of 1000, burn-in period of 5000 iterations and collect 5000 iterations, storing values of `pF`, `pM` and the log odds ratio `lambda`.

```
the_data <- list("yF" = 75, "nF" = 151,
                 "yM" = 39, "nM" = 93,
                 "mu0" = 0, "phi0" = 0.001, "phi" = 2)

posterior <- run.jags(modelString,
                 data = the_data,
                 n.chains = 1,
                 monitor = c("pF", "pM", "lambda"),
                 adapt = 1000,
                 burnin = 5000,
                 sample = 5000)
```

Since the main goal is to learn about the association structure in the table, Figure 9.18 displays a density estimate of the posterior draws of the log odds ratio λ. A reference line at $\lambda = 0$ is drawn on the graph which corresponds to the case where $p_M = p_L$. What is the probability that women are more likely than men to make more visits to Facebook? This is directly answered by computing the posterior probability $Prob(\lambda < 0 \mid data)$ that is computed to be 0.874. Based on this computation, one concludes that it is very probable that women have a greater tendency than men to visit Facebook frequently.

```
post <- data.frame(posterior$mcmc[[1]])
post %>%
  summarize(Prob = mean(lambda < 0))
     Prob
1 0.874
```

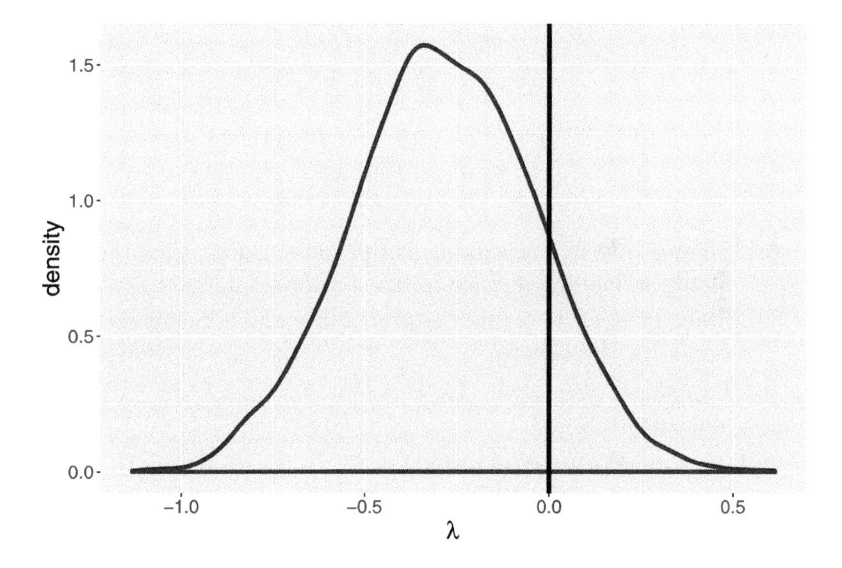

FIGURE 9.18
Posterior density estimate of simulated draws of log odds ratio λ for visits to Facebook example. A vertical line is drawn at the value $\lambda = 0$ corresponding to no association between gender and visits to Facebook.

In the end-of-chapter exercises, the reader will be asked to perform further explorations with this two proportion model.

9.8 Exercises

1. **Normal and Cauchy Priors**

   In the example in Section 9.1.2, it was assumed that the prior for the average snowfall μ was normal with mean 10 inches and standard deviation 3 inches.

   (a) Confirm that the 25th and 75th percentiles of this prior are equal to 8 and 12 inches, respectively.

   (b) Show that under this normal prior, it is unlikely that the mean μ is at least as large as 26.75 inches.

   (c) Confirm that a Cauchy distribution with location 10 inches and scale parameter 2 inches also has 25th and 75th percentiles equal to 8 and 12 inches, respectively.

2. **A Random Walk**

   The following matrix represents the transition matrix for a random walk on the integers $\{1, 2, 3, 4, 5\}$.

$$P = \begin{bmatrix} .2 & .8 & 0 & 0 & 0 \\ .2 & .2 & .6 & 0 & 0 \\ 0 & .2 & .4 & .2 & 0 \\ 0 & 0 & .6 & .2 & .2 \\ 0 & 0 & 0 & .8 & .2 \end{bmatrix}$$

   (a) Suppose one starts walking at the state value 4. Find the probability of landing at each location after a single step.

   (b) Starting at value 4, find the probability of landing at each location after three steps.

   (c) Explain what is means for this Markov chain to be irreducible and aperiodic.

3. **A Random Walk (continued)**

   Consider the random walk Markov chain described in Exercise 2.

   (a) Suppose one starts at the location 1. Using an R script with the `sample()` function (see example script Section 9.2.3), simulate 1000 steps of the Markov chain using the probabilities given in the transition matrix. Store the locations of the walk in a vector.

   (b) Compute the relative frequencies of the walker in the five states from the simulation output. From this computation, guess at the value of the stationary distribution vector w.

   (c) Confirm that your guess is indeed the stationary distribution by using the matrix computation w %*% P.

4. **Weird Weather**

Suppose a city in Alaska has interesting weather. The four possible weather states are "sunny" (SU), "rainy" (R), "cloudy" (C), and "snow" (SN). If it is sunny one day, it is equally likely to be rainy, cloudy, and snow on the next day. If is currently rainy, then the probabilities of sunny, rain, cloudy, and snow on the next day are respectively 1/2, 1/6, 1/6, and 1/6. The following matrix gives the transitions of weather from one day to the next day.

$$\begin{array}{c} \\ \text{SU} \\ \text{R} \\ \text{C} \\ \text{SN} \end{array} \begin{array}{c} \begin{array}{cccc} \text{SU} & \text{R} & \text{C} & \text{SN} \end{array} \\ \begin{pmatrix} 0 & 1/3 & 1/3 & 1/3 \\ 1/2 & 1/6 & 1/6 & 1/6 \\ 0 & 1/4 & 1/2 & 1/4 \\ 0 & 1/4 & 1/4 & 1/2 \end{pmatrix} \end{array}$$

(a) If the weather is rainy today, find the probability that is rainy two days later.

(b) Starting with a sunny day, write an R script to simulate 1000 days of weather using this Markov chain.

(c) Find the relative frequencies of the four states. Are these values approximately the stationary distribution of the Markov chain?

5. **Ehrenfest Urn Model**

Grinstead and Snell (2006) describe a model used to explain diffusion of gases. One version of this model is described in the setting of two urns that contain a total of four balls. A state is the number of balls in the first urn. There are five possible states 0, 1, 2, 3, and 4. At each step, one ball is chosen at random and moved from the urn it is located to the other urn. The transition matrix for this Markov chain is shown below:

$$P = \begin{bmatrix} 0 & 1 & 0 & 0 & 0 \\ 1/4 & 0 & 3/4 & 0 & 0 \\ 0 & 1/2 & 0 & 1/2 & 0 \\ 0 & 0 & 3/4 & 0 & 1/4 \\ 0 & 0 & 0 & 1 & 0 \end{bmatrix}$$

(a) Starting at state 1, find the probabilities of each state after two steps.

(b) Starting at state 1, find the probabilities of each state after three steps.

(c) Explain why this Markov chain is not aperiodic.

(d) Does a stationary distribution exist for this Markov chain? Why or why not?

6. **Metropolis Sampling in a Random Walk**

   Suppose the variable X takes on values from 1 to 9 with respective probabilities that are proportional to the values 9, 7, 5, 3, 1, 3, 5, 7, 9. This probability distribution displayed in Figure 9.19 has a "bathtub" shape.

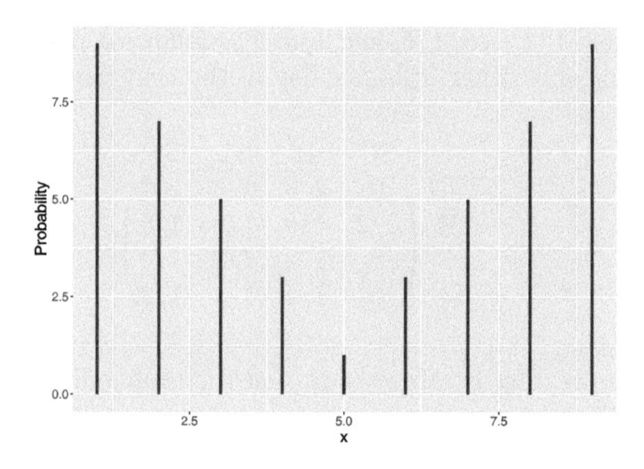

FIGURE 9.19
Bathtub shaped probability distribution.

   (a) Write an R function that computes this probability distribution for any value of X.
   (b) Using the Metropolis algorithm described in Section 9.3.1 as programmed in the function `random_walk()`, simulate 10,000 draws from this probability distribution starting at the value $X = 2$.
   (c) Collect the simulated draws and find the relative frequencies of the values 1 through 9. Compare these approximate probabilities with the exact probabilities.

7. **Metropolis Sampling of a Binomial Distribution**

   (a) Using the Metropolis algorithm described in Section 9.3 as programmed in the function `random_walk()`, simulate 1000 draws from a binomial distribution with parameters $n = 20$ and $p = 0.3$.
   (b) Collect the simulated draws and find the relative frequencies of the values 0 through 20. Compare these approximate probabilities with the exact probabilities.
   (c) Using the simulated values, estimate the mean μ and standard deviation σ of the distribution and compare these estimates with the known values of μ and σ of a binomial distribution.

8. **Metropolis Sampling - Poisson-Gamma Model**

   Suppose we observe $y_1, ..., y_n$ from a Poisson distribution with mean λ, and the parameter λ has a Gamma(a, b) distribution. The posterior density is proportional to

$$\pi(\lambda \mid y_1, \cdots, y_n) \propto \left[\prod_{i=1}^{n} \exp(-\lambda)\lambda^{y_i}\right] \left[\lambda^{a-1}\exp(-b\lambda)\right].$$

   (a) Write a function to compute the logarithm of the posterior density. Assume that one observes the sample 2, 5, 10, 5, 6, and the prior parameters are $a = b = 1$.

   (b) Use the `metropolis()` function in Section 9.3.3 to collect 1000 draws from the posterior distribution. Use a starting value of $\lambda = 5$ and a neighborhood scale value of $C = 2$.

   (c) Inspect MCMC diagnostic graphs to assess if the simulated sample approximates the posterior density of λ.

9. **Metropolis Sampling from a Bimodal Distribution**

   Suppose we observe a random sample $y_1, ..., y_n$ from a Cauchy distribution with location θ and scale parameter 1 with density

$$f(y_i \mid \theta) = \frac{1}{\pi\left[1 + (y_i - \theta)^2\right]}. \tag{9.38}$$

   If a uniform prior is placed on θ, then the posterior density of θ is proportional to

$$\pi(\theta \mid y_1, \cdots, y_n) \propto \prod_{i=1}^{n} \frac{1}{\pi\left[1 + (y_i - \theta)^2\right]} \tag{9.39}$$

   If we observe the values 3, 6, 7, 8, 15, 14, 16, 17, Figure 9.20 displays the bimodal shape of the posterior density.

   (a) Write a function to compute the logarithm of the posterior density.

   (b) Using the `metropolis()` function in Section 9.3.3, collect a simulated sample of 1000 from the posterior distribution. Run the sampler twice, once using a starting value of $\theta = 10$ and a neighborhood scale value of $C = 3$, and a second time with the same starting value and a scale value of $C = 0.2$.

   (c) By inspecting MCMC diagnostic graphs, which value of C appears to result in a simulated sample that is a better approximation to the posterior distribution? Explain.

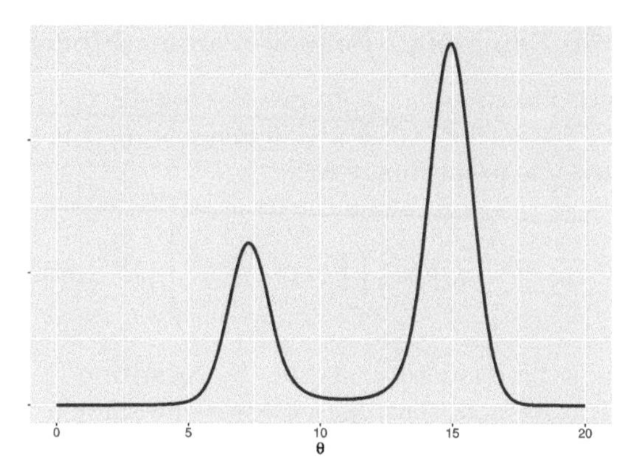

FIGURE 9.20
Posterior density of location parameter with Cauchy sampling.

10. **Gibbs Sampling - Poisson-Gamma Model**

    Suppose a single observation Y conditional on λ is Poisson with mean λ, and λ has a Gamma(a, b) prior with density equal to

    $$\pi(\lambda) = \frac{b^a}{\Gamma(a)} \lambda^{a-1} \exp(-b\lambda).$$

    (a) Write down the joint density of Y and λ.
    (b) Identify the conditional distribution Y conditional on λ, and the conditional distribution of λ conditional on $Y = y$.
    (c) Use the information from part (b) to construct a Gibbs sampling algorithm to sample from the joint distribution of (Y, λ).
    (d) Write an R function to implement one cycle of Gibbs sampling, and run 1000 iterations of Gibbs sampling for the case where $a = 3$ and $b = 3$.
    (e) By integration, find the marginal density of Y. Compare the exact values of the marginal density with the simulated draws of Y found using Gibbs sampling.

11. **Gibbs Sampling - Coin Flips**

    Suppose one observes the outcomes of four fair coin flips $W_1, ..., W_4$ where $W_i = 1$ if the outcome is heads and $W_i = 0$ otherwise. Let $X = W_1 + W_2 + W_3$ denote the number of heads in the first three flips and $Y = W_2 + W_3 + W_4$ is the number of heads in the last three flips. The joint probability of X and Y is given in Table 9.3.

    (a) Find the conditional distribution $f(x \mid Y = 1)$.
    (b) Find the conditional distribution $f(y \mid X = 2)$.

TABLE 9.3
The joint probability mass function $f(x, y)$ of the number of heads in the first three flips X and the number of heads in the last three flips Y in four tosses of a fair coin.

		Y			
		0	1	2	3
	0	1/16	1/16	0	0
X	1	1/16	3/16	2/16	0
	2	0	2/16	3/16	1/16
	3	0	0	1/16	1/16

(c) Describe how Gibbs sampling can be used to simulate from the joint distribution of X and Y.

(d) Using the `gibbs_discrete` function in Section 9.5.1, simulate 1000 iterations of Gibbs sampling using this probability distribution. By tabulating the (X, Y) output and computing relative frequencies, confirm that the relative frequencies are good approximations to the actual probabilities.

12. **Normal Sampling with Both Parameters Unknown**

The heights in inches of 20 college women were collected, observing the following measurements:

47	64	61	61	63	61	64	66	63	67
63.5	65	62	64	61	56	63	65	64	59

Suppose one assumes that the normal mean and precision parameters are independent with μ distributed Normal(62, 1) and ϕ distributed gamma with parameters $a = 1$ and $b = 1$.

(a) Using the `gibbs_normal()` function in Section 9.5.3, collect a sample of 5000 from the joint posterior distribution of (μ, ϕ).

(b) Find a 90% interval estimate for the standard deviation $\sigma = 1/\sqrt{\phi}$.

(c) Suppose one is interested in estimating the 90th percentile of the height distribution $P_{90} = \mu + 1.645\sigma$. Collect simulated draws from the posterior of P_{90} and construct a density estimate.

13. **Normal Sampling with Both Parameters Unknown (continued)**

In Exercise 12, one learned about the mean and precision of the heights by use of a Gibbs sampling algorithm. Use JAGS and the `runjags` package to collect MCMC draws from this model. Write a JAGS script for this normal sampling problem and use the `run.jags()` function. Answer questions from parts (c) and (d) from Exercise 12. (Note that the sample JAGS script in Section 9.7.1 returns samples of μ and σ.)

14. **Normal Sampling with Both Parameters Unknown (continued)**

    If we graph the height data from Exercise 12, we see one usually small height value, 47. We want to determine if this minimum height is consistent with the fitted model.

    (a) Write a function to simulate a sample of size 20 from the posterior predictive distribution. You can use either the `gibbs_normal()` function in Section 9.5.3 or the JAGS sample script in Section 9.7.1 to generate a sample from the posterior distribution of (μ, ϕ) or (μ, σ). For each sample, compute the minimum value $T(\tilde{y})$.
    (b) Repeat the procedure 1000 times, collecting a sample of the predictive distribution of the minimum observation.
    (c) Graph the predictive distribution. From comparing the observed minimum height with this distribution, what can you conclude about the suitability of the model?

15. **Comparing Proportions**

    In Section 9.7.4, the problem of comparing proportions of high numbers of visits to Facebook from male and female students was considered.

    (a) Using the same prior, use JAGS to take a simulated sample of size 5000 from the posterior of p_F and p_M. Construct a 90% probability interval estimate for the difference in proportions $\delta = p_W - p_M$.
    (b) Use the same simulated sample to perform inferences about the ratio of proportions $R = p_W / p_M$. Construct a density estimate of R and construct a 90% probability interval estimate.

16. **Comparing Poisson Rates**

    Suppose the number of customers y_j arriving at a bank during a half-hour period in the morning is Poisson with mean λ_M, and the number of customers w_j arriving in an afternoon half-hour period is Poisson with mean λ_A. Suppose one observes the counts 3, 3, 6, 3, 2, 3, 7, 6 for the morning periods, and the counts 11, 3, 9, 10, 10, 5, 8, 7 for the afternoon periods. Assume that λ_M and λ_A have independent Gamma(1, 1) priors. Use JAGS to obtain a simulated sample from the joint posterior of (λ_M, λ_A) and use the output to obtain a 90% posterior interval estimate for the ratio of means $R = \lambda_A / \lambda_M$.

17. **Normal Sampling with a Cauchy Prior**

    In Section 9.4, we considered the problem of estimating the mean snowfall amount in Buffalo with a Cauchy prior. The sample mean $\bar{y}$ is normal with mean μ and standard error se and μ is Cauchy with location 10 and scale 2. In our problem, $\bar{y} = 26.785$ and $se = 3.236$. Write a JAGS script for this Bayesian model. Use the `run.jags()` function to simulate 1000 draws of the posterior distribution for μ. Compute the posterior mean and posterior standard deviation for μ.

18. **Normal Sampling with a Cauchy Prior (continued)**

    In Exercise 17, we used JAGS to simulate values from the posterior of μ from a single MCMC chain. Instead use two chains with the different starting values of $\mu = 0$ and $\mu = 50$. Run JAGS with two chains and estimate the posterior mean and posterior standard deviation using output from each of the two chains. Based on the output, comment on the sensitivity of the MCMC run with the choice of the starting value.

19. **Bivariate Normal**

    Section 6.7 introduced the bivariate normal distribution. Suppose we wish to use Gibbs sampling to simulate from this distribution. In the following assume (X, Y) is bivariate normal with parameters $(\mu_X, \mu_Y, \sigma_X, \sigma_Y, \rho)$.

    (a) Using results from Section 6.7, identify the two conditional distributions $f(x \mid y)$ and $f(y \mid x)$ and write down a Gibbs sampling algorithm for simulating from the joint distribution of (X, Y).
    (b) Write an R function to simulate a sample from the distribution using Gibbs sampling.
    (c) Assume $\mu_X = 0, \mu_Y = 0, \sigma_X = 1, \sigma_Y = 1, \rho = 0.5$ and run the simulation for 1000 iterations. Compare the means, standard deviations, and correlation computed from the simulation with the true values of the parameters.
    (d) Repeat part (c) using the correlation value $\rho = 0.95$ and again compare the simulation estimates with the true values. Explain why Gibbs sampling does not appear to work as well in this situation.

20. **A Normal Mixture Model**

    Consider a three-component mixture distribution, where the density for x has the form

$$f(x) = 0.45 \times \phi(x, -3, 1/3) + 0.1 \times \phi(x, 0, 1/3) + 0.45 \times \phi(x, 3, 1/3), \tag{9.40}$$

where $\phi(x, \mu, \sigma)$ is the normal density with mean μ and standard deviation σ. Consider the following two ways of simulating from this mixture density.

Approach 1: Monte Carlo: Introduce a "mixture component indicator", δ, an unobserved latent variable. The variable z is equal to 1, 2, and 3 with respective probabilities 0.45, 0.1, and 0.45. The density for x conditional on z is normal where $[x \mid z = 1] \sim \text{Normal}(-3, 1/3)$, $[x \mid z = 2] \sim \text{Normal}(0, 1/3)$, and $[x \mid z = 3] \sim \text{Normal}(3, 1/3)$.

One simulates x by first simulating a value of z from its discrete distribution and then simulating a value of x from the corresponding conditional distribution. By repeating this method, one obtains a Monte Carlo simulated sample from the exact mixture distribution.

Approach 2: Gibbs Sampling: An alternative way of simulating from the mixture density is based on Gibbs sampling. Introduce the latent variable z and consider the two conditional distributions $[x \mid z]$ and $[z \mid x]$. The conditional distribution $[x \mid z]$ will be a normal density where the normal parameters depend on the value of the latent variable. The conditional distribution $[z \mid x]$ is discrete on the values 1, 2, 3 where the probabilities are proportional to $0.45 \times \phi(x, -3, 1/3)$, $0.1 \times \phi(x, 0, 1/3)$, $0.45 \times \phi(x, 3, 1/3)$ respectively.

Write R scripts to use both the Monte Carlo and Gibbs sampling methods to simulate 1000 draws from this mixture density.

21. **A Normal Mixture Model – MCMC Diagnostics**

Figure 9.21 displays histograms of simulated draws from the mixture distribution using the Monte Carlo and Gibbs sampling algorithms, and the exact mixture density is overlaid on top. It is clear from the figure that the Gibbs sampling does not appear to perform as well as the Monte Carlo method in simulating from this distribution. Using MCMC diagnostic graphs, explore the Gibbs sampling output. Are there particular features in these diagnostic graphs that would indicate problems in the convergence of the Gibbs sampling algorithm?

22. **Change Point Analysis**

The World Meteorological Association collects data on tropical storms, and scientists want to find out whether the distribution of storms changed over time, and if so, when. Data on the number of storms per year has been collected for n years, and let y_i be the number of storms in year i, where $i = 1, \cdots, n$. Let M be the year in which the distribution of Y changes, where $M \in \{1, \cdots, n-1\}$.

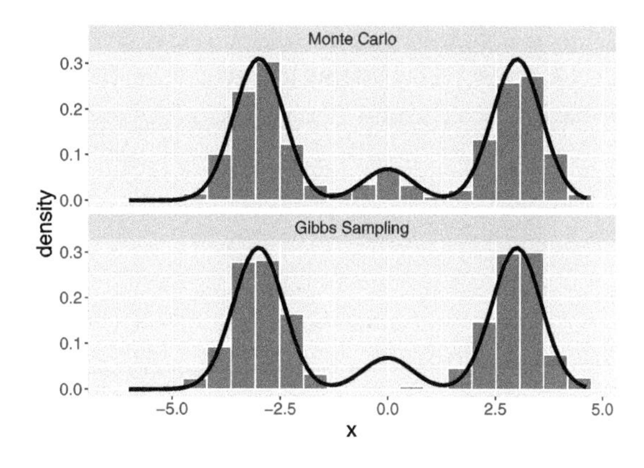

FIGURE 9.21
Histogram of 1000 samples of μ from the Monte Carlo and Gibbs sampling algorithms.

A reasonable sampling model for Y is:

$$\begin{aligned} y_i \mid \lambda_1, M &\sim \text{Poisson}(\lambda_1), \; i = 1, \cdots, M; \\ y_i \mid \lambda_2, M &\sim \text{Poisson}(\lambda_2), \; i = M+1, \cdots, n. \end{aligned}$$

Suppose one gives a uniform prior for M over integers from 1 to $n-1$ to represent complete uncertainty about change point:

$$M \mid \lambda_1, \lambda_2 \sim \text{Discrete}(\frac{1}{n-1}, \cdots, \frac{1}{n-1}), \;\; M \in \{1, \cdots, n-1\}.$$

Equivalently, you can think of the uniform prior as:

$$Prob(M = m) = \frac{1}{n-1}, \;\; M \in \{1, \cdots, n-1\}.$$

Recall that gamma distributions are conjugate prior distributions for Poisson data model. Suppose one uses independent conjugate gamma priors for λ_1 and λ_2:

$$\begin{aligned} \lambda_1 \mid a_1, b_1 &\sim \text{Gamma}(a_1, b_1), \\ \lambda_2 \mid a_2, b_2 &\sim \text{Gamma}(a_2, b_2). \end{aligned}$$

(a) Write the joint posterior distribution, $\pi(\lambda_1, \lambda_2, M \mid y_1, \cdots, y_n)$, up to a constant.

(b) Find the full conditional posterior distribution for λ_1 and λ_2. Write the name of the distributions and expressions for their parameter values.

(c) Find the full conditional posterior distribution for M, which should be a discrete distribution over $m = 1, \cdots, n-1$.

(d) Describe how you would design a Gibbs sampling to simulate posterior draws of the set of parameters, $(\lambda_1, \lambda_2, M)$.

10

Bayesian Hierarchical Modeling

10.1 Introduction

10.1.1 Observations in groups

Chapters 7, 8, and 9 make an underlying assumption about the source of data: observations are assumed to be identically and independently distributed (*i.i.d.*) following a single distribution with one or more unknown parameters. In Chapter 7, the binomial data model is based on the assumptions that a student's chance of preferring dining out on Friday is the same for all students, and the dining preferences of different students are independent. To refresh your memory, recall the four conditions of a binomial experiment: a fixed number of trials, only two outcomes, a fixed success probability, and independent trials. In Chapter 8, the normal sampling model is based on the assumptions that Roger Federer's time-to-serves are independent observations following a single normal distribution with an unknown mean μ and known standard deviation σ. That is, $Y_i \overset{i.i.d.}{\sim} \textrm{Normal}(\mu, \sigma)$. Similarly in Chapter 9, the underlying assumption is that the snowfall amounts in Buffalo for the month of January for the last 20 years follow the same $\textrm{Normal}(\mu, \sigma)$ distribution with both parameters unknown.

In many situations, treating observations as *i.i.d.* from the same distribution with the same parameter(s) is not sensible. In our dining out example, dining preferences for students may be different from dining preferences of senior citizens, so it would not make sense to use a single success probability for a combined group of students and senior citizens. In a similar fashion, if one considered time-to-serve data for a group of tennis players, then it would not be reasonable to use a single normal distribution with a single mean to represent these data – the mean time-to-serve for a quick-serving player would likely be smaller than the mean time-to-serve for a slower player. For many applications, some observations share characteristics, such as age or player, that distinguish them from other observations, therefore multiple distinct groups are observed.

10.1.2 Example: standardized test scores

As a new example, consider a study in which students' scores of a standardized test such as the SAT are collected from five different senior high schools in a given year. Suppose a researcher is interested in learning about the mean SAT score. Since five different schools participated in this study and students' scores might vary from school to school, it makes sense for the researcher to learn about the mean SAT score for each school and compare students' mean performance across schools.

To start modeling this education data, it is inappropriate to use Y_i as the random variable for the SAT score of student i ($i = 1, \cdots, n$, where n is the total number of students from all five schools) since this ignores the inherent grouping of the observations. Instead, the researcher adds a school label j to Y_i to reflect the grouping. Let Y_{ij} denote the SAT score of student i in school j, where $j = 1, \cdots, 5$, and $i = 1, \cdots, n_j$, where n_j is the number of students in school j, and $n = \sum_{j=1}^{5} n_j$.

Since SAT scores are continuous, the normal sampling model is a reasonable choice for a data distribution. Within school j, one assumes that SAT scores are *i.i.d.* from a normal data model with a mean and standard deviation depending on the school. Specifically, one assumes a school-specific mean μ_j and a school-specific standard deviation σ_j for the normal data model for school j. Combining the information for the five schools, one has

$$Y_{ij} \overset{i.i.d.}{\sim} \text{Normal}(\mu_j, \sigma_j), \tag{10.1}$$

where $j = 1, \cdots, 5$ and $j = 1, \cdots, n_j$.

10.1.3 Separate estimates?

One approach for handling this group estimation problem is find separate estimates for each school. One focuses on the observations in school j,$\{Y_{1j}, Y_{2j}, \cdots, Y_{n_j j}\}$, choose a prior distribution $\pi(\mu_j, \sigma_j)$ for the mean and the standard deviation parameters, follow the Bayesian inference procedure in Chapter 9 and obtain posterior inference on μ_j and σ_j. If one assumes that the prior distributions on the individual parameters for the schools are independent, one is essentially fitting five separate Bayesian models and one's inferences about one particular school will be independent of the inferences on the remaining schools.

This "separate estimates" approach may be reasonable, especially if the researcher thinks the means and the standard deviations from the five normal models are completely unrelated to each other. That is, one's prior beliefs about the parameters of the SAT score distribution in one school are unrelated to the prior beliefs about the distribution parameters in another school.

To see if this assumption is reasonable, let us consider a thought experiment for the school testing example. Suppose you are interested in learning about the mean SAT score μ_N for school N. You may not be familiar with the distribution of SAT scores and it would be difficult to construct an informative prior for μ_N. But suppose that you are told that the students from another school, call it school M, average 1200 on their SAT scores. That information would likely influence your prior on μ_N, since now you have some general idea about SAT scores. This means that your prior beliefs about the mean SAT scores μ_N and μ_M are not independent – some information about one school's mean SAT scores would change your prior on the second school's mean SAT score. So in many situations, this independence assumption would be questionable.

10.1.4 Combined estimates?

Another way to handle this group estimation problem is to ignore the fact that there is a grouping variable and estimate the parameters in the combined sample. In our school example, one ignores the school variable and simply assumes that the SAT scores $Y_i's$ are distributed from a single normal population with mean μ and standard deviation σ. Here, $i = 1, \cdots, n$ where n is the total number of students from all five schools.

If ones ignores the grouping variable, then the inference procedure described in Chapter 9 can be used. One constructs a prior for the parameters μ and σ and use Gibbs sampling to obtain a simulated sample from the posterior distribution of (μ, σ).

Using this approach, one is effectively ignoring any differences between the five schools. Although it is reasonable to assume some similarity in the SAT scores across different schools, one probably does not believe that the schools are indistinguishable. In fact, state officials assume the schools have distinct features such as student bodies with different socioeconomic statuses so that SAT scores from different schools can be substantially different. In some states in the United States, all schools are ranked on different criteria which reflects the belief that schools are different with respect to student achievement.

10.1.5 A two-stage prior leading to compromise estimates

If one applies the "separate estimates" approach, one performs separate analyses on the different groups, and one ignores any prior knowledge about the similarity between the groups. On the other extreme, the "combined estimates" approach ignores the grouping variable and assumes that the groups are identical with respect to the response variable SAT score. Is there an alternative approach that compromises between the separate and combined estimate methods?

Let us return to the model Normal(μ_j, σ_j) where μ_j is the parameter representing the mean SAT score of students in school j. For simplicity of discussion it is assumed the standard deviation σ_j of the j-th school is known. Consider the collection of five mean parameters, $\{\mu_1, \mu_2, \mu_3, \mu_4, \mu_5\}$ representing the means of the five schools' SAT scores. One believes that the μ_j's are distinct, because each μ_j depends on the characteristics of school j, such as size and socioeconomic status. But one also believes that the mean parameters are similar in size. Imagine if you were given some information about the location of one mean, say μ_j, then this information would influence your beliefs about the location of another mean μ_k. One wishes to construct a prior distribution for the five mean parameters that reflects the belief that $\mu_1, \mu_2, \mu_3, \mu_4$, and μ_5 are related or similar in size. This type of "similarity" prior allows one to combine the SAT scores of the five schools in the posterior distribution in such a way to obtain compromise estimates of the separate mean parameters.

The prior belief in similarity of the means is constructed in two stages.

Stage 1 The prior distribution for the j-th mean, μ_j is normal, where the mean and standard deviation parameters are shared among all μ_j's:

$$\mu_j \mid \mu, \tau \sim \text{Normal}(\mu, \tau), \; j = 1, ..., 5. \tag{10.2}$$

Stage 2 In the Stage 1 specification, the parameters μ and τ are unknown. So this stage assigns the parameters a prior density π.

$$\mu, \tau \sim \pi(\mu, \tau). \tag{10.3}$$

Several comments can be made about this two-stage prior.

- Specifying the same prior distribution for all μ_j's at Stage 1 does not say that the μ_j's are the same value. Instead, Stage 1 indicates that the μ_j's *a priori* are related and come from the same distribution. If the prior distribution Normal(μ, τ) has a large standard deviation (that is, if τ is large), the μ_j's can be very different from each other *a priori*. On the other hand, if the standard deviation τ is small, the μ_j's will be very similar in size.

- To follow up the previous comment, if one considers the limit of the Stage 1 prior as the standard deviation τ approaches zero, the group means μ_j will be identical. Then one is in the "combined groups" situation where one is pooling the SAT data to learn about a single population. At the other extreme, if one allows the standard deviation τ of the Stage 1 prior to approach infinity, then one is saying that the group means $\mu_1, ..., \mu_5$ are unrelated and that leads to the separate estimates situation.

- In the school testing example, this prior Normal(μ, τ) distribution is a model about all μ_j's in the U.S., i.e. the population of SAT score means corresponding to all schools in the United States. The five schools in the dataset represent a sample from all schools in the U.S.

- Since μ and τ are parameters in the prior distribution, they are called hyperparameters. Learning about μ and τ provides information about the population of μ_j's. Naturally in Bayesian inference, one learns about μ and

τ by specifying a hyperprior distribution and performing inference based on the posterior distribution. In this example, inferences about μ and τ tell us about the location and spread of the population of mean SAT scores of schools in the U.S.

To recap, one models continuous outcomes in groups through the school-specific sampling density in Equation (10.1) and the common normal prior distribution in Equation (10.2) for the mean parameters. An important and appealing feature of this approach is learning simultaneously about each school (group) and learning about the population of schools (groups). Specifically in the current setup, the model simultaneously estimates the means for the schools (the μ_j's) and the variation among the means (μ and τ). It will be seen that the hierarchical model posterior estimates for one school borrows information from other schools. This process is often called "partial pooling" information among groups.

From the structural point of view, due to the two stages of the model, this approach is called hierarchical or multilevel modeling. In essence, hierarchical modeling takes into account information from multiple levels, acknowledging differences and similarities among groups. In the posterior analysis, one learns simultaneously about each group and learns about the population of groups by pooling information across groups.

In this chapter, hierarchical modeling is described in two situations that extend the Bayesian models for one proportion and one normal mean described in Chapters 7 and 8, respectively. Section 10.2 introduces hierarchical normal modeling using a sample of ratings of animation movies released in 2010; and Section 10.3 describes hierarchical beta-binomial modeling with an example of deaths after heart attack. In each section, we motivate the consideration of hierarchical models, outline the model structure, and implement model inference through Markov chain Monte Carlo simulation.

10.2 Hierarchical Normal Modeling

10.2.1 Example: ratings of animation movies

MovieLens is a website which provides personalized movie recommendations from users who create accounts and rate movies that they have seen. Based on such information, MovieLens works to build a custom preference profile for each user and provide movie recommendations. MovieLens is run by Group-Lens Research, a research laboratory at the University of Minnesota that has made MovieLens rating datasets available to the public. GroupLens Research regularly updates these datasets on its website and the datasets are useful for new research, education and development initiatives.

In one study, a sample from the MovieLens database was collected on movie ratings for eight different animation movies released in 2010. There are a total of 55 movie ratings, where a rating is is for a particular animation movie completed by a MovieLens user. The ratings are likely affected by the quality of the movie itself, as some movies are generally favored by the audience while others might be less favored. Therefore there exists a natural grouping of the 55 ratings by the movie title.

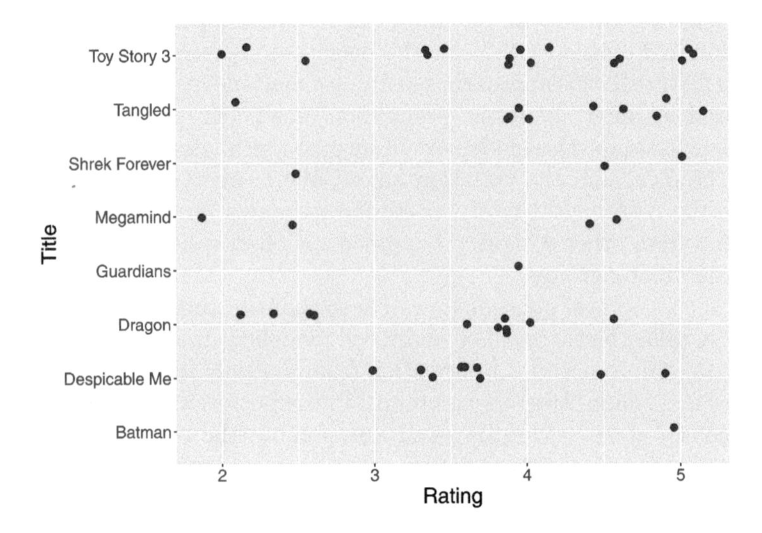

FIGURE 10.1
Jittered dotplot of the ratings for the eight animation movies.

Figure 10.1 displays a jittered dotplot of the ratings grouped by movie title and Table 10.1 lists the sample mean, sample standard deviation, and the number of ratings for each title. Note the variability in the sample sizes – *Toy Story 3* received 16 ratings and *Legend of the Guardians* and *Batman: Under the Red Hood* only received a single rating. For a movie with only one observed rating, such as *Legend of the Guardians* and *Batman: Under the Red Hood*, it would be difficult to learn much about its mean rating. Here it is desirable to improve the estimate of its mean rating by using rating information from similar movies.

10.2.2 A hierarchical Normal model with random σ

In this situation it is reasonable to develop a model for the movie ratings where the grouping variable is the movie title. We index ratings by two subscripts, where Y_{ij} denotes the i-th rating for the j-th movie title ($j = 1, \cdots, 8$).

What sampling model should be used for the movie ratings? Since the ratings are continuous, it is reasonable to use the normal data model described in Chapter 8. Recall that a normal model has two parameters, the mean and the standard deviation. Based on previous reasoning, the mean parameter is

TABLE 10.1
The movie title, the mean rating, the standard deviation of the ratings, and the number of ratings.

Movie Title	Mean	SD	N
Batman: Under the Red Hood	5.00		1
Despicable Me	3.72	0.62	9
How to Train Your Dragon	3.41	0.86	11
Legend of the Guardians	4.00		1
Megamind	3.38	1.31	4
Shrek Forever After	4.00	1.32	3
Tangled	4.20	0.89	10
Toy Story 3	3.81	0.96	16

assumed to be movie-specific, so μ_j will represent the mean of the ratings for movie j. Thinking about the standard deviation parameter, should the standard deviation also be movie-specific, where σ_j represents the standard deviation of the ratings for movie j? Or can we assume a common value of the standard deviation, say σ, across movies? For simplicity and ease of illustration, a common and shared unknown standard deviation σ is assumed for all normal models. This is a simplified version of random σ_j's — the more flexible hierarchical model with random σ_j's will be left as an end-of-chapter exercise.

One begins by writing down the sampling distributions for the ratings of the eight movies. Recall that Y_{ij} denotes the i-th rating of movie j, where μ_j denote the mean of the normal model for movie j, and σ denote the shared standard deviation of the normal models across different movies. In our notation, n_j represents the number of ratings for movie j.

- Sampling, for $j = 1, \cdots, 8$ and $i = 1, \cdots, n_j$:

$$Y_{ij} \mid \mu_j, \sigma \overset{i.i.d.}{\sim} \text{Normal}(\mu_j, \sigma). \tag{10.4}$$

The next task is to set up a prior distribution for the eight mean parameters, $\{\mu_1, \mu_2, \cdots, \mu_8\}$ and the shared standard deviation parameter σ. Focus first on the prior distribution for the mean parameters. Since these movies are all animations, it is reasonable to believe that the mean ratings are similar across movies. So one assigns each mean rating the same normal prior distribution at the first stage:

- Prior for μ_j, $j = 1, \cdots, 8$:

$$\mu_j \mid \mu, \tau \sim \text{Normal}(\mu, \tau). \tag{10.5}$$

As discussed in Section 10.1, this prior allows for a flexible method for pooling information across movies. If the prior distribution has a large standard deviation (e.g. a large value of τ), the μ_j's are very different from each other *a priori*, and one would have modest pooling of the eight sets of ratings. If instead this prior has a small standard deviation (e.g. a small value of τ), the μ_j's are very similar *a priori* and one would essentially be pooling the ratings to get an estimate at each of the μ_j. This shared prior Normal(μ, τ) distribution among the μ_j's simultaneously estimates both a mean for each movie (the μ_j's) and also lets us learn about variation among the movies by the parameter τ.

The hyperparameters μ and τ are treated as random since we are unsure about the degree of pooling of the eight sets of ratings. In typical practice, one specifies weakly informative hyperprior distributions for these "second-stage" parameters, indicating that one has little prior knowledge about the locations of these parameters. After observing data, inference is performed about μ and τ based on their posterior distributions. The posterior on the mean parameter μ is informative about an "average" mean rating and the posterior on τ lets one know about the variation among the μ_j's in the posterior.

Treating μ and τ as random, one arrives at the following hierarchical model.

- Sampling: for $j = 1, \cdots, 8$ and $i = 1, \cdots, n_j$:

$$Y_{ij} \mid \mu_j, \sigma \overset{i.i.d.}{\sim} \text{Normal}(\mu_j, \sigma). \tag{10.6}$$

- Prior for μ_j, Stage 1: μ_j, $j = 1, \cdots, 8$:

$$\mu_j \mid \mu, \tau \sim \text{Normal}(\mu, \tau). \tag{10.7}$$

- Prior for μ_j, Stage 2:

$$\mu, \tau \sim \pi(\mu, \tau). \tag{10.8}$$

In our model $\pi(\mu, \tau)$ denotes an arbitrary joint hyperprior distribution for the Stage 2 hyperparameters μ and τ. When the MovieLens ratings dataset is analyzed, the specification of this hyperprior distribution will be described.

To complete the model, one needs to specify a prior distribution for the standard deviation parameter, σ. As discussed in Chapter 9, when making inference about the standard deviation in a normal model, one uses a gamma prior on the precision (the inverse of the variance), for example,

- Prior for σ:

$$1/\sigma^2 \mid a_\sigma, b_\sigma \quad \sim \quad \text{Gamma}(a_\sigma, b_\sigma) \tag{10.9}$$

One assigns a known gamma prior distribution for $1/\sigma^2$, with fixed hyperparameter values a_σ and b_σ. In some situations, one may consider the situation where a_σ and b_σ are random and assign hyperprior distributions for these unknown hyperparameters.

Before continuing to the graphical representation and simulation by MCMC using JAGS, it is helpful to contrast the two-stage prior distribution for $\{\mu_j\}$ and the one-stage prior distribution for σ. The hierarchical model specifies a common prior for the means μ_j's which induces sharing of information across ratings from different movies. On the other hand, the model uses a shared σ for all movies which also induces sharing of information, though different from the sharing induced by the two-stage prior distribution for $\{\mu_j\}$.

What is the difference between the two types of sharing? For the means $\{\mu_j\}$, we have discussed that specifying a common prior distribution for different μ_j's pools information across the movies. One is simultaneously estimating both a mean for each movie (the μ_j's) and the variation among the movies (μ and τ). For the standard deviation σ, the hierarchical model also pools information across movies. However, all of the observations are combined in the estimation of σ. Since separate values of σ_j's are not assumed, one cannot learn about the differences and similarities among the σ_j's. If one is interested in pooling information across movies for the σ_j's, one needs to allow random σ_j's, and specify a two-stage prior distribution for these parameters. Interested readers are encouraged to try out this approach as an end-of-chapter exercise.

Graphical representation of the hierarchical model

An alternative way of expressing this hierarchical model uses the following graphical representation.

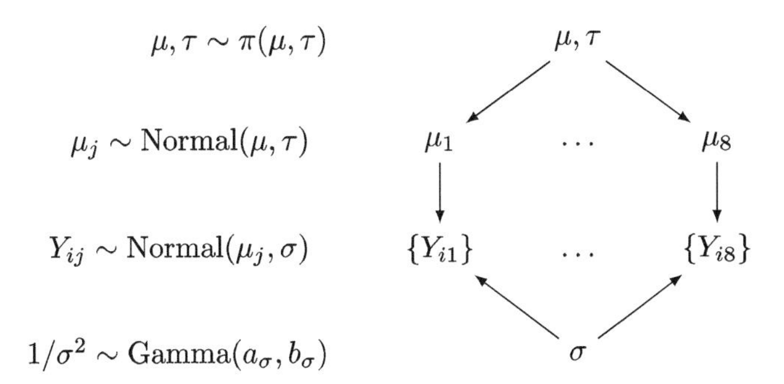

In the middle section of the graph, $\{Y_{ij}\}$ represents the collection of random variables for all ratings of movie j, and the label to the left indicates the assumed normal sampling distribution. The two parameters in the normal sampling density, μ_j and σ, are connected from above and below, with arrows pointing from the parameters to the random variables.

The upper section of the graph focuses on the μ_j's. All means follow the same prior, a normal distribution with mean μ and standard deviation τ. Therefore, arrows come from the common hyperparameters μ and τ to each

μ_j. Since μ and τ are random, these second-stage parameters are associated with the prior label $\pi(\mu, \tau)$.

The lowest section of the graph is about σ, or to be precise, $1/\sigma^2$. If one wants to allow hyperparameters a_σ and b_σ to be random as well, the lower part of the graph grows further, in a similar manner as the upper section for μ_j.

Second-stage prior

The hierarchical normal model presented in Equations (10.6) through (10.9) has not specified the hyperprior distribution $\pi(\mu, \tau)$. How does one construct a prior on these second-stage hyperparameters?

Recall that μ and τ are parameters for the normal prior distribution for $\{\mu_j\}$ the collection of eight different normal sampling means. The mean μ and standard deviation τ in this normal prior distribution reflect respectively the mean and spread of the mean ratings across eight different movies.

Following the discussion in Section 9.5.3, a typical approach for normal models is to assign two independent prior distributions — a normal distribution for the mean μ and a gamma distribution for the precision $1/\tau^2$. Such a specification facilitates the use of the Gibbs sampling due to the availability of the conditional posterior distributions of both parameters (see the details of this work in Section 9.5.3). Using this approach, the density $\pi(\mu, \tau)$ is replaced by the two hyperprior distributions below.

- The hyperprior for μ and τ:

$$\mu \mid \mu_0, \gamma_0 \sim \text{Normal}(\mu_0, \gamma_0) \tag{10.10}$$
$$1/\tau^2 \mid a, b \sim \text{Gamma}(a_\tau, b_\tau) \tag{10.11}$$

The task of choosing a prior for (μ, τ) reduces to the problem of choosing values for the four hyperparameters μ_0, γ_0, a_τ, and b_τ. If one believes that μ is located around the value of 3 and is not very confident of this choice, the set of values $\mu_0 = 3$ and $\gamma_0 = 1$ could be chosen. As for τ, one chooses a weakly informative prior with $a_\tau = b_\tau = 1$, as Gamma(1, 1). Moreover, to choose a prior for σ, let $a_\sigma = b_\sigma = 1$ to have the weakly informative Gamma(1, 1) prior.

10.2.3 Inference through MCMC

With the specification of the prior, the complete hierarchical model is described as follows:

- Sampling: for $j = 1, \cdots, 8$ and $i = 1, \cdots, n_j$:

$$Y_{ij} \mid \mu_j, \sigma_j \overset{i.i.d.}{\sim} \text{Normal}(\mu_j, \sigma_j) \tag{10.12}$$

- Prior for μ_j, Stage 1: for $j = 1, \cdots, 8$:

$$\mu_j \mid \mu, \tau \sim \text{Normal}(\mu, \tau) \tag{10.13}$$

- Prior for μ_j, Stage 2: the hyperpriors:

$$\mu \sim \text{Normal}(3, 1) \tag{10.14}$$
$$1/\tau^2 \sim \text{Gamma}(1, 1) \tag{10.15}$$

- Prior for σ:

$$1/\sigma^2 \sim \text{Gamma}(1, 1) \tag{10.16}$$

If one uses JAGS for simulation by MCMC, one writes out the model section by following the model structure above closely. Review Section 9.7 for an introduction and a description of several examples of JAGS.

Describe the model by a script

R The first step in using the JAGS software is to write the following script defining the hierarchical model. The model is saved in the character string `modelString`.

```
modelString <-"
model {
## sampling
for (i in 1:N){
   y[i] ~ dnorm(mu_j[MovieIndex[i]], invsigma2)
}
## priors
for (j in 1:J){
   mu_j[j] ~ dnorm(mu, invtau2)
}
invsigma2 ~ dgamma(a_s, b_s)
sigma <- sqrt(pow(invsigma2, -1))
## hyperpriors
mu ~ dnorm(mu0, g0)
invtau2 ~ dgamma(a_t, b_t)
tau <- sqrt(pow(invtau2, -1))
}
"
```

In the sampling part of the script, note that the loop goes from `1` to `N`, where `N` is the number of observations with index `i`. However, because now `N` observations are grouped according to movies, indicated by `j`, one needs to create one vector, `mu_j` of length eight, and use `MovieIndex[i]` to grab the corresponding `mu_j` based on the movie index.

In the priors part of the script, the loop goes from `1` to `J`, and `J = 8` in the current example. Inside the loop, the first line corresponds to the prior distribution for `mu_j`. Due to a commonly shared `sigma`, `invsigma2` follows `dgamma(a_g, b_g)` outside of the loop. In addition, `sigma <- sqrt(pow(invsigma2, -1))` is added to help track `sigma` directly.

Finally in the hyperpriors section of the script, one specifies the normal hyperprior for `mu`, a gamma hyperprior for `invtau2`. Keep in mind that the arguments in the `dnorm` in JAGS are the mean and the precision. If one is interested instead in the standard deviation parameter `tau`, one could return it in the script by using `tau <- sqrt(pow(invtau2, -1))`, enabling the tracking of its MCMC chain in the posterior inferences.

Define the data and prior parameters

R After one has defined the model script, the next step is to provide the data and values for parameters of the prior. In the R script below, a list `the_data` contains the vector of observations, the vector of movie indices, the number of observations, and the number of movies. It also contains the normal hyperparameters `mu0` and `g0`, and two sets of gamma hyperparameters (`a_t` and `b_t`) for `invtau2`, and (`a_s` and `b_s`) for `invsigma2`.

```
y <- MovieRatings$rating
MovieIndex <- MovieRatings$Group_Number
N <- length(y)
J <- length(unique(MovieIndex))
the_data <- list("y" = y, "MovieIndex" = MovieIndex,
                 "N" = N, "J" = J,
                 "mu0" = 3, "g0" = 1,
                 "a_t" = 1, "b_t" = 1,
                 "a_s" = 1, "b_s" = 1)
```

Generate samples from the posterior distribution

R One uses the `run.jags()` function in the `runjags` R package to generate posterior samples by using the MCMC algorithms in JAGS. The script below runs one MCMC chain with 1000 iterations in the adapt period (preparing for MCMC), 5000 iterations of burn-in and an additional set of 5000 iterations to be run and collected for inference. By using `monitor = c("mu", "tau", "mu_j", "sigma")`, one collects the values of all parameters in the model. In the end, the output variable `posterior` contains a matrix of simulated draws.

```
posterior <- run.jags(modelString,
                      n.chains = 1,
                      data = the_data,
```

```
                     monitor = c("mu", "tau", "mu_j", "sigma"),
                     adapt = 1000,
                     burnin = 5000,
                     sample = 5000)
```

MCMC diagnostics and summarization

R In any implementation of MCMC sampling, diagnostics are crucial to perform to ensure convergence. To perform some MCMC diagnostics in our example, one uses the `plot()` function, specifying the variable to be checked by the `vars` argument. For example, the script below returns four diagnostic plots (trace plot, empirical PDF, histogram, and autocorrelation plot) in Figure 10.2 for the hyperparameter τ. Note that the trace plot only includes 5000 iterations in sample, although its index starts from adapt (1000 adapt + 5000 burn-in). The trace plot and autocorrelation plot suggest good mixing of the chain, therefore indicating convergence of the MCMC chain for τ.

```
plot(posterior, vars = "tau")
```

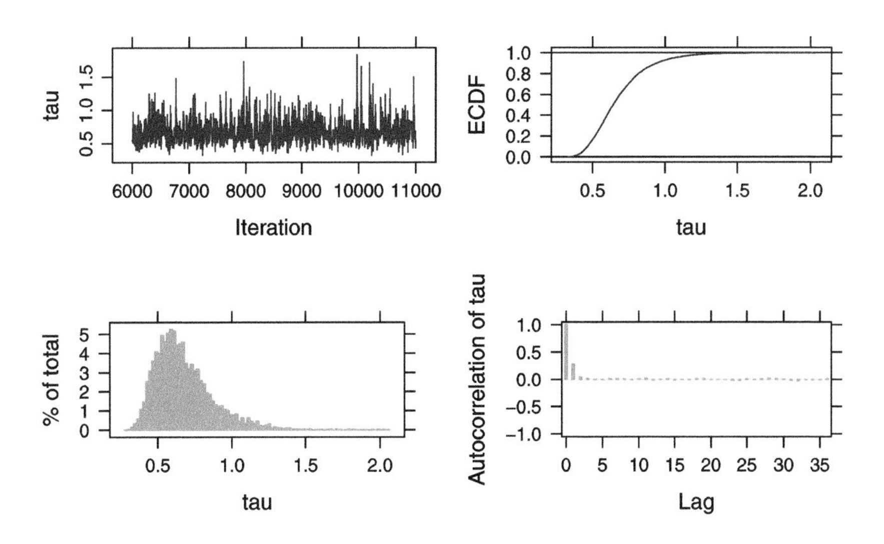

FIGURE 10.2
Diagnostic plots of simulated draws of τ using the JAGS software with the `runjags` package.

In practice MCMC diagnostics should be performed for all parameters to justify the overall MCMC convergence. In our example, the above diagnostics should be implemented for each of the eleven parameters in the model: $\mu, \tau, \mu_1, \mu_2, \cdots, \mu_8$, and σ. Once diagnostics are done, one reports posterior summaries of the parameters using `print()`. Note that these summaries are based on the 5000 iterations from the sample period, excluding the adapt and burn-in iterations.

```
print(posterior, digits = 3)
        Lower95 Median Upper95  Mean     SD Mode   MCerr
mu         3.19   3.78    4.34  3.77  0.286   -- 0.00542
tau       0.357  0.638    1.08 0.677    0.2   -- 0.00365
mu_j[1]    2.96   3.47    3.99  3.47  0.262   -- 0.00376
mu_j[2]    3.38   3.81    4.25  3.82  0.221   -- 0.00313
mu_j[3]    3.07   3.91    4.75  3.91  0.425   -- 0.00677
mu_j[4]    3.21   3.74    4.31  3.74  0.285   -- 0.00428
mu_j[5]    3.09   4.15    5.43  4.18  0.588   --  0.0115
mu_j[6]     2.7   3.84    4.99  3.85  0.576   -- 0.00915
mu_j[7]    2.74   3.53    4.27  3.51  0.388   -- 0.00595
mu_j[8]    3.58   4.12    4.66  4.12  0.276   -- 0.00423
sigma     0.763   0.92    1.12  0.93 0.0923   -- 0.00142
```

One performs various inferential summaries and inferences based on the output. For example, the movies *How to Train Your Dragon* (corresponding to μ_1) and *Megamind* (corresponding to μ_7) have the lowest average ratings with short 90% credible intervals, (2.96, 3.99) and (2.74, 4.27) respectively, whereas *Legend of the Guardians: The Owls of Ga'Hoole* (corresponding to μ_6) also has a low average rating but with a wider 90% credible interval (2.70, 4.99). The differences in the width of the credible intervals stem from the sample sizes: there are eleven ratings for *How to Train Your Dragon*, four ratings for *Megamind*, and only a single rating for *Legend of the Guardians: The Owls of Ga'Hoole*. The smaller the sample size, the larger the variability in the inference, even if one pools information across groups.

Among the movies with high average ratings, *Batman: Under the Red Hood* (corresponding to μ_5) is worth noting. This movie's average rating μ_5 has the largest median value among all μ_j's, at 4.15, and also a wide 90% credible interval, (3.09, 5.43). *Batman: Under the Red Hood* also received one rating in the sample resulting in a wide credible interval.

Shrinkage

Recall that the two-stage prior in Equations (10.7) to (10.8) specifies a shared prior Normal(μ, τ) for all μ_j's which facilitates simultaneous estimation of the movie mean ratings (the μ_j's), and estimation of the variation among the movie mean ratings through the parameters μ and τ. The posterior mean of the rating for a particular movie μ_j shrinks the observed mean rating towards an average rating. Figure 10.3 displays a shrinkage plot which illustrates the movement of the observed sample mean ratings towards an average rating.

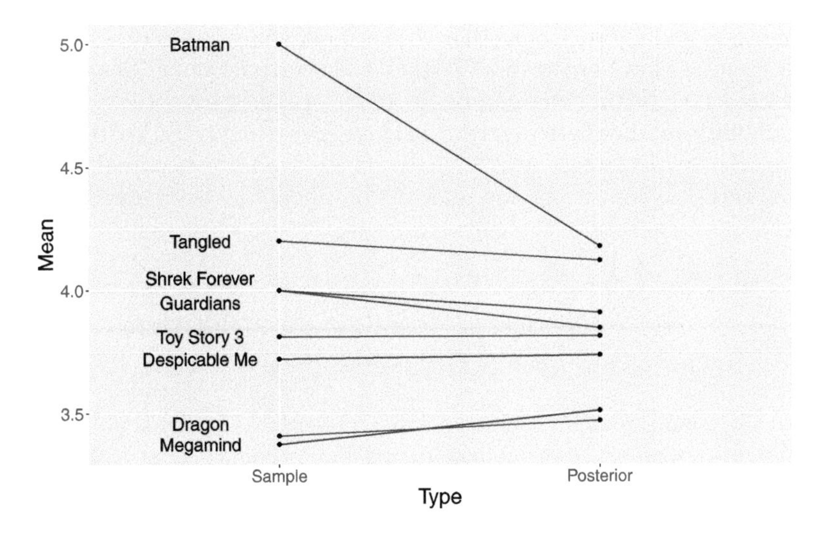

FIGURE 10.3
Shrinkage plot of sample means and posterior means of movie ratings for eight movies.

The left side of Figure 10.3 plots the sample movie rating means and lines connect the sample means to the corresponding posterior means (i.e. means of the posterior draws of μ_j). The shrinkage effect is obvious for the movie *Batman: Under the Red Hood* which corresponds to the dot at the value 5.0 on the left. This movie only received one rating of 5.0 and its mean rating μ_5 shrinks to the value 4.178 on the right, which is still the highest posterior mean among the nine movie posterior means. A large shrinkage is desirable for a movie with a small number of ratings such as *Batman: Under the Red Hood.* For a movie with a small sample size, information about other ratings of similar movies helps to produce a more reasonable estimate at the "true" average movie rating. The amount of shrinkage is more modest for movies with larger sample sizes. Furthermore, by pooling ratings across movies, one is able to estimate the standard deviation σ of the ratings. Without this pooling, one would be unable to estimate the standard deviation for a movie with only one rating.

Sources of variability

As discussed in Section 10.1, the prior distribution Normal(μ, τ) is shared among the means μ_j's of all groups in a hierarchical normal model, and the hyperparameters μ and τ provide information about the population of μ_j's. Specifically, the standard deviation τ measures the variability among the μ_j's. When the hierarchical model is estimated through MCMC, summaries from the simulation draws from the posterior of τ provide information about this source of variation after analyzing the data.

There are actually two sources for the variability among the observed Y_{ij}'s. At the sampling level of the model, the standard deviation σ measures variability of the Y_{ij} within the groups. In contrast, the parameter τ measures the variability in the measurements between the groups. When the hierarchical model is fit through MCMC, summaries from the marginal posterior distributions of σ and τ provide information about the two sources of variability.

$$Y_{ij} \overset{i.i.d.}{\sim} \text{Normal}(\mu_j, \sigma) \text{ [within-group variability]} \tag{10.17}$$
$$\mu_j \mid \mu, \tau \sim \text{Normal}(\mu, \tau) \text{ [between-group variability]} \tag{10.18}$$

The Bayesian posterior inference in the hierarchical model is able to compare these two sources of variability, taking into account the prior belief and the information from the data. One initially provides prior beliefs about the values of the standard deviations σ and τ through gamma distributions. In the MovieLens ratings application, weakly informative priors of Gamma$(1, 1)$ are assigned to both σ and τ. These prior distributions assume *a priori* the within-group variability, measured by σ, is believed to be the same size as the between-group variability measured by τ.

What can be said about these two sources of variability after the estimation of the hierarchical model? As seen in the output of `print(posterior, digits = 3)`, the 90% credible interval for σ is (0.763, 1.12) and the 90% credible interval for τ is (0.357, 1.08). After observing the data, the within-group variability in the measurements is estimated to be larger than the between-group variability.

To compare these two sources of variation one computes the fraction $R = \frac{\tau^2}{\sigma^2+\tau^2}$ from the posterior samples of σ and τ. The interpretation of R is that it represents the fraction of the total variability in the movie ratings due to the differences between groups. If the value of R is close to 1, most of the total variability is attributed to the between-group variability. On the other side, if R is close to 0, most of the variation is within groups and there is little significant difference between groups.

R Sample code shown below computes simulated values of R from the MCMC output. A density plot of R is shown in Figure 10.4.

```
tau_draws <- as.mcmc(posterior, vars = "tau")
sigma_draws <- as.mcmc(posterior, vars = "sigma")
R <- tau_draws ^ 2 / (tau_draws ^ 2 + sigma_draws ^ 2)
```

A 95% credible interval for R is (0.149, 0.630). Since much of the posterior probability of R is located below the value 0.5, this confirms that the variation between the mean movie rating titles is smaller than the variation of the ratings within the movie titles in this example.

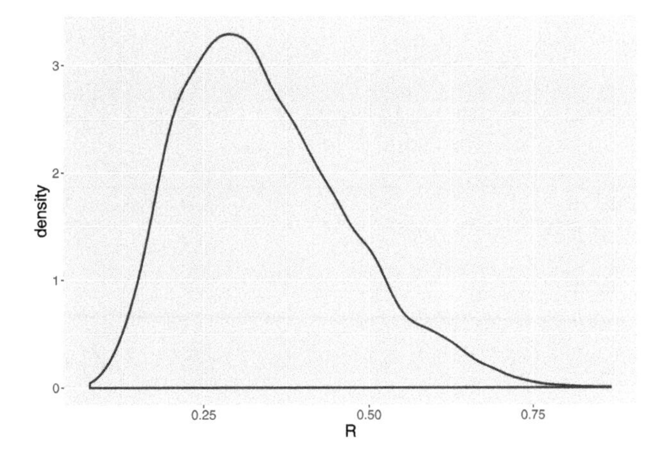

FIGURE 10.4
Density plot of the ratio $R = \frac{\tau^2}{\sigma^2+\tau^2}$ from the posterior samples of τ and σ.

10.3 Hierarchical Beta-Binomial Modeling

10.3.1 Example: Deaths after heart attacks

The New York State (NYS) Department of Health collects and releases data on mortality after acute myocardial infarction (AMI), commonly known as a heart attack. Its 2015 report was the initial public data release by the NYS Department of Health on risk-adjusted mortality outcomes for AMI patients at hospitals across the state. We focus on 13 hospitals in Manhattan, New York City, with the goal of learning about the percentages of deaths resulting from heart attacks in hospitals cited below. Table 10.2 records for each hospital the number of heart attack cases, the corresponding number of resulted deaths, and their computed percentage of resulted deaths.

10.3.2 A hierarchical beta-binomial model

Treating "cases" as trials and "deaths" as successes, the binomial sampling model is a natural choice for this data, and the objective is to learn about the death probability p of the hospitals. If one looks at the actual death percentages in Table 10.2, some hospitals have much higher death rates than other hospitals. For example, the highest death rate belongs to Mount Sinai Roosevelt, at 13.043% which is more than four times the rate of Harlem Hospital Center at 2.857%. If one assumes a common probability p for all thirteen hos-

TABLE 10.2
The number of heart attack cases, the number of resulted deaths, and the percentage of resulted deaths of 13 hospitals in New York City - Manhattan in 2015. NYP stands for New York Presbyterian.

Hospital	Cases	Deaths	Death %
Bellevue Hospital Center	129	4	3.101
Harlem Hospital Center	35	1	2.857
Lenox Hill Hospital	228	18	7.894
Metropolitan Hospital Center	84	7	8.333
Mount Sinai Beth Israel	291	24	8.247
Mount Sinai Hospital	270	16	5.926
Mount Sinai Roosevelt	46	6	13.043
Mount Sinai St. Luke's	293	19	6.485
NYU Hospitals Center	241	15	6.224
NYP Allen Hospital	105	13	12.381
NYP Columbia Presbyterian Center	353	25	7.082
NYP New York Weill Cornell Center	250	11	4.400
NYP Lower Manhattan Hospital	41	4	9.756

pitals, this model does not allow for possible differences between the death rates among these hospitals.

On the other hand, if one creates thirteen separate binomial sampling models, one for each hospital, and conducts separate inferences, one loses the ability to use potential information about the death rate from hospital j when making inference about that of a different hospital i. Since these are all hospitals in Manhattan, New York City, they may share attributes in common related to death rates from heart attack. The separate modeling approach does not allow for the sharing of information across hospitals.

A hierarchical model provides a compromise between the combined and separate modeling approaches. In Section 10.2, a hierarchical normal density was used to model mean rating scores from different movies. In this setting, one builds a hierarchical model by assuming the hospital death rate parameters *a priori* come from a common distribution. Specifically, one builds a hierarchical model based on a common beta distribution that generalizes the beta-binomial conjugate model described in Chapter 7. This modeling setup provides posterior estimates that partially pool information among hospitals

Let Y_i denote the number of resulted deaths from heart attack, n_i the number of heart attack cases, and p_i the death rate for hospital i. The sampling density for Y_i for hospital i is a binomial distribution with n_i and p_i, as in Equation (10.19). Suppose that the proportions $\{p_i\}$ independently follow the same conjugate beta prior distribution, as in Equation (10.20). So the sampling and first stage of the prior of our model is written as follows:

- Sampling, for $i, \cdots, 13$:

$$Y_i \sim \text{Binomial}(n_i, p_i) \tag{10.19}$$

- Prior for p_i, $i = 1, \cdots, 13$:

$$p_i \sim \text{Beta}(a, b) \tag{10.20}$$

Note that the hyperparameters a and b are shared among all hospitals. If a and b are known values, then the posterior inference for p_i of hospital i is simply another beta distribution by conjugacy (review material in Chapter 7 if needed):

$$p_i \mid y_i \sim \text{Beta}(a + y_i, b + n_i - y_i). \tag{10.21}$$

In the general situation where the hyperparameters a and b are unknown, a second stage of the prior $\pi(a, b)$ needs to specified for these hyperparameters. With this specification, one arrives at the hierarchical model below.

- Sampling, for $i, \cdots, 13$:

$$Y_i \sim \text{Binomial}(n_i, p_i) \tag{10.22}$$

- Prior for p_i, Stage1: for $i = 1, \cdots, 13$:

$$p_i \sim \text{Beta}(a, b) \tag{10.23}$$

- Prior for p_i, Stage 2: the hyperprior:

$$a, b \sim \pi(a, b) \tag{10.24}$$

We use $\pi(a, b)$ to denote an arbitrary distribution for the joint hyperprior distribution for a and b. When we start analyzing the New York State heart attack death rate dataset, the specification of this hyperprior distribution $\pi(a, b)$ will be described.

Graphical representations of the hierarchical model

Below is a sketch of a graphical representation of the hierarchical beta-binomial model.

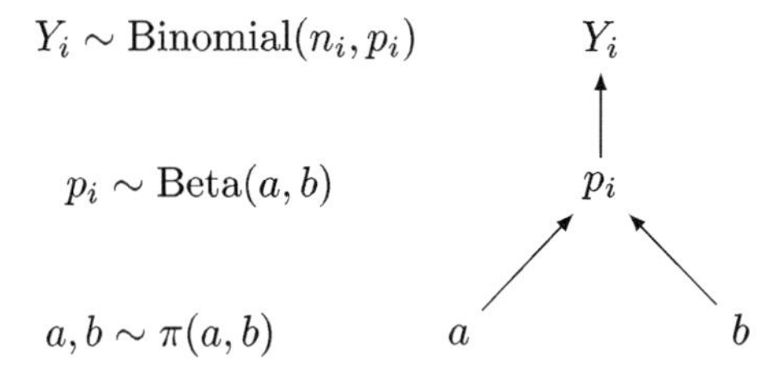

Focusing on the graph on the right, one sees that the upper section of the graph represents the sampling density, with the arrow directing from p_i to Y_i. Here the start of the arrow is the parameter and the end of the arrow is the random variable. The lower section of the graph represents the prior, with arrows directing from a and b to p_i. In this case, the start of the arrow is the hyperparameter and the end of the arrow is the parameter. On the left side of the display, the sampling density, prior and hyperprior distributional expressions are written next to the graphical representation.

In the situation where the beta parameters a and b are known constants, the graphical representation changes to the beta-binomial conjugate model displayed below.

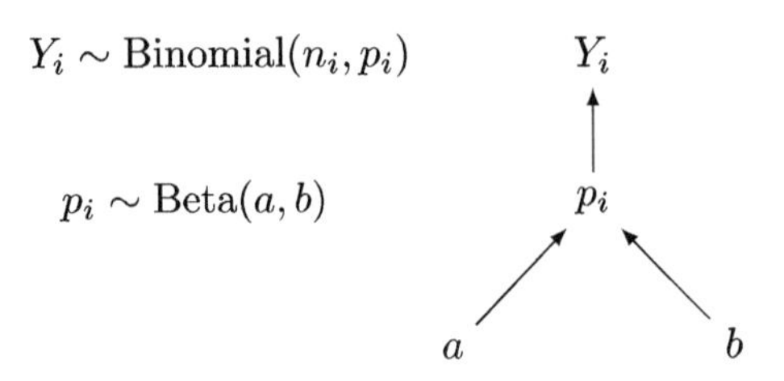

To illustrate another graphical representation, we display below the one for the separate models approach in the hospitals death rate application where a fully specified beta prior is specified for each death rate. The separate models are represented by thirteen graphs, one for each hospital. This graphical structure shows clearly the separation of the subsamples and the resulting separation of the corresponding Bayesian posterior distributions.

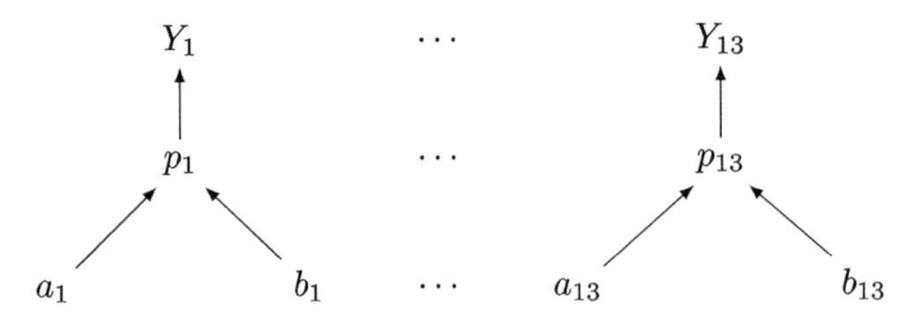

In comparing graphical representations for hierarchical models, the interested reader might notice that the structure for the hierarchical beta-binomial model looks different from the ones in Section 10.2 for the hierarchical normal models. In this chapter, one is dealing with one-parameter models (recall that beta-binomial is an example of one-parameter models; other examples include gamma-Poisson), whereas the normal models in Section 10.2 involve two parameters. Typically, when working with one-parameter models, one starts from the top with the sampling density, then next writes down the priors and continues with the hyperpriors. When there are multiple parameters, one needs to be careful in describing the graphical structure. In fact, for a large number of parameters, a good graphical representation might not be feasible. In that case, one writes a representation that focuses on the key parts of the model.

Also note that there is no unique way of sketching a graphical representation, as long as the representation is clear and shows the relationship among the random variables, parameters and hyperparameters with the arrows in the correct directions.

10.3.3 Inference through MCMC

In this section the application of JAGS script for simulation by MCMC is illustrated for the hierarchical beta-binomial models for the New York State heart attach death rate dataset. Before this is done, we discuss the specification of the hyperprior density $\pi(a, b)$ for the hyperparameters a and b for the common beta prior distribution for the proportions p_i's.

Second-stage prior

In Chapter 7, the task was to specify the values of a and b for a single beta curve Beta(a, b) and the beta shape parameter values were selected by trial-and-error using the `beta.select()` function in the `ProbBayes` package. In this hierarchical model setting, the shape parameters a and b are random and the goal is learn about these parameters from its posterior distribution.

In this prior construction, it is helpful to review some facts on beta curves from Chapter 7. For a Beta(a, b) prior distribution for a proportion p, one considers the parameter a as the prior count of "successes", the parameter b as the prior count of "failures", and the sum $a + b$ represents the prior sample size. Also the expectation of Beta(a, b) is $\frac{a}{a+b}$. From these facts, a more natural parameterization of the hyperprior distribution $\pi(a, b)$ is $\pi(\mu, \eta)$, where $\mu = \frac{a}{a+b}$ is the hyperprior mean and $\eta = a + b$ is the hyperprior sample size. One rewrites the hyperprior distribution in terms of the new parameters μ and η as follows:

$$\mu, \eta \sim \pi(\mu, \eta), \tag{10.25}$$

where $a = \mu\eta$ and $b = (1 - \mu)\eta$. These expressions are useful in writing the JAGS script for the hierarchical beta-binomial Bayesian model.

A hyperprior is constructed from the (μ, η) representation. Assume μ and η are independent which means that one's beliefs about the prior mean are independent of the beliefs about the prior sample size. The hyperprior expectation μ is the mean measure for p_i, the average death rate across 13 hospitals. If one has little prior knowledge about the expectation μ, one assigns this parameter a uniform prior which is equivalent to a Beta$(1, 1)$ prior.

To motivate the prior choice for the hyperparameter sample size η, consider the case where the hyperparameter values are known. If y^* and n^* are respectively the number of deaths and number of cases for one hospital, then the posterior mean of death rate parameter p^* is given by

$$E(p^* \mid y^*) = \frac{y^* + \mu\eta}{n^* + \eta}. \tag{10.26}$$

With a little algebra, the posterior mean is rewritten as

$$E(p^* \mid y^*) = (1 - \lambda)\frac{y^*}{n^*} + \lambda\mu, \tag{10.27}$$

where λ is the shrinkage fraction

$$\lambda = \frac{\eta}{n^* + \eta}. \tag{10.28}$$

The parameter λ falls in the interval (0, 1) and represents the degree of shrinkage of the posterior mean away from the sample proportion y^*/n^* towards the prior mean μ.

Suppose one believes *a priori* that, for a representative sample size n^*, the shrinkage λ is uniformly distributed on (0, 1). By performing a transformation, this implies that the prior density for the prior sample size η has the form

$$\pi(\eta) = \frac{n^*}{(n^* + \eta)^2}, \ \eta > 0. \tag{10.29}$$

Equivalently, the logarithm of η, $\theta = \log \eta$, has a logistic distribution with location $\log n^*$ and scale 1. We represent this distribution as Logistic($\log n^*, 1$), with pdf:

$$\pi(\theta) = \frac{e^{-(\theta - \log n^*)}}{(1 + e^{-(\theta - \log n^*)})^2}. \tag{10.30}$$

With this specification of the hyperparameter distribution, one writes down the complete hierarchical model as follows:

- Sampling, for $i, \cdots, 13$:

$$Y_i \sim \text{Binomial}(n_i, p_i) \tag{10.31}$$

- Prior for p_i, Stage 1: for $i = 1, \cdots, 13$:

$$p_i \sim \text{Beta}(a, b) \tag{10.32}$$

- Prior for p_i, Stage 2:

$$\mu \sim \text{Beta}(1, 1), \tag{10.33}$$
$$\log \eta \sim \text{Logistic}(\log n^*, 1) \tag{10.34}$$

where $a = \mu\eta$ and $b = (1 - \mu)\eta$.

Writing the JAGS script

R Following this model structure above, one writes out the model section of the JAGS script for the hierarchical beta-binomial model. The model script is saved in `modelString`.

```
modelString <-"
model {
## likelihood
for (i in 1:N){
   y[i] ~ dbin(p[i], n[i])
}
## priors
for (i in 1:N){
   p[i] ~ dbeta(a, b)
}
## hyperpriors
a <- mu*eta
b <- (1-mu)*eta
mu ~ dbeta(mua, mub)
eta <- exp(logeta)
logeta ~ dlogis(logn, 1)
}
"
```

In the sampling part of the script, the loop goes from `1` to `N`, where `N` is the total number of observations, with index `i`. Another loop going from `1` to `N` is needed for the priors as each `p[i]` follows the same `dbeta(a, b)` distribution. The hyperpriors section uses the new parameterization of the Beta(a, b) distribution in terms of `mu` and `eta`. Here one expresses the hyperparameters `a` and `b` in terms of the new hyperparameters `mu` and `eta`, and then assigns to the parameters `mu` and `logeta` the independent distributions `dbeta(mua, mub)` and `dlogist(logn, 1)`, respectively. One also needs to transform `logeta` to `eta`. The values of `mua`, `mub`, and `logn` are assigned together with the data in the setup of JAGS, following Equation (10.33) and Equation (10.34).

Define the data and prior parameters

R Following the usual implementation of JAGS, the next step is to define the data and provide values for the parameters of the prior. In the script below, a list `the_data` contains the vector of death counts in `y`, the vector of hearth attack cases in `n`, the number of observations `N`, the values of `mua`, `mub`, and `logn`. Note that we are setting $\log n^* = \log(100)$ which indicates that *a priori* we believe the shrinkage $\lambda = \eta/(\eta + 100)$ is uniformly distributed on (0, 1).

```
y <- deathdata$Deaths
n <- deathdata$Cases
N <- length(y)
the_data <- list("y" = y, "n" = n, "N" = N,
                 "mua" = 1, "mub" = 1,
                 "logn" = log(100))
```

Generate samples from the posterior distribution

The `run.jags()` function is used to generate samples by MCMC in JAGS following the sample script below. It runs one MCMC chain with 1000 iterations in the adapt period, 5000 iterations of burn-in and an additional set of 5000 iterations to be run and collected for inference. One keeps tracks of all parameters in the model by using the argument `monitor = c("p", "mu", "logeta")`. The output of the MCMC runs is the variable `posterior` containing a matrix of simulated draws.

```
posterior <- run.jags(modelString,
                      n.chains = 1,
                      data = the_data,
                      monitor = c("p", "mu", "logeta"),
                      adapt = 1000,
                      burnin = 5000,
                      sample = 5000)
```

MCMC diagnostics and summarization

As usual, it is important to perform MCMC diagnostics to ensure convergence of the simulated sample. The `plot()` function returns diagnostics plots of a designated parameter. For brevity, the diagnostics for a are performed and results shown in Figure 10.5. Readers should implement MCMC diagnostics for all parameters in the model.

```
plot(posterior, vars = "logeta")
```

After the diagnostics are performed, one reports posterior summaries of the parameters using `print()`. Note that these summaries are based on the 5000 iterations from the sampling period (excluding the adapt and burn-in periods).

```
print(posterior, digits = 3)
       Lower95 Median Upper95   Mean     SD Mode    MCerr
p[1]    0.0314 0.0602  0.0847 0.0593 0.0138   -- 0.000619
p[2]    0.0312  0.066   0.095 0.0654 0.0156   -- 0.000496
```

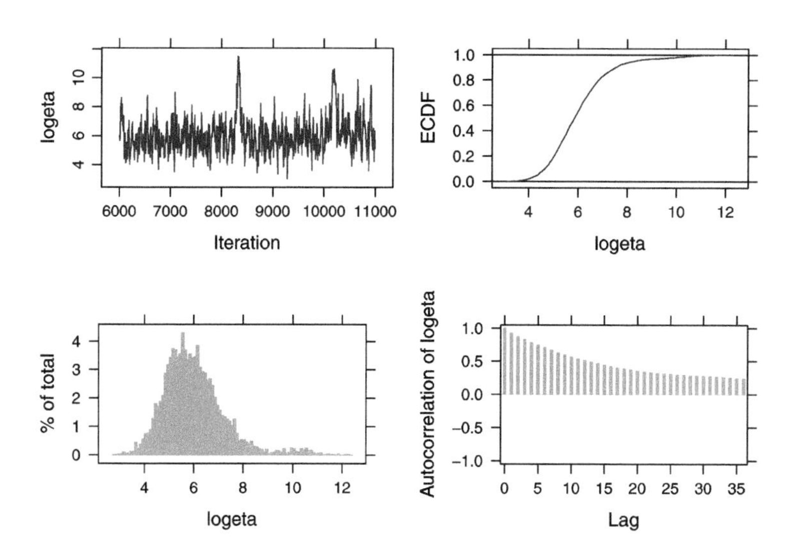

FIGURE 10.5
Diagnostics plots of simulated draws of $\log \eta$ using the JAGS software with the `run.jags` package.

```
p[3]     0.0515 0.0731     0.1 0.0741  0.0122   -- 0.000398
p[4]      0.044 0.0726   0.105  0.074  0.0155   -- 0.000486
p[5]     0.0553 0.0756     0.1 0.0765  0.0116   -- 0.000348
p[6]     0.0435 0.0655  0.0871 0.0655  0.0111   --  0.00042
p[7]     0.0466 0.0765   0.119 0.0797  0.0191   -- 0.000717
p[8]     0.0473 0.0683  0.0889 0.0683  0.0104   -- 0.000277
p[9]     0.0442 0.0669  0.0879 0.0671  0.0111   -- 0.000301
p[10]    0.0544 0.0811   0.122 0.0845  0.0178   -- 0.000732
p[11]    0.0521 0.0704  0.0934 0.0711  0.0103   -- 0.000279
p[12]    0.0369   0.06  0.0818 0.0596  0.0116   -- 0.000504
p[13]    0.0444 0.0729   0.113 0.0752  0.0176   -- 0.000593
mu       0.0576 0.0705  0.0881 0.0714 0.00788   -- 0.000375
logeta     3.63   5.84    8.38   6.01    1.26   --    0.107
```

From the posterior output, one evaluates the effect of information pooling in the hierarchical model. Figure 10.6 displays a shrinkage plot showing how the sample proportions are shrunk towards the overall death rate. Two of the lines in the figure are labelled to indicate the the death rates for the hospitals Mount Sinai Roosevelt and and NYP Allen Hospital. Mount Sinai Roosevelt's death rate of $6/46 = 0.13043$ exceeds the rate of NYP Allen of $13/105 = 0.12381$, but the figure shows the posterior death rate of NYP

Allen exceeds the posterior death rate of Mount Sinai Roosevelt. Due to the relatively small sample size, one has less confidence in the 0.13043 death rate of Mount Sinai and this rate is shrunk significantly towards the overall death rate in the hierarchical posterior analysis.

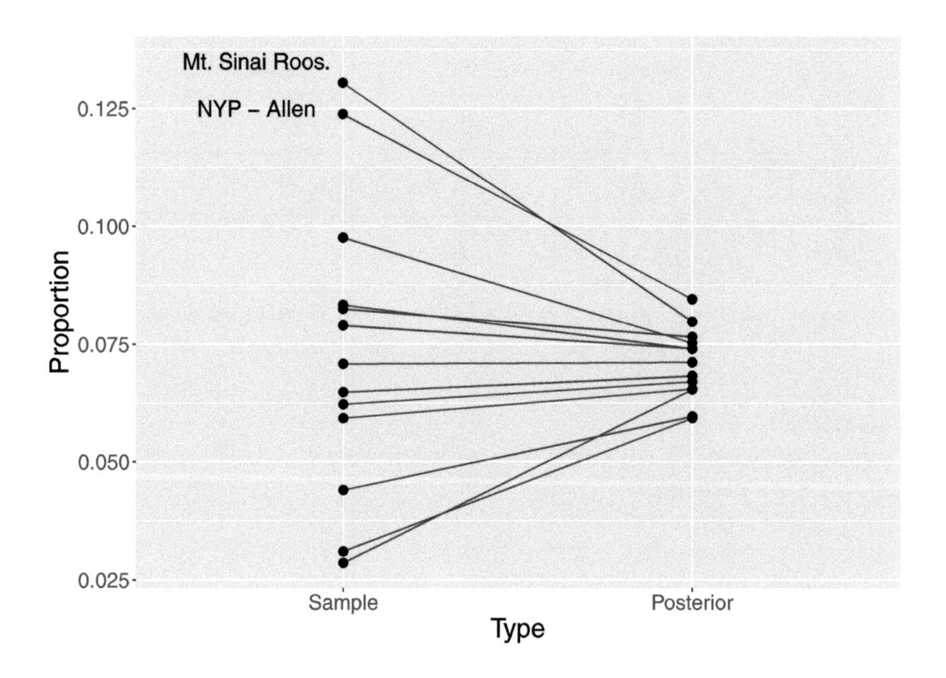

FIGURE 10.6
Shrinkage plot of sample proportions and posterior means of proportions of resulted heart attack deaths of 13 hospitals. The death rates of two particular hospitals are labeled. Due to the varying sample sizes, Mt Sinai Roosevelt has a higher observed death rate than NYP Allen, but NYP Allen has a higher posterior proportion than Mt Sinai Roosevelt.

To compare the posterior densities of the different p_i, one displays the density estimates in a single graph as in Figure 10.7. Because of the relatively large number of parameters, such plots are difficult to read. Combining the graph and the output above, one sees that p_7 and and p_{10} have the largest median values with large standard deviations. One makes inferential statements such as Mount Sinai Roosevelt's (corresponding to p_7) death rate of heart attack cases has a posterior 90% credible interval of (0.0466, 0.119), the highest among the 13 hospitals in the dataset.

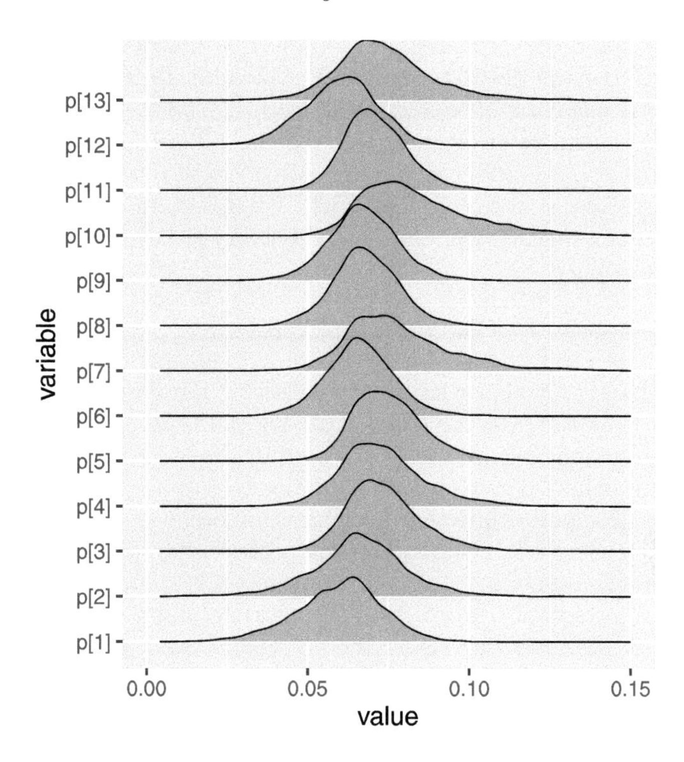

FIGURE 10.7
Density plots of simulated draws of p_i's using the JAGS software with the `run.jags` package.

Comparison of hospitals

R One uses this MCMC output to compare the death rates of two hospitals directly, for example, NYP Columbia Presbyterian Center and NYP New York Weill Cornell Center corresponding respectively to p_{11} and p_{12}. One collects the vector of simulated values of the difference of the death rates ($\delta = p_{11} - p_{12}$) by subtracting the sets of simulated proportion draws. From the simulated values of the difference in proportions `diff`, one estimates the probability that $p_{11} > p_{12}$ is positive.

```
p11draws <- as.mcmc(posterior, vars = "p[11]")
p12draws <- as.mcmc(posterior, vars = "p[12]")
diff = p11draws - p12draws
sum(diff > 0)/5000
[1] 0.7872
```

A 78.72% posterior probability of $p_{11} > p_{12}$ indicates strong posterior evidence that the the death rate of NYP Columbia Presbyterian Center is higher than that of NYP New York Weill Cornell Center.

Generally, when one presents a table such as Table 10.2, one is interested in ranking the 13 hospitals from best (smallest death rate) to worst (largest death rate). A particular hospital, say Bellevue Hospital Center, is interested in its rank among the 13 hospitals. The probability Bellevue has rank 1 is the posterior probability

$$P(p_1 < p_2, ..., p_1 < p_{13} \mid y), \tag{10.35}$$

and this probability is approximated by collecting the posterior draws where the simulated value of p_1 is the smallest among the 13 simulated proportions. Likewise, one computes from the MCMC output the probability that Bellevue has rank 2 through 13. These rank probabilities are displayed in Figure 10.8 for two hospitals. The probability that Bellevue is the best hospital with respect to death rate is 0.25 and by summing several probabilities, the probability that Bellevue is ranked among the top three hospitals is 0.54. In contrast, from Figure 10.8, the rank of Harlem Hospital is less certain since the probability distribution is relatively flat across the 13 possible rank values. This is not surprising since this particular hospital had only 35 cases, compared to 129 cases at Bellevue.

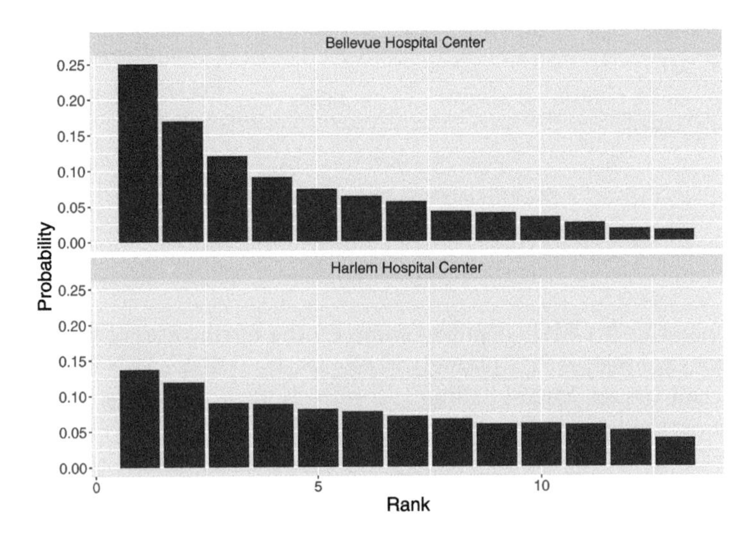

FIGURE 10.8
Posterior probabilities of rank for two hospitals.

From a patient's perspective, she would be interested in learning the identity of the hospital that is ranked best among the 13. For each simulation draw of $p_1, ..., p_{13}$, one identifies the hospital with the smallest simulated value. By collecting this information over the 5000 draws, one computes the posterior probability that each hospital is ranked first. These probabilities are displayed in Figure 10.9. The identity of the best hospital is not certain, but the top three hospitals are Bellevue, NYP New York Weill Cornell Center, and Harlem with respective probabilities 0.250, 0.220, and 0.137 of being the best.

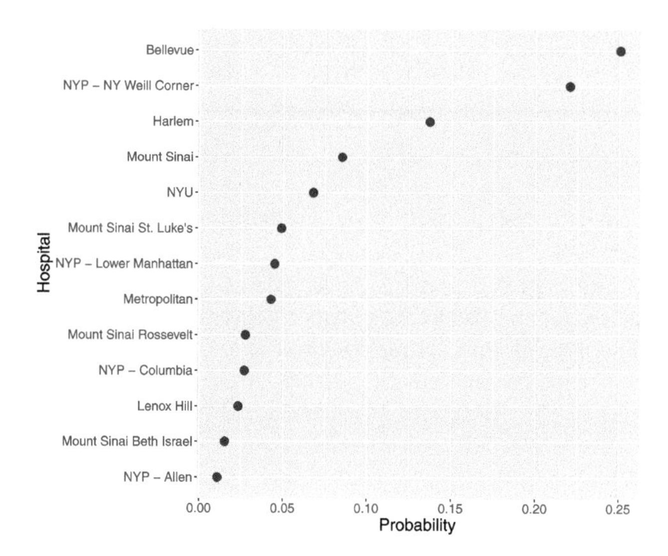

FIGURE 10.9
Posterior probabilities of the hospital that was ranked first.

10.4 Exercises

1. **Time-to-Serve for Six Tennis Players**

   Table 10.3 displays the sample size n_j and the mean time-to-serve $\bar{y}_j$ (in seconds) for six professional tennis players. Assume that the sample mean for the i-th player $\bar{y}_i$ is normally distributed with mean μ_i and standard deviation $\sigma/\sqrt{n_i}$ where we assume $\sigma = 5.5$ seconds.

   (a) (Separate estimate) Suppose one is interested in estimating Murray's mean time-to-serve μ_1 using only Murray's time-to-serve data. Assume that one's prior beliefs about μ_1 are represented by a normal density with mean 20 and standard deviation 10 seconds. Use results from Chapter 8 to find the

TABLE 10.3
Number of serves and mean time-to-serve for six professional tennis players.

Player	n	$\bar{y}$
Murray	731	23.56
Simon	570	18.07
Federer	491	16.21
Ferrer	456	21.70
Isner	403	22.32
Kyrgios	274	14.11

posterior distribution of μ_1 and construct a 90% interval estimate for μ_1.

(b) (Combined estimate) Suppose instead that one believes that there are no differences between players and $\mu_1 = ... = \mu_6 = \mu$. The overall mean time-to-serve is $\bar{y} = 19.9$ seconds with a combined sample size $n = 2925$. Assuming that μ has a Normal(20,10) prior, find the posterior distribution of μ and construct a 90% interval estimate for μ.

(c) Which approach, part (a) or part (b), seems more reasonable in this situation? Explain.

2. **Time-to-Serve for Six Tennis Players (continued)**

Suppose one wishes to estimate the mean time-to-serve values for the six players by the following hierarchical model. Remember that we are assuming $\sigma = 5.5$ seconds.

$$\begin{aligned}
\bar{y}_i &\sim \text{Normal}(\mu_j, \sigma/\sqrt{n_i}), \;\; i = 1, ..., 6. \\
\mu_i &\sim \text{Normal}(\mu, \tau), \;\; i = 1, ..., 6. \\
\mu &\sim \text{Normal}(20, 1/0.001), \\
1/\tau^2 &\sim \text{Gamma}(0.1, 0.1).
\end{aligned}$$

(a) Use JAGS to simulate a sample of size 1000 from the posterior distribution from this hierarchical model, storing values of the means $\mu_1, ..., \mu_6$.

(b) Construct a 90% interval estimate for each of the means.

(c) Compare the 90% estimate for Murray with the separate and combined interval estimates from Exercise 1.

3. **Random σ_j's for Movie Ratings**

In Section 10.2.2, consider the situation where the standard deviations of the ratings differ across movies, so σ_j represents the standard deviation of the ratings for movie j.

(a) Write out the likelihood, the prior distributions, and hyperprior distributions for this varying means and varying standard deviations model.

(b) Discuss the implications of specifying varying σ_j's by comparing this hierarchical model to the developed model in Section 10.2.2.

(c) What prior distributions do you choose for σ_j's? Why?

(d) Carry out the simulation by MCMC using JAGS. Report and discuss the findings.

4. **Smoothing Counts**

A general issue in statistical inference is how to handle situations where there are zero observed counts in a sample. This exercise illustrates several Bayesian modeling approaches to this problem.

(a) Suppose one is learning about the probability p a particular player successively makes a three-point shot in basketball. One assigns a uniform prior for p. This player attempts 10 shots and one observes $y = 0$ successes. Derive the posterior density of p and compute the posterior mean.

(b) Suppose that you are learning about the probabilities p_1, p_2, p_3, p_4, p_5 of five players making three-throw shots. You assign the following hierarchical prior on the probabilities. You assume $p_1, ..., p_5$ are independently identically distributed beta with shape parameters α and α, and at the second stage, you assign α a uniform prior on $(0, 1)$. Write down a graphical representation of this hierarchical model.

(c) In part (b), suppose that each player attempts 10 shots and you observe 0, 2, 3, 1, 3 successes for the five players. Use JAGS to obtain posterior samples from the parameters $\alpha, p_1, p_2, p_3, p_4, p_5$. Compute the posterior means of α and p_1 and compare your probability estimates with the estimate of p using the single-stage prior in part (a).

5. **Schedules and Producers in Korean Drama Ratings**

The Korean entertainment industry has been continuously booming. The global audience for K-drama is exploding across Asia and even spreading to other parts of the world, notably the U.S. and Europe. This surge of Korean cultural popularity is called "Hallyu", literally meaning the "Korean wave". K-dramas are popular on multiple streaming websites in the U.S., such as Hulu, DramaFever, and even Netflix.

How are K-dramas being rated in Korea? How are the producing company and broadcasting schedule affecting the drama ratings? In one study, data were collected on 101 K-dramas from 2014 to 2016.

Each drama was produced by one of the three main producers, and was broadcasted in one of four different times of the week. The ratings of dramas were collected from the AGB Nielsen Media Research Group[1]. In particular, the national AGB TV ratings of each drama were recorded.

The data is stored in `KDramaData.csv`. Table 10.4 provides information about the variables in the complete dataset.

TABLE 10.4
Table of the variables in K-dramas application.

Name	Variable Information
Drama Name	Name of drama
Schedule	1 = Mon. and Tue., 2 = Wed. and Thu., 3 = Fri., 4 = Sat. and Sun.
Producer	1 = Seoul Broadcasting System, 2 = Korean Broadcasting System, 3 = Munhwa Broadcasting Corporation
Viewership	AGB national TV ratings, in percentage
Date	Month, day, year

(a) Explore the ratings graphically by schedule and by producer.

(b) Explain how the ratings differ by schedule and by producer. Are there particular days when the ratings are high or low? Does one producer tend to have larger ratings than the other producer?

(c) Choose a subset of the `KDramaData.csv` dataset for a particular producer. Develop a hierarchical model to make inference about the mean ratings of dramas across different schedules. Discuss your conclusions and the advantage of using hierarchical modeling in this situation.

6. **Homework Hours for Five Schools**

To compare weekly hours spent on homework by students, data is collected from a sample of five different schools. The data is stored in `HWhours5schools.csv`.

(a) Explore the weekly hours spent on homework by students from the five schools. Do the school-specific means seem significantly different from each other? What about their variances?

[1]AGB Nielsen Media Research Group is one of the biggest companies that measures audiences' television ratings. At the time the data were collected, AGB Nielsen analyzed viewing of 2020 households from five major and five medium-sized cities in South Korea to determine TV ratings.

(b) Set up a hierarchical model with common and unknown σ in the likelihood, as in Section 10.2.2. Write out the likelihood, the prior distributions and the hyperprior distributions.

(c) Use JAGS to obtain posterior samples of the parameters in the hierarchical model. Perform appropriate MCMC diagnostics.

(d) Compute posterior means and 95% credible intervals for every school mean. Compute the posterior probability that the mean hour in school 1 is higher than that of school 2. Discuss your findings.

(e) Compute and summarize the posterior distribution of the ratio $R = \frac{\tau^2}{\tau^2+\sigma^2}$. Comment on the evidence of between-school variability for this data..

7. **Heart Attack Deaths - New York City**

In Section 10.3, the heart attack deaths dataset of thirteen hospitals in Manhattan, New York City were analyzed using a hierarchical beta-binomial model. A complete dataset of heart attack death information of 45 hospitals in all 5 boroughs of New York City (Manhattan, the Bronx, Brooklyn, Queens, and Staten Island) is stored in `DeathHeartAttackDataNYCfull.csv`. Table 10.5 lists the variables and their description.

TABLE 10.5
The list of variables in the New York City heart attack deaths dataset and their description.

Variable	Description
Hospital	Name of hospital
Borough	Borough location of hospital
Type	N = Non-PCI hospital; P = PCI hospital
Cases	Number of heart attack cases
Deaths	Number of deaths among heart attack cases

Note: PCI = Percutaneous coronary intervention, also known as coronary angioplasty, is a nonsurgical procedure that improves blood flow to your heart.

(a) Write out the complete hierarchical beta-binomial model for the subset of thirteen hospitals in Brooklyn. Sketch a graphical representation and discuss how to choose priors and hyperpriors.

(b) Use JAGS to obtain posterior samples of the parameters in the hierarchical model. Perform appropriate MCMC diagnostics.

(c) Compute the posterior probability that the death rate of Kings County Hospital Center is higher than that of the Kingsbrook Jewish Medical Center. Report and discuss your findings.

8. **Heart Attack Deaths - New York City (continued)**

   Develop a hierarchical beta-binomial model for the subset of sixteen hospitals in the Bronx and Queens. Instead of allowing a p_i for each hospital i in the subset, allow a p_B to be shared among hospitals in the Bronx, and a p_Q to be shared among hospitals in Queens.

   (a) How does the hierarchical beta-binomial model change from the specification in Exercise 7? Write out the complete hierarchical beta-binomial model, sketch a graphical representation. Discuss how to choose priors and hyperpriors.
   (b) Use JAGS to obtain posterior samples of the parameters in the hierarchical model. Perform appropriate MCMC diagnostics.
   (c) Compute the posterior probability that the death rate of all hospitals in the Bronx is higher than that of all hospitals in Queens. Report and discuss your findings.

9. **Heart Attack Deaths - New York City (continued)**

   Develop a hierarchical beta-binomial model for the complete dataset of 45 hospitals in New York City. Instead of allowing a p_i for each hospital i in the subset, allow a p_P to be shared among hospitals of Type P, and a p_N to be shared among hospitals of Type N.

   (a) Write out the complete hierarchical beta-binomial model, sketch a graphical representation. Discuss how to choose priors and hyperpriors.
   (b) Use JAGS to obtain posterior samples of the parameters in the hierarchical model. Perform appropriate MCMC diagnostics.
   (c) Compute the posterior probability that the death rates of all hospitals of Type P are higher than those of all hospitals of Type N. Report and discuss your findings.
   (d) Can you develop a hierarchical beta-binomial model for all 45 hospitals in New York City that takes into account Borough and Type? Describe how you would design the hierarchical model, write JAGS script to obtain posterior samples of the parameters and discuss any findings from your work.

10. **Hierarchical Gamma-Poisson Modeling - Marriage Rates in Italy**

    Annual marriage counts per 1000 of the population in Italy from 1936 to 1951 were collected and recorded in Table 10.6. Can we learn something about Italians' marriage rates during this 16-year period? The dataset is stored in `marriage_counts.csv`.

    (a) Recall that with count data, a common conjugate model is the gamma-Poisson model, introduced in Section 8.8. Write out the likelihood, the prior distribution, and its posterior distribution under the gamma-Poisson model.

TABLE 10.6
The year and the marriage counts per 1000 of the population in Italy from 1936 to 1951.

Year	Count	Year	Count
1936	7	1944	5
1937	9	1945	7
1938	8	1946	9
1939	7	1947	10
1940	7	1948	8
1941	6	1949	8
1942	6	1950	8
1943	5	1951	7

(b) Observations are considered *i.i.d.* in the model in part (a). Figure 10.10 plots the marriage rates in Italy across years. Discuss whether the *i.i.d.* assumption is reasonable.

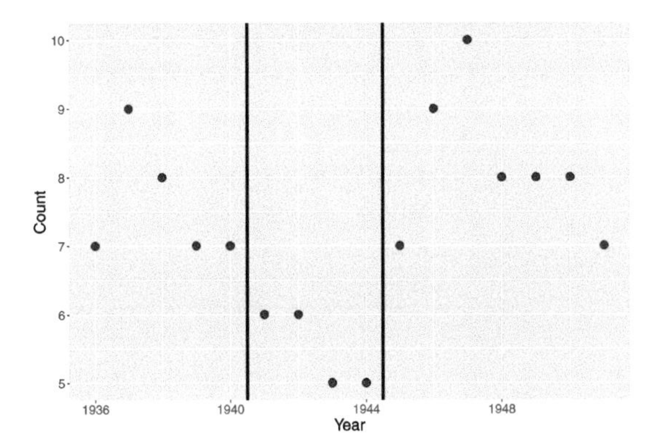

FIGURE 10.10
Dotplot of marriage rates in Italy from 1936 to 1951.

(c) Suppose one believes that the mean marriage rate differs across the three time periods shown in Figure 10.10. Using this belief, model the Italian marriage rates in a hierarchical approach. Write out the likelihood, the prior distributions, and any hyperprior distributions under a hierarchical gamma-Poisson model.

(d) Sketch a graphical representation of the hierarchical gamma-Poisson model.

(e) Simulate posterior draws by MCMC using JAGS. Perform MCMC diagnostics and make sure your MCMC has converged.

(f) Do you see clear differences between the three rate parameters in the posterior? Report and discuss your findings.

11. **Hierarchical Gamma-Poisson Modeling - Fire Calls in Pennsylvania**

Table 10.7 displays the number of fire calls and the number of building fires for ten zip codes in Montgomery County, Pennsylvania from 2015 through 2019. This data is currently described as "Emergency - 911 Calls" from `kaggle.com`. Suppose that the number of building fires for the j-th zip code is Poisson with mean $n_j \lambda_j$, where n_j and λ_j are respectively the number of fire calls and rate of building fires for the j-th zip code.

TABLE 10.7
The number of fire calls and building fires for ten zip codes in Montgomery County, Pennsylvania.

Zip Code	Fire Calls	Building Fires
18054	266	12
18103	1	0
19010	1470	59
19025	246	11
19040	1093	47
19066	435	26
19116	2	0
19406	2092	113
19428	2025	73
19474	4	1

(a) Suppose that the building fire rates $\lambda_1, ..., \lambda_{10}$ follow a common Gamma(α, β) distribution where the hyperparameters α and β follow weakly informative distributions. Use JAGS to simulate a sample of size 5000 from the joint posterior distribution of all parameters of the model.

(b) The individual estimates of the building rates for zip codes 18054 and 19010 are $12/266$ and $59/1470$, respectively. Contrast these estimates with the posterior means of the rates λ_1 and λ_3.

(c) The parameter $\mu = \alpha/\beta$ represents the mean building fire rates across zip codes. Construct a density estimate of the posterior distribution of μ.

(d) Suppose that the county has 50 fire calls to the zip code 19066. Use the simulated predictive distribution to construct a 90% predictive interval for the number of building fires.

12. **Hierarchical Gamma-Exponential Modeling - Times Between Traffic Accidents**

    Exercise 20 in Chapter 8 describes the exponential distribution, which is often used as a model for time between events, such as traffic accidents. The exercise also describes the gamma distribution as a conjugate prior choice for the exponential data model. 10 times between traffic accidents are collected: 1.5, 15, 60.3, 30.5, 2.8, 56.4, 27, 6.4, 110.7, 25.4 (in minutes).

    (a) Suppose the 10 collected times are observed at 4 different locations, shown in Table 10.8. Using this information, model the times between traffic accidents in a hierarchical approach. Write out the likelihood, the prior distributions, and any hyperprior distributions under a hierarchical gamma-exponential model.

    (b) Sketch a graphical representation of the hierarchical gamma-exponential model.

    (c) Simulate posterior draws by MCMC using JAGS. Perform MCMC diagnostics and make sure your MCMC has converged.

    (d) Do you see clear differences between the rate of traffic accidents at the 5 locations? Report and discuss your findings.

TABLE 10.8
The time between traffic accidents and recorded location.

Time	Location	Time	Location
1.5	1	15	1
60.3	2	30.5	2
2.8	3	56.4	3
27	4	6.4	4
110.7	5	25.4	5

13. **Bird Survey Trend Estimates**

    The North American Breeding Bird Survey (BBS) is a yearly survey to monitor the bird population. Regression models were used to estimate the change in population size for many species of birds between 1966 to 1999. For each of 28 particular grassland species of birds, Table 10.9 displays the trend estimate $\hat{\beta}_i$ and the corresponding standard error σ_i. This data is stored in the data file `BBS_survey.csv`. Assume that the trend estimates are independent with $\hat{\beta}_i \sim \text{Normal}(\beta_i, \sigma_i)$ where we assume that the standard errors $\{\sigma_i\}$ are known.

    (a) Suppose one assumes that the population trend estimates are equal, that is, $\beta_1 = ... \beta_{28} = \beta$. Using JAGS to simulate from

TABLE 10.9
Trend estimate $\hat{\beta}_i$ and associated standard error σ_i for 28 grassland species birds.

Species Name	Trend	SE
Upland Sandpiper	0.76	0.39
Long-billed Curlew	-0.77	1.01
Mountain Plover	-1.05	2.24
Greater Prairie-Chicken	-2.54	2.33
Sharp-tailed Grouse	-0.92	1.43
Ring-necked Pheasant	-1.06	0.32
Northern Harrier	-0.80	4.00
Ferruginous Hawk	3.52	1.31
Common Barn Owl	-2.00	2.14
Short-eared Owl	-6.23	4.55
Burrowing Owl	1.00	2.74
Horned Lark	-1.89	0.22
Bobolink	-1.25	0.31
Eastern Meadlowlark	-2.69	0.17
Western Meadowlark	-0.75	0.17
Chestnut-col Longspur	-1.36	0.68
McCown's Longspur	-9.29	8.27
Vesper Sparrow	-0.61	0.24
Savannah Sparrow	-0.34	0.29
Baird's Sparrow	-2.04	1.48
Grasshopper Sparrow	-3.73	0.47
Henslow's Sparrow	-4.82	2.50
LeConte's Sparrow	0.91	0.95
Cassin's Sparrow	-2.10	0.51
Dickcissel	-1.46	0.28
Lark Bunting	-3.74	2.30
Sprague's Pipit	-5.62	1.34
Sedge Wren	3.18	0.73

the posterior distribution of β assuming a weakly informative prior on β. Find the posterior mean and posterior standard deviation of β and compare your answers to the trend estimates and standard errors in Table 10.9.

(b) Next assume that the population trend estimates $\beta_1 = ... \beta_{28}$ are a random sample from a normal distribution with mean μ and standard deviation τ. Assuming weakly informative priors on μ and τ, use JAGS to simulate from the posterior distribution of all parameters. Find the posterior means of the $\{\beta_j\}$ and compare your estimates with the trend estimates in Table 10.9.

14. **Predicting Baseball Batting Averages**

    The data file `batting_2018.csv` contains batting data for every player in the 2018 Major League Baseball season. The variables `AB.x` and `H.x` in the dataset contain the number of at-bats (opportunities) and number of hits of each player in the first month of the baseball season. One assumes that y_i, the number of hits of the i-th player is Binomial(n_i, p_i) where n_i is the number of at-bats and p_i is the probability of a hit.

    (a) Select a random sample of 20 players from the dataset.

    (b) Assume that the hitting probabilities $\{p_i\}$ have a common Beta(a, b) prior where $a = \eta\mu$ and $b = \eta(1 - \mu)$. Assume that the hyper parameters η and μ are independent where μ is Uniform(0, 1) and $\log(\eta)$ has a logistic distribution with mean $\log(50)$ and scale 1.

    (c) Use a JAGS script similar to what is presented in Section 10.3.3, draw a sample of 5000 from the posterior distribution, monitoring values of the $\{p_i\}$, μ, and $\log(\eta)$.

    (d) Compare unpooled, pooled, and hierarchical estimates of the $\{p_i\}$ in predicting the batting averages in the remainder of the season.

15. **Estimating Kidney Cancer Death Rates**

    This exercise is a variation of an activity described in Gelman and Nolan (2017). Suppose one is interested in estimating the kidney cancer death rates for the ten Ohio counties displayed in Table 10.10. Suppose the true death rates $\theta_1, ..., \theta_{10}$ are a sample from a Gamma(α, β) distribution. The observed number of deaths y_j in the jth county is assumed to be Poisson$(n_j\theta_j)$ where n_j is the population size.

TABLE 10.10
Populations of ten Ohio counties from recent census estimates.

County	Population	County	Population
Cuyahoga	1,243,857	Jackson	32,384
Gallia	29,979	Knox	61,893
Hamilton	816,684	Noble	14,354
Henry	27,086	Seneca	55,207
Holmes	43,892	Van Wert	28,281

(a) Assuming $\alpha = 27, \beta = 58{,}000$, simulate ten true cancer rates $\theta_1, ..., \theta_{10}$ from a Gamma(α, β) distribution. For each county, simulate the number of deaths in all counties. (Use the following R code.)

```
true_rates <- rgamma(10, shape = 27, rate = 58000)
pop_size <- c(1243857, 29979, 816684, 27086, 43892,
32384, 61893, 14354, 55207, 28281)
deaths <- rpois(10, lambda = pop_size * true_rates)
```

(b) Compute the observed death rates $\{y_j/n_j\}$. Identify the counties with the lowest and highest death rates.

(c) Using JAGS, fit a hierarchical model to the data assuming weakly informative gamma priors on the parameters α and β. Simulate a sample of 10,000 draws from the posterior distribution and compute the posterior means of the $\{\theta_j\}$.

(d) Identify the counties with the lowest and highest posterior means of the true rates. Compare these "best" and "worst" counties with the best and worst counties identified in part (b).

Exercises 16 to 20 concern additional Bayesian hierarchical models with more complicated structures. These exercises are here to help the reader gain familiarity of working with joint posterior distribution and deriving full conditional posterior distributions. These skills are essential to creating one's own Metropolis and Gibbs sampling algorithms instead of using JAGS.

16. **Inference for the Binomial N parameter**

Suppose that we want inference about an unknown number of animals N in a fixed-size population. On five separate days, we take photographs of some areas where they reside, and count the number of animals in the photos $(y_1, \ldots, y_5)$. Suppose further that each animal has a constant probability θ of appearing in a photograph and that appearances are independent across animals and days. A reasonable model for such data is a binomial distribution, $y_i \mid N, \theta \sim \text{Binomial}(N, \theta)$. In our setting, neither the number of trials N nor the probability θ are known.

To get a posterior distribution for N and θ, we propose the following system of models (Raftery 1988):

$$
\begin{aligned}
y_i \mid N, \theta &\sim \text{Binomial}(N, \theta) \\
N \mid \theta, \lambda &\sim \text{Poisson}(\lambda/\theta) \\
\pi(\lambda, \theta) &\propto 1/\lambda,
\end{aligned}
$$

where $\lambda > 0$ is a continuous random variable introduced to help with computations.

(a) Write down the joint posterior distribution, $\pi(N, \theta, \lambda \mid y_1, \ldots, y_5)$, up to a multiplicative constant.

(b) Find an expression for the conditional distribution, $\pi(\lambda \mid y_1, \ldots, y_5, N, \theta)$. Write the name of the distribution and expressions for its parameter values.

(c) Find the posterior distribution $\pi(N, \theta \mid y_1, \ldots, y_5)$ by integrating $\pi(N, \theta, \lambda \mid y_1, \ldots, y_5)$ with respect to λ. You don't need to name the distribution; just write its mathematical form.

(d) Find the conditional distribution, $\pi(\theta \mid y_1, \ldots, y_5, N)$. Write the name of the distribution and expressions for its parameter values.

17. **Successes and Failures in Tests**

A standard model for success or failure in testing situations is the item response model, also called the Rasch model. Suppose that J persons are given a test with K items. For $j = 1, \ldots, J$ and $k = 1, \ldots, K$, let $y_{jk} = 1$ if person j gets item k correct, and let $y_{jk} = 0$ otherwise. The Rasch model is

$$p(Y_{jk} = 1 \mid \pi_{jk}) = \text{Bernoulli}(\pi_{jk}) \quad (10.36)$$

$$\log\left(\frac{\pi_{jk}}{1 - \pi_{jk}}\right) = \alpha_j - \beta_k. \quad (10.37)$$

Here, α_j represents the ability of person j, and β_k represents the difficulty of item k. For a Bayesian version of the Rasch model, we use the hierarchical model distributions,

$$\begin{aligned} \alpha_j &\sim \text{Normal}(0, \sqrt{1/\tau}), \text{ for } j = 1, \ldots, J \\ \beta_k &\sim \text{Normal}(\mu, \sqrt{1/\phi}), \text{ for } k = 1, \ldots, K \end{aligned}$$

where τ and σ are precisions. For prior distributions, we use

$$\begin{aligned} \tau &\sim \text{Gamma}(a, b), \\ \phi &\sim \text{Gamma}(c, d), \\ \mu &\sim \text{Normal}(0, e), \end{aligned}$$

for known positive constants (a, b, c, d, e). We intend to run an MCMC to estimate the posterior distributions of all parameters. This problem asks you to outline some of the MCMC steps.

(a) Write the joint posterior distribution of $\pi(\alpha_1, \ldots, \alpha_J, \beta_1, \ldots, \beta_K, \tau, \phi, \mu \mid \{y_{jk}\})$, up to a constant.

(b) Write the steps you'd take to sample μ given $(\alpha_1, \ldots, \alpha_J, \beta_1, \ldots, \beta_K, \tau, \phi, \{y_{jk}\})$. If you can use a Gibbs step, write the name of the full conditional posterior distribution for μ and its parameter values. If you use a Metropolis step, write an expression for the acceptance probability and suggest a family of proposal distributions.

(c) Write the steps you'd take to sample ϕ given $(\alpha_1, \ldots, \alpha_J, \beta_1, \ldots, \beta_K, \tau, \mu, \{y_{jk}\})$. If you can use a Gibbs step, write the name of the full conditional posterior distribution for ϕ and its parameter values. If you use a Metropolis step, write an expression for the acceptance probability and suggest a family of proposal distributions.

(d) Write the steps you'd take to sample τ given $(\alpha_1, \ldots, \alpha_J, \beta_1, \ldots, \beta_K, \phi, \mu, \{y_{jk}\})$. If you use a Gibbs step, write the name of the full conditional distribution for τ. If you use a Metropolis step, write an expression for the acceptance probability and suggest a family of proposal distributions.

18. **Success and Failures in Tests (continued)**

Continuing from Exercise 17.

(a) Write the steps you'd take to sample β_k given $(\alpha_1, \ldots, \alpha_J, \beta_1, \ldots \beta_{k-1}, \beta_{k+1}, \ldots, \beta_K, \tau, \phi, \mu, \{y_{jk}\})$. If you can use a Gibbs step, write the name of the full conditional posterior distribution for β_k and its parameter values. f you use a Metropolis step, write an expression for the acceptance probability and suggest a family of proposal distributions.

(b) Write the steps you'd take to sample α_j given $(\alpha_1, \ldots, \alpha_{j-1}, \alpha_{j+1}, \ldots, \alpha_J, \beta_1, \ldots, \beta_K, \tau, \phi, \mu, \{y_{jk}\})$. If you use a Gibbs step, write the name of the full conditional distribution for α_j. If you use a Metropolis step, write an expression for the acceptance probability and suggest a family of proposal distributions.

19. **Success and Failures in Tests (continued)**

Suppose that you have 1000 approximately uncorrelated draws of the parameters from the joint posterior distribution of the Rasch model in Exercise 17. Describe how you would do the following tasks.

(a) Find the posterior probability that the variability in peoples' abilities exceeds the variability in item difficulty.

(b) Find the item in the test that appears to be the most difficult, and attach a posterior probability that it in fact is the most difficult among all K items.

(c) Perform a posterior predictive check of the model.

20. **AR(1) Models in Finance and Macroeconomics**

A common model in finance and macroeconomics is the AR(1) model. Suppose that we have n measurements ordered in time. For $j = 1, \ldots, n$, let y_j be the measurement at time j. Suppose we consider the measurement at time 1 as known (not a random variable).

Then, for $j = 2, \ldots, n$, a typical AR(1) model is $y_j = \beta y_{j-1} + \epsilon_j$ where $\epsilon_j \sim \text{Normal}(0, \sigma)$. Equivalently, we have

$$p(y_j \mid y_{j-1}, \ldots, y_1, \beta, \sigma^2) = \text{Normal}(\beta y_{j-1}, \sigma) \text{ for } j = 2, \ldots, n. \tag{10.38}$$

Note that what happens at time j only depends on what happened at time $j - 1$. For prior distributions, we will use

$$\begin{aligned} 1/\sigma^2 &\sim \text{Gamma}(a, b), \\ \beta &\sim \text{Normal}(c, d), \end{aligned}$$

for known positive constants (a, b, c, d). We intend to run an MCMC sampler to estimate the posterior distribution of (β, σ^2). Whenever possible, we will sample directly from full conditionals. This problem asks you to outline some of the MCMC steps, and to make a prediction for a future observation.

(a) Write the kernel of the joint distribution of $\pi(\beta, \sigma^2 \mid y_1, \ldots, y_n)$. [Hint: write $p(y_2, \ldots, y_n \mid y_1, \beta, \sigma^2) = p(y_n \mid y_{n-1}, \ldots, y_2, y_1, \beta, \sigma^2) p(y_{n-1} \mid y_{n-2}, \ldots, y_2, y_1, \beta, \sigma^2) \cdots p(y_2 \mid y_1, \beta, \sigma^2)$.]

(b) Write the steps you'd take to sample σ^2 given $(\beta, y_1, \ldots, y_n)$. If you use a Gibbs step, write the name of the full conditional distribution for σ and its parameter values. If you use a Metropolis step, write an expression for the acceptance probability and suggest a family of proposal distributions.

(c) Write the steps you'd take to sample β given $(\sigma, y_1, \ldots, y_n)$. If you use a Gibbs step, write the name of the full conditional distribution for β and its parameter values. If you use a Metropolis step, write an expression for the acceptance probability and suggest a family of proposal distributions.

(d) Describe how you would make a 95% posterior interval for the future value of Y_{n+2}.

11

Simple Linear Regression

11.1 Introduction

For continuous response variables such as Roger Federer's time-to-serve data in Chapter 8 and snowfall amounts in Buffalo, New York in Chapter 9, normal sampling models have been applied. The basic underlying assumption in a normal sampling model is that observations are identically and independently distributed (i.i.d.) according to a normal density, as in $Y_i \overset{i.i.d.}{\sim} \textrm{Normal}(\mu, \sigma)$.

Adding a predictor variable

When continuous responses are observed, it is common that other variables are recorded that may be associated with the primary response measure. In the Buffalo snowfall example, one may also observe the average temperature in winter season and one believes that the average season temperature is associated with the corresponding amount of snowfall. For the tennis example, one may believe that the time-to-serve measurement is related to the rally length of the previous point. Specifically, a long rally in the previous point may be associated with a long time-to-serve in the current point.

In Chapter 9, a normal curve was used to model the snowfalls $Y_1, ..., Y_n$ for n winters,

$$Y_i \mid \mu, \sigma \overset{i.i.d.}{\sim} \textrm{Normal}(\mu, \sigma), \; i = 1, \cdots, n. \tag{11.1}$$

The model in Equation (11.1) assumes that each winter snowfall follows the same normal density with mean μ and σ. From a Bayesian viewpoint, one assigns prior distributions for μ and σ and bases inferences about these parameters from the posterior distribution.

However when the average temperature in winter i, x_i, is also available, one might wonder if the snowfall amount Y_i can be explained by the average temperature x_i in the same winter. One typically calls x_i a predictor variable as one is interested in predicting the snowfall amount Y_i from the value of x_i. How does one extend the basic normal sampling model in Equation (11.1) to study the possible relationship between the average temperature and the snowfall amount?

An observation-specific mean

The model in Equation (11.1) assumes a common mean μ for each Y_i. Since one wishes to introduce a new variable x_i specific to winter i, the model in Equation (11.1) is adjusted to Equation (11.2) where the common mean μ is replaced by a winter specific mean μ_i .

$$Y_i \mid \mu_i, \sigma \overset{ind}{\sim} \text{Normal}(\mu_i, \sigma), \;\; i = 1, \cdots, n. \tag{11.2}$$

Note that the observations $Y_1, ..., Y_n$ are no longer identically distributed since they have different means, but the observations are still independent which is indicated by *ind* written over the distributed $\sim$ symbol in the formula.

Linear relationship between the mean and the predictor

One basic approach for relating a predictor x_i and the response Y_i is to assume that the mean of Y_i, μ_i, is a linear function of x_i. This linear relationship is written as

$$\mu_i = \beta_0 + \beta_1 x_i, \tag{11.3}$$

for $i = 1, \ldots, n$. In Equation (11.3), each x_i is a known constant (that is why a small letter is used for x) and β_0 and β_1 are unknown parameters. As one might guess, these intercept and slope parameters are random. One assigns a prior distribution to (β_0, β_1) and performs inference by summarizing the posterior distribution of these parameters.

In this model, the linear function $\beta_0 + \beta_1 x_i$ is interpreted as the *expected* snowfall amount when the average temperature is equal to x_i. The intercept β_0 represents the expected snowfall when the winter temperature is $x_i = 0$. The slope parameter β_1 gives the increase in the expected snowfall when the temperature x_i increases by one degree. It is important to note that the linear relationship in Equation (11.3) with parameters β_0 and β_1 describes the association between the mean μ_i and the predictor x_i. This linear relationship is a statement about the expected or average snowfall amount μ_i, not the *actual* snowfall amount Y_i.

Linear regression model

Substituting Equation (11.3) into the model in Equation (11.2), one obtains the linear regression model.

$$Y_i \mid \beta_0, \beta_1, \sigma \overset{ind}{\sim} \text{Normal}(\beta_0 + \beta_1 x_i, \sigma), \;\; i = 1, \cdots, n. \tag{11.4}$$

This is a special case of a normal sampling model, where the Y_i independently follow a normal density with observation specific mean $\beta_0 + \beta_1 x_i$ and common standard deviation σ. Since there is only a single predictor x_i, this model is commonly called the simple linear regression model.

One restates this regression model as

$$Y_i = \mu_i + \epsilon_i, i = 1, \cdots, n, \tag{11.5}$$

where the mean response $\mu_i = \beta_0 + \beta_1 x_i$ and the residuals $\epsilon_1, ..., \epsilon_n$ are *i.i.d.* from a normal distribution with mean 0 and standard deviation σ. In the context of our example, this model says that the snowfall for a particular season Y_i is a linear function of the average season temperature x_i plus a random error ϵ_i that is normal with mean 0 and standard deviation σ.

The simple linear regression model is displayed in Figure 11.1. The line in the graph represents the equation $\beta_0 + \beta_1 x$ for the mean response $\mu = E(Y)$. The actual response Y is equal to $\beta_0 + \beta_1 x + \epsilon$ where the random variable ϵ is distributed normal with mean 0 and standard deviation σ. The normal curves (drawn sideways) represent the locations of the response Y for three distinct values of the predictor x. The parameter σ represents the deviation of the response Y about the mean value $\beta_0 + \beta_1 x$. One is interested in learning about the parameters β_0 and β_1 that describe the line and the standard deviation σ which describes the deviations of the random response about the line.

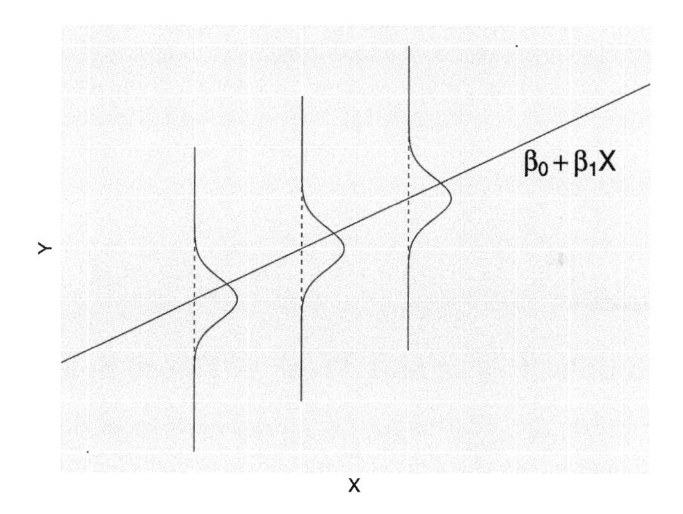

FIGURE 11.1
Display of linear regression model. The line represents the unknown regression line $\beta_0 + \beta_1 x$ and the normal curves (drawn sideways) represent the distribution of the response Y about the line.

In the linear regression model, the observation Y_i is random, the predictor x_i is a fixed constant and the unknown parameters are β_0, β_1, and σ. Using the Bayesian paradigm, a joint prior distribution is assigned to $(\beta_0, \beta_1, \sigma)$. After the response values $Y_i = y_i, i = 1, ..., n$ are observed, one learns about the parameters through the posterior distribution. An MCMC algorithm will be used to simulate a posterior sample, and using the simulation sample, one makes inferences about the expected response $\beta_0 + \beta_1 x$ for a specific value of the predictor x. Also, one will be able to assess the sizes of the errors by summarizing the posterior density of the standard deviation σ.

In our snowfall example, one is interested in learning about the relationship between the average temperature and the mean snowfall that is described by the linear model $\mu = \beta_0 + \beta_1 x$. If the posterior probability that $\beta_1 < 0$ is large, that indicates that lower average temperatures will likely result in larger mean snowfall. Also one is interested in using this model for prediction. If given the average winter temperature in the following season, can one predict the Buffalo snowfall? This question is addressed by use of the posterior predictive distribution of a future snowfall $\tilde{Y}$. Using the usual computing strategy, one simulates a large sample of values from the posterior predictive distribution and finds an interval that contains $\tilde{Y}$ with a prescribed probability.

In this chapter, regression is introduced in Section 11.2 by a dataset containing several characteristics of 24 house sales in an area in Ohio. In this example, one is interested in predicting the price of a house given the house size and Section 11.3 presents a simple linear regression model to explain this relationship. The practice of standardizing variables will be introduced which is helpful in the process of assigning an informative prior on the regression parameters. Inference through MCMC is presented in Section 11.6 and methods for performing Bayesian inferences with simple linear regression are illustrated in Section 11.7.

11.2 Example: Prices and Areas of House Sales

Zillow is an online real estate database company that collects information on 110 million homes across the United States. Data is collected from a random sample of 24 houses for sale in the Findlay, Ohio area during October 2018. For each house, the dataset contains the selling price (in \$1000) and size (in 1000 square feet). Table 11.1 displays the first five observations of the dataset.

Suppose one is interested in predicting a house's selling price from its size. In this example, one is treating price as the response variable and size as the single predictor. Figure 11.2 constructs a scatterplot of price (y-axis) against the size (x-axis) for the houses in the sample. This figure shows a positive relationship between the size and the price of a house sale, suggesting that the house sale price increases as the house size increases. Can one quantify this relationship through a Bayesian linear regression model? In particular, is there sufficient evidence that there is a positive association among the population of all homes? Can one predict the sale price of a home given its size?

TABLE 11.1
The house index, price (in \$1000), and size (in 1000 sq feet) of 5 house sales in Findlay, Ohio area during October 2018. The random sample contains 24 house sales.

Index	Price (\$1000)	Size (1000 sq feet)
1	167	1.625
2	236	1.980
3	355	2.758
4	148	1.341
5	93	1.465

11.3 A Simple Linear Regression Model

The house sale example can be fit into the linear regression model framework. It is assumed the response variable, the price of a house sale, is a continuous variable is distributed as a normal random variable. Specifically, the price Y_i for house i, is normally distributed with mean μ_i and standard deviation σ.

$$Y_i \mid \mu_i, \sigma \overset{ind}{\sim} \textrm{Normal}(\mu_i, \sigma), \tag{11.6}$$

where $i = 1, \cdots, n$, where $n = 24$ is the number of homes in the dataset. The *ind* over $\sim$ in Equation (11.6) indicates that each response Y_i independently follows its own normal density. Moreover, unlike the house-specific mean μ_i, a common standard deviation σ is shared among all responses Y_i's.

Since one believes the size of the house is helpful in understanding a house's price, one represents the mean price μ_i as a linear function of the house size x_i depending on two parameters β_0 and β_1.

$$\mu_i = \beta_0 + \beta_1 x_i \tag{11.7}$$

How does one interpret the intercept and slope parameters? The intercept β_0 gives the expected price μ_i for a house i that has zero square feet ($x_i = 0$). This is not a meaningful parameter since no house (not even a tiny house) has zero square feet. The slope parameter β_1 gives the change in the expected price μ_i, when the size x_i of house i increases by 1 unit, i.e., increases by 1000 square feet.

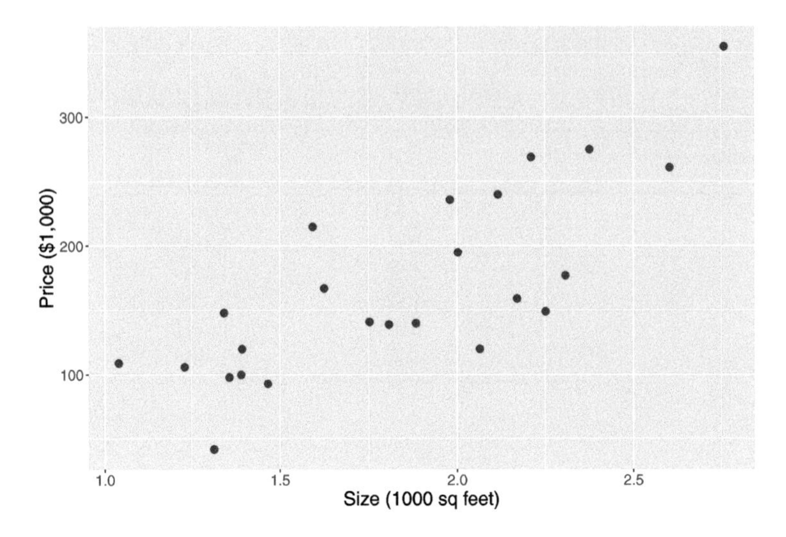

FIGURE 11.2
Scatterplot of price against size of house sales.

11.4 A Weakly Informative Prior

In some situations, the user has limited prior information about the location of the regression parameters or the standard deviation. To implement the Bayesian approach, one has to assign a prior distribution, but it is desirable in this situation to assign a prior that has little impact on the posterior distribution.

Suppose that one's beliefs about the regression coefficients (β_0, β_1) are independent from one's opinion about the standard deviation σ. Then the joint prior density for the parameters $(\beta_0, \beta_1, \sigma)$ is written as

$$\pi(\beta_0, \beta_1, \sigma) = \pi(\beta_0, \beta_1)\pi(\sigma).$$

The choice of weakly informative priors on (β_0, β_1) and σ are described in separate sections.

Prior on the intercept β_0 and slope β_1

If one assumes independence of one's opinion about the intercept and the slope, one represents the joint prior $\pi(\beta_0, \beta_1)$ as the product of priors $\pi(\beta_0)\pi(\beta_1)$, and it is convenient to use normal priors. So it is assumed $\beta_0 \sim \text{Normal}(\mu_0, s_0)$ and $\beta_1 \sim \text{Normal}(\mu_1, s_1)$.

The choice of the standard deviation s_j in the normal prior reflects how confident the person believes in a prior guess of β_j. If one has little information about the location of a regression parameter, then the choice of the prior guess μ_j is not that important and one chooses a large value for the prior standard deviation s_j. So the regression intercept and slope are each assigned a normal prior with a mean of 0 and standard deviation equal to the large value of 100.

Prior on sampling standard deviation σ

In the current regression model, one assumes that $Y_i \sim \text{Normal}(\beta_0 + \beta_1 x_i, \sigma)$ and σ represents the variability of the house price about the regression line. It is typically hard to specify informative beliefs about a standard deviation than a mean parameter such as $\beta_0 + \beta_1 x$. So following the suggestions from Chapter 9 and Chapter 10, one assigns a weakly informative prior for the standard deviation σ. A gamma prior for the precision parameter $\phi = 1/\sigma^2$ with small values of the shape and rate parameters, say $a = 1$ and $b = 1$, was seen in those chapters to represent weak prior information, and a similar prior is assigned in this regression setting.

$$\phi = 1/\sigma^2 \sim \text{Gamma}(1, 1).$$

11.5 Posterior Analysis

In the sampling model one has that $Y_1, ..., Y_n$ are independent with $Y_i \sim \text{Normal}(\beta_0 + \beta_1 x_i, \sigma)$. Suppose the pairs $(x_1, y_1), ..., (x_n, y_n)$ are observed. The likelihood is the joint density of these observations viewed as a function of $(\beta_0, \beta_1, \sigma)$. For convenience, the standard deviation σ is reexpressed as the precision $\phi = 1/\sigma^2$.

$$\begin{aligned} L(\beta_0, \beta_1, \phi) &= \prod_{i=1}^{n} \left[\frac{\sqrt{\phi}}{\sqrt{2\pi}} \exp\left\{ -\frac{\phi}{2}(y_i - \beta_0 - \beta_1 x_i)^2 \right\} \right] \\ &\propto \phi^{\frac{n}{2}} \exp\left\{ -\frac{\phi}{2} \sum_{i=1}^{n} (y_i - \beta_0 - \beta_1 x_i)^2 \right\} \end{aligned} \tag{11.8}$$

By multiplying the likelihood by the prior for (β_0, β_1, ϕ), one obtains an expression for the posterior density.

$$\begin{aligned} \pi(\beta_0, \beta_1, \phi \mid y_1, \cdots, y_n) &\propto \phi^{\frac{n}{2}} \exp\left\{ -\frac{\phi}{2} \sum_{i=1}^{n} (y_i - \beta_0 - \beta_1 x_i)^2 \right\} \\ &\times \exp\left\{ -\frac{1}{2s_0^2}(\beta_0 - \mu_0)^2 \right\} \exp\left\{ -\frac{1}{2s_1^2}(\beta_1 - \mu_1)^2 \right\} \\ &\times \phi^{a-1} \exp(-b\phi) \end{aligned} \tag{11.9}$$

Since this is not a familiar probability distribution, one needs to use an MCMC algorithm to obtain simulated draws from the posterior.

11.6 Inference through MCMC

R It is convenient to draw an MCMC sample from a regression model using the JAGS software. One attractive feature of JAGS is that it is straightforward to transpose the statement of the Bayesian model (sampling density and prior) directly to the JAGS model script.

Describe the model by a script

The first step in using JAGS is writing the following script defining the linear regression model, saving the script in the character string `modelString`.

```
modelString <-"
model {
## sampling
for (i in 1:N){
   y[i] ~ dnorm(beta0 + beta1*x[i], invsigma2)
}
## priors
beta0 ~ dnorm(mu0, g0)
beta1 ~ dnorm(mu1, g1)
invsigma2 ~ dgamma(a, b)
sigma <- sqrt(pow(invsigma2, -1))
}
"
```

In the sampling section of the script, the loop goes from `1` to `N`, where `N` is the number of observations with index `i`. Recall that the normal distribution `dnorm` in JAGS is stated in terms of the mean and precision, and so the variable `invsigma2` corresponds to the normal sampling precision. The variable `sigma` is defined in the prior section of the script so one can track the simulated values of the standard deviation σ. Also the variables `g0` and `g1` correspond to the precisions of the normal prior densities for `beta0` and `beta1`.

Define the data and prior parameters

The next step is to provide the observed data and the values for the prior parameters. In the R script below, a list `the_data` contains the vector of sale prices, the vector of house sizes, and the number of observations. This list also contains the means and precisions of the normal priors for `beta0` and `beta1`, and the values of the two parameters `a` and `b` of the gamma prior for `invsigma2`. The prior standard deviations of the normal priors on `beta0` and `beta1` are both 100, and so the corresponding precision values of `g0` and `g1` are both $1/100^2 = 0.0001$.

```
y <- PriceAreaData$price
x <- PriceAreaData$newsize
N <- length(y)
the_data <- list("y" = y, "x" = x, "N" = N,
                 "mu0" = 0, "g0" = 0.0001,
                 "mu1" = 0, "g1" = 0.0001,
                 "a" = 1, "b" = 1)
```

Generate samples from the posterior distribution

The `run.jags()` function in the `runjags` package generates posterior samples by the MCMC algorithm using the JAGS software. The script below runs one MCMC chain with an adaption period of 1000 iterations, a burn-in period of 5000 iterations, and an additional set of 5000 iterations to be run and collected for inference. By using the argument `monitor = c("beta0", "beta1", "sigma")`, one keeps tracks of all three model parameters. The output variable `posterior` contains a matrix of simulated draws.

```
posterior <- run.jags(modelString,
                      n.chains = 1,
                      data = the_data,
                      monitor = c("beta0", "beta1", "sigma"),
                      adapt = 1000,
                      burnin = 5000,
                      sample = 5000)
```

MCMC diagnostics and summarization

Using JAGS one obtains 5000 posterior samples for the vector of parameters. Below the first 10 posterior samples are displayed for the triplet $(\beta_0, \beta_1, \sigma)$. Note that the index starts from 6001 since 6000 samples were already generated in the adaption and burn-in periods.

```
     beta0 beta1 sigma
6001 -17.62 103.3 40.68
6002 -21.35 107.3 44.92
6003 -34.34 114.0 37.11
6004 -42.06 110.5 51.84
```

```
6005 -47.71 111.4 62.63
6006 -47.49 113.9 53.80
6007 -18.85 106.0 50.92
6008 -28.50 114.8 42.71
6009 -32.10 105.1 47.41
6010 -37.41 119.3 45.88
```

To obtain valid inferences from the posterior draws from the MCMC simulation, convergence of the MCMC chain is necessary. The `plot()` function with the argument input `vars` returns four diagnostic plots (trace plot, empirical CDF, histogram and autocorrelation plot) for the specified parameter. For example, Figure 11.3 shows the diagnostic plots for the intercept parameter β_0 by the following command.

```
plot(posterior, vars = "beta0")
```

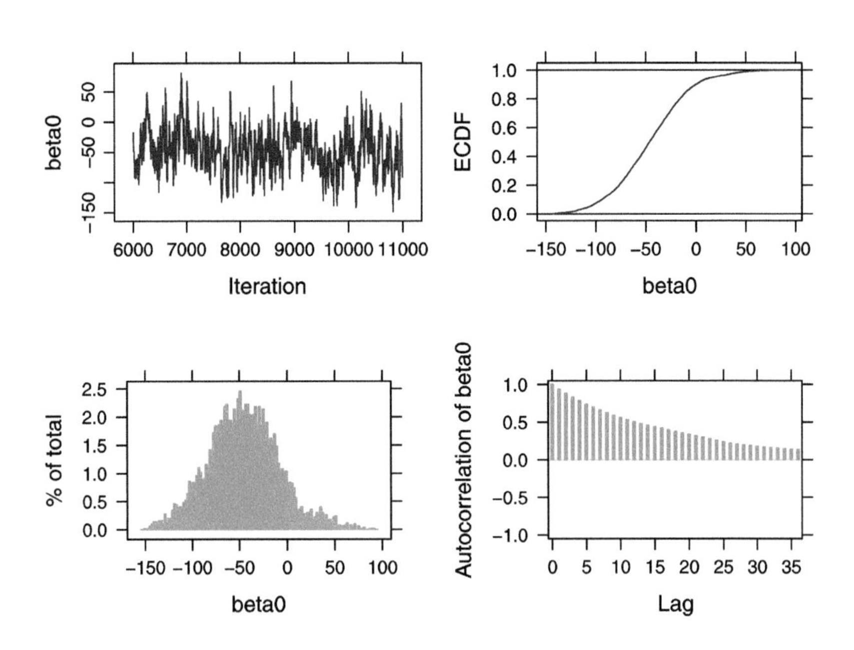

FIGURE 11.3
MCMC diagnostics plots for the regression intercept parameter β_0.

The upper left trace plot shows good MCMC mixing for the 5000 simulated draws of β_0. The lower right autocorrelation plot indicates close to zero correlation between adjacent posterior draws of β_0. Overall these indicate convergence of the MCMC chain for β_0. In usual practice, one should perform these diagnostics for all three parameters in the model.

Figure 11.4 displays a scatterplot of the simulated draws of the regression parameters β_0 and β_1. It is interesting to note the strong negative correlation in these parameters. If one assigned informative independent priors on β_0 and β_1, these prior beliefs would be counter to the correlation between the two parameters observed in the data.

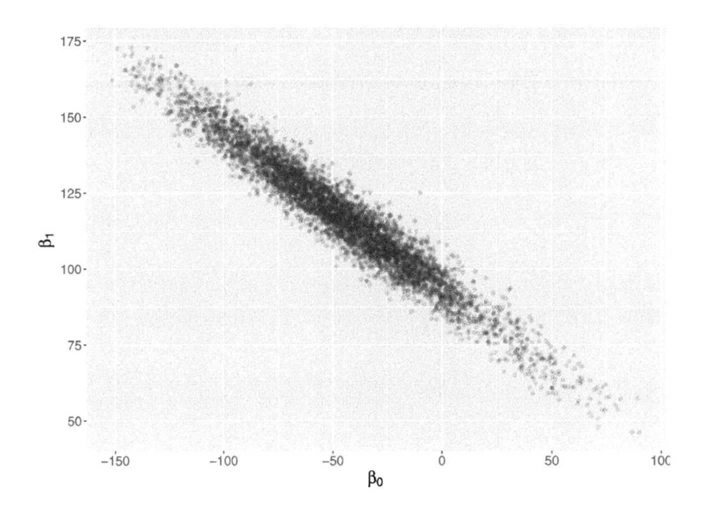

FIGURE 11.4
Scatterplot of posterior draws of the intercept and slope parameters β_0 and β_1.

Posterior summaries of the parameters are obtained by use of the `print(posterior, digits = 3)` command. Note that these summaries are based on the 5000 iterations from the sampling period excluding the samples from the adaption and burn-in periods.

```
print(posterior, digits = 3)
      Lower95 Median Upper95  Mean    SD Mode MCerr
beta0    -122  -46.2    31.4 -45.7 37.6   --  2.98
beta1    78.7    117     159   117   20   --  1.65
sigma    33.2     45    59.3  45.7 6.93   -- 0.157
```

Then intercept parameter β_0 does not have a useful interpretation, so values of these particular posterior summaries will not be interpreted. The summaries of the slope β_1 indicate a positive slope with a posterior median of 117 and a 90% credible interval (78.7, 159). That is, with every 1000 square feet increase of the house size, the house price increases by $117,000. In addition, this increase in the house price falls in the interval ($78,700, $159,000) with

90% posterior probability. The posterior median of the standard deviation σ is the large value 45 or $45,000 which indicates that there are likely additional variables than house size that determine the price.

11.7 Bayesian Inferences with Simple Linear Regression

11.7.1 Simulate fits from the regression model

The intercept β_0 and slope β_1 determine the linear relationship between the mean of the response Y and the predictor x.

$$E(Y) = \beta_0 + \beta_1 x. \tag{11.10}$$

Each pair of values (β_0, β_1) corresponds to a line $\beta_0 + \beta_1 x$ in the space of values of x and y. If one finds the posterior mean of these coefficients, say $\tilde{\beta_0}$ and $\tilde{\beta_1}$, then the line

$$y = \tilde{\beta_0} + \tilde{\beta_1} x$$

corresponds to a "best" line of fit through the data.

This best line represents a most likely value of the line $\beta_0 + \beta_1 x$ from the posterior distribution. One learns about the uncertainty of this line estimate by drawing a sample of J rows from the matrix of posterior draws of (β_0, β_1) and collecting the line estimates

$$\tilde{\beta_0}^{(j)} + \tilde{\beta_1}^{(j)} x, j = 1, ..., J.$$

R Using the R script below, one produces a graph showing the best line of fit (solid line) and ten simulated fits from the posterior as in Figure 11.5.

```
post <- as.mcmc(posterior)
post_means <- apply(post, 2, mean)
post <- as.data.frame(post)
ggplot(PriceAreaData, aes(newsize, price)) +
  geom_point(size=3) +
  geom_abline(data=post[1:10, ],
              aes(intercept=beta0, slope=beta1),
              alpha = 0.5) +
  geom_abline(intercept = post_means[1],
              slope = post_means[2],
              size = 2) +
  ylab("Price") + xlab("Size") +
  theme_grey(base_size = 18, base_family = "")
```

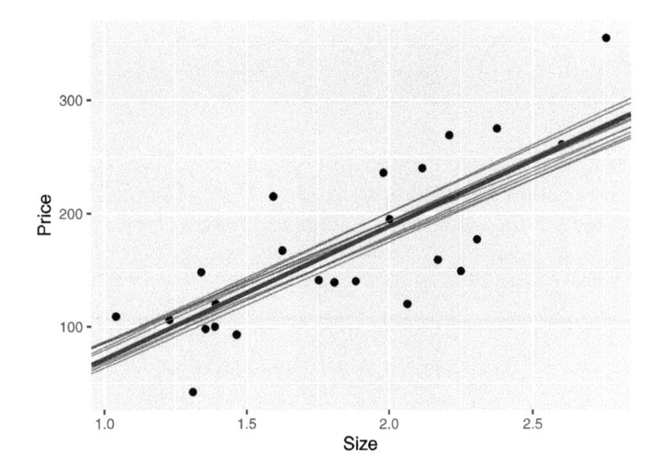

FIGURE 11.5
Scatterplot of the (size, price) data with the best line of fit (solid line) and ten simulated fits $\beta_0 + \beta_1 x$ from the posterior distribution.

From Figure 11.5, since there is inferential uncertainty about the intercept β_0 and slope β_1, one sees variation among the ten fits from the posterior of the linear regression line $\beta_0 + \beta_1 x$. This variation about the best-fitting line is understandable since the size of our sample of data is the relatively small value of 24. A larger sample size would help to reduce the posterior variation for the intercept and slope parameters and result in posterior samples of fits that are more tightly clustered about the best fitting line in Figure 11.5.

11.7.2 Learning about the expected response

In regression modeling, one may be interested in learning about the expected response $E(Y)$ for a specific value of the predictor x. In the house sale example, one may wish to learn about the expected house price for a specific value of the house size. Since the expected response $E(Y) = \beta_0 + \beta_1 x$ is a linear function of the intercept and slope parameters, one obtains a simulated sample from the posterior of $\beta_0 + \beta_1 x$ by computing this function on each of the simulated pairs from the posterior of (β_0, β_1).

R For example, suppose one is interested in the expected price $E(Y)$ for a house with a size of 1, i.e. $x = 1$ (1000 sq feet). In the R script below, one simulates 5000 draws from the posterior of the expected house prices, $E[Y]$ from the 5000 posterior samples of the pair (β_0, β_1).

```
size <- 1
mean_response <- post[, "beta0"] + size * post[, "beta1"]
```

This process is repeated for the four sizes $x = 1.2, 1.6, 2.0, 2.4$ (1200 sq feet, 1600 sq feet, 2000 sq feet, and 2400 sq feet). Let $E(Y \mid x)$ denote the expected price for a house with size x. Figure 11.6 displays density plots of the simulated posterior samples for the expected prices $E(Y \mid 1.2)$, $E(Y \mid 1.6)$, $E(Y \mid 2.0)$, $E(Y \mid 2.4)$ for these four house sizes.

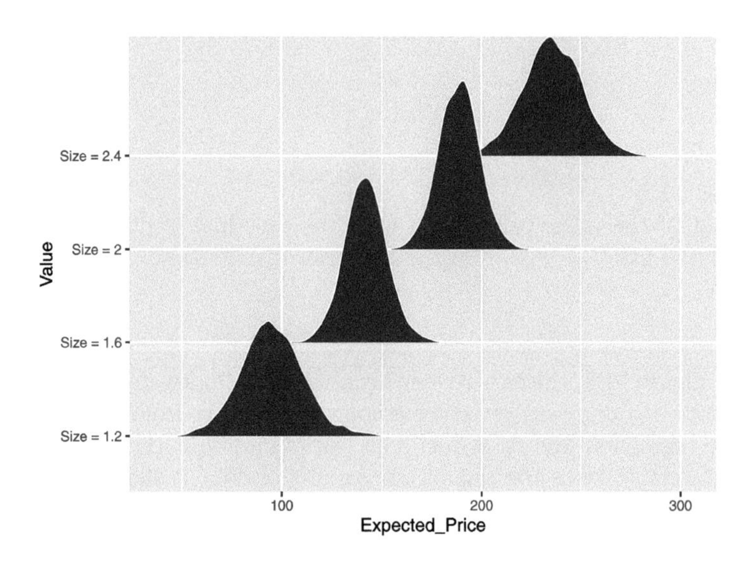

FIGURE 11.6
Density plots of the simulated draws of the posterior expected house price for four different values of the house size.

The R output below provides summaries of the posterior of the expected price for each of the four values of the house size. From this output, one sees that, for a house of size of 1.2 (1200 sq feet), the posterior median of the expected price is \$94,500, and the probability that the expected price falls between \$69,800 and \$121,000 is 90%.

```
  Value         P05   P50   P95
  <chr>       <dbl> <dbl> <dbl>
1 Size = 1.2   69.8  94.5   121
2 Size = 1.6 125    142     159
3 Size = 2   172    189     205
4 Size = 2.4 211    236     260
```

11.7.3 Prediction of future response

Learning about the regression model and values of the expected response values focuses on the deterministic linear relationship between x and $E[Y]$ through the intercept β_0 and the slope β_1, as shown in Equation (11.10). The variability among the fitted lines in Figure 11.5 and the variability among the simulated house price for fixed size in Figure 11.6 reflect the variability in the posterior draws of β_0 and β_1.

However, if one wants to predict future values for a house sale price Y given its size x, one needs to go one step further to incorporate the sampling model in the simulation process.

$$Y_i \mid \beta_0, \beta_1, \sigma \overset{ind}{\sim} \text{Normal}(\beta_0 + \beta_1 x_i, \sigma) \tag{11.11}$$

As shown in Equation (11.11), the sampling model of Y is a normal with a mean expressed as a linear combination of β_0 and β_1 and a standard deviation σ. To obtain a predicted value of Y given $x = x_i$, one first simulates the expected response from $\beta_0 + \beta_1 x_i$, and then simulates the predicted value of Y_i from the sampling model: $Y_i \sim \text{Normal}(E[Y_i], \sigma)$. Below is a diagram for the prediction process for an observation where its house size is given as x, and predicted value denoted as $\tilde{y}^{(s)}$ for iteration s. Here the simulation size S is 5000 as there are 5000 posterior samples of each of the three parameters.

$$
\begin{array}{rcl}
\text{simulate } E[y]^{(1)} = \beta_0^{(1)} + \beta_1^{(1)} x & \rightarrow & \text{sample } \tilde{y}^{(1)} \sim \text{Normal}(E[y]^{(1)}, \sigma^{(1)}) \\
\text{simulate } E[y]^{(2)} = \beta_0^{(2)} + \beta_1^{(2)} x & \rightarrow & \text{sample } \tilde{y}^{(2)} \sim \text{Normal}(E[y]^{(2)}, \sigma^{(2)}) \\
 & \vdots & \\
\text{simulate } E[y]^{(S)} = \beta_0^{(S)} + \beta_1^{(S)} x & \rightarrow & \text{sample } \tilde{y}^{(S)} \sim \text{Normal}(E[y]^{(S)}, \sigma^{(S)})
\end{array}
$$

R The R function `one_predicted()` obtains a simulated sample of the predictive distribution of the house price given a value of the house size. First one uses the posterior sample of (β_0, β_1) to obtain a posterior sample of the "linear response" $\beta_0 + \beta_1 x$. Then one simulates draws of the future observation by simulating from a normal distribution with mean $\beta_0 + \beta_1 x$ and standard deviation σ, where draws of σ are taken from its posterior distribution.

```
one_predicted <- function(x){
  lp <- post[ , "beta0"] +  x * post[ , "beta1"]
  y <- rnorm(5000, lp, post[, "sigma"])
  data.frame(Value = paste("Price =", x),
             Predicted_Price = y)
}
```

This process is repeated for each of the house sizes $x = 1.2, 1.6, 2.0, 2.4$ (1200 sq feet, 1600 sq feet, 2000 sq feet, and 2400 sq feet). Figure 11.7 displays density estimates of these simulated samples from the predictive distributions of the house price. Comparing Figure 11.7 with Figure 11.6, note that the predictive distributions are much wider than the posterior distributions on the expected response. This is what one would anticipate, since the predictive distribution incorporates two types of uncertainty – the inferential uncertainty in the values of the regression line $\beta_0 + \beta_1 x$ and the predictive uncertainty expressed in the sampling density of the response y with standard deviation σ.

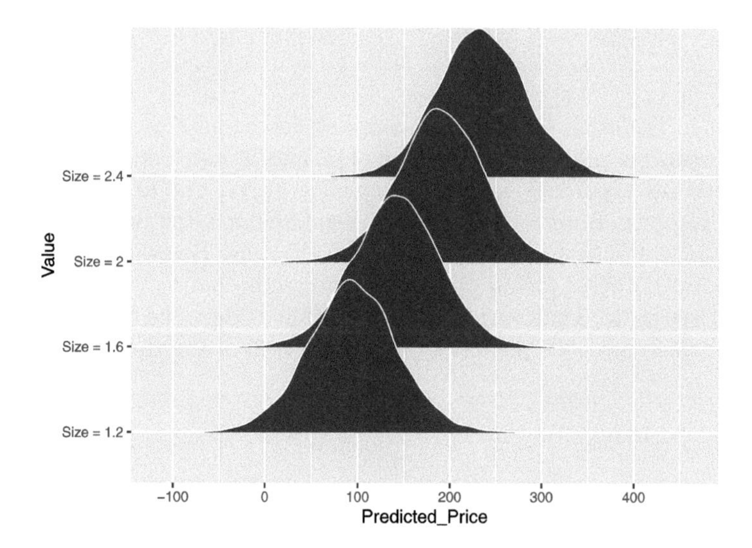

FIGURE 11.7
Density plots of the simulated draws of the predicted house price for four different values of the house size.

To reinforce this last point, the R output below displays the 5th, 50th, and 95th percentiles of the predictive distribution of the house price for each of the four values of the house size. One saw earlier that a 90% interval estimate for the expected price for a house with $x = 1.2$ was given by (69.8, 121). Below one sees that a 90% prediction interval for the price of the same house size is $(15.5, 175)$. The prediction interval is substantially wider than the posterior interval estimate. This is true since the predictive distribution incorporates the sizable uncertainty in the house price given the house size represented by the sampling standard deviation σ.

```
Value        P05   P50   P95
<chr>      <dbl> <dbl> <dbl>
```

```
1 Size = 1.2  15.5  94.4   175
2 Size = 1.6  64.5 142     219
3 Size = 2   110   189     266
4 Size = 2.4 157   234     315
```

11.7.4 Posterior predictive model checking

Simulating replicated datasets

The posterior predictive distribution is used to predict the value of a house's price for a particular house size. It is also helpful in judging the suitability of the linear regression model. The basic idea is that the observed response values should be consistent with predicted responses generated from the fitted model.

In our example, one observed the house size x and the house price y for a sample of 24 houses. Suppose one simulates a sample of prices for a sample of 24 houses with the same sizes from the posterior predictive distribution. This is implemented in two steps.

1. Values of the parameters $(\beta_0, \beta_1, \sigma)$ are simulated from the posterior distribution – call these simulated values $(\beta_0^*, \beta_1^*, \sigma^*)$.

2. A sample $\{y_1^R, ..., y_n^R\}$ is simulated where the sample size is $n = 24$ and y_i^R is Normal(μ_i^*, σ^*), where $\mu_i^* = \beta_0^* + \beta_1^* x_i$.

This is called a *replicated* sample from the posterior predictive distribution since one is using the same sample size and covariate values as the original dataset.

For our example, this simulation process was repeated eight times, where each iteration produces a sample $(x_i, y_i^R), i = 1, ..., 24$. Scatterplots of these eight replicated samples are displayed in Figure 11.8. The observed sample is also displayed in this figure.

The question one wants to ask is: Do the scatterplots of the simulated replicated samples resemble the scatterplot of the observed data? Since the x values are the same for the observed and replicated datasets, one focuses on possible differences in the observed and replicated response values. Possibly, the sample prices display more variation than the replicated prices, or perhaps the sample prices have a particular outlier or other feature that is not present in the replicated prices.

In the examination of these scatterplots, the distribution of the observed responses does not seem markably different from the distribution of the response in the simulated replicated datasets. Therefore in this brief examination, one does not see any indication of model misfit – the observed (x, y) data seems consistent with replicated data generated from the posterior predictive distribution.

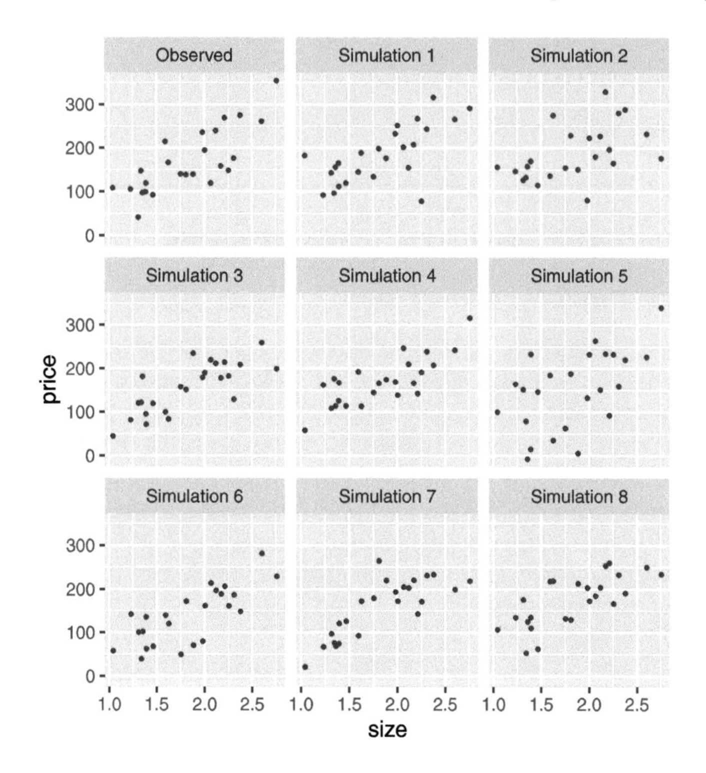

FIGURE 11.8
Scatterplots of observed and eight replicated datasets from the posterior predictive distribution.

Predictive residuals

In linear regression, one typically explores the residuals that are the deviations of the observations $\{y_i\}$ from the fitted regression model. The posterior prediction distribution is used to define a suitable Bayesian residual.

Consider the observed point (x_i, y_i). One asks the question – is the observed response value y_i consistent with predictions $\tilde{y}_i$ of this observation from the fitted model? One simulates predictions $\tilde{y}_i$ from the posterior predictive distribution in two steps:

1. One simulates $(\beta_0, \beta_1, \sigma)$ from the posterior distribution.
2. One simulates $\tilde{y}_i$ from a normal distribution with mean $\beta_0 + \beta_1 x_i$ and standard deviation σ.

By repeating this process many times, one has a sample of values $\{\tilde{y}_i\}$ from the posterior predictive distribution.

To see how close the observed response y_i is to the predictions $\{\tilde{y}_i\}$, one computes the predictive residual

$$r_i = y_i - \tilde{y}_i. \tag{11.12}$$

If this predictive residual is away from zero, that indicates that the observation is not consistent with the linear regression model. Remember that $\tilde{y}_i$, and therefore the predictive residual r_i is random. So one constructs a 90% interval estimate for the predictive residual r_i and says that the observation is unusual if the predictive residual interval estimate does not include zero.

Figure 11.9 displays a graph of the 90% interval estimates for the predictive residuals $\{r_i\}$ plotted against the size variable. A horizontal line at the value 0 is displayed and we look for intervals that are located on one side of zero. One notices that a few of the intervals barely overlap zero – this indicates that the corresponding points (x_i, y_i) are somewhat inconsistent with the fitted regression model.

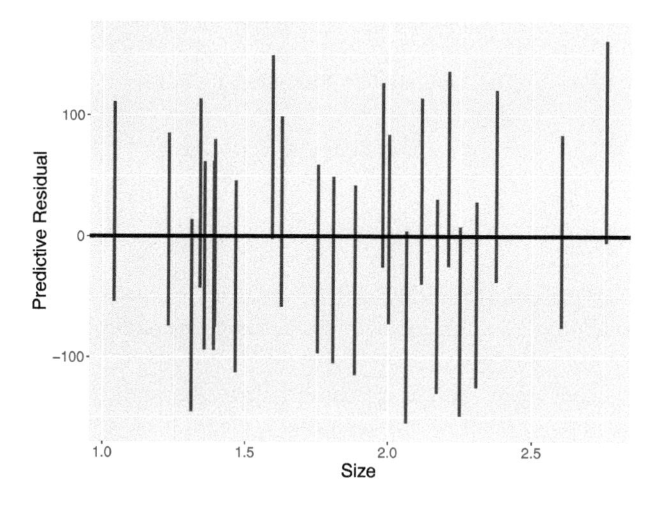

FIGURE 11.9
Display of predictive residuals. Each line covers 90% of the probability of the predictive residual $y_i - \tilde{y}_i$.

11.8 Informative Prior

One challenge in a Bayesian analysis is the construction of a prior that reflects beliefs about the parameters. In the usual linear function representation in

Equation (11.7), thinking about prior beliefs can be difficult since the intercept β_0 does not have a meaningful interpretation. To make the regression parameters β_0 and β_1 easier to interpret, one considers standardizing the response and predictor variables. With this standardization, the task of constructing informative priors will be facilitated.

11.8.1 Standardization

Standardization is the process of putting different variables on similar scales. As we can see in Figure 11.2, the house size variable ranges from 1.0 to over 2.5 (in 1000 sq feet), while the price variable ranges from below 50 to over 350 (in \$1000). The standardization process works as follows: for each variable, calculate the sample mean and the sample standard deviation, and then for each observed value of the variable, subtract the sample mean and divide by the sample standard deviation.

For example, let y_i be the observed sale price and x_i be the size of a house. Let $\bar{y}$ and $\bar{x}$ denote the sample means and s_y and s_x denote the sample standard deviations for the y_i's and x_i's, respectively. Then the standardized variables y_i^* and x_i^* are defined by the following formula.

$$y_i^* = \frac{y_i - \bar{y}}{s_y}, \; x_i^* = \frac{x_i - \bar{x}}{s_x}. \tag{11.13}$$

In R, the function `scale()` performs standardization.

```
PriceAreaData$price_standardized <- scale(PriceAreaData$price)
PriceAreaData$size_standardized <- scale(PriceAreaData$newsize)
```

A standardized value represents the number of standard deviations that the value falls above or below the mean. For example, if $x_i^* = -2$, then this house size is two standard deviations below the mean of all house sizes, and a value $y_i = 1$ indicates a sale price that is one standard deviation larger than the mean. Figure 11.10 constructs a scatterplot of the original (x, y) data (top) and the standardized (x^*, y^*) data (bottom). Note that the ranges of the standardized scores for the x^* and y^* are similar – both sets of standardized scores fall between -2 and 2. Also note that the association patterns of the two graphs agree which indicates that the standardization procedure has no impact on the relationship of house size with the sale price.

One advantage of standardization of the variables is that it provides more meaningful interpretations of the regression parameters β_0 and β_1. The linear regression model with the standardized variables is written as follows:

$$Y_i^* \mid \mu_i^*, \sigma \overset{ind}{\sim} \text{Normal}(\mu_i^*, \sigma), \tag{11.14}$$

$$\mu_i^* = \beta_0 + \beta_1 x_i^*. \tag{11.15}$$

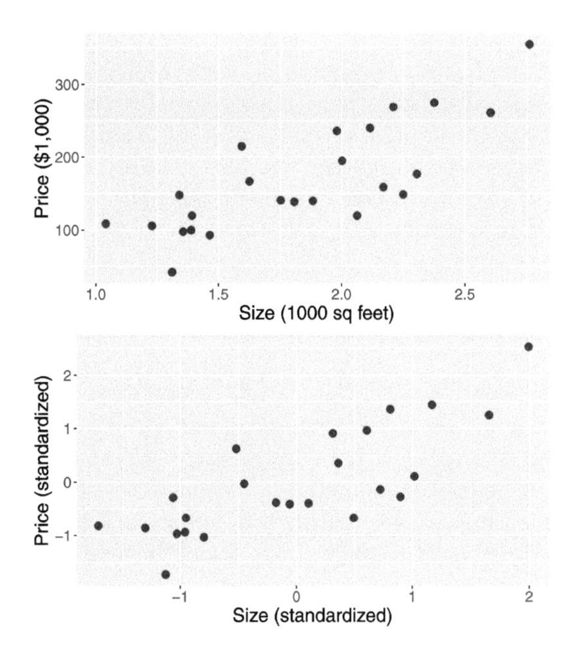

FIGURE 11.10
Two scatterplots of price against size of house sales: both variables unstandardized (top) and both variables standardized (bottom).

The intercept parameter β_0 now is the expected standardized sale price for a house where $x_i^* = 0$ corresponding to a house of average size. The slope β_1 gives the change in the expected standardized sale price μ_i^* when the standardized size x_i^* increases by 1 unit, or when the size variable increases by one standard deviation. In addition, when the variables are standardized, the slope β_1 can be shown equal to the correlation between x_i and y_i. So this slope provides a meaningful measure of the the linear relationship between the standardized predictor x_i^* and the expected standardized response μ_i^*. A positive value β_1 indicates a positive linear relationship between the two variables, and the absolute value of β_1 indicates the strength of the relationship.

11.8.2 Prior distributions

As in the weakly informative prior case, assume that the three parameters β_0, β_1 and σ are independent so the joint prior is factored into the marginal components.

$$\pi(\beta_0, \beta_1, \sigma) = \pi(\beta_0)\pi(\beta_1)\pi(\sigma).$$

Then the task of assigning a joint prior simplifies to the task of assigning priors separately to each of the three parameters. The process of assigning an informative prior is described for each parameter.

Prior on the intercept β_0

After the data is standardized, recall that the intercept β_0 represents the expected standardized sale price given a house of average size (i.e. $x_i^* = 0$). If one believes a house of average size will also have an average price, then a reasonable guess of β_0 is zero. One can give a normal prior for β_0 with mean $\mu_0 = 0$ and standard deviation s_0:

$$\beta_0 \sim \text{Normal}(0, s_0).$$

The standard deviation s_0 in the normal prior reflects how confident one believes in the guess of $\beta_0 = 0$. For example, if one specifies $\beta_0 \sim \text{Normal}(0, 1)$, this indicates that a price of a house of average size could range from one standard deviation below to one standard deviation above the average price. Since this is a wide range, one is stating that he or she is unsure that a house of average size will have an average price. If one instead is very sure of the guess that $\beta_0 = 0$, one could choose a smaller value of s_0.

Prior on the slope β_1

For standardized data, the slope β_1 represents the correlation between the house size and the sale price. One represents one's belief about the location of β_1 by means of a normal prior.

$$\beta_1 \sim \text{Normal}(\mu_1, s_1),$$

For this prior, μ_1 represents one's best guess of the correlation and s_1 represents the sureness of this guess. For example, if one lets β_1 be Normal$(0.7, 0.15)$, this means that one's best guess of the correlation is 0.7 and one is pretty certain that the correlation falls between $0.7 - 0.15$ and $0.7 + 0.15$. If one is not very sure of the guess of 0.7, one could choose a larger value of s_1.

Prior on σ

It is typically harder to specify informative beliefs about a standard deviation than a mean parameter such as $\beta_0 + \beta_1 x$. So it seems reasonable to assign a weakly informative prior for the sampling error standard deviation σ. A gamma prior for the precision parameter $\phi = 1/\sigma^2$ with small values of the shape and rate parameters, say $a = 1$ and $b = 1$, can represent weak prior information in this regression setting.

$$1/\sigma^2 \sim \text{Gamma}(1, 1).$$

To summarize, the informative prior distribution for (β_0, β_1, σ) is defined as follows.

$$\pi(\beta_0, \beta_1, \sigma) = \pi(\beta_0)\pi(\beta_1)\pi(\sigma), \quad (11.16)$$
$$\beta_0 \sim \text{Normal}(0, 1), \quad (11.17)$$
$$\beta_1 \sim \text{Normal}(0.7, 0.15), \quad (11.18)$$
$$1/\sigma^2 \sim \text{Gamma}(1, 1). \quad (11.19)$$

11.8.3 Posterior Analysis

R One again uses the JAGS software to simulate from the posterior distribution of the parameters. The `modelString` is written in the same way as in Section 11.6.

Since the data have been standardized, one needs to do some initial preliminary work before the MCMC implementation. First, in R, one defines new variables `price_standardized` and `size_standardized` that are standardized versions of the original `price` and `newsize` variables.

```
PriceAreaData$price_standardized <- scale(PriceAreaData$price)
PriceAreaData$size_standardized <- scale(PriceAreaData$newsize)
```

Then the variables `y` and `x` in `modelString` now correspond to the standardized data. Also in the definition of the `the_data` list, we enter the mean and precision values of the informative priors placed on the regression intercept and slope. Remember that one needs to convert the prior standard deviations s_0 and s_1 to the corresponding precision values.

```
y <- as.vector(PriceAreaData$price_standardized)
x <- as.vector(PriceAreaData$size_standardized)
N <- length(y)
the_data <- list("y" = y, "x" = x, "N" = N,
                 "mu0" = 0, "g0" = 1,
                 "mu1" = 0.7, "g1" = 44.4,
                 "a" = 1, "b" = 1)
```

With the redefinition of the standardized variables `y` and `x`, the same JAGS script `modelString` is used to define the posterior distribution. As before, the `run.jags()` function is run, collecting a sample of 5000 draws from $(\beta_0, \beta_1, \sigma)$.

```
posterior2 <- run.jags(modelString,
                       n.chains = 1,
                       data = the_data,
                       monitor = c("beta0", "beta1", "sigma"),
                       adapt = 1000,
                       burnin = 5000,
                       sample = 5000)
```

Comparing posteriors for two priors

R To understand the influence of the informative prior, one can contrast this posterior distribution with a posterior using a weakly informative prior. Suppose one assumes that β_0, β_1, and σ are independent with $\beta_0 \sim \text{Normal}(0, 100)$, $\beta_1 \sim \text{Normal}(0.7, 100)$ and $\phi = 1/\sigma^2 \sim \text{Gamma}(1, 1)$. This prior differs from the informative prior in that large values are assigned to the standard deviations, reflecting weak information about the location of the regression intercept and slope.

```
the_data <- list("y" = y, "x" = x, "N" = N,
                 "mu0" = 0, "g0" = 0.0001,
                 "mu1" = 0.7, "g1" = 0.0001,
                 "a" = 1, "b" = 1)
```

```
posterior3 <- run.jags(modelString,
                       n.chains = 1,
                       data = the_data,
                       monitor = c("beta0", "beta1", "sigma"),
                       adapt = 1000,
                       burnin = 5000,
                       sample = 5000)
```

Figure 11.11 displays density estimates of the simulated posterior draws of the slope parameter β_1 under the informative and weakly informative prior distributions. Note that the informative prior posterior has less spread than the weakly informative prior posterior. This is to be expected since the informative prior adds more information about the location of the slope parameter. In addition, the informative prior posterior shifts the weakly informative prior posterior towards the prior belief that the slope is close to the value 0.7.

After viewing Figure 11.11, one would expect the posterior interval estimate for the slope β_1 to be shorter with the informative prior. We had earlier found that the 90% interval estimate for β_1 to be (0.551, 0.959) with the informative prior. The 90% interval for the slope with the weakly informative prior is (0.501, 1.08) which is about 40% longer than the interval using the informative prior.

```
print(posterior2, digits = 3)
      Lower95   Median Upper95     Mean    SD Mode   MCerr
beta0  -0.267 0.000358   0.276 0.000372 0.138   -- 0.00195
beta1   0.551    0.751   0.959    0.749 0.104   -- 0.00147
sigma   0.498     0.67   0.878    0.682 0.102   -- 0.00154
```

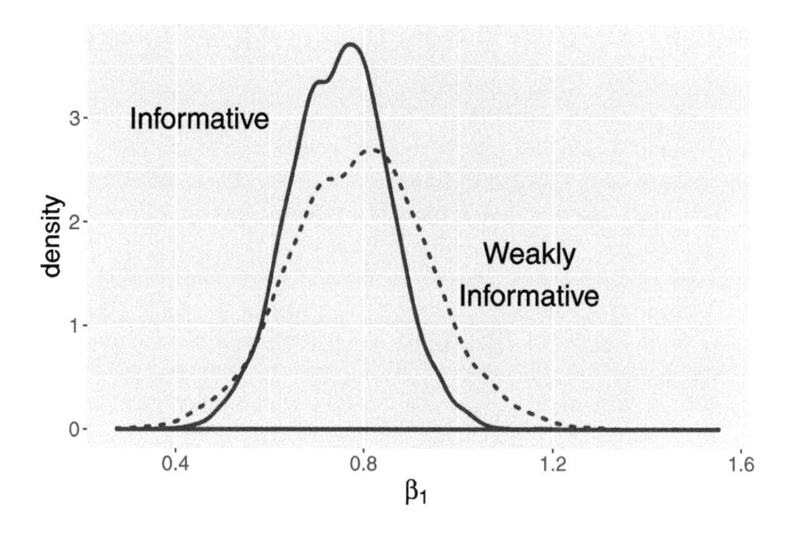

FIGURE 11.11
Density plots of posterior distributions of regression slope parameter β_1 using informative and weakly informative prior distributions.

```
print(posterior3, digits = 3)
      Lower95   Median Upper95     Mean    SD Mode    MCerr
beta0  -0.273 0.000362   0.281 0.000421 0.141   -- 0.00199
beta1   0.501    0.794    1.08    0.792 0.146   -- 0.00207
sigma   0.502    0.677   0.894    0.688 0.105   -- 0.00163
```

11.9 A Conditional Means Prior

In this chapter, we have illustrated two methods for constructing a prior on the parameters of a regression model. The first method reflects weakly informative prior beliefs about the parameters, and the second method assesses an informative prior on the regression parameters on a model on standardized data. In this section, a third method is described for representing prior beliefs on a regression model on the original data. This approach assesses a prior on $(\beta_0, \beta_1, \sigma)$ indirectly by stating prior beliefs about the expected response value conditional on specific values of the predictor variable.

Learning about a gas bill from the outside temperature

A homeowner will typically have monthly payments on basic utilities such as water, natural gas, and electricity. One particular homeowner observes that her monthly natural gas bill seems to vary across the year. The bill is larger for colder months and smaller for warmer months. That raises the question: can one accurately predict one's monthly natural gas bill from the outside temperature?

To address this question, the homeowner collects the mostly gas bills in dollars and the average mostly outside temperatures for all twelve months in a particular year. Figure 11.12 displays a scatterplot of the temperatures and bill amounts. Note that the month bill appears to decrease as a function of the temperature. This motivates consideration of the linear regression model

$$Y_i \mid \beta_0, \beta_1, \sigma \sim \text{Normal}(\beta_0 + \beta_1 x_i, \sigma), \tag{11.20}$$

where x_i and y_i are respectively the average temperature (degrees in Fahrenheit) and the bill amount (in dollars) in month i, and $(\beta_0, \beta_1, \sigma)$ are the unknown regression parameters.

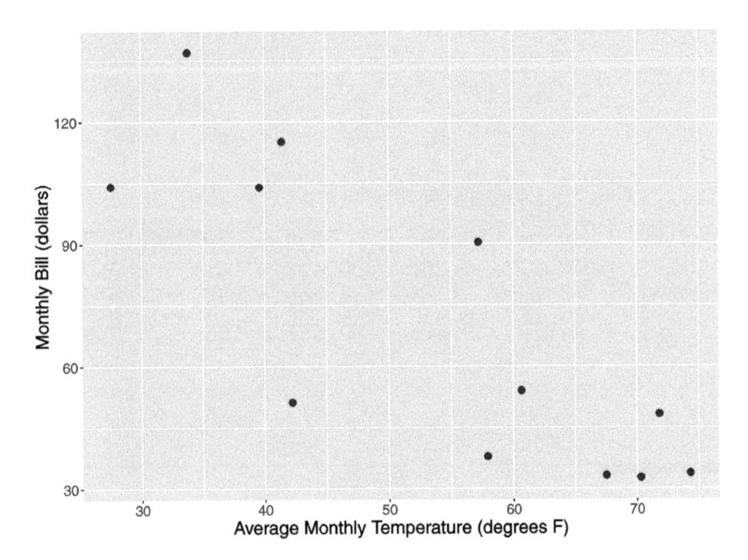

FIGURE 11.12
Scatterplot of average temperatures and gas bills for twelve payments.

A conditional means prior

To construct a prior, first assume that one's beliefs about the regression parameters (β_0, β_1) are independent of the beliefs on the standard deviation

σ and so the joint prior can be factored into the marginal densities:

$$\pi(\beta_0, \beta_1, \sigma) = \pi(\beta_0, \beta_1)\pi(\sigma).$$

With the unstandardized data, it is difficult to think directly about plausible values of the intercept β_0 and slope β_1 and also how these regression parameters are related. But it may be easier to formulate prior opinion about the mean values

$$\mu_i^* = \beta_0 + \beta_1 x_i^*, \tag{11.21}$$

for two specified values of the predictor x_1^* and x_2^*. The conditional means approach proceeds in two steps.

1. For the first predictor value x_1^* construct a normal prior for the mean value μ_1^*. Let the mean and standard deviation values of this prior be denoted by m_1 and s_1, respectively.
2. Similarly, for the second predictor value x_2^* construct a normal prior for the mean value μ_2^* with respective mean and standard deviation m_2 and s_2.

If one assumes that one's beliefs about the conditional means are independent, then the joint prior for the vector (μ_1^*, μ_2^*) has the form

$$\pi(\mu_1^*, \mu_2^*) = \pi(\mu_1^*)\pi(\mu_2^*).$$

This prior on the two conditional means implies a bivariate normal prior on the regression parameters. The two conditional means μ_1^* and μ_2^* were written above as a function of the regression parameters β_0 and β_1. By solving these two equations for the regression parameters, one expresses each parameter as a function of the conditional means:

$$\beta_1 = \frac{\mu_2^* - \mu_1^*}{x_2 - x_1}, \tag{11.22}$$

$$\beta_0 = \mu_1^* - x_1 \left(\frac{\mu_2^* - \mu_1^*}{x_2 - x_1} \right). \tag{11.23}$$

Note that both the slope β_0 and β_1 are linear functions of the two conditional means μ_1^* and μ_2^* and this implies that β_0, β_1 will have a bivariate normal distribution.

Regression analysis of the gas bill example

The process of constructing a conditional means prior is illustrated for our gas bill example. Consider two different temperature values, say 40 degrees and 60 degrees, and, for each temperature, construct a normal prior for the expected monthly bill. After some thought, the following priors are assigned.

- If $x = 40$, the mean bill $\mu_1^* = \beta_0 + \beta_1(40)$ is normal with mean \$100 and standard deviation \$20. This statement indicates that one believes the average gas bill will be relatively high during a cold month averaging 40 degrees.
- If $x = 60$, the mean bill $\mu_2^* = \beta_0 + \beta_1(100)$ is normal with mean \$50 and standard deviation \$15. Here the month's average temperature is warmer and one believes the gas cost will average \$50 lower than in the first scenario.

By assuming independence of our prior beliefs about the two means, we have

$$\pi(\mu_1^*, \mu_2^*) = \phi(\mu_1^*, 100, 20)\phi(\mu_2^*, 50, 15), \tag{11.24}$$

where $\phi(y, \mu, \sigma)$ denotes the normal density with mean μ and standard deviation σ.

The prior on the two means is an indirect way of assessing a prior on the regression parameters β_0 and β_1. One simulates pairs (β_0, β_1) from the prior distribution by simulating values of the means μ_1^* and μ_2^* from independent normal distributions and applying Equation (11.22) and Equation (11.23).

R Simulated draws from the prior are conveniently produced using the JAGS software. The prior is specified for the conditional means by two applications of the `dnorm()` function and the regression parameters are defined as functions of the conditional means.

```
modelString = "
model{
beta1 <- (mu2 - mu1) / (x2 - x1)
beta0 <- mu1 - x1 * (mu2 - mu1) / (x2 - x1)
mu1 ~ dnorm(m1, s1)
mu2 ~ dnorm(m2, s2)
}"
```

Figure 11.13 displays 1000 simulated draws of (β_0, β_1) from the the conditional means prior. It is interesting to note that although the conditional means μ_1^* and μ_2^* are independent, the implied prior on the regression coefficients indicates that β_0 and β_1 are strongly negatively correlated.

The conditional means approach is used to indirectly specify a prior on the regression vector $\beta = (\beta_0, \beta_1)$. To complete the prior, one assigns the precision parameter $\phi = 1/\sigma^2$ a gamma prior with parameters a and b. Then the prior density on all parameters has the form

$$\pi(\beta_0, \beta_1, \sigma) = \pi_{CM}(\beta_0, \beta_1)\pi(\sigma),$$

where π_{CM} is the conditional means prior.

Using this conditional means prior and the gas bill data, one also uses JAGS to simulate from the posterior distribution of $(\beta_0, \beta_1, \sigma)$. In the exercises, the reader will have the opportunity to perform inference about the regression line. In addition, there will be an opportunity to compare inferences using conditional means and weakly informative priors.

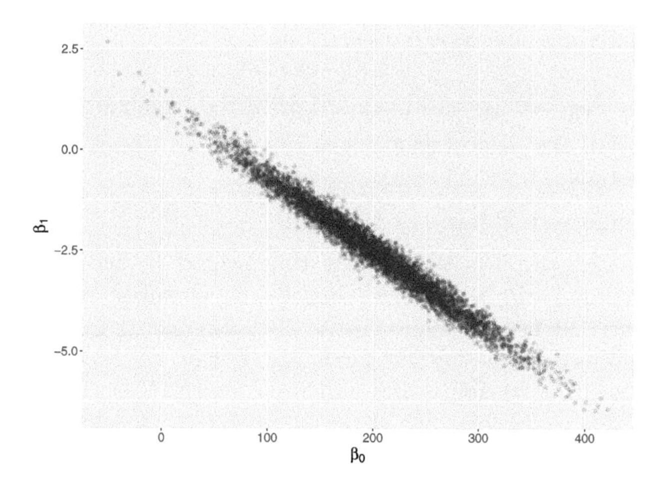

FIGURE 11.13
Scatterplot of simulated draws of the regression parameters (β_0, β_1) from the conditional means prior.

11.10 Exercises

1. **Linear Regression Model**

   Suppose one is interested in predicting a person's height y_i in cm from his or her arm span x_i in cm. Let μ_i denote the mean $\mu_i = E(Y_i \mid x_i)$. Are the following linear models? If not, explain why.

   (a) $\mu_i = \beta_0 + x_i/\beta_1$
   (b) $\mu_i = \beta_1 x_i$
   (c) $\mu_i = \beta_0 + \beta_1 x_i$
   (d) $\mu_i = \exp(\beta_0 + \beta_1 x_i)$

2. **Linear Regression Model**

   Suppose a researcher collects some daily weather data for Denver, Colorado for several winter months. She considers the regression model $Y_i \mid \mu_i, \sigma \sim \text{Normal}(\mu_i, \sigma)$ where

   $$\mu_i = \beta_0 + \beta_1 x_i,$$

   and x_i and y_i are respectively the observed temperature (in degrees Fahrenheit) and snowfall (in inches) on data collected on the i-th day.

   (a) Interpret the intercept parameter β_0.

(b) Interpret the slope parameter β_1.

(c) Suppose $\beta_0 = 5$, $\beta_1 = -0.2$, and $\sigma = 0.2$. If the temperature is 10 degrees, use this model to predict the amount of snowfall.

(d) With the same assumptions in part (c), find a 90% interval estimate for the amount of snowfall.

3. **Pythagorean Result in Baseball**

Table 11.2 displays the average runs scored R, the average runs allowed RA, the number of wins W, the number of losses L for 15 National League teams in the 2018 baseball season. This data is contained in the datafile `pythag2018.csv`. By the Pythagorean formula, if $y = \log(W/L)$ and $x = \log(R/RA)$, then approximately $y = \beta x$ for some slope parameter β. Consider the model $Y_i \sim \text{Normal}(\beta x_i, \sigma)$ where x_i and y_i are the values of $\log(R/RA)$ and $\log(W/L)$ for the i-th team.

(a) Suppose one assumes β and σ are independent where $\beta \sim \text{Normal}(0, 10)$ and the precision $\phi = 1/\sigma^2$ is gamma with parameters 0.001 and 0.001. Write down the expression for the joint posterior density of (β, σ).

(b) Using JAGS, write a script defining the Bayesian model and use MCMC to simulate a sample of 1000 draws from the posterior distribution.

(c) From the simulated output, construct 90% interval estimates for the slope β and for the standard deviation σ.

TABLE 11.2
Average runs scored (R), average runs scored against (RA), wins (W) and losses (L) for 15 National League teams (Tm).

Tm	R	RA	W	L	Tm	R	RA	W	L
MIL	4.6	4	96	67	ARI	4.3	4	82	80
CHC	4.7	4	95	68	PHI	4.2	4.50	80	82
LAD	4.9	3.70	92	71	NYM	4.2	4.40	77	85
COL	4.8	4.60	91	72	SFG	3.7	4.30	73	89
ATL	4.7	4.10	90	72	CIN	4.3	5.10	67	95
STL	4.7	4.30	88	74	SDP	3.8	4.70	66	96
PIT	4.3	4.30	82	79	MIA	3.7	5	63	98
WSN	4.8	4.20	82	80					

4. **Pythagorean Result in Baseball (continued)**

(a) Suppose a team scores on average 4.5 runs and allows, on average, 4.0 runs, so $x = \log(R/RA) = \log(4.5/4) = 0.118$. Simulate 1000 draws from the posterior distribution of $\mu = \beta x$.

(b) If $x = 0.118$, use the work from part (a) to simulate repeated draws of the posterior predictive distribution of $y = \log(W/L)$ and use the output to construct a 90% prediction interval for y.

(c) From the interval found in part (a), find a 90% prediction interval for the number of wins in a 162 game season.

5. **Pythagorean Result in Baseball (continued)**

A traditional least-squares of the model $y = \beta x$ can be found by use of the `lm()` function for the 2018 team data as follows.

```
fit <- lm(I(log(W / L)) ~ 0 + I(log(R / RA)),
                data = pythag2018)
summary(fit)
```

The "Estimate" value is the least-squares estimate of the slope β and the "Std. Error" provides an estimate at the sampling error of this estimate. Compare these estimates with the Bayesian posterior mean and standard deviation of β using a weakly informative prior.

6. **Height and Arm Span**

A person's arm span is strongly related to his or her height. To investigate this relationship, arm spans and heights were measured (in cm) for a sample of 20 students and stored in the file `arm_height.csv`. (This data was simulated using statistics from Mohanty, Babu, and Nair (2001).) Consider the regression model $Y_i \sim \text{Normal}(\mu_i, \sigma)$ where $\mu_i = \beta_0 + \beta_1 x_i$, and y_i and x_i are respectively the height and arm span for the i-th student.

(a) Suppose that one assigns a weakly informative prior to the vector $(\beta_0, \beta_1, \sigma)$ where β_0 and β_1 are independent normal with mean 0 and standard deviation 10, and the precision $\phi = 1/\sigma^2$ is gamma with parameters 0.1 and 0.1. Use JAGS to obtain a simulated sample from the posterior distribution. Find the posterior means of the regression intercept and slope and interpret these posterior means in the context of the problem.

(b) Rescale the heights and arm spans and consider the alternative regression model $Y_i^* \sim \text{Normal}(\mu_i, \sigma)$ where $\mu_i = \beta_0 + \beta_1 x_i^*$ and x_i^* and y_i^* are the rescaled measurements found by subtracting the respective means and dividing by the respectively standard deviations, i.e. standardized. By using similar weakly informative priors as in part (a), use JAGS to simulate from the joint posterior distribution. Find the posterior means of the regression parameters for this rescaled problem and interpret the means.

7. **Height and Arm Span (continued)**

   Consider the problem of learning about a student height using his or her arm span where the measurements are both standardized and one assigns weakly informative priors on the parameters.

   (a) Consider a student whose arm span is one standard deviation above the mean so $x_i^* = 1$. Using the simulated sample from the posterior distribution, find the posterior mean and 90% interval estimate for the expected rescaled height.
   (b) For the same value $x_i^* = 1$, simulate a sample from the posterior predictive distribution of a future standardized height y_i^*. Estimate the mean and construct a 90% prediction interval for y_i^*.
   (c) Compare the intervals computed in parts (a) and (b) and explain how they are different.

8. **Serving Size and Calories of Sandwiches**

   McDonald restaurant publishes nutritional information on the sandwiches served. Table 11.3 displays the serving size (in grams) and the calories for some sandwiches. (This data is available from the file `mcdonalds.csv`.). One is interested in the model $Y_i \sim \text{Normal}(\mu_i, \sigma)$, where $\mu_i = \beta_0 + \beta_1 x_i$ and y_i and x_i are respectively the calories and serving size for the i-th sandwich.

TABLE 11.3
Serving size (grams) and calories for some McDonalds sandwiches.

Sandwich	Size	Calories
Hamburger	105	260
Cheeseburger	119	310
Double Cheeseburger	173	460
Quarter Pounder with Cheese	199	510
Double Quarter Pounder with Cheese	280	730
Big Mac	219	560
Big N' Tasty	232	470
Filet-O-Fish	141	400
McChicken	147	370
Premium Grilled Chicken Classic Sandwich	229	420
Premium Crispy Chicken Classic Sandwich	232	500

   (a) Using a suitable weakly informative prior for the regression parameters and the standard deviation, use JAGS to obtain a simulated sample from the joint posterior distribution.

(b) Construct a graph and 95% interval estimate for the regression slope β_1. Is there sufficient evidence to say that sandwiches with larger serving sizes have more calories?

(c) For a sandwich with serving size 200 grams, simulate a sample from the predictive distribution of the number of calories. Construct a 95% prediction interval for the number of calories.

9. **Serving Size and Calories of Sandwiches (continued)**

   In Exercise 8, one obtained a simulated sample from the posterior distribution of $(\beta_0, \beta_1, \sigma)$ using a weakly informative prior.

   (a) Suppose one is interested in learning about the expected calories μ for a sandwich where the serving size is 300 grams. From the simulated sample from the posterior, construct a sample from the posterior of μ and construct a 90% interval estimate.

   (b) Suppose one is interested in learning about the value of the serving size x^* such that the mean calorie value $\beta_0 + \beta_1 x^*$ is equal to 500. First write the serving size x^* as a function of β_0 and β_1. Then use this representation to find a 90% interval estimate for the serving size x^*.

10. **Movie Sales**

    Table 11.4 displays the weekend and gross sales, in millions of dollars, for ten popular movies released in 2017. This data is contained in the file `movies2017.csv`. Suppose one is interested in studying the relationship between the two variables by fitting the model

$$Y_i \mid \beta_0, \beta_1, \sigma \sim \text{Normal}(\beta_0 + \beta_i x_i, \sigma),$$

    where y_i and x_i are respectively the gross sales and weekend sales for the i-th movie.

TABLE 11.4
Weekend and gross sales (in millions of dollars) for ten popular movies released in 2017.

Movie	Weekend	Gross
Beauty and the Beast	174	504
The Fate of the Furious	99	226
Despicable Me 3	72	264
Spider-Man: Homecoming	117	334
Guardians of the Galaxy, Vol 2.	147	389
Thor: Ragnarok	122	315
Wonder Woman	103	413
Pirates of the Caribbean	63	173
It	123	327
Justice League	94	229

(a) Assuming weekly informative priors on the regression parameters and the standard deviation, use JAGS to simulate from the joint posterior distribution of $(\beta_0, \beta_1, \sigma)$.
(b) Construct a scatterplot of the simulated draws of β_0 and β_1.
(c) From your output, is there significant evidence that weekend sales is a useful predictor of gross sales? Explain.
(d) From the simulated draws, construct a 80% interval estimate for the average gross sales for all movies that has \$100 million weekend sales.
(e) Suppose you are interested in predicting the gross sales for a single movie that has \$100 million weekend sales. Construct a 80% prediction interval for the gross sales.

11. **Fog Index and Complex Words of Books**

The `amazon.com` website used to provide "text statistics" for many of the books it sold. For a particular book, the website displayed the "fog index", the number of years of formal education required to read and understand a passage of text, and the "complex words", the percentage of words in the book with three or more syllables. Table 11.5 displays the complex words and fog index for a selection of 21 popular books. (This data is contained in the file `book_stats.csv`.) Suppose one is interested in predicting the fog index of a book given its complex words measurement. Using suitable weakly informative priors on the parameters, fit a simple regression model by simulating 5000 draws from the joint posterior distribution. Use this output to construct a 90% prediction interval for a book whose complex word measurement is 10.

12. **Distances of Batted Balls**

Figure 11.14 displays a scatterplot of the launch speed (mph) and distance traveled (feet) for batted balls hit by the baseball player Mike Trout during the 2018 season. This data is contained in the file `trout20.csv`.

(a) In R, rescale both the explanatory and response variables so each has mean 0 and standard deviation 1.
(b) Consider the regression model on the rescaled data: $Y_i^* \sim$ Normal(μ_i^*, σ) where

$$\mu_i^* = \beta_0 + \beta_1 x_i^*,$$

where y_i^* and x_i^* are respectively the rescaled distance traveled and rescaled launch speed of the i-th batted ball. Suppose one has little prior knowledge about the values of the parameters β_0, β_1 and σ. One assumes that the parameter are independent where β_0 and β_1 are assigned normal priors with mean 0 and standard deviation 10, and the precision $\phi = 1/\sigma^2$ is

TABLE 11.5
Measures of complex words and fog indexes for a selection of 20 books.

Book	Complex Words	Fog Index
A Million Little Pieces	4.00	5.70
The Five People You Meet in Heaven	6.00	6.60
The Glass Castle	6.00	8.40
The Mermaid Chair	7.00	8.20
The Kite Runner	7.00	7.10
Marley & Me	8.00	9.20
Memoirs of a Geisha	8.00	10.10
In Cold Blood	10.00	9.80
Moneyball	10.00	10.30
Jim Cramer's Real Money	10.00	11.70
The Da Vinci Code	12.00	9.10
Power of Thinking Without Thinking	12.00	11.60
A Mathematician at the Ballpark	12.00	10.20
Misquoting Jesus	13.00	16.10
The Tipping Point	13.00	12.60
Freakonomics	14.00	11.10
Curve Ball	14.00	10.10
The World is Flat	15.00	15.00
Confessions of an Economics Hit Man	15.00	12.80
Collapse	17.00	18.00
Ordinal Data Modeling	24.00	14.60

gamma with parameters 0.1 and 0.1. Use JAGS to obtain a simulated sample of 5000 draws from the posterior distribution of $(\beta_0, \beta_1, \sigma)$.

(c) From the simulated output, construct a density estimate and 90% interval estimate for the slope parameter β_1.

(d) Using the output, obtain simulated draws of the expected standardized distances when the standardized speed is equal to one. Construct a 90% interval estimate for the expected distance.

(e) Obtain a simulated sample from the predicted standardized distance when the standardized speed is equal to one. Construct a 90% prediction interval and compare this interval with the interval constructed in part (c).

13. **Distance of Batted Balls (continued)**

Consider the use of the regression model on the standardized data $Y_i^* \sim \text{Normal}(\mu_i^*, \sigma)$ where

$$\mu_i^* = \beta_0 + \beta_1 x_i^*,$$

where y_i^* and x_i^* are respectively the rescaled distance traveled and rescaled launch speed of the i-th batted ball.

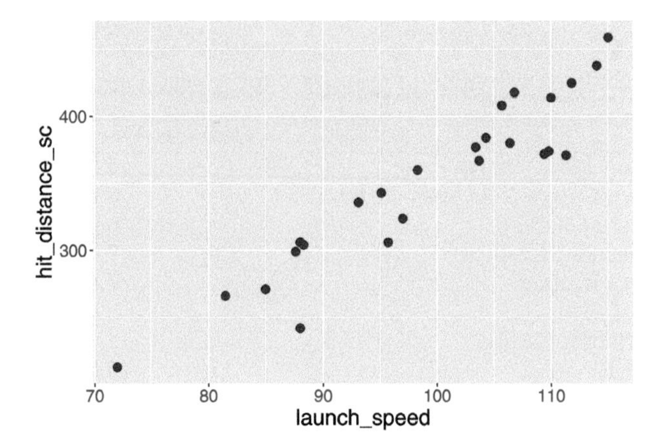

FIGURE 11.14
Scatterplot of launch speed and distance traveled for batted balls of Mike Trout during the 2018 season.

(a) Suppose one believes that if Trout hits a ball with average speed, his expected distance is normal with mean 350 feet and standard deviation 10 feet. In addition, one believe the correlation between launch speed and distance is normal with mean 0.9 and a standard deviation of 0.02. Construct a prior distribution on the vector of parameters (β_0, β_1) that reflects this information.

(b) Use JAGS with this informative prior to obtain a simulated sample from the posterior distribution of all parameters.

(c) Construct a 90% interval estimate for the slope parameter β_1 and compare your answer with the interval estimate constructed in part (c) of Exercise 12.

14. **Gas Bill**

The dataset `gas2017.csv` contains the average temperatures and gas bills in dollars for twelve months for a particular homeowner. This chapter described the use of a conditional means prior to construct a prior distribution on the regression vector (β_0, β_1) in Section 11.9.

(a) Using this conditional means prior and a weakly informative Gamma(1, 1) prior on $\phi = 1/\sigma^2$, use JAGS to simulate 5000 draws from the posterior distribution of $(\beta_0, \beta_1, \sigma)$. Construct a 90% interval estimate for the regression slope β_1.

(b) Rerun this analysis using a suitable weakly informative prior on all the parameters. Simulate 5000 draws and construct a 90% interval estimate for the slope β_1.

(c) Compare your interval estimates for parts (a) and (b).

15. **Gas Bill (continued)**

A traditional maximum likelihood fit of the model $Y_i \mid \beta_0, \beta_1, \sigma \sim \text{Normal}(\beta_0 + \beta_1 x, \sigma)$ can be found by use of the `lm()` function.

```
fit <- lm(Bill ~ Temp, data = gas2017)
summary(fit)
```

The output provides estimates of the regression coefficients β_0 and β_1 and the Residual standard error provides an estimate at the standard deviation σ. Compare these estimates with the posterior means of β_0, β_1, σ found in Exercise 14.

16. **Conditional Means Prior**

Suppose you are interested in predicting the cost of purchasing groceries based on the number of items purchased. You consider the regression model $Y_i \mid \beta_0, \beta_1, \sigma \sim \text{Normal}(\beta_0 + \beta_1 x_i, \sigma)$, where x_i and y_i are respectively the number of grocery items and the total cost (in dollars) of the i-th purchase. Use a conditional mean prior using the following information.

Let $\mu_1^* = \beta_0 + \beta_1(10)$ and $\mu_2^* = \beta_0 + \beta_1(30)$ denote the expected cost of purchasing 10 and 30 grocery items, respectively. You assume that your beliefs about μ_1^* and μ_2^* are independent where $\mu_1^* \sim \text{Normal}(20, 5)$ and $\mu_2^* \sim \text{Normal}(70, 5)$.

(a) Use JAGS to simulate 1000 draws from the prior distribution of (β_0, β_1).

(b) Construct a scatterplot of the values of β_0 and β_1 and describe the relationship that you see in the scatterplot.

(c) Suppose you believe that your prior beliefs about the regression parameters are too strong. Choose a new conditional means prior that reflects this belief.

17. **Olympic Swimming Times**

The dataset `olympic_butterfly.csv` contains the winning time in seconds for the men's and women's 100 m butterfly race for the Olympics from 1964 through 2016. Suppose we focus on the women's times. If y_i and x_i denote respectively the winning time for the women's 100 m butterfly and year for the i-th Olympics, consider the use of the regression model

$$Y_i \mid \beta_0, \beta_1, \sigma \sim \text{Normal}(\beta_0 + \beta_1(x_i - 1964), \sigma).$$

(a) Give interpretations for the regression parameters β_0 and β_1.

(b) Assuming weakly informative priors for all parameters, use JAGS to simulate 5000 values from the joint posterior distribution.

(c) Suppose one is interested in predicting the winning time for the women's 100 m butterfly the 2020 Olympics. Simulate 5000 draws from the posterior predictive distribution and construct a 90% prediction interval for this winning time.

18. **Olympic Swimming Times (continued)**

One way to judge the suitability of the linear model $\mu_i = \beta_0 + \beta_1(x_i - 1964)$ for the Olympics race data is to look for a pattern in the predictive residuals $r_i = y_i - \hat{y}_i$.

(a) Using the draws from the posterior distribution of $(\beta_0, \beta_1, \sigma)$, simulate a sample from the posterior predictive distribution of the future observation $\tilde{y}_i$ for all i using the algorithm described in Section 11.7.4.

(b) Compute the sample of predictive residuals r_i and find 90% interval estimates for each i.

(c) Construct a graph of the 90% intervals for r_i against x_i.

(d) Comment on any lack of fit of the linear model from looking at the residual graph.

19. **Priors for Two-Group Model**

Returning to a tennis example described in Section 8.3, suppose one is interested in comparing the time-to-serve for two tennis servers Roger Federer and Rafael Nadal. One collects the time to serve y_i for both players for many serves. One assumes that Y_i is distributed Normal(μ_i, σ) where

$$\mu_i = \beta_0 + \beta_1 x_i,$$

where x_i is an indicator of the server where $x_i = 0$ if Federer is serving and $x_i = 1$ if Nadal is serving. In this setting, β_0 is the mean time to serve for Federer and β_1 is the increase in mean serving time for Nadal.

(a) Construct a reasonable prior for the intercept parameter β_0.

(b) If you believe that Nadal is significantly slower than Federer in his time-to-serve, construct a prior for β_1 that reflects this belief.

(c) Suppose that the range of serving times is about 3 seconds. Construct a prior for σ that reflects this knowledge.

(d) Construct a joint prior for $(\beta_0, \beta_1, \sigma)$ using the priors constructed in parts (a), (b), and (c).

20. **Two-Group Model (continued)**

    In Exercise 19, you constructed an informative prior for $(\beta_0, \beta_1, \sigma)$ for the regression model for the time-to-serve measurements for two tennis servers. The data file **two_players_time_to_serve.csv** contain measurements for 100 serves for the players Roger Federer and Rafael Nadal. Use JAGS to obtain a simulated sample of the posterior distribution using your prior and this data. Construct a 90% interval estimate for the regression slope β_1 that measures the differences in the mean time to serve for the two players.

12

Bayesian Multiple Regression and Logistic Models

12.1 Introduction

In Chapter 11, we introduced simple linear regression where the mean of a continuous response variable was represented as a linear function of a single predictor variable. In this chapter, this regression scenario is generalized in several ways. In Section 12.2, the multiple regression setting is considered where the mean of a continuous response is written as a function of several predictor variables. Methodology for comparing different regression models is described in Section 12.3. The second generalization considers the case where the response variable is binary with two possible responses in Section 12.4. Here one is interested in modeling the probability of a particular response as a function of an predictor variable. Although these situations are more sophisticated, the Bayesian methodology for inference and prediction follows the general approach described in the previous chapters.

12.2 Bayesian Multiple Linear Regression

12.2.1 Example: expenditures of U.S. households

The U.S. Bureau of Labor Statistics (BLS) conducts the Consumer Expenditure Surveys (CE) through which the BLS collects data on expenditures, income, and tax statistics about households across the United States. Specifically, this survey provides information on the buying habits of U.S. consumers. The summary, domain-level statistics published by the CE are used for both policy-making and research, including the most widely used measure of inflation, the Consumer Price Index (CPI). In addition, the CE has measurements of poverty that determine thresholds for the U.S. Government's Supplemental Poverty Measure.

The CE consists of two surveys. The Quarterly Interview Survey, taken each quarter, aims to capture large purchases (such as rent, utilities, and vehicles), containing approximately 7000 interviews. The Diary Survey, administrated on an annual basis, focuses on capturing small purchases (such as food, beverages, tobacco), containing approximately 14,000 interviews of households.

The CE publishes public-use microdata (PUMD), and a sample of the Quarterly Interview Survey in 2017 1st quarter is collected from the PUMD. This sample contains 1000 consumer units (CU), and provides information of the CU's total expenditures in last quarter, the amount of CU income before taxes in past 12 months, and the CU's urban or rural status. Table 12.1 provides the description of each variable in the CE sample.

TABLE 12.1
Variable description for CE sample.

Variable	Description
Expenditure	Continuous; CU's total expenditures in last quarter
Income	Continuous; the amount of CU income before taxes in past 12 months
UrbanRural	Binary; the urban or rural status of CU: 1 = Urban, 2 = Rural

Suppose someone is interested in predicting a CU expenditure based on his or her urban or rural status and its income before taxes. In this example, one is treating expenditure as the response variable and the other two variables as predictors. To proceed, one needs to develop a model to express the relationship between expenditure and the other two predictors *jointly*. This requires extending the simple linear regression model introduced in Chapter 11 to the case with multiple predictors. This extension is known as multiple linear regression – the word "multiple" indicates two or more predictors are present in the regression model. This section describes how to set up a multiple linear regression model, how to specify prior distributions for regression coefficients of multiple predictors, and how to make Bayesian inferences and predictions in this setting.

Recall in Chapter 11, the mean response μ_i was expressed as a linear function of the single continuous predictor x_i depending on an intercept parameter β_0 and a slope parameter β_1:

$$\mu_i = \beta_0 + \beta_1 x_i.$$

In particular, the slope parameter β_1 is interpreted as the change in the expected response μ_i, when the predictor x_i of record i increases by a single unit. In the household expenditures example, not only there are multiple predictors, but the predictors are of different types including one continuous predictor (income), and one binary categorical (rural or urban status) predictor. As Chapter 11 focused on continuous-valued predictors, the interpretation of a regression coefficient for a binary categorical predictor is an important topic for discussion in this section.

12.2.2 A multiple linear regression model

Similar to a simple linear regression model, a multiple linear regression model assumes an observation-specific mean μ_i for the i-th response variable Y_i.

$$Y_i \mid \mu_i, \sigma \overset{ind}{\sim} \text{Normal}(\mu_i, \sigma), \; i = 1, \cdots, n. \tag{12.1}$$

In addition, it assumes that the mean of Y_i, μ_i, is a linear function of all predictors. In general, one writes

$$\mu_i = \beta_0 + \beta_1 x_{i,1} + \beta_2 x_{i,2} + \cdots + \beta_r x_{i,r}, \tag{12.2}$$

where $\mathbf{x}_i = (x_{i,1}, x_{i,2}, \cdots, x_{i,r})$ is a vector of r known predictors for observation i, and $\beta = (\beta_0, \beta_1, \cdots, \beta_r)$ is a vector of unknown regression parameters (coefficients), shared among all observations.

For studies where all r predictors are continuous, one interprets the intercept parameter β_0 as the expected response μ_i for observation i, where all of its predictors take values of 0 (i.e. $x_{i,1} = x_{i,2} = \cdots = x_{i,r} = 0$). One can also interpret the slope parameter β_i ($j = 1, 2, \cdots, r$) as the change in the expected response μ_i, when the j-th predictor, $x_{i,j}$, of observation i increases by a single unit while all remaining $(r-1)$ predictors stay constant.

However in the household expenditures example from the CE data sample, not all predictors are continuous. The urban or rural status variable is a binary categorical variable, taking a value of 1 if the CU is in an urban area, and taking value of 2 if the CU is in a rural area. It is possible to consider the variable as continuous and interpret the associated regression coefficient as the change in the expected response μ_i when the CU's urban or rural status changes by one unit from urban to rural (corresponding to change from one to two). But it is much more common to consider this variable as a binary categorical variable that classifies the observations into two distinct groups: the urban group and the rural group. It will be seen that this classification puts an emphasis on the difference of the expected responses between the two distinct groups.

Consequently, consider the construction of a new indicator variable in place of the binary variable. This new indicator variable takes a value of 0 if the CU is in an urban area, and a value of 1 if the CU is in a rural area. To understand the implication of this indicator variable, it is helpful to consider a simplified regression model with a single predictor, the binary indicator for rural area x_i. This simple linear regression model expresses the linear relationship as

$$\mu_i = \beta_0 + \beta_1 x_i = \begin{cases} \beta_0, & \text{the urban group;} \\ \beta_0 + \beta_1, & \text{the rural group.} \end{cases} \tag{12.3}$$

The expected response μ_i for CUs in the urban group is given by β_0, and the expected response μ_i for CUs in the rural group is $\beta_0 + \beta_1$. In this case β_1 represents the change in the expected response μ_i from the urban group to the rural group. That is, β_1 represents the effect of being a member of the rural group.

Before continuing, there is a need for some data transformation. Both the expenditure and income variables are highly skewed, and both variables have more even distributions if we apply logarithm transformations. So the response variable will be the logarithm of the CU's total expenditure and the continuous predictor will be the logarithm of the CU 12-month income. Figure 12.1 displays scatterplots of log income and log expenditure where the two panels correspond to urban and rural residents. Note that in each panel there appears to be a positive association between log income and log expenditure.

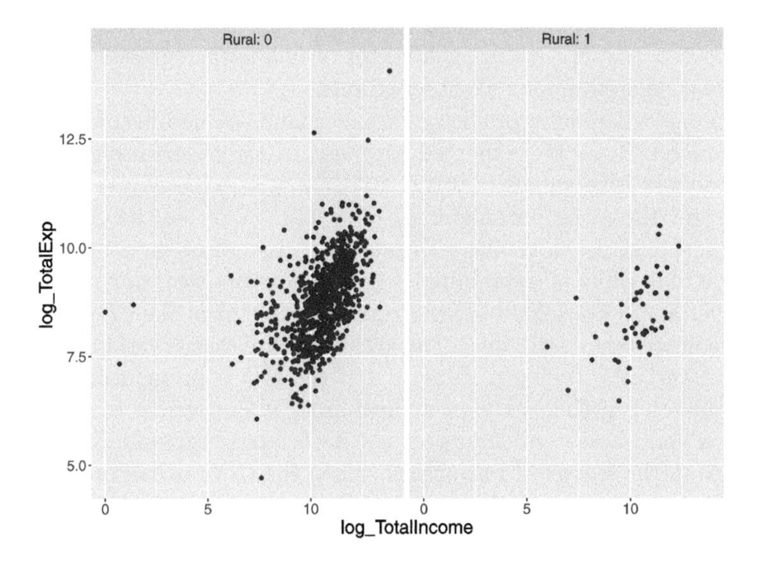

FIGURE 12.1
Scatterplot of log total income and log total expenditure for the urban and rural groups.

Now the the data transformations are completed, one is ready to set up a multiple linear regression model for the log expenditure response including one continuous predictor and one binary categorical predictor. The expected response μ_i is expressed as a linear combination of the log income variable and the rural indicator variable.

$$\mu_i = \beta_0 \quad + \quad \beta_1 x_{i,income} + \beta_2 x_{i,rural}. \tag{12.4}$$

The multiple linear regression model is written as

$$Y_i \mid \beta_0, \beta_1, \beta_2, \sigma \overset{ind}{\sim} \text{Normal}(\beta_0 \quad + \quad \beta_1 x_{i,income} + \beta_2 x_{i,rural}, \sigma), \tag{12.5}$$

where $\mathbf{x}_i = (x_{i,income}, x_{i,rural})$ is a vector of predictors and σ is the standard deviation in the normal model shared among all responses Y_i's.

The regression parameters have clear interpretations. The intercept parameter β_0 is the expected log expenditure when both the remaining variables are 0's: $x_{i,income} = x_{i,rural} = 0$. This intercept represents the mean log expenditure for an urban CU with a log income of 0.

The regression slope coefficient β_1 is associated with the continuous predictor variable, log income. This slope β_1 can be interpreted as the change in the expected log expenditure when the predictor log income of record i increases by one unit, while all other predictors stay unchanged.

The remaining regression coefficient β_2 represents the change in the expected log expenditure compared relative to the expected log expenditure of the associated reference category, while all other predictors stay unchanged. In other words, β_2 is the change in the expected log expenditure of a rural CU comparing to an urban CU, when the two CUs have the same log income.

With an understanding of the meaning of the regression coefficients, one can now proceed to a description of a prior and MCMC algorithm of this multiple linear regression model. Note that one needs to construct a prior distribution for the set of parameters $(\beta_0, \beta_1, \beta_2, \sigma)$. We begin by describing the weakly informative prior approach and the subsequent MCMC inference.

12.2.3 Weakly informative priors and inference through MCMC

In situations where the data analyst has limited prior information about the regression parameters or the standard deviation, it is desirable to assign a prior that has little impact on the posterior. Similar to the weakly informative prior for simple linear regression described in Chapter 11, one assigns a weakly informative prior for a multiple linear regression model using standard functional forms. Assuming independence, the prior density for the set of parameters $(\beta_0, \beta_1, \beta_2, \sigma)$ is written as a product of the component densities:

$$\pi(\beta_0, \beta_1, \beta_2, \sigma) = \pi(\beta_0)\pi(\beta_1)\pi(\beta_2)\pi(\sigma),$$

where β_0 is Normal(m_0, s_0), β_1 is Normal(m_1, s_1), β_2 is Normal(m_2, s_2), and the precision parameter $\phi = 1/\sigma^2$, the inverse of the variance σ^2, is Gamma(a, b).

If one has little information about the location of the regression parameters β_0, β_1, and β_2, one assigns the respective prior means to be 0 and the prior standard deviations to be large values, say 20. In similar fashion, if little knowledge exists about the location of the sampling standard deviation σ, one assigns small values for the hyperparameters, a and b, say $a = b = 0.001$, for the Gamma prior placed on the precision $\phi = 1/\sigma^2$.

One uses the JAGS software to draw MCMC samples from this multiple linear regression model. The process of using JAGS mimics the general approach used in earlier chapters.

Describe the model by a script

R The first step in using JAGS writes the following script defining the multiple linear regression model, saving the script in the character string `modelString`.

```
modelString <-"
model {
## sampling
for (i in 1:N){
   y[i] ~ dnorm(beta0 + beta1*x_income[i] +
                beta2*x_rural[i], invsigma2)
}
## priors
beta0 ~ dnorm(mu0, g0)
beta1 ~ dnorm(mu1, g1)
beta2 ~ dnorm(mu2, g2)
invsigma2 ~ dgamma(a, b)
sigma <- sqrt(pow(invsigma2, -1))
}
"
```

In the sampling section of the script, the iterative loop goes from `1` to `N`, where `N` is the number of observations with index `i`. Recall that the normal distribution `dnorm` in JAGS is stated in terms of the mean and the precision and the variable `invsigma2` corresponds to the normal sampling precision. The variable `sigma` is defined in the prior section of the script so one can track the simulated values of the standard deviation σ. Also the variables `m0`, `m1`, `m2` correspond to the means, and `g0`, `g1`, `g2` correspond to the precisions of the normal prior densities for the three regression parameters.

Define the data and prior parameters

The next step is to provide the observed data and the values for the prior parameters. In the R script below, a list `the_data` contains the vector of log expenditures, the vector of log incomes, the indicator variables for the categories of the binary categorical variable, and the number of observations. This list also contains the means and precisions of the normal priors for `beta0` through `beta2` and the values of the two parameters `a` and `b` of the gamma prior for invsigma2. The prior mean of the normal priors on the individual regression coefficients is 0, for mu0 through mu2. The prior standard deviations of the normal priors on the individual regression coefficients are 20, and so the corresponding precision values are $1/20^2 = 0.0025$ for g0 through g2.

```
y <- as.vector(CEsample$log_TotalExp)
x_income <- as.vector(CEsample$log_TotalIncome)
x_rural <- as.vector(CEsample$Rural)
N <- length(y)
the_data <- list("y" = y, "x_income" = x_income,
                 "x_rural" = x_rural, "N" = N,
                 "mu0" = 0, "g0" = 0.0025,
                 "mu1" = 0, "g1" = 0.0025,
                 "mu2" = 0, "g2" = 0.0025,
                 "a" = 0.001, "b" = 0.001)
```

Generate samples from the posterior distribution

The run.jags() function in the runjags package generates posterior samples by the MCMC algorithm using the JAGS software. The script below runs one MCMC chain with an adaption period of 1000 iterations, a burn-in period of 5000 iterations, and an additional set of 20,000 iterations to be run and collected for inference. By using the argument monitor = c("beta0", "beta1", "beta2", "sigma"), one keeps tracks of all four model parameters. The output variable posterior contains a matrix of simulated draws.

```
posterior <- run.jags(modelString,
                      n.chains = 1,
                      data = the_data,
                      monitor = c("beta0", "beta1",
                                  "beta2", "sigma"),
                      adapt = 1000,
                      burnin = 5000,
                      sample = 20000)
```

MCMC diagnostics

To obtain valid inferences from the posterior draws from the MCMC simulation, one should assess convergence of the MCMC chain. The plot() function with the argument input vars returns four diagnostic plots (trace plot, empirical CDF, histogram and autocorrelation plot) for the specified parameter. For example, Figure 12.2 shows the diagnostic plots for the slope parameter β_1 for the log income predictor using the following code.

```
plot(posterior, vars = "beta1")
```

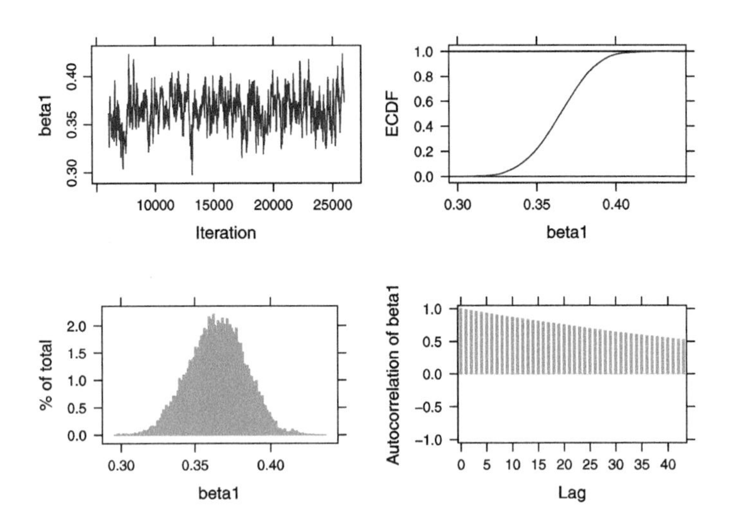

FIGURE 12.2
MCMC diagnostics plots for the regression slope parameter β_1 for the log income predictor.

The upper left trace plot shows MCMC mixing for the 20,000 simulated draws of β_1. In this example, the lower right autocorrelation plot indicates relatively large correlation values between adjacent posterior draws of β_1. In this particular example, since the mixing was not great, it was decided to take a larger sample of 20,000 draws to get good estimates of the posterior distribution. In usual practice, one should perform these diagnostics for all parameters in the model.

Summarization of the posterior

Posterior summaries of the parameters are obtained by use of the `print(posterior, digits = 3)` command. Note that these summaries are based on the 20,000 iterations from the sampling period excluding the samples from the adaption and burn-in periods.

```
print(posterior, digits = 3)
      Lower95 Median Upper95   Mean     SD Mode    MCerr
beta0    4.59   4.95    5.36   4.95  0.201   --   0.0166
beta1   0.328  0.365     0.4  0.365 0.0188   --  0.00155
beta2  -0.482 -0.267 -0.0476 -0.269  0.112   --  0.00112
sigma   0.735  0.769   0.802  0.769 0.0172   -- 0.000172
```

One way to determine if the two variables are useful predictors is to inspect the location of the 90% probability intervals. The interval estimate for β_1 (corresponding to log income) is (0.328, 0.400) and the corresponding estimate for β_2 (corresponding to the rural variable) is $(-0.482, -0.048)$. Neither interval covers zero, thus indicating that both log income and the rural variables are helpful in predicting log expenditure.

Several types of summaries of the posterior distribution are illustrated. Suppose one is interested in learning about the expected log expenditure. From the regression model, the mean log expenditure is equal to

$$\beta_0 + \beta_1 x_{income} \tag{12.6}$$

for urban CUs, and equal to

$$\beta_0 + \beta_1 x_{income} + \beta_2 \tag{12.7}$$

for rural CUs. Figure 12.3 displays simulated draws from the posterior of the expected log expenditure superposed over the scatterplots of log income and log expenditure for the urban and rural cases. Note that there is more variation in the posterior draws for the rural units – this is reasonable since only a small portion of the data came from rural units.

Figure 12.4 displays the posterior density of the mean log expenditure for the predictor pairs (log Income = 9, Rural = 1), (log Income = 9, Rural = 0), (log Income = 12, Rural = 1), and (log Income = 12, Rural = 0). It is pretty clear from this graph that log income is the more important predictor. For both urban and rural CUs, the log total expenditure is much larger for log income = 12 than for log income = 9. Given a particular value of log expenditure, the log expenditure is slightly higher for urban (Rural = 0) compared to rural units.

12.2.4 Prediction

A related problem is to predict a CU's log expenditure for a particular set of predictor values. Let $\tilde{Y}$ denote the future response value for the expenditure for given values of income x^*_{income} and rural value x^*_{rural}. One represents the posterior predictive density of $\tilde{Y}$ as

$$f(\tilde{Y} = \tilde{y} \mid y) = \int f(\tilde{y} \mid y, \beta, \sigma)\pi(\beta, \sigma \mid y) d\beta, \tag{12.8}$$

where $\pi(\beta, \sigma|y)$ is the posterior density and $f(\tilde{Y} = \tilde{y} \mid y, \beta, \sigma)$ is the normal sampling density which depends on the predictor values.

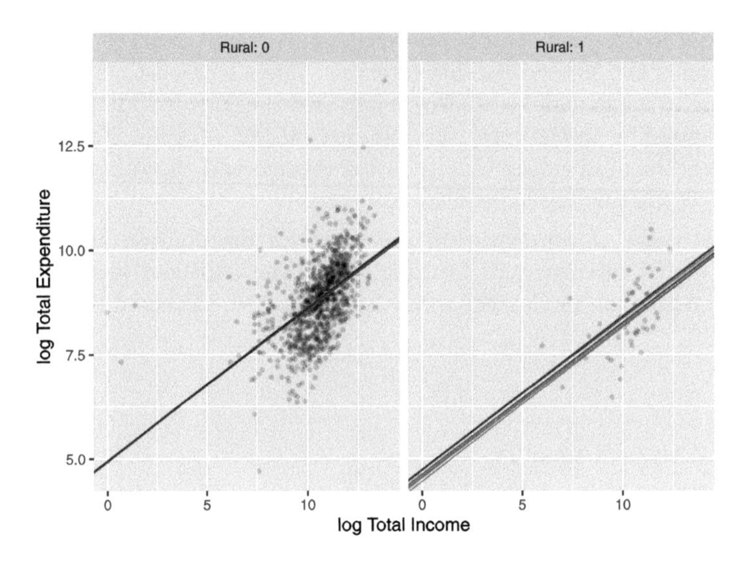

FIGURE 12.3
Scatterplot of log income and log expenditure for the urban and rural groups. The superposed lines represent draws from the posterior distribution of the expected response.

R Since we have already produced simulated draws from the posterior distribution, it is straightforward to simulate from the posterior predictive distribution. One simulates a single draw from $f(\tilde{Y} = \tilde{y} \mid y)$ by first simulating a value of (β, σ) from the posterior – call this draw $(\beta^{(s)}, \sigma^{(s)})$. Then one simulates a draw of $\tilde{Y}$ from a normal density with mean $\beta_0^{(s)} + \beta_1^{(s)} x^*_{income} + \beta_2^{(s)} x^*_{rural}$ and standard deviation $\sigma^{(s)}$. By repeating this process for a large number of iterations, the function `one_predicted()` simulates a sample from the posterior prediction distribution for particular predictor values x^*_{income} and x^*_{rural}.

```
one_predicted <- function(x1, x2){
  lp <- post[ , "beta0"] +  x1 * post[ , "beta1"] +
    x2 * post[, "beta2"]
  y <- rnorm(5000, lp, post[, "sigma"])
  data.frame(Value = paste("Log Income =", x1,
                           "Rural =", x2),
             Predicted_log_TotalExp = y)
}
df <- map2_df(c(12, 12),
              c(0, 1), one_predicted)
```

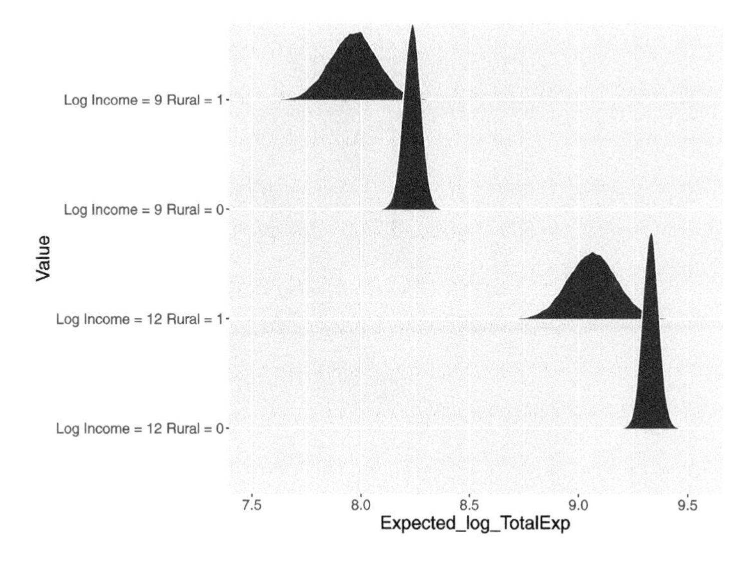

FIGURE 12.4
Posterior distributions of the expected log expenditure for units with different income and rural variables.

This procedure is implemented for the two sets of predictor values (log income, rural) = (12, 1) and (log income, rural) = (12, 0). Figure 12.5 displays density estimates of the posterior predictive distributions for the two cases. Comparing Figures 12.4 and 12.5, note the increased width of the prediction densities relative to the expected response densities. One confirms this by computing interval estimates. For example, for the values (log income, rural) = (12, 1), a 90% interval for the expected log expenditure is (8.88, 9.25) and the 90% interval for the predicted log expenditure for the same predictor values is (7.81, 10.34).

12.3 Comparing Regression Models

When one fits a multiple regression model, there is a list of inputs, i.e. potential predictor variables, and there are many possible regression models to fit depending on what inputs are included in the model. In the household expenditures example, there are two possible inputs, the log total income and the rural/urban status and there are $2 \times 2 = 4$ possible models depending on the inclusion or exclusion of each input. When there are many inputs, the number of possible regression models can be quite large and so there needs to be

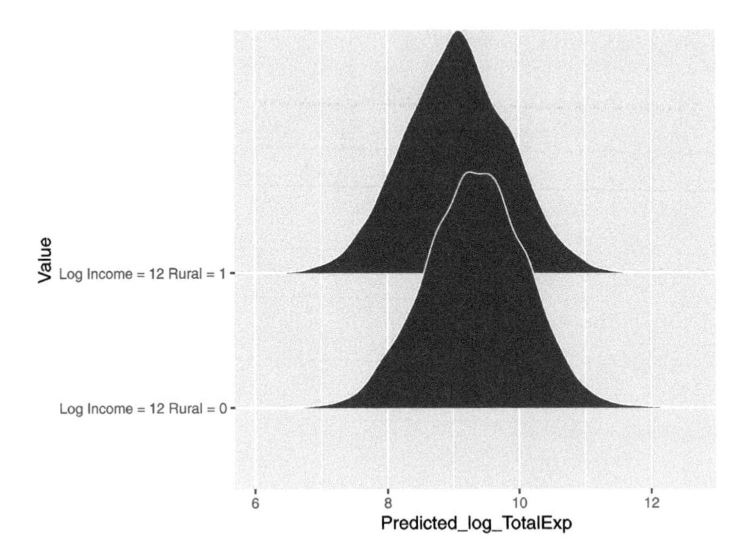

FIGURE 12.5
Predictive distributions of the log expenditure for units with different income and rural variables.

some method for choosing the best regression model. A simple example will be used to describe what is meant by a best model and then a general method is outlined for selecting between models.

Learning about a career trajectory

To discuss model selection in a simple context, consider a baseball modeling problem that will be more thoroughly discussed in Chapter 13. One is interested in seeing how a professional athlete ages during his or her career. In many sports, an athletic enters his or her professional career at a modest level of performance, gets better until a particular age when peak performance is achieved, and then decreases in the level of performance until retirement. One can use a regression model to explore the pattern of performance over age – this pattern is typically called the athletic's career trajectory.

We focus on a particular great historical baseball player Mike Schmidt who played in Major League Baseball from 1972 through 1989. Figure 12.6 first displays a scatterplot of the rate that Schmidt hit home runs as a function of his age. If y_i denotes Schmidt's home run rate during the i-th season when his age was x_i, Figure 12.6 further overlays fits from the following three career trajectory models:

- Model 1 - Linear:

$$Y_i \mid \beta_0, \beta_1, x_i, \sigma \sim \textrm{Normal}(\beta_0 + \beta_1(x_i - 30), \sigma).$$

- Model 2 - Quadratic:

$$Y_i \mid \beta_0, \beta_1, \beta_2, x_i, \sigma \sim \text{Normal}(\beta_0 + \beta_1(x_i - 30) + \beta_2(x_i - 30)^2, \sigma).$$

- Model 3 - Cubic:

$$Y_i \mid \beta_0, \beta_1, \beta_2, \beta_3 x_i, \sigma \sim \text{Normal}(\beta_0 + \beta_1(x_i - 30) + \beta_2(x_i - 30)^2 + \beta_3(x_i - 30)^3, \sigma).$$

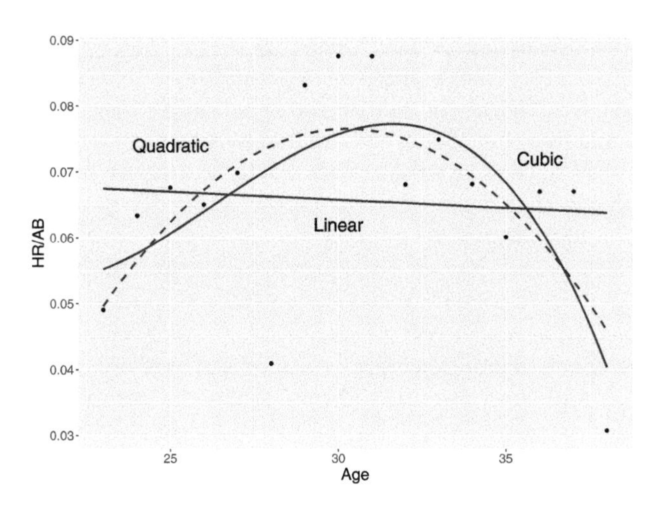

FIGURE 12.6
Scatterplot of age and home run rate for Mike Schmidt. Fits from linear, quadratic, and cubic models are overlaid.

Model 1 says that Schmidt's true home run performance is a linear function of his age, Model 2 says that his home run performance follows a parabolic shape, and Model 3 indicates that his performance follows a cubic curve. Based on the earlier comments about the knowledge of shapes of career trajectories, the linear function of age given in Model 1 does not appear suitable in reflecting the "down, up, down" trend that we see in the scatterplot. The fits of Models 2 and 3 appear to be similar in appearance, but there are differences in the interpretation of the fits. The quadratic fit (Model 2) indicates that Schmidt's peak performance occurs about the age of 30 while the cubic fit (Model 3) indicates that his peak performance occurs around the age of 33. How can we choose between the two models?

Underfitting and overfitting

In model building, there are two ways of misspecifying a model that we call "underfitting" and "overfitting" that are described in the context of this career trajectory example. First, it is important to include all useful inputs in the model to explain the variation in the response variable. Failure to include relevant inputs in the model will result in underfitting. In our example, age is the predictor variable and the possible inputs are age, age^2, and age^3. If we use Model 1 which includes only the input age, this particular model appears to underfit the data since this model does not reflect the increasing and decreasing pattern in the home run rates that we see in Figure 12.6.

At the other extreme, one should be careful not to include too many inputs in the model. When one includes more inputs in our regression model than needed, one has overfitting. Model 3 possibly overfits the data, since it may not be necessary to represent a player's trajectory by a cubic curve – perhaps a quadratic curve is sufficient. In an extreme situation, by increasing the degree of the polynomial function of age, one can find a fitted curve that goes through most of the points in the scatterplot. This would be a severe case of overfitting since it is unlikely that a player's true career trajectory is represented by a polynomial of a high degree.

Cross-validation

How does one choose a suitable regression model that avoids the underfitting and overfitting problems described above? A general method of comparing models is called cross-validation. In this method, one partitions the dataset into two parts – the training and testing components. One initially fits each regression model to the training dataset. Then one uses each fitted model to predict the response variable in the testing dataset. The model that is better in predicting observations in the future testing dataset is the preferred model.

Let's describe how one implements cross-validation for our career trajectory example. In the example, Mike Schmidt had a total of 8170 at-bats for 13 seasons. One randomly divides these 8170 at-bats into two datasets – 4085 of the at-bats (and the associated home run and age variables) are placed in a training dataset and the remaining at-bats become the testing dataset. Let $\{(x_i^{(1)}, y_i^{(1)})\}$ denote the age and home run rate variables from the training dataset and $\{(x_i^{(2)}, y_i^{(2)})\}$ denote the corresponding variables from the testing dataset.

Suppose one considers the use of Model 1 where the home run rate $Y_i^{(1)} \sim$ Normal(μ_i, σ) where the mean rate is $\mu_i = \beta_0 + (\beta_1 - 30)x_i^{(1)}$. One places a weakly informative prior on the vector of parameters $(\beta_0, \beta_1, \sigma)$ and define the likelihood using the training data. One uses JAGS to simulate from the posterior distribution and obtain the fitted regression

$$\mu = \tilde{\beta}_0 + (\tilde{\beta}_1 - 30)x,$$

where $\tilde{\beta}_0$ and $\tilde{\beta}_1$ are the posterior means of the regression intercept and slope respectively.

One now uses this fitted regression to predict values of the home run rate from the testing dataset. One could simulate predictions from the posterior predictive distribution, but for simplicity, suppose one is interested in making a single prediction. For the i-th value of age $x_i^{(2)}$ in the testing dataset, our best prediction of the i-th home run rate from Model 1 would be $\tilde{y}_i^{(2)}$ where

$$\tilde{y}_i^{(2)} = \tilde{\beta}_0 + (\tilde{\beta}_1 - 30)x_i^{(2)}.$$

If one performs this computation for all ages, one obtains a set of predictions $\{\tilde{y}_i^{(2)}\}$ that one would like to be close to the actual home run rates $\{y_i^{(2)}\}$ in the training dataset. It is unlikely that the prediction will be on target so one considers the prediction error that is the difference between the prediction and the response $|\tilde{y}_i^{(2)} - y_i^{(2)}|$. One measures the closeness of the predictions by computing the sum of squared prediction errors (SSPE):

$$SSPE = \sum (\tilde{y}_i^{(2)} - y_i^{(2)})^2. \tag{12.9}$$

The measure $SSPE$ describes how well the fitted model predicts home run rates from the training dataset. One uses this measure to compare predictions from alternative regression models. Specifically, suppose each of the regression models (Model 1, Model 2, and Model 3) is fit to the training dataset and each of the fitted models is used to predict the home run rates of the testing dataset. Suppose the sum of squared prediction errors for the three fitted models are $SSPE_1$, $SSPE_2$ and $SSPE_3$. The best model is the model corresponding to the smallest value of $SSPE$. If this model turns out to be Model 2, then we say that Model 2 is best in that it is best in predicting home run rates in a future or out-of-sample dataset.

Approximating cross-validation by DIC

The cross validation method of assessing model performance can be generally applied in many situations. However, there are complications in implementing cross validation in practice. One issue is how the data should be divided into the training and testing components. In our example, the data was divided into two datasets of equal size, but it is unclear if this division scheme is best in practice. Another issue is that the two datasets were divided using a random mechanism. The problem is that the predictions and the sum of squared prediction errors can depend on the random assignment of the two groups. That raises the question – is it necessary to perform cross validation to compare the predictive performance of two models?

A best regression model is the one that provides the best predictions of the response variable in an out-of-sample or future dataset. Fortunately, it is not necessary in practice to go through the cross-validation process. It is possible to compute a measure, called the Deviance Information Criterion or DIC, from the simulated draws from the posterior distribution that approximates a model's out-of-sample predictive performance. The description and derivation of the DIC measure is outside of the scope of this text – a brief description of this method is contained in the appendices. But we illustrate the use of DIC measure for the career trajectory example. It can be applied generally and is helpful for comparing the predictive performance of several Bayesian models.

Example of model comparison

To illustrate the application of DIC, let's return to the career trajectory example. As usual practice, JAGS will be used to fit a specific Bayesian model. To fit the quadratic model M_2, one writes the following JAGS model description.

At the sampling stage, the home run rates `y[i]` are assumed to be a quadratic function of the ages `x[i]`, and at the prior stage, the regression coefficients `beta0`, `beta1`, `beta2`, and the precision `phi` are assigned weakly informative priors. The variable `the_data` is a list containing the observed home run rates, ages, and sample size.

```
modelString = "
model {
for (i in 1:N){
   y[i] ~ dnorm(mu[i], phi)
   mu[i] <- beta0 + beta1 * (x[i] - 30) +
            beta2 * pow(x[i] - 30, 2)
}
beta0 ~ dnorm(0, 0.001)
beta1 ~ dnorm(0, 0.001)
beta2 ~ dnorm(0, 0.001)
phi ~ dgamma(0.001, 0.001)
}
"
d <- filter(sluggerdata,
            Player == "Schmidt", AB >= 200)
the_data <- list(y = d$HR / d$AB,
                 x = d$Age,
                 N = 16)
```

The model is fit by the `run.jags()` function. To compute DIC, it is necessary to run multiple chains, which is indicated by the argument `n.chains = 2` that two chains will be used.

```
post2 <- run.jags(modelString,
                  n.chains = 2,
                  data = the_data,
                  monitor = c("beta0", "beta1",
                              "beta2", "phi"))
```

To compute DIC, the `extract.runjags()` function is applied on the runjags object `post2`. In `Penalized deviation`, output is the value of DIC computed on the simulated MCMC output.

```
extract.runjags(post2, "dic")
Mean deviance:  -88.98
penalty 4.817
Penalized deviance: -84.17
```

The value of DIC $= -84.17$ for this single quadratic regression model is not meaningful, but one compares values of DIC for competing models. Suppose one wishes to compare models M_1, M_2, M_3 and a quartic regression where one represents the home run rate as a polynomial of fourth degree of the age. For each model, a JAGS script is written where the regression coefficients and the precision parameter are assigned weakly informative priors. The `run.jags()` function is applied to produce a posterior sample and the `extract.runjags()` with the `"dic"` argument to extract the value of DIC. Table 12.2 displays the values of DIC for the four regression models. The best model is the model with the smallest value of DIC. Looking at the values in Table 12.2, one sees that the quadratic model has the smallest value of -84.2. The interpretation is that the quartic model is best in the sense that it will provide the best out-of-sample predictions.

TABLE 12.2
DIC values for four regression models fit to Mike Schmidt's home run rates.

Model	DIC
Linear	−80.4
Quadratic	−84.2
Cubic	−82.1
Quartic	−79.0

12.4 Bayesian Logistic Regression

12.4.1 Example: U.S. women labor participation

The University of Michigan Panel Study of Income Dynamics (PSID) is the longest running longitudinal household survey in the world. The study began in 1968 with a nationally representative sample of over 18,000 individuals living in 5000 families in the United States. Information on these individuals and their descendants has been collected continuously, including data covering employment, income, wealth, expenditures, health, marriage, childbearing, child development, philanthropy, education, and numerous other topics.

The PSID 1976 survey has attracted particular attention since it interviewed wives in the households directly in the previous year. The survey provides helpful self-reporting data sources for studies of married women's labor supply. A sample includes information on family income exclusive of wife's income (in \$1000) and the wife's labor participation (yes or no). This PSID sample contains 753 observations and two variables. Table 12.3 provides the description of each variable in the PSID sample.

TABLE 12.3
The variable descriptions for the PSID sample.

Variable	Description
LaborParticipation	Binary; labor participation of gift: 1 = year, 0 = no
FamilyIncome	Continuous; family income exclusive of wife's income, in \$1000, 1975 U.S. dollars

Suppose one is interested in predicting a wife's labor participation status from the family income exclusive of her income. In this example, one is treating labor participation as the response variable and the income variable as a predictor. Furthermore, the response variable is not continuous, but binary – either the wife is working or she is not. To analyze a binary response such as labor participation, one is interested in estimating the probability of a labor participation (yes) as a function of the predictor variable, family income exclusive of her income. This requires a new model that can express the probability of a yes as a function of the predictor variable.

Figure 12.7 displays a scatterplot of the family income against the labor participation status. Since the labor participation variable is binary, the points are jittered in the vertical direction. From this graph, we see that roughly half of the wives are working and it is difficult to see if the family income is predictive of the participation status.

Recall in Chapter 11, when one had a continuous-valued response variable and a single continuous predictor, the mean response μ_i was expressed as a linear function of the predictor through an intercept parameter β_0 and a slope parameter β_1:

$$\mu_i = \beta_0 + \beta_1 x_i. \tag{12.10}$$

Moreover it is reasonable to use a normal regression model where the response Y_i is Normally distributed where the mean μ_i with a linear function as in Equation (12.10).

$$Y_i \mid \mu_i, \sigma \overset{ind}{\sim} \text{Normal}(\mu_i, \sigma), \;\; i = 1, \cdots, n.$$

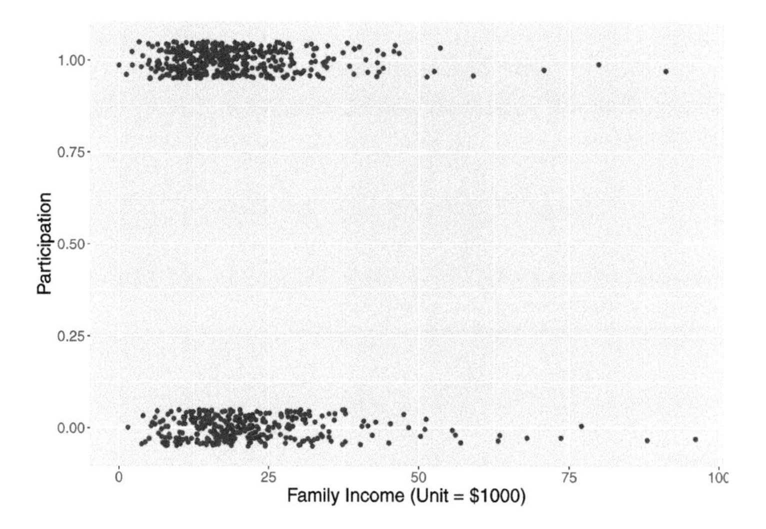

FIGURE 12.7
Scatterplot of the family income against the wife's labor participation. Since the participation value is binary, the points have been jittered in the vertical direction.

However, such a normal density setup is not sensible for this labor participation example. For a binary response Y_i, the mean is a probability μ_i that falls in the interval from 0 to 1. Thus the model $\mu_i = \beta_0 + \beta_1 x_i$ is not sensible since the linear component $\beta_0 + \beta_1 x_i$ is on the real line, not in the interval [0, 1].

In the upcoming subsections, it is described how to construct a regression model for binary responses using a linear function. In addition, this section describes how to interpret regression coefficients, how to specify prior distributions for these coefficients, and simulate posterior samples for these models.

12.4.2 A logistic regression model

Recall in Chapter 1 and Chapter 7, the definition of odds was introduced – an odds is the ratio of the probability of some event will take place over the probability that the event will not take place. The notion of odds will be used in how one represents the probability of the response in the regression model.

In the PSID example, let p_i be the probability of labor participation of married woman i, and the corresponding odds of participation is $\frac{p_i}{1-p_i}$. The probability p_i falls in the interval [0, 1] and the odds is a positive real number. If one applies the logarithm transformation on the odds, one obtains a quantity, called a log odds or logit, that can take both negative and positive values on the real line. One obtains a linear regression model for a binary response by writing the logit in terms of the linear predictor.

The binary response Y_i is assumed to have a Bernoulli distribution with probability of success p_i.

$$Y_i \mid p_i \overset{ind}{\sim} \text{Bernoulli}(p_i), \;\; i = 1, \cdots, n. \tag{12.11}$$

The logistic regression model writes that the logit of the probability p_i is a linear function of the predictor variable x_i:

$$\text{logit}(p_i) = \log\left(\frac{p_i}{1 - p_i}\right) = \beta_0 + \beta_1 x_i. \tag{12.12}$$

It is more challenging to interpret the regression coefficients in a logistic model. In simple linear regression with one predictor, the interpretation of the intercept and the slope is relatively straightforward, as the linear function is directly assigned to the mean μ_i. With the logit function as in Equation (12.12), one sees the the regression coefficients β_0 and β_1 are directly related to the log odds $\log\left(\frac{p_i}{1-p_i}\right)$ instead of p_i.

For example, the intercept β_0 is the log odds $\log\left(\frac{p_i}{1-p_i}\right)$ for observation i when the predictor takes a value of 0. In the PSID example, it refers to the log odds of labor participation of a married woman, whose family has 0 family income exclusive of her income.

The slope β_1 refers to the change in the expected log odds of labor participation of a married woman who has an additional \$1000 family income exclusive of her own income.

By rearranging the logistic regression Equation (12.12), one expresses the regression as a nonlinear equation for the probability of success p_i:

$$\begin{aligned} \log\left(\frac{p_i}{1 - p_i}\right) &= \beta_0 + \beta_1 x_i \\ \frac{p_i}{1 - p_i} &= \exp(\beta_0 + \beta_1 x_i) \\ p_i &= \frac{\exp(\beta_0 + \beta_1 x_i)}{1 + \exp(\beta_0 + \beta_1 x_i)}. \end{aligned} \tag{12.13}$$

Equation (12.13) shows that the logit function guarantees that the probability p_i lies in the interval [0, 1].

With these building blocks, one proceeds to prior specification and MCMC posterior inference of this logistic regression model. Note that a prior distribution is needed for the set of regression coefficient parameters: (β_0, β_1). In the next subsections, a conditional means prior approach is explored in this prior construction and the subsequent MCMC inference.

12.4.3 Conditional means priors and inference through MCMC

A conditional means prior can be constructed in a straightforward manner for logistic regression with a single predictor. This type of prior was previously constructed in Chapter 11 for a normal regression problem in the gas bill example. A weakly informative prior can always be used when little prior information is available. In contrast, the conditional means prior allows the data analyst to incorporate useful prior information about the probabilities at particular observation values.

The task is to construct a prior on the vector of regression coefficients $\beta = (\beta_0, \beta_1)$. Since the linear component $\beta_0 + \beta_1 x$ is indirectly related to the probability p, it is generally difficult to think directly about plausible values of the intercept β_0 and slope β_1 and think about the relationship between these regression parameters. Instead of constructing a prior on β directly, a conditional means prior indirectly specifies a prior by constructing priors on the probability values p_1 and p_2 corresponding to two predictor values x_1^* and x_2^*. By assuming independence of one's beliefs about p_1^* and p_2^*, this implies a prior on the probability vector (p_1^*, p_2^*). Since the regression coefficients β_0 and β_1 are functions of the probability values, this process essentially specifies a prior on the vector β.

A conditional means prior

To construct a conditional means prior, one considers two values of the predictor x_1^* and x_2^* and constructs independent beta priors for the corresponding probabilities of success.

1. For the first predictor value x_1^*, construct a beta prior for the probability p_1^* with shape parameters a_1 and b_1.

2. Similarly, for the second predictor value x_2^*, construct a beta prior for the probability p_2^* with shape parameters a_2 and b_2.

If one's beliefs about the probabilities p_1^* and p_2^* are independent, the joint prior for the vector (p_1^*, p_2^*) has the form

$$\pi(p_1^*, p_2^*) = \pi(p_1^*)\pi(p_2^*).$$

The prior on (p_1^*, p_2^*) implies a prior on the regression coefficient vector (β_0, β_1). First write the two conditional probabilities p_1^* and p_2^* as function of the regression coefficient parameters β_0 and β_1, as in Equation (12.13). By solving these two equations for the regression coefficient parameters, one expresses each regression parameter as a function of the conditional probabilities.

$$\beta_1 = \frac{\text{logit}(p_1^*) - \text{logit}(p_2^*)}{x_1^* - x_2^*}, \tag{12.14}$$

$$\beta_0 = \log\left(\frac{p_1^*}{1 - p_1^*}\right) - \beta_1 x_1^*. \tag{12.15}$$

Let's illustrate constructing a conditional means prior for our example. Consider two different family incomes (exclusive of the wife's income), say \$20,000 and \$80,000 (predictor is in \$1000 units). For each family income, a beta prior is constructed for the probability of the wife's labor participation. As in Chapter 7, a beta prior is assessed by specifying two quantiles of the prior distribution and finding the values of the shape parameters that match those specific quantile values.

- Consider the labor participation probability p_1^* for the value $x = 20$, corresponding to a \$20,000 family income. Suppose one believes the median of this probability is 0.10 and the 90th percentile is equal to 0.2. Using the R function `beta_select()` this belief is matched to a beta prior with shape parameters 2.52 and 20.08.

- Next, consider the participation probability p_2^* for the value $x = 80$, corresponding to a \$80,000 family income. The median and 90th percentile of this probability are thought to be 0.7 and 0.8, respectively, and this information is matched to a beta prior with shape parameters 20.59 and 9.01.

Figure 12.8 illustrates the conditional means prior for this example. Each bar displays the 90% interval estimate for the participation probability for a particular value of the family income.

Assuming independence of the prior beliefs about the two probabilities, one represents the joint prior density function for (p_1^*, p_2^*) as the product of densities

$$\pi(p_1^*, p_2^*) = \pi_B(p_1^*, 2.52, 20.08)\pi_B(p_2^*, 20.59, 9.01), \tag{12.16}$$

where $\pi_B(y, a, b)$ denotes the beta density with shape parameters a and b.

As said earlier, this prior distribution on the two probabilities implies a prior distribution on the regression coefficients. To simulate pairs (β_0, β_1) from the prior distribution, one simulates values of the means p_1^* and p_2^* from independent beta distributions in Equation (12.16), and applies the expressions in Equation (12.14) and Equation (12.15). One then obtains prior draws of the regression coefficient pair (β_0, β_1). Figure 12.9 displays a scatterplot of the simulated pairs (β_0, β_1) from the prior. Note that, although the two probabilities p_1^* and p_2^* have independent priors, the implied prior on the regression coefficient vector β indicates strong negative dependence between the intercept β_0 and the slope β_1.

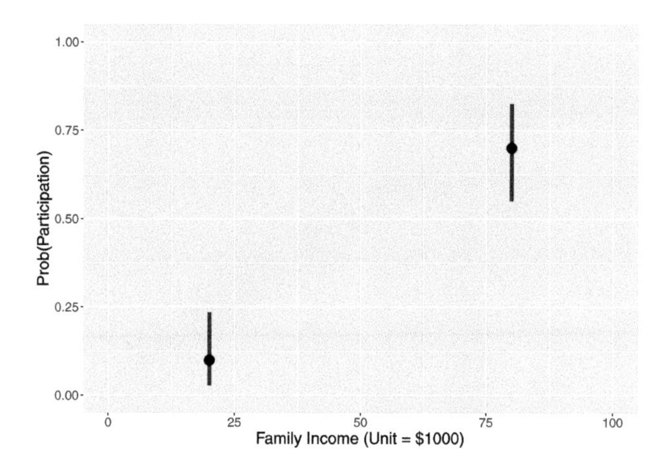

FIGURE 12.8
Illustration of the conditional means prior. Each line represents the limits of a 90% interval for the prior for the probability of participation for a specific family income value.

Inference using MCMC

Once the prior on the regression coefficients is defined, it is straightforward to simulate from the Bayesian logistic model by MCMC and the JAGS software.

The JAGS script

R As usual, the first step in using JAGS is writing a script defining the logistic regression model, and saving the script in the character string `modelString`.

```
modelString <-"
model {
## sampling
for (i in 1:N){
   y[i] ~ dbern(p[i])
   logit(p[i]) <- beta0 + beta1*x[i]
}
## priors
beta1 <- (logit(p1) - logit(p2)) / (x1 - x2)
beta0 <- logit(p1) - beta1 * x1
p1 ~ dbeta(a1, b1)
p2 ~ dbeta(a2, b2)
}
"
```

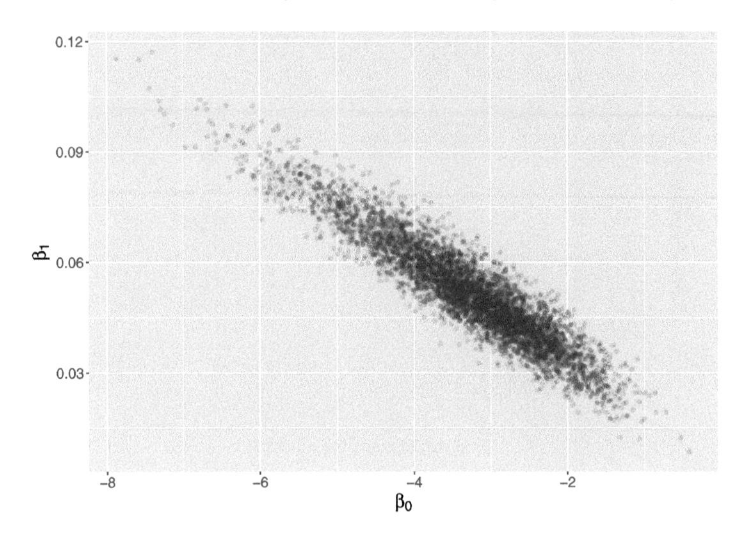

FIGURE 12.9
Scatterplot of simulated draws of the regression parameters for the conditional means prior for the logistic model.

In the sampling section of the script, the loop goes from `1` to `N`, where `N` is the number of observations with index `i`. Since $Y_i \mid p_i \overset{ind}{\sim} \text{Bernoulli}(p_i)$, one uses `dbern()` for `y[i]`. In addition, since $\text{logit}(p_i) = \beta_0 + \beta_1 x_i$, `logit()` is written for establishing this linear relationship.

In the prior section of the script, one expresses `beta0` and `beta1` according to the expressions in Equation (12.14) and Equation (12.15), in terms of `p1`, `p2`, `x1`, and `x2`. One also assign beta priors to `p1` and `p2`, according to the conditional means prior discussed previously. Recall that the beta distribution is represented by `dbeta()` in the JAGS code where the arguments are the associated shape parameters.

Define the data and prior parameters

The next step is to provide the observed data and the values for the prior parameters. In the R script below, a list `the_data` contains the vector of binary labor participation status values, the vector of family incomes (in \$1000), and the number of observations. It also contains the shape parameters for the beta priors on p_1^* and p_2^* and the values of the two incomes, x_1^* and x_2^*.

```
y <- as.vector(LaborParticipation$Participation)
x <- as.vector(LaborParticipation$FamilyIncome)
N <- length(y)
the_data <- list("y" = y, "x" = x, "N" = N,
```

```
                "a1" = 2.52, "b1" = 20.08,
                "a2" = 20.59, "b2" = 9.01,
                "x1" = 20, "x2" = 80)
```

Generate samples from the posterior distribution

The `run.jags()` function in the `runjags` package generates posterior samples by the MCMC algorithm using the JAGS software. The script below runs one MCMC chain with an adaption period of 1000 iterations, a burn-in period of 5000 iterations, and an additional set of 5000 iterations to be simulated. By using the argument `monitor = c("beta0", "beta1")`, one keeps tracks of the two regression coefficient parameters. The output variable `posterior` contains a matrix of simulated draws.

```
posterior <- run.jags(modelString,
                      n.chains = 1,
                      data = the_data,
                      monitor = c("beta0", "beta1"),
                      adapt = 1000,
                      burnin = 5000,
                      sample = 5000)
```

MCMC diagnostics and summarization

Once the simulated values are found, one applies several diagnostic procedures to check if the simulations appear to converge to the posterior distribution. Figures 12.10 and 12.11 display MCMC diagnostic plots for the regression parameters β_0 and β_1. From viewing these graphs, it appears that there is a small amount of autocorrelation in the simulated draws and the draws appear to have converged to the posterior distributions.

By use of the `print()` function, posterior summaries are displayed for the regression parameters. One primary question is whether the family income is predictive of the labor participation status and so the key parameter of interest is the regression slope β_1. From the output, one sees that the posterior median for β_1 is -0.0052 and a 90% interval estimate is $(-0.0143, 0.0029)$. This tells us several things. First, since the regression slope is negative, there is a negative relationship between family income and labor participation – wives from families with larger income (exclusive of the wife's income) tend not to work. Second, this relationship does not appear to be strong since the value 0 is included in the 90% interval estimate.

```
print(posterior, digits = 3)
      Lower95   Median Upper95     Mean      SD Mode     MCerr
beta0   0.101    0.358    0.59     0.36   0.125   --   0.00214
```

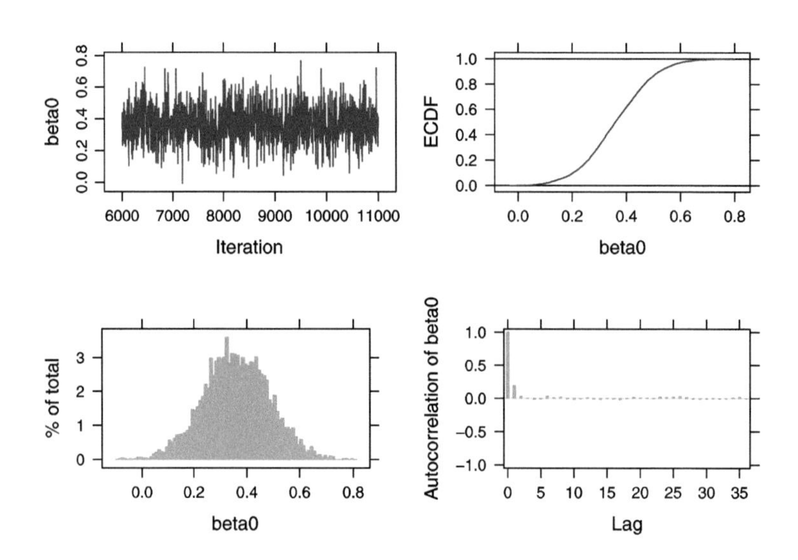

FIGURE 12.10
MCMC diagnostics plots for the logistic regression intercept parameter β_0.

```
beta1 -0.0143 -0.00524 0.00285 -0.00532 0.00438   -- 7.69e-05
```

One difficulty in interpreting a logistic regression model is that the linear component $\beta_0 + \beta_1 x$ is on the logit scale. It is easier to understand the fitted model when one expresses the model in terms of the probability of participation p_i:

$$p_i = \frac{\exp(\beta_0 + \beta_1 x_i)}{1 + \exp(\beta_0 + \beta_1 x_i)}. \tag{12.17}$$

For a specific value of the predictor x_i, it is straightforward to simulate the posterior distribution of the probability p_i. If $(\beta_0^{(s)}, \beta_1^{(s)})$ represents a simulated draw from the posterior of β, and one computes $p_i^{(s)}$ using Equation (12.13) from the simulated draw, then $p_i^{(s)}$ is a simulated draw from the posterior of p_i.

This process was used to obtain simulated samples from the posterior distribution of the probability p_i for the income variable values 10, 20, ..., 70. In Figure 12.12 the posterior medians of the probabilities p_i are displayed as a line graph and 90% posterior interval estimates are shown as vertical bars. The takeaway message from this figure is that the probability of labor participation is close to one-half and this probability slightly decreases as the family income increases. Also note that the length of the posterior interval

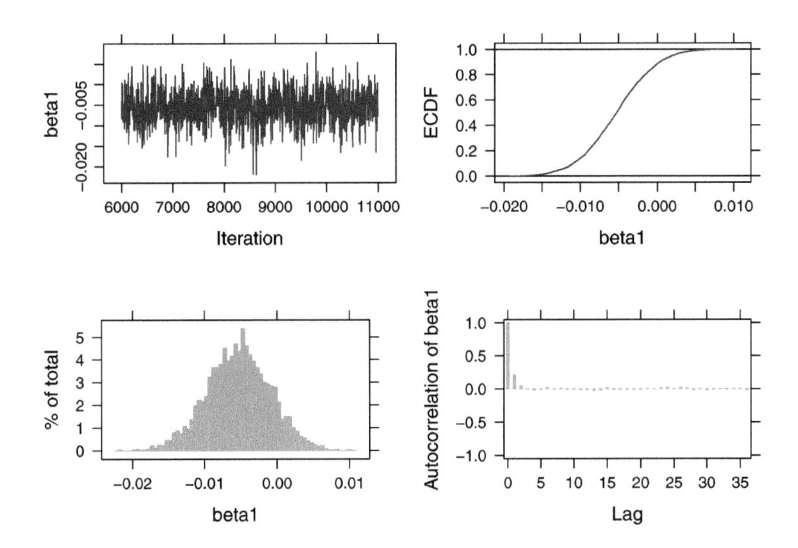

FIGURE 12.11
MCMC diagnostics plots for the logistic regression intercept parameter β_1.

estimate increases for larger family incomes – this is expected since much of the data is for small income values.

12.4.4 Prediction

We have considered learning about the probability p_i of labor participation for a specific income value x_i^*. A related problem is to predict the fraction of labor participation for a sample of n women with a specific family income. If $\tilde{y}_i$ represents the number of women who work among a sample of n with family income x_i, then one would be interested in the posterior predictive distribution of the fraction $\tilde{y}_i/n$.

One represents this predictive density of $\tilde{y}_i$ as

$$f(\tilde{Y}_i = \tilde{y}_i \mid y) = \int \pi(\beta \mid y) f(\tilde{y}_i, \beta) d\beta, \tag{12.18}$$

where $\pi(\beta \mid y)$ is the posterior density of $\beta = (\beta_0, \beta_1)$ and $f(\tilde{y}_i, \beta)$ is the binomial sampling density of $\tilde{y}_i$ conditional on the regression vector β.

A strategy for simulating the predictive density is implemented similar to what was done in the linear regression setting. Suppose that one focuses on the predictor value x_i^* and one wishes to consider a future sample of $n = 50$ of women with that income level. The simulated draws from the posterior distribution of β are stored in a matrix `post`. For each of the simulated parameter

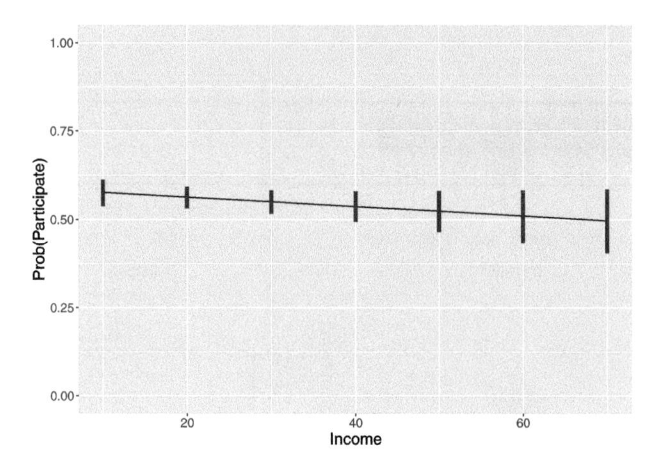

FIGURE 12.12
Posterior interval estimates for the probability of labor participation for seven values of the income variable.

draws, one computes the probability of labor participation $p^{(s)}$ for that income level – these values represent posterior draws of the probability $\{p^{(s)}\}$. Given those probability values, one simulates binomial samples of size $n = 50$ where the probabilities of success are given by the simulated $\{p^{(s)}\}$ – the variable $\tilde{y}$ represents the simulated binomial variable. By dividing $\tilde{y}$ by n, one obtains simulated proportions of labor participation for that income level. Each group of simulated draws from the predictive distribution of the labor proportion is summarized by the median, 5th, and 95th percentiles.

In the following R script, the function `prediction_interval()` obtains the quantiles of the prediction distribution of $\tilde{y}/n$ for a fixed income level, and the `sapply()` function computes these predictive quantities for a range of income levels. Figure 12.13 graphs the predictive median and interval bounds against the income variable. By comparing Figure 12.12 and Figure 12.13, note that one is much more certain about the probability of labor participation than the fraction of labor participation in a future sample of 50.

```
prediction_interval <- function(x, post, n = 20){
    lp <- post[, 1] + x * post[, 2]
    p <- exp(lp) / (1 + exp(lp))
    y <- rbinom(length(p), size = n, prob = p)
    quantile(y / n,
             c(.05, .50, .95))
  }
out <- sapply(seq(10, 70, by = 10),
                 prediction_interval, post, n = 50)
```

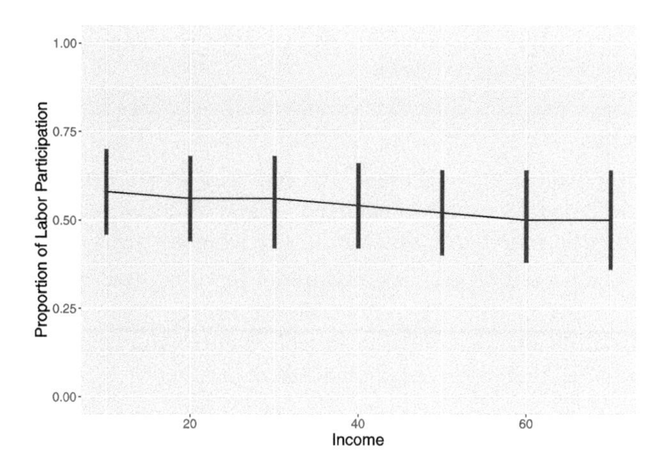

FIGURE 12.13
Prediction intervals for the fraction of labor participation of a sample of size $n = 50$ for seven values of the income variable.

12.5 Exercises

1. **Olympic Swimming Times**

   The dataset `olympic_butterfly.csv` contains the winning time in seconds for the men's and women's 100 m butterfly race for the Olympics from 1964 through 2016. Let y_i and x_i denote the winning time and year for the i-th Olympics. In addition, let w_i denote an indicator variable that is 1 for the women's race and 0 for the men's race. Consider the regression model $Y_i \sim \text{Normal}(\mu_i, \sigma)$, where the mean is given by

   $$\mu_i = \beta_0 + \beta_1(x_i - 1964) + \beta_2 w_i.$$

   (a) Interpret the parameter β_0 in terms of the winning time in the race.
   (b) Interpret the parameter $\beta_0 + \beta_2$.
   (c) Interpret the parameter $\beta_0 + 8\beta_1$.
   (d) Interpret the parameter $\beta_0 + 8\beta_1 + \beta_2$.

2. **Olympic Swimming Times (continued)**

   Consider the regression model for the 100 m butterfly race times described in Exercise 1. Suppose the regression parameters $\beta_0, \beta_1, \beta_2$ and the precision parameter $\phi = 1/\sigma^2$ are assigned weakly informative priors.

(a) Using JAGS, sample 5000 draws from the joint posterior distribution of all parameters.

(b) Construct 90% interval estimates for each of the regression coefficients.

(c) Based on your work, describe how the mean winning time in the butterfly race has changed over time. In addition, describe how the men times differ from the women times.

(d) Construct 90% interval estimates for the predictive residuals $r_i = y_i - \tilde{y}_i$ where $\tilde{y}_i$ is simulated from the posterior predictive distribution. Plot these interval estimates and comment on any interval that does not include zero.

3. **Olympic Swimming Times (continued)**

For the 100 m butterfly race times described in Exercise 1 consider the regression model where the mean race time has the form

$$\mu_i = \beta_0 + \beta_1(x_i - 1964) + \beta_2 w_i + \beta_3(x_i - 1964)w_i,$$

where x_i denotes the year for the i-th Olympics and w_i denote an indicator variable that is 1 for the women's race and 0 for the men's race.

(a) Write the expression for the mean time for the men's race, and for the mean time for the women's race. Using this expressions, interpret the parameters β_2 and β_3.

(b) Using weakly informative priors for all parameters, use JAGS to draw a sample of 5000 draws from the joint posterior distribution.

(c) Based on your work, is there evidence that the regression model between year and mean race time differs between men and women?

4. **Prices of Personal Computers**

What factors determine the price of a personal computer in the early days? A sample of 500 personal computer sales was collected from 1993 to 1995 in the United States. In addition to the sale price (price in U.S. dollars of 486 PCs), information on clock speed in MHz, size of hard drive in MB, size of RAM in MB, and name of the manufacturer (e.g. IBM, COMPAQ) was collected. The dataset is in `ComputerPriceSample.csv`. Suppose one considers the regression model $Y_i \sim \text{Normal}(\mu_i, \sigma)$ where

$$\mu_i = \beta_0 + \beta_1 x_{i1} + \beta_2 x_{i2},$$

y_i is the sale price, x_{i1} is the clock speed, and x_{2i} is the logarithm of the hard drive size.

(a) Using a weakly informative prior on $\beta = (\beta_0, \beta_1, \beta_2)$ and σ, use JAGS to produce a simulated sample of size 5000 from the posterior distribution on (β, σ).

(b) Obtain 95% interval estimates for β_1 and β_2.

(c) On the basis of your work, are both clock speed and hard drive size useful predictors of the sale price?

5. **Prices of Personal Computers (continued)**

(a) Suppose a consumer is interested in a computer with a clock speed of 33 MHz and a 540 MB hard drive (so $\log 450 = 6.1$). Simulate 5000 draws from the expected selling price $\beta_0 + \beta_1 x_1 + \beta_2$ for computer with this clock speed and hard drive size. Construct a 90% interval estimate for the expected sale price.

(b) Instead suppose the consumer wishes to predict the selling price of a computer with this clock speed and hard drive size. Simulate 5000 draws from the posterior predictive distribution and use these simulated draws to find a 90% prediction interval.

6. **Salaries for Professors**

A sample contains the 2008-09 nine-month academic salary for Assistant Professors, Associate Professors and Professors in a college in the U.S. The data were collected as part of the on-going effort of the college's administration to monitor salary differences between male and female faculty members. In addition to the nine-month salary (in U.S. dollars), information on gender, rank (Assistant Professor, Associate Professor, Professor), discipline (A is "theoretical" and B is "applied"), years since PhD, and years of service were collected. The dataset is in `ProfessorSalary.csv`. Suppose that the salary of the i-th professor, y_i, is distributed normal with mean μ_i and standard deviation σ, where the mean is given by

$$\mu_i = \beta_0 + \beta_1 x_{i1} + \beta_2 x_{i2},$$

where x_{i1} is the years of service and x_{i2} is the gender (where 1 corresponds to male and 0 to female).

(a) Assuming a weakly informative prior on β and σ, use JAGS to simulate a sample of 5000 draws from the posterior distribution on (β, σ).

(b) Simulate 1000 draws from the posterior of $\beta_0 + 10\beta_1$, the mean salary among all female professors with 10 years of service.

(c) Simulate 1000 draws from the posterior of the mean salary of male professors with 10 years of service $\beta_0 + 10\beta_1 + \beta_2$.

(d) By comparing the intervals computed in parts (b) and (c), is there a substantial difference in the mean salaries of male and female professors with 10 years of service?

7. **Salaries for Professors (continued)**

   (a) Suppose the college is interested in predicting the salary of a female professor with 10 years of service. By simulating 5000 draws from the posterior predictive distribution, construct a 90% prediction interval for this salary.

   (b) Use a similar method to obtain a 90% prediction interval for the salary of a male professor with 10 years of service.

8. **Graduate School Admission**

   What factors determine admission to graduate school? In a study, data on 400 graduate school admission cases was collected. Admission is a binary response, with 0 indicating not admitted, and 1 indicating admitted. Moreover, the applicant's GRE score, and undergraduate grade point average (GPA) are available. The dataset is in `GradSchoolAdmission.csv` (GRE score is out of 800). Let p_i denote the probability that the i-th student is admitted. Consider the logistic model

$$\log\left(\frac{p_i}{1-p_i}\right) = \beta_0 + \beta_1 x_{i1} + \beta_2 x_{i2},$$

   where x_{1i} and x_{2i} are respectively the GRE score and the GPA for the i-th student.

   (a) Assuming weakly informative priors on β_0, β_1, and β_2, write a JAGS script defining the Bayesian model.

   (b) Take a sample of 5000 draws from the posterior distribution of $\beta = (\beta_0, \beta_1, \beta_2)$.

   (c) Consider a student with a 550 GRE score and a GPA of 3.50. Construct a 90% interval estimate for the probability that this student is admitted to graduate school.

   (d) Construct a 90% interval estimate for the probability a student with a 500 GRE score and a 3.2 GPA is admitted to graduate school.

9. **Graduate School Admission (continued)**

   Consider the logistic model described in Exercise 8 where the logit probability of being admitted to graduate school is a linear function of GRE score and GPA. It is assumed that JAGS is used to obtain a simulated sample from the posterior distribution of the regression vector.

   (a) Consider a student with a 580 GRE score. Construct 90% posterior interval estimates for the probability that this student achieves admission for GPA values equally spaced from 3.0 to 3.8. Graph these posterior interval estimates as a function of the GPA.

(b) Consider a student with a 3.4 GPA. Find 90% interval estimates for the probability this student is admitted for GRE score values equally spaced from 520 to 700. Graph these interval estimates as a function of the GRE score.

10. **Personality Determinants of Volunteering**

In a study of the personality determinants of volunteering for psychological research, a subject's neuroticism (scale from Eysenck personality inventory), extraversion (scale from Eysenck personality inventory), gender, and volunteering status were collected. One intends to find out what personality determinants affect a person's volunteering choice. The dataset is in `Cowles.csv`. Let p_i denote the probability that the i-th subject elects to volunteer. Consider the logistic model

$$\log\left(\frac{p_i}{1-p_i}\right) = \beta_0 + \beta_1 x_{i1} + \beta_2 x_{i2},$$

where x_{1i} and x_{2i} are respectively the neuroticism and extraversion measures for the i-th subject.

(a) Assuming weakly informative priors on β_0, β_1, and β_2, write a JAGS script defining the model and draw a sample of 5000 draws from the posterior distribution of $\beta = (\beta_0, \beta_1, \beta_2)$.

(b) By inspecting the locations of the posterior distributions of β_1 and β_2, which personality characteristic is most important in determining a person's volunteering choice?

(c) Let $O = p/(1-p)$ denote the odds of volunteering. Construct a 90% interval estimate for the odds a student with a neuroticism score of 12 and an extraversion score of 13 will elect to volunteer.

11. **The Divide by Four Rule**

Suppose one considers the logistic model $\log\left(\frac{p}{1-p}\right) = \beta_0 + \beta_1 x$. This model is rewritten as

$$p = \frac{\exp(\beta_0 + \beta_1 x)}{1 + \exp(\beta_0 + \beta_1 x)}.$$

(a) Show that the derivative of p with respect to x is written as

$$\frac{dp}{dx} = p(1-p)\beta_1.$$

(b) Suppose the probability is close to the value 0.5. Using part (a), what is the approximate derivative of p with respect to x in this region?

(c) Fill in the blank in the following sentence. In this logistic model, the quantity $\beta_1/4$ can be interpreted as the change in the ________ when x increases by one unit.

(d) Suppose one is interested in fitting the logistic model $\log \frac{p}{1-p} = \beta_0 + \beta_1 x$ where x is the number of study hours and p is the probability of passing an exam. One obtains the fitted model

$$\log \frac{\hat{p}}{1 - \hat{p}} = -1 + 0.2x.$$

Using your work in parts (b) and (c), what is the (approximate) change in the fitted pass probability if a student studies an additional hour for the exam?

12. **Football Field Goal Kicking**

The data file `football_field_goal.csv` contains data on field goal attempts for professional football kickers. Focus on the kickers who played during the 2015 season. Let y_i denote the response (success or failure) of a field goal attempt from x_i yards. One is interested in fitting the logistic model

$$\log \frac{p_i}{1 - p_i} = \beta_0 + \beta_1 x_i,$$

where p_i is the probability of a successful attempt.

(a) Using weakly informative priors on β_0 and β_1, use JAGS to take a simulated sample from the posterior distribution of (β_0, β_1).

(b) Suppose a kicker is attempting a field goal from 40 yards. Construct a 90% interval estimate for the probability of a success.

(c) Suppose instead that one is interested in estimating the yardage x^* where the probability of a success is equal to 0.8. First express the yardage x^* as a function of β_0 and β_1, and then find a 90% interval estimate for x^*.

(d) Suppose 50 field goals are attempted at a distance of 40 yards. Simulate from the posterior predictive distribution to construct a 90% interval estimate for the number of successful attempts.

13. **Predicting Baseball Batting Averages**

The data file `batting_2018.csv` contains batting data for every player in the 2018 Major League Baseball season. The variables `AB.x` and `H.x` in the dataset contain the number of at-bats (opportunities) and number of hits of each player in the first month of the baseball season. The variables `AB.y` and `H.y` in the dataset contain the at-bats and hits of each player for the remainder of the season.

Take a random sample of size 50 from `batting_2018.csv`. Suppose one is interested in predicting the players' batting averages

$H.y/AB.y$ for the remainder of the season. Consider the following three estimates:

- Individual Estimate: Use the player's first month batting average $H.x/AB.x$.
- Pooled Estimate: Use the pooled estimate $\sum H.x / \sum AB.x$.
- Compromise Estimate: Use the shrinkage estimate

$$\frac{AB.x}{AB.x+135}\frac{H.x}{AB.x}+\frac{135}{AB.x+135}\frac{\sum H.x}{\sum AB.x}.$$

For your sample, compute values of the individual, pooled, and compromise estimates. For each set of estimates, compute the sum of squared prediction errors, where the prediction error is defined to be the difference between the estimate and the batting average in the remainder of the season. Which estimate do you prefer? Why?

14. **Predicting Baseball Batting Averages (continued)**

In Exercise 13, for the i-th player in the sample of 50 one observes the number of hits y_i (variable `H.x`) distributed binomial with sample size n_i (variable `AB.x`) and probability of success p_i. Consider the logistic model

$$\log\left(\frac{p_i}{1-p_i}\right)=\gamma_i.$$

Use JAGS to simulate from the following three models:

(a) Individual Model: Assume the γ_i values are distinct and assign each parameter a weakly informative normal distribution.

(b) Pooled Model: Assume that $\gamma_1 = ... = \gamma_{50} = \gamma$ and assign the single γ parameter a weakly informative normal distribution.

(c) Partially Pooled Hierarchical Model: Assume that $\gamma_i \sim Normal(\mu, \tau)$ where μ and the precision $P = 1/\tau^2$ are assigned weakly informative distributions.

(d) Focus on a particular player corresponding to the index k. Contrast 90% interval for estimates for p_k using the individual, pooled, and partially pooled hierarchical models fit in parts (a), (b), and (c).

15. **Comparing Career Trajectory Models**

In Section 12.3, the Deviance Information Criterion (DIC) was used to compare four regression models for Mike Schmidt's career trajectory of home run rates. By fitting the model using JAGS and using the `extract.runjags()` function, find the DIC values for fitting the linear, cubic, and quartic models and compare your answers with the values in Table 12.2. For each model, assume that the regression parameters and the precision parameter have weakly informative priors.

16. **Comparing Models for the CE Sample Example**

For the Consumer Expenditure Survey (CE) example, the objective was to learn about a CU's expenditure based on the person's income and his or her urban or rural status. There are four possible regression models depending on the inclusion or exclusion of each predictor. Use JAGS to fit each of the possible models and compute the value of DIC. For each model, assume that the regression parameters and the precision parameter have weakly informative priors. By comparing the DIC values, decide on the most appropriate model and compare your results with the discussion in Section 12.2.

17. **Grades in a Calculus Class**

Suppose one is interested in how the grade in a calculus class depends on the grade in the prerequisite math course. One is interested in fitting the logistic model

$$\log\left(\frac{p_i}{1-p_i}\right) = \beta_0 + \beta_1 x_i,$$

where p_i is the probability of an A of the ith student and x_i represents the grade of the ith student in the previous math class (1 if an A was received, and 0 otherwise).

(a) Suppose one believes a Beta(12, 8) prior reflects the belief about the probability of an A for a student who has received an A in the previous math, and a Beta(5, 15) prior reflects the belief about the probability of an A for a student who has not received an A in the previous course. Use JAGS to simulate 1000 draws from the prior of (β_0, β_1).

(b) Data for 100 students is contained in the data file `calculus.grades.csv`. Use JAGS to simulate 5000 draws from the posterior of (β_0, β_1).

(c) Construct a 90% interval estimate for β_1. Is there evidence that the grade in the prerequisite math course is helpful in explaining the grade in the calculus class?

18. **Grades in a Calculus Class (continued)**

The traditional way of fitting the logistic model in Exercise 17 is by maximum likelihood. The variables `grade` and `prev.grade` contain the relevant variables in the data frame `calculus.grades`. The maximum likelihood is achieved by the function `glm` with the `family = binomial` option.

```
fit <- glm(grade ~ prev.grade, data = calculus.grades,
           family = binomial)
summary(fit)
```

Look at the estimates and associated standard errors of the regression coefficients and contrast these values with the posterior means and standard deviations from the informative prior Bayesian analysis in Exercise 17.

19. **Logistic Model to Compare Proportions**

    In Exercise 19 of Chapter 7, one was comparing proportions of science majors for two years at some liberal arts colleges. One can formulate this problem in terms of logistic regression. Let y_i denote the number of science majors out of a sample of n_i for the ith year. One assumes that y_i is distributed Binomial(n_i, p_i) where p_i satisfies the logistic model

    $$\log\left(\frac{p_i}{1-p_i}\right) = \beta_0 + \beta_1 x_i,$$

    where $x_i = 0$ for year 2005 and $x = 1$ for year 2015.

    (a) Assuming that β_0 and β_1 are independent with weakly informative priors, use JAGS to simulate a sample of 5000 from the posterior distribution. (In the JAGS script, the `dbin(p, n)` denotes the Binomial distribution with probability `p` and sample size `n`.)
    (b) Find a 90% interval estimate for β_1.
    (c) Use the result in (b) to describe how the proportion of science majors has changed (on the logit scale) from 2005 to 2015,

20. **Separation in Logistic Regression**

    Consider data in Table 12.4 that gives the number of class absences and the grade (1 for passing and 0 for failure) for ten students. If p_i denotes the probability the ith student passes the class, then consider the logistic model

    $$\log\left(\frac{p_i}{1-p_i}\right) = \beta_0 + \beta_1 x_i,$$

    where x_i is the number of absences.

TABLE 12.4
Number of absences and grades for ten students.

Student	Absences	Grade	Student	Absences	Grade
1	0	1	6	2	1
2	0	1	7	2	1
3	0	1	8	5	0
4	1	1	9	8	0
5	1	1	10	10	0

(a) Using the `glm()` function as shown in Exercise 18, find maximum likelihood estimates of β_0 and β_1.

(b) Comment on the output of implementing the `glm()` function. (The strange behavior is related to the problem of separation in logistic research.) Do some research on this topic and describe why one is observing this unusual behavior.

(c) By use of a weakly informative prior, use JAGS to simulate a sample of 5000 from the posterior distribution.

(d) Compute posterior means and standard deviations of β_0 and β_1 and compare your results with the traditional fit in part (a).

13

Case Studies

13.1 Introduction

This chapter provides several illustrations of Bayesian modeling that extend some of the models described in earlier chapters. Mosteller and Wallace (1963), in one of the early significant Bayesian applications, explore the frequencies of word use in the well-known *Federalist Papers* to determine the authorship between Alexander Hamilton and James Madison. Section 13.2 revisits this word use application. This study raises several interesting explorations such as determining a suitable sampling distribution and finding suitable ways of comparing the word use of several authors.

In sports, teams are very interested in learning about the pattern of increase and decrease in the performance of a player, commonly called a career trajectory. A baseball player is believed to reach a level of peak performance at age of 30, although this "peak age" may vary between players. Section 13.3 illustrates the use of a hierarchical model to simultaneously estimate the career trajectories for a group of baseball players using on-base percentage as the measure of performance.

Suppose a class is taking a multiple choice exam where there are two groups of students. Some students are well-prepared and are familiar with the exam content and other students have not studied and will essentially guess at the answers to the exam questions. Section 13.4 introduces a latent class model that assumes that the class consists of two groups of students with different success rates and the group identifications of the students are unknown. In the posterior analysis, one learns about the location of the two success rates and the group classifications of the students. Using this latent class framework, the *Federalist Papers* example is revisited and the frequencies of particular filler words is used to learn about the true author identity of some disputed authorship *Federalist Papers.*

13.2 Federalist Papers Study

13.2.1 Introduction

The *Federalist Papers* were a collection of articles written in the late 18th century by Alexander Hamilton, James Madison and John Jay to promote the ratification of the United States Constitution. Some of these papers are known to be written by Hamilton, other papers were clearly written by Madison, and the true authorship of some of the remaining papers has been in doubt.

In one of the early significant applied Bayesian papers, Mosteller and Wallace (1963) illustrate the use of Bayesian reasoning in solving the authorship problem. They focused on the frequencies of word counts. Since the topic of the article may influence the frequencies of words used, Mosteller and Wallace were careful to focus on counts of so-called filler words such as "an", "of", and "upon" that are not influenced by the topics of the articles.

In this case study, the use of different sampling distributions is described to model word counts in a group of *Federalist Papers.* The Poisson distribution is perhaps a natural choice for modeling a group of word counts, but it will be seen that the Poisson can not accommodate the spread of the distribution of word counts. This motivates the use of a negative binomial sampling distribution and this model will be used to compare rates of use of some filler words by Hamilton and Madison.

13.2.2 Data on word use

To begin our study, let's look at the occurrences of the word "can" in all of the *Federalist Papers* authored by Alexander Hamilton or James Madison. Table 13.1 shows the format of the data. For each paper, the total number of words, the number of occurrences of the word "can" and the rate of use of this word per 1000 words are recorded.

TABLE 13.1
Portion of the data table counting the number of words and occurrences of the word "can" in 74 *Federalist Papers.*

	Name	Total	Count	Rate	Authorship
1	Federalist No. 1	1622	3	1.85	Hamilton
2	Federalist No. 10	3008	4	1.33	Madison
3	Federalist No. 11	2511	5	1.99	Hamilton
4	Federalist No. 12	2171	2	0.92	Hamilton
5	Federalist No. 13	970	4	4.12	Hamilton
6	Federalist No. 14	2159	9	4.17	Madison

Figure 13.1 displays parallel jittered dotplots of the rates (per 1000 words) of "can" for the Madison and Hamilton papers. Note the substantial variability in the rates across papers. But it appears that this is a slight tendency for Hamilton to use this particular word more frequently than Madison. Later in this section we will formally perform inference about the ratio of the true rates of use of "can" for the two authors.

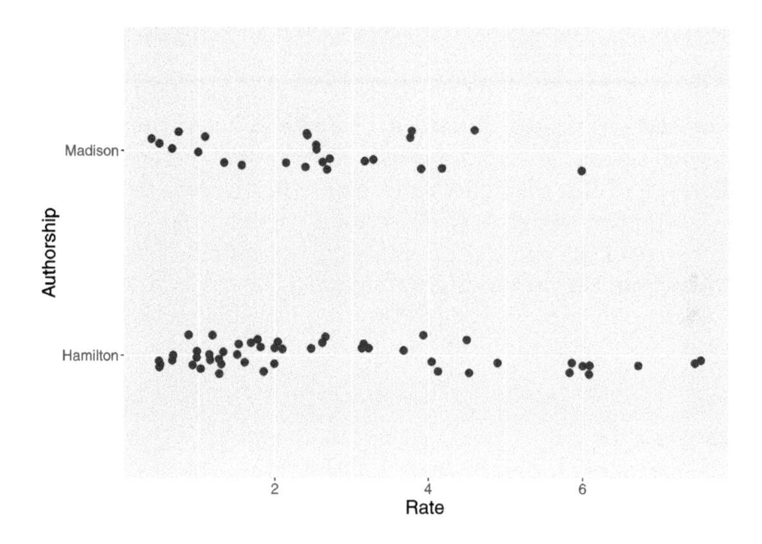

FIGURE 13.1
Observed rates of use of the word "can" in *Federalist Papers* authored by Hamilton and Madison.

13.2.3 Poisson density sampling

Consider first the word use of all of the *Federalist Papers* written by Hamilton. The initial task is to find a suitable sampling distribution for the counts of a particular function word such as "can". Since Poisson is a popular sampling distribution for counts, it is initially assumed that for the i-th paper the count y_i of the word "can" has a Poisson density with mean $n_i\lambda/1000$ where n_i is the total number of words and λ is the true rate of the word among 1000 words. There are N papers in total. Using the Poisson density expression, one writes

$$f(Y_i = y_i \mid \lambda) = \frac{(n_i\lambda/1000)^{y_i} \exp(-n_i\lambda/1000)}{y_i!}. \tag{13.1}$$

Assuming independence of word use between papers, the likelihood function is the product of Poisson densities

$$L(\lambda) = \prod_{i=1}^{N} f(y_i \mid \lambda), \tag{13.2}$$

and the posterior density of λ is given by

$$\pi(\lambda \mid y_1, \cdots, y_N) \propto L(\lambda)\pi(\lambda), \tag{13.3}$$

where $\pi()$ is the prior density.

R Suppose one knows little about the true rate of "can"s and to reflect this lack of information, one assigns λ a gamma density with parameters $\alpha = 0.001$ and $\beta = 0.001$. Recall in Section 8.8 in Chapter 8, a gamma prior is conjugate to a Poisson sampling model. A JAGS script is written to specify this Bayesian model and by use of the `run.jags()` function, one obtains a simulated sample of 5000 draws from the posterior distribution.

```
modelString = "
model{
## sampling
for (i in 1:N) {
   y[i] ~ dpois(n[i] * lambda / 1000)
}
## prior
lambda ~ dgamma(0.001, 0.001)
}
"
```

When one observes count data such as these, one general concern is *overdispersion*. Do the observed counts display more variability than one would anticipate with the use of this Poisson sampling model? One can check for overdispersion by use of a posterior predictive check. First one simulates one replicated dataset from the posterior predictive distribution. This is done in two steps: 1) one simulates a value of λ from the posterior distribution; 2) given the simulated value $\lambda = \lambda^*$, one simulates counts $y_1^R, ..., y_N^R$ from independent Poisson distribution with means $n_1\lambda^*/1000, ..., n_N\lambda^*/1000$. Given a replicated dataset of counts $\{y_i^R\}$, one computes the standard deviation. In this setting a standard deviation is a reasonable choice of a testing function since one is concerned about the variation or spread in the data.

```
one_rep <- function(i){
  lambda <- post[i]
  sd(rpois(length(y), n * lambda / 1000))
}
sapply(1:5000, one_rep) -> SD
```

One repeats this process 5000 times, obtaining 5000 replicated datasets from the posterior predictive distribution and 5000 values of the standard deviation. Figure 13.2 displays a histogram of the standard deviations from the predictive distribution and the standard deviation of the observed counts $\{y_i\}$ is displayed as a vertical line. Note that the observed standard deviation is very large relative to the standard deviations of the counts from the predictive distribution. The takeaway is that there is more variability in the observed counts of "can"s than one would predict from the Poisson sampling model.

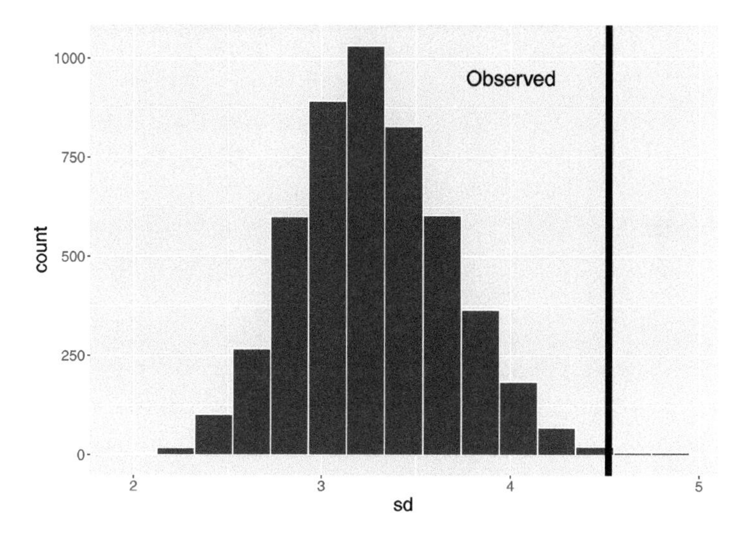

FIGURE 13.2
Histogram of standard deviations from 5000 replicates from the posterior predictive distribution from the Poisson sampling model. The observed standard deviation is displayed as a vertical line.

13.2.4 Negative binomial sampling

In the previous section, we presented evidence that the observed counts of "can" from a group of *Federalist Papers* of Alexander Hamilton were overdispersed in that there was more variability in the counts than predicted by the Poisson sampling model. One way of handling this overdispersion issue to find an alternative sampling density for the counts that is able to accommodate this additional variation.

One popular alternative density is the negative binomial density. Recall that y_i represents the number of "can"s in the i-th *Federalist Papers*. Conditional on parameters α and β, one assigns y_i the negative binomial density defined as

$$f(Y_i = y_i \mid \alpha, \beta) = \frac{\Gamma(y_i + \alpha)}{\Gamma(\alpha)} p_i^{\alpha} (1 - p_i)^{y_i}, \tag{13.4}$$

where

$$p_i = \frac{\beta}{\beta + n_i/1000}. \tag{13.5}$$

One can show that this density is a natural generalization of the Poisson density. The mean count is given by $E(y_i) = \mu_i$ where

$$\mu_i = \frac{n_i}{1000} \frac{\alpha}{\beta}. \tag{13.6}$$

Recall that the mean count for y_i the Poisson model was $n_i \lambda/1000$, so the ratio α/β is playing the same role as λ – one can regard α/β as the true rate of the particular word per 1000 words.

One can show that the variance of the count y_j is given by

$$Var(y_i) = \mu_i \left(1 + \frac{n_i}{1000\beta}\right). \tag{13.7}$$

The variance for the Poisson model is equal to μ_i, so the negative binomial model has the extra multiplicative term $\left(1 + \frac{n_i}{1000\beta}\right)$. So the negative binomial family is able to accommodate the additional variability in the counts $\{y_i\}$.

The posterior analysis using a negative binomial density is straightforward. The counts $y_1, ..., y_N$ are independent negative binomial with parameters α and β and the likelihood function is equal to

$$L(\alpha, \beta) = \prod_{i=1}^{N} f(y_i \mid \alpha, \beta). \tag{13.8}$$

If little is known a priori about the locations of the positive parameter values α and β, then it reasonable to assume the two parameters are independent and assign to each α and β a gamma density with parameters 0.001 and 0.001. Then the posterior density is given by

$$\pi(\alpha, \beta \mid y_1, \cdots, y_N) \propto L(\alpha, \beta)\pi(\alpha, \beta) \tag{13.9}$$

where $\pi(\alpha, \beta)$ is the product of gamma densities.

R One simulates the posterior with negative binomial sampling using JAGS. The negative binomial density is represented by the JAGS function `dnegbin()` with parameters `p[i]` and `alpha`. In the JAGS script below, note that one first defines `p[i]` in terms of the parameter `beta` and the sample size `n[i]`, and then expresses the negative binomial density in terms of `p[i]` and `alpha`.

```
modelString = "
model{
## sampling
for(i in 1:N){
   p[i] <- beta / (beta + n[i] / 1000)
   y[i] ~ dnegbin(p[i], alpha)
}
## priors
mu <- alpha / beta
alpha ~ dgamma(.001, .001)
beta ~ dgamma(.001, .001)
}
"
```

We earlier made a statement that the Negative Binomial density can accommodate the extra variability in the word counts. One can check this statement by a posterior predictive check. One replication of the posterior predictive checking method is implemented in the R function `one_rep()`. We start with a simulated value (α^*, β^*) from the posterior distribution. Then we simulated a replicated dataset $y_1^R, ..., y_N^R$ where y_i^R has a negative binomial distribution with parameters α^* and $\beta^*/(\beta^* + n_i/1000)$. Then we compute the standard deviation of the $\{y_i^R\}$.

```
one_rep <- function(i){
  p <- post$beta[i] / (post$beta[i] + n / 1000)
  sd(rnbinom(length(y), size = post$alpha[i], prob = p))
}
```

By repeating this algorithm for 5000 iterations, one has 5000 draws of the standard deviation of samples from the predictive distribution stored in the R vector `SD`.

```
sapply(1:5000, one_rep) -> SD
```

Figure 13.3 displays a histogram of the standard deviations of samples from the predictive distribution and the observed standard deviation of the counts is shown as a vertical line. In this case the observed standard deviation value is in the middle of the predictive distribution. The interpretation is that predictions with a negative binomial sampling model are consistent with the spread in the observed word counts.

Now that the negative binomial model seems reasonable, one performs inferences about the mean use of the word "can" in Hamilton essays. The parameter $\mu = \alpha/\beta$ represents the true rate of use of this word per 1000 words. Figure 13.4 displays MCMC diagnostic plots for the parameter μ. The trace plot and autocorrelation plot indicate good mixing and so one believes the

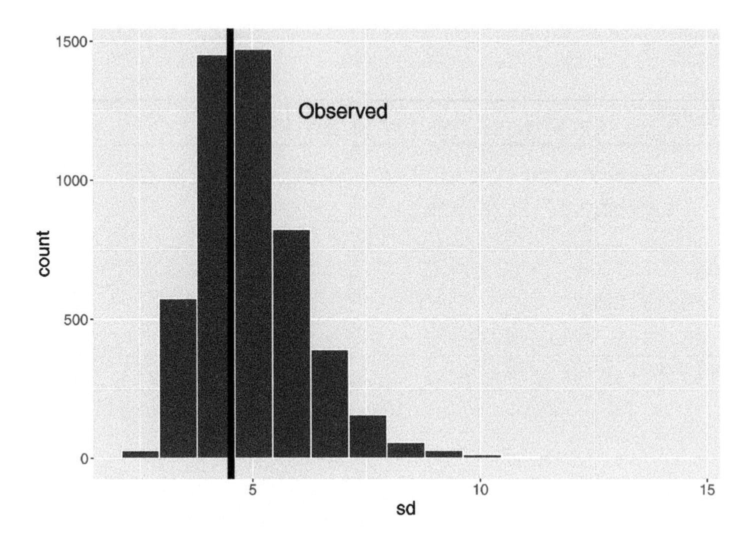

FIGURE 13.3
Histogram of standard deviations from 5000 replicates from the posterior predictive distribution in the negative binomial sampling model. The observed standard deviation is displayed as a vertical line.

histogram in the lower-left section represents the marginal posterior density for μ. A 90% posterior interval estimate for the rate of "can" is (2.20, 3.29).

13.2.5 Comparison of rates for two authors

Recall that the original problem was to compare the word use of Alexander Hamilton with that of James Madison. Suppose we collect the counts $\{y_{1i}\}$ of the word "can" in the *Federalist Papers* authored by Hamilton and the counts $\{y_{2i}\}$ of "can" in the *Federalist Papers* authored by Madison. The general problem is to compare the true rates per 1000 words of the two authors.

Since a negative binomial sampling model appears to be suitable in the one-sample situation, we extend this in a straightforward away to the two-sample case. The Hamilton counts $y_{11}, ..., y_{1N_1}$, conditional on parameters α_1 and β_1 are assumed to be independent negative binomial, where y_{1i} is negative binomial(p_{1i}, α_1) with

$$p_{1i} = \frac{\beta_1}{\beta_1 + n_{1i}/1000}, \tag{13.10}$$

and $\{n_{1i}\}$ are the word counts for the Hamilton essays. Similarly, the Madison counts $y_{21}, ..., y_{2N_2}$, conditional on parameters α_2 and β_2 are assumed to be

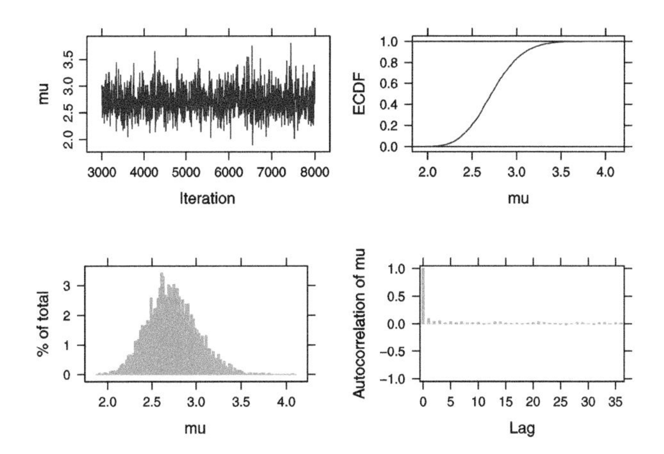

FIGURE 13.4
MCMC diagnostic plots for the rate $\mu = \alpha/\beta$ of use of the word "can" in Hamilton essays.

independent negative binomial, where y_{2i} is Negative Binomial(p_{2i}, α_2) with

$$p_{2i} = \frac{\beta_2}{\beta_2 + n_{2i}/1000}, \tag{13.11}$$

and $\{n_{2i}\}$ are the word counts for the Madison essays. The focus will be to learn about μ_M/μ_H, the ratio of the rates (per 1000 words) of use of the word "can" of the two authors, where $\mu_M = \alpha_2/\beta_2$ and $\mu_H = \alpha_1/\beta_1$.

Assume that the observed counts of word "can" of the two authors are independent. Moreover, assume that the prior distributions of the parameters (α_1, β_1) and (α_2, β_2) are independent. Then the posterior distribution is given, up to an unknown proportionality constant, by

$$\pi(\alpha_1, \beta_1, \alpha_2, \beta_2 \mid \{y_{1i}\}, \{y_{12}\}) \propto \prod_{k=1}^{2} \left(\prod_{i=1}^{n_{ki}} f(y_{ki} \mid \alpha_k, \beta_k) \pi(\alpha_k, \beta_k) \right). \tag{13.12}$$

We assume that the user has little prior information about the location of the negative binomial parameters and we assume they are independent with each parameter assigned a gamma prior with parameters 0.001 and 0.001.

R The posterior sampling is implemented using the JAGS software. The model description script is an extension of the previous script for a single negative binomial sample. Note that the `ratio` parameter is defined to be the ratio of the word rates for the two samples.

```
modelString = "
model{
## sampling
for(i in 1:N1){
   p1[i] <- beta1 / (beta1 + n1[i] / 1000)
   y1[i] ~ dnegbin(p1[i], alpha1)
}
for(i in 1:N2){
   p2[i] <- beta2 / (beta2 + n2[i] / 1000)
   y2[i] ~ dnegbin(p2[i], alpha2)
}
## priors
alpha1 ~ dgamma(.001, .001)
beta1 ~ dgamma(.001, .001)
alpha2 ~ dgamma(.001, .001)
beta2 ~ dgamma(.001, .001)
ratio <- (alpha2 / beta2) / (alpha1 / beta1)
}
"
```

Since the focus is to compare the word use of the two authors, Figure 13.5 displays MCMC diagnostics for the ratio of "can" rates $R = \mu_M/\mu_H$. Note that most of the posterior probability of R is found in an interval about the value one. From the simulated draws, one finds the posterior median is 0.92 and a 95% probability interval for R is found to be (0.71, 1.19). Since this interval contains the value one, there is no significant evidence to conclude that Hamilton and Madison have different rates of use of the word "can".

13.2.6 Which words distinguish the two authors?

In the previous section, it was found that the word "can" was not a helpful discriminator between the essays written by Hamilton and the essays written by Madison. However, other words may be useful in this discrimination task. Following suggestions in Mosteller and Wallace (1963), the previous two-sample analysis was repeated for each of the following words: also, an, any, by, can, from, his, may, of, on, there, this, to, and upon. For a given word, the counts of occurrence of that word were collected for each of the essays authored by Hamilton and Madison. For each word, we focus on inferences about the parameter R, the ratio of mean rates of the particular word by Madison and Hamilton. A ratio value of $R > 1$ indicates that Madison was a more frequent user of the word, and a ratio value $R < 1$ indicates that Hamilton used it more frequently. Fourteen separate two-sample analyses were conducted and the posterior distributions of R were summarized by posterior medians and 95% probability intervals.

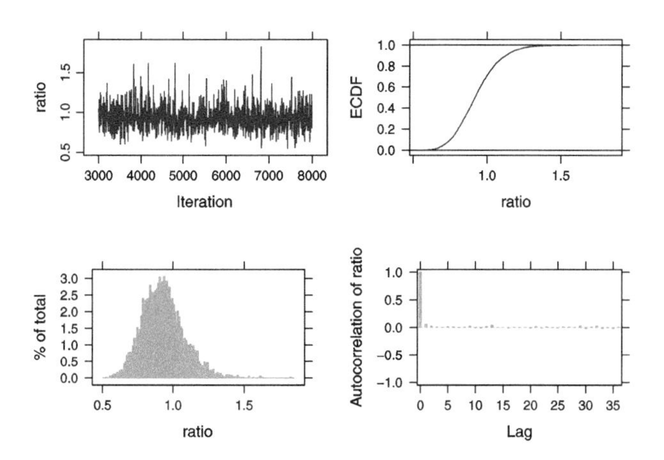

FIGURE 13.5
MCMC diagnostic plots for the ratio of rates μ_M/μ_H of use of the word "can" *Federalist Papers* essays written by Hamilton and Madison.

Figure 13.6 displays the locations of the posterior medians and interval estimates for all of the 14 analyses. Intervals that are completely on one side of the value $R = 1$ indicate that one author was more likely to use that particular word. Looking at the figure, one sees that the words "upon", "to", "this", "there", "any", and "an" were more likely be used by Hamilton, and the words "on", "by", and "also" were more likely be used by Madison. The posterior intervals for the remaining words ("may", "his", "from", "can", and "also") cover the value one, and so one cannot say from these data that one author was more likely to use those particular words.

13.3 Career Trajectories

13.3.1 Introduction

For an athlete in a professional sport, his or her performance typically begins at a small level, increases to a level in the middle of his or her career where the player has peak performance, and then decreases until the player's retirement. This pattern of performance over a player's career is called the *career trajectory.* A general problem in sports is to predict future performance of a player and one relevant variable in this prediction is the player's age. Due to the ready availability of baseball data, it is convenient to study career trajec-

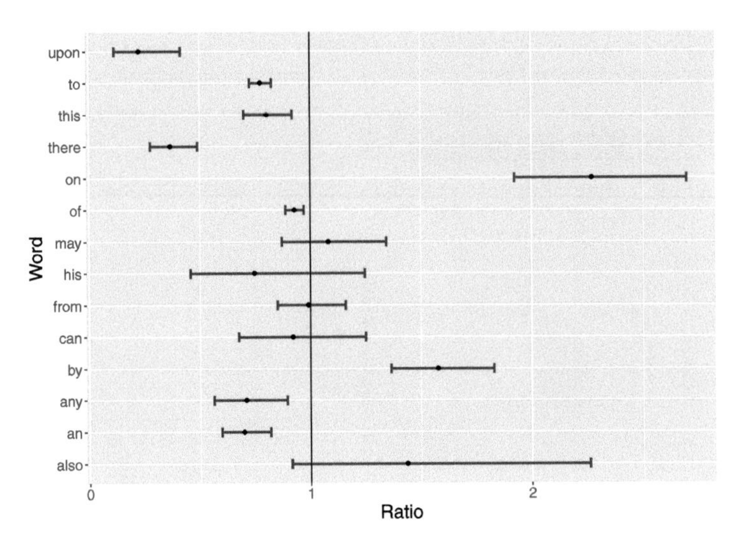

FIGURE 13.6
Display of posterior median and 95% interval estimates for the ratio of rates μ_H/μ_M for 14 different words in *Federalist Papers* essays written by Hamilton and Madison.

tories for baseball players, although the methodology will apply to athletes in other sports.

13.3.2 Measuring hitting performance in baseball

Baseball is a bat and ball game first played professionally in the United States in the mid 19th century. Players are measured by their ability to hit, pitch, and field, and a wide variety of statistical measures have been developed. One of the more popular measures of batting performance is the on-base percentage or OBP. A player comes to bat during a *plate appearance* and it is desirable for the batter to get on base. The OBP is defined to be the fraction of plate appearances where the batter reaches a base. As an example, during the 2003 season, Chase Utley had 49 on-base events in 152 plate appearances and his OBP was $49/152 = 0.322$.

13.3.3 A hitter's career trajectory

A baseball player typically plays between 5 to 20 years in Major League Baseball (MLB), the top-tier professional baseball league in the United States. In this case study, we explore career trajectories of the OBP measure of baseball players as a function of their ages. To illustrate a career trajectory, consider Chase Utley who played in the Major Leagues from 2003 through 2018. Figure

13.7 displays Utley's OBP as a function of his age for all of the seasons of his career. A quadratic smoothing curve is added to the scatterplot. One sees that Utley's OBP measure increases until about age 30 and then steadily decreases towards the end of his career.

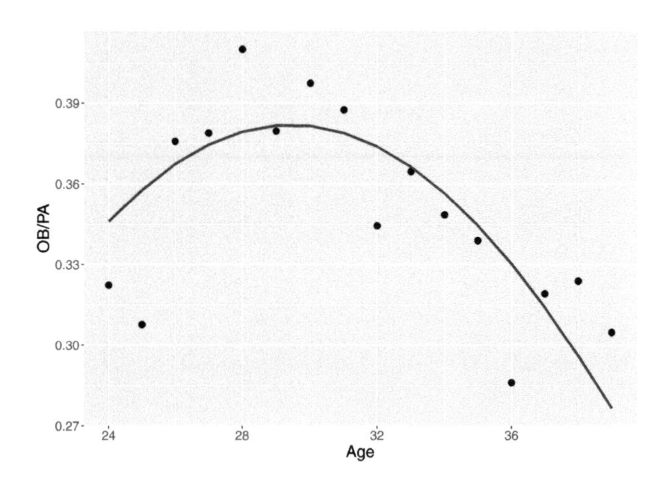

FIGURE 13.7
Career trajectory of Chase Utley's on-base percentages. A quadratic smoothing curve is added to the plot.

Figure 13.8 displays the career trajectory of OBP for another player Josh Phelps who had a relatively short baseball career. In contrast, Phelps does not have a clearly defined career trajectory. In fact, Phelps' OBP values appear to be relatively constant from ages 24 to 30 and the quadratic smoothing curve indicates that Phelps had a minimum OBP at age 26. The purpose of this case study is to see if one can improve the career trajectory smooth of this player by a hierarchical Bayesian model that combines data from a number of baseball players. Recall in Chapter 10, we have seen how hierarchical Bayesian models have the pooling effect that could borrow information from other groups to improve the estimation of one group, especially for groups with small sample size.

13.3.4 Estimating a single trajectory

First we consider learning about a single hitter's OBP career trajectory. Let y_j denote the number of on-base events in n_j plate appearances during a hitter's j-th season. It is reasonable to assume that y_j has a Binomial distribution with parameters n_j and probability of success p_j. One represents the logit of

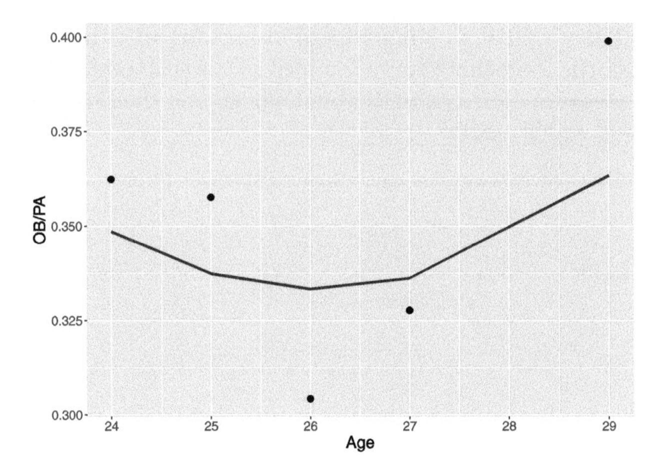

FIGURE 13.8
Career trajectory of Josh Phelps' on-base percentages. A quadratic smoothing curve is added to the plot.

the success probability as a quadratic function of the player's age:

$$\log\left(\frac{p_j}{1-p_j}\right) = \beta_0 + \beta_1(x_j - 30) + \beta_2(x_j - 30)^2, \qquad (13.13)$$

where x_j represents the age of the player in the j-th season.

Note that the age value is centered by 30 in the logistic model – this is done for ease of interpretation. The intercept β_0 is an estimate of the player's OBP performance at age 30. Specific functions of the regression vector $\beta = (\beta_0, \beta_1, \beta_2)$ are of specific interest in this application.

- The quadratic function reaches its largest value at

$$h_1(\beta) = 30 - \frac{\beta_1}{2\beta_2}.$$

  This is the age where the player is estimated to have his peak on-base performance during his career.

- The maximum value of the curve, on the logistic scale, is

$$h_2(\beta) = \beta_0 - \frac{\beta_1^2}{4\beta_2}.$$

  The maximum value of the curve on the probability scale is

$$p_{max} = \exp(h_2(\beta))/(1 + \exp(h_2(\beta))). \qquad (13.14)$$

  The parameter p_{max} is the estimated largest OBP of the player over his career.

- The coefficient β_2, typically a negative value, tells us about the degree of curvature in the quadratic function. If a player has a large value of β_2, this indicates that he more rapidly reaches his peak level and more rapidly decreases in ability until retirement. One simple interpretation is that β_2 represents the change in OBP from his peak age to one year later.

It is straightforward to fit this Bayesian logistic model using the JAGS software. Suppose one has little prior information about the location of the regression vector β. Then one assumes the regression coefficients are independent with each coefficient assigned a normal prior with mean 0 and precision 0.0001. The posterior density of β is given, up to an unknown proportionality constant, by

$$\pi(\beta \mid \{y_j\}) \propto \prod_j \left(p_j^{y_j}(1 - p_j)^{n_j - y_j}\right) \pi(\beta), \tag{13.15}$$

where p_j is defined by the logistic model and $\pi(\beta)$ is the prior density.

R The JAGS model script is shown below. The `dbin()` function is used to define the binomial distribution and the `logit()` function describes the log odds reexpression.

```
modelString = "
model {
## sampling
for (j in 1:N){
   y[j] ~ dbin(p[j], n[j])
   logit(p[j]) <- beta0 + beta1 * (x[j] - 30) +
               beta2 * (x[j] - 30) * (x[j] - 30)
}
## priors
beta0 ~ dnorm(0, 0.0001)
beta1 ~ dnorm(0, 0.0001)
beta2 ~ dnorm(0, 0.0001)
}
"
```

The JAGS software is used to simulate a sample from the posterior distribution of the regression vector β. From this sample, it is straightforward to learn about any function of the regression vector of interest. To illustrate, one performs inference about the peak age function $h_1(\beta)$ by computing this function on the simulated β draws – the output is a posterior sample from the peak age function. In a similar fashion, one obtains a sample from the posterior of the maximum value function p_{max} by computing this function on the simulated β values. Figure 13.9 displays density estimates of the simulated values of $h_1(\beta)$ and p_{max}. From this graph, one sees that Utley's peak performance was most likely achieved at age 29, although there is uncertainty about this most likely peak age. Also the posterior of the peak value p_{max} indicates that Utley's peak on-base probability ranged from 0.38 and 0.40.

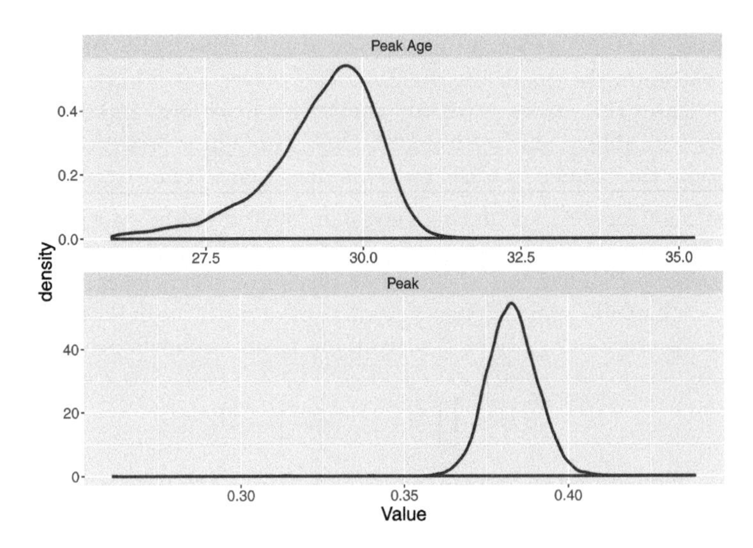

FIGURE 13.9
Density estimates of the peak age and peak for logistic model on Chase Utley's trajectory.

13.3.5 Estimating many trajectories by a hierarchical model

We have focused on estimating the career trajectory of a single baseball player such as Chase Utley. But there are many baseball players and it is reasonable to want to simultaneously estimate the career trajectories for a group of players. As an example, suppose one focuses on the Major League players who were born in the year 1978 and had at least 1000 career at-bats. Figure 13.10 displays scatterplots of age and OBP with quadratic smoothing curves for the 36 players in this group. Looking at these curves, one notices that many of the curves follow a familiar concave down shape with the player achieving peak performance near an age of 30. But for some players, especially for those players who played a small number of seasons, note that the trajectories have different shapes. Some trajectories are relatively constant over the age variable and other trajectories have an unusual concave up appearance.

In this situation, it may be desirable to partially pool the data from the 36 players using a hierarchical model to obtain improved trajectory estimates for all players. For the i-th player, one observes the on-base events $\{y_{ij}\}$ where y_{ij} is binomial with sample size n_{ij} and probability of on-base success p_{ij}. The logit of the on-base probability for the i-th player during the j-th season is given by

$$\log\left(\frac{p_{ij}}{1-p_{ij}}\right) = \beta_{i0} + \beta_{i1}(x_{ij} - 30) + \beta_{i2}(x_{ij} - 30)^2, \tag{13.16}$$

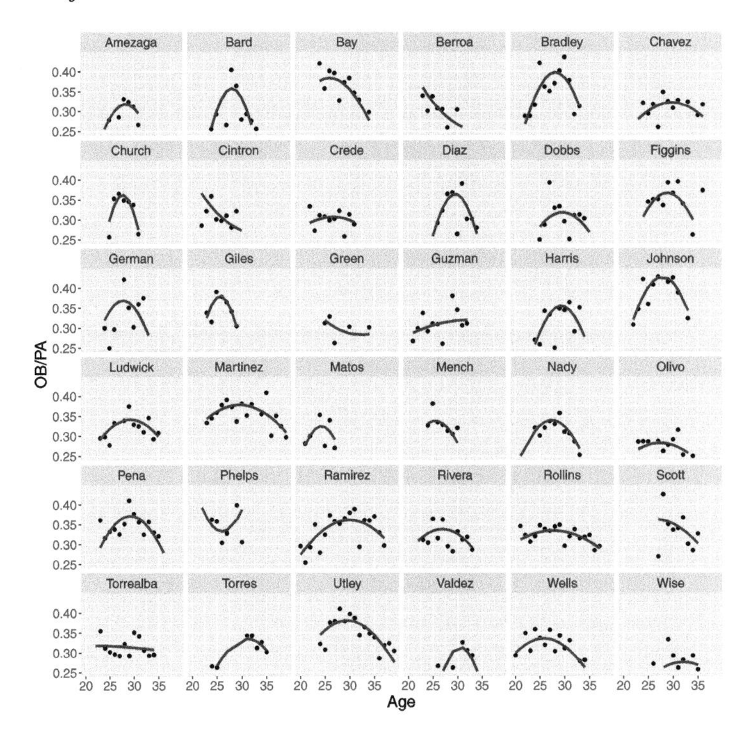

FIGURE 13.10
Career trajectories and individual quadratic fits for all players born in the year 1978 and having at least 1000 career at-bats.

where x_{ij} is the age of the i-th player during the j-th season. If $\beta_i = (\beta_{i0}, \beta_{i1}, \beta_{i2})$ represents the vector regression coefficients for the i-th player, then one is interested in estimating the regression vectors $(\beta_1, ..., \beta_N)$ for the N players in the study.

One constructs a two-stage prior on these regression vectors. In Chapter 10, one assumed that the normal means were distributed according to a common normal distribution. In this setting, since each regression vector has three components, at the first stage of the prior, one assumes that $\beta_1, ..., \beta_N$ are independently distributed from a common multivariate normal distribution with mean vector μ_β and precision matrix τ_β. Then, at the second stage, vague prior distributions are assigned to the unknown values of μ_β and τ_β.

R In our application, there are $N = 36$ players, so one is estimating $36 \times 3 = 108$ regression parameters together with unknown parameters in the prior distributions of μ_β and τ_β at the second stage. Fortunately the JAGS script defining this model is a straightforward extension of the JAGS script for a logistic regression model for a single career trajectory. The variable `player` indicates the player number, and the variables `beta0[i]`, `beta1[i]`, and `beta2[i]` represent the logistic regression parameters for the i-th player. The vector `B[j, 1:3]` represents a vector of parameters for one player and `mu.beta` and `Tau.B` represent respectively the second-stage prior mean vector and precision matrix values. The variables `mean`, `prec`, `Omega` are specified parameters that indicate weak information about the parameters at the second stage.

```
modelString = "
model {
## sampling
for (i in 1:N){
   y[i] ~ dbin(p[i], n[i])
   logit(p[i]) <- beta0[player[i]] +
                  beta1[player[i]] * (x[i] - 30) +
                  beta2[player[i]] * (x[i] - 30) * (x[i] - 30)
}
## priors
for (j in 1:J){
   beta0[j] <- B[j,1]
   beta1[j] <- B[j,2]
   beta2[j] <- B[j,3]
   B[j,1:3] ~ dmnorm (mu.beta[], Tau.B[,])
}
mu.beta[1:3] ~ dmnorm(mean[1:3],prec[1:3 ,1:3 ])
Tau.B[1:3 , 1:3] ~ dwish(Omega[1:3 ,1:3 ], 3)
}
"
```

After JAGS is used to simulate from the posterior distribution of this hierarchical model, a variety of inferences are possible. The player trajectories $\beta_1, ..., \beta_{36}$ are a sample from a normal distribution with mean μ_β. Figure 13.11 displays draws of the posterior of μ_β expressed (using equation (13.16)) as probabilities over a grid of age values from 23 to 37. The takeaway if that the career trajectories appear to be centered about 29.5 – a typical MLB player in this group peaks in on-base performance about age 29.5.

By combining data across players, the Bayesian hierarchical model is helpful in borrowing information for estimating the career trajectories of players with limited career data. This is illustrated in Figure 13.12 that shows individual and hierarchical posterior mean fits of the career trajectories for two players. For Chase Utley, the two fits are very similar since Utley's career trajectory was well-estimated just using his data. In contrast, we saw that Phelps had an unrealistic concave up individual estimated trajectory. In the hierarchical model, this career trajectory is corrected to be more similar to the concave down trajectory for most players.

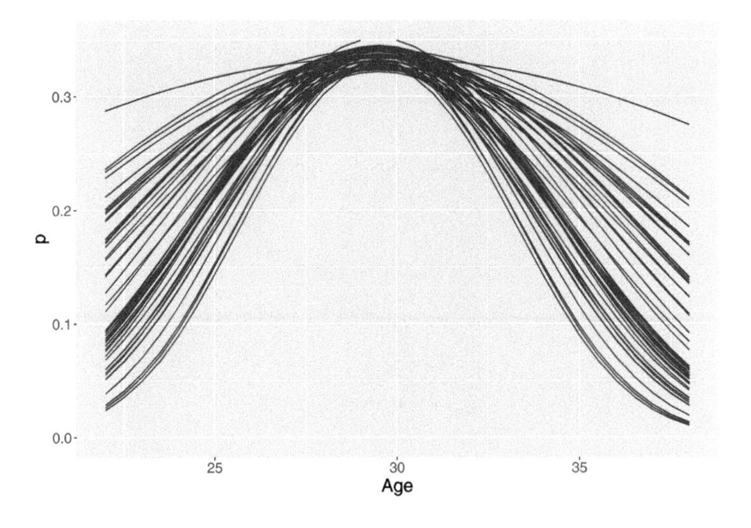

FIGURE 13.11
Samples from the posterior distribution of the mean trajectory μ_β.

13.4 Latent Class Modeling

13.4.1 Two classes of test takers

Suppose thirty people are given a 20-question true or false exam and the number of correct responses for all people are graphed in Figure 13.13. From this figure note that test takers 1 through 10 appear to have a low level of knowledge about the subject matter as their scores are centered around 10. The remaining test takers 11 through 30 seem to have a higher level of knowledge as their scores range from 15 to 20.

Are there really two groups of test takers, a random-guessing group and a knowledgeable group? If so, how can one separate the people in the two ability groups, and how can one make inferences about the correct rate for each group? Furthermore, can one be sure that two ability groups exist? Is it possible to have more than two groups of people by ability level?

The above questions relate to the classification of observations and the number of classes. In the introduction of hierarchical models in Chapter 10, there was a natural grouping of the observations. For example, in the animation movie ratings example in Chapter 10, each rating was made on one animation movie, so grouping based on movie is natural, and the group assignment of the observations was known. It was then reasonable to specify a two-stage prior where the rating means shared the same prior distribution at the first stage.

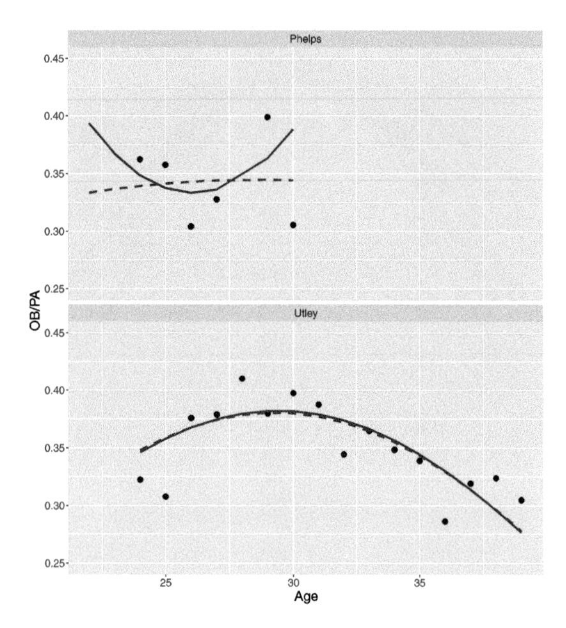

FIGURE 13.12
Individual (solid line) and hierarchical (dashed line) fits of the career trajectories for Josh Phelps and Chase Utley.

In contrast, in the true or false exam example, since the group assignment is not known, it not possible to proceed with a hierarchical model with a common prior at the first stage. In this testing example one believes the people fall in two ability groups, however one does not observe the actual classification of the people into groups. So it is assumed that there exists *latent* or unobserved classification of observations. The class assignments of the individuals are unknown and can be treated as random parameters in our Bayesian approach.

If two classes exist, the class assignment parameter for the i-th observation z_i is unknown and assumed to follow a Bernoulli distribution with probability π belonging to the first class, i.e. $z_i = 1$. With probability $1 - \pi$ the i-th observation belongs to the second class, i.e. $z_i = 0$.

$$z_i \mid \pi \sim \text{Bernoulli}(\pi). \tag{13.17}$$

If one believes there are more than two classes, the class assignment parameter follows a multinomial distribution. For ease of description of the model, we focus on the two classes situation.

Once the class assignment z_i is known for observation i, the response variable Y_i follows a data model with a group-specific parameter. In the case of a true/false exam where the outcome variable Y_i is the number of correct answers, the binomial model is a good choice for a sampling model. The

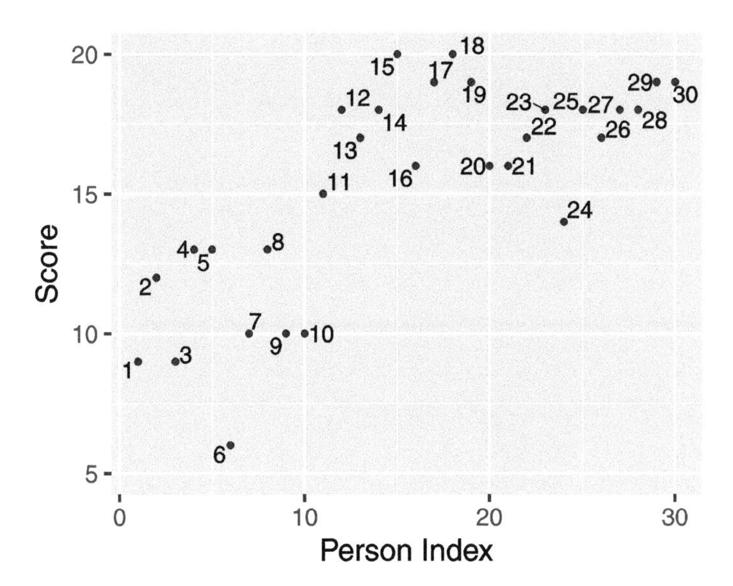

FIGURE 13.13
Scatterplot of test scores of 20 test takers. The number next to each point is the person index.

response variable Y_i conditional on the class assignment variable z_i is assigned a binomial distribution with probability of success p_{z_i}.

$$Y_i = y_i \mid z_i, p_{z_i} \sim \text{Binomial}(20, p_{z_i}). \tag{13.18}$$

One writes the success probability p_{z_i} with subscript z_i since this probability is class-specific. For the guessing group, the number of correct answers is Binomial with parameter p_1, and for the knowledgeable group the number of correct answers is Binomial with parameter p_0.

This model for responses to a true/false with unknown ability levels illustrates latent class modeling. The fundamental assumption is that two latent classes exist, and each latent class has its own sampling model with class-specific parameters. All n observations belong to one of the two latent classes and each observation is assigned to the latent classes one and two with respective probabilities π and $(1 - \pi)$. From Equation (13.17), once the latent class assignment is determined, the outcome variable y_i follows a class-specific data model as in Equation (13.18).

The tree diagram below illustrates the latent class model.

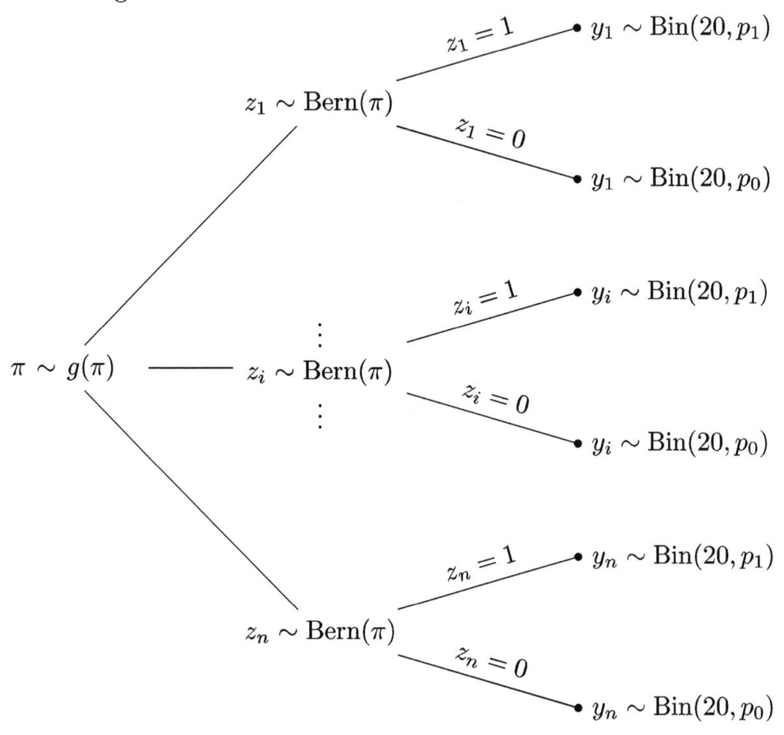

To better understand this latent class model, consider a thought experiment where one simulates outcomes $y_1, \cdots, y_n$ from this model.

- Step 1: First simulate the latent class assignments of the n test takers. One samples n values, $z_1, \cdots, z_n$, from a Bernoulli distribution with probability π. Once the latent class assignments are simulated, one has partitioned the test takers into the random-guessing group where $z_i = 1$ and the knowledgeable group where $z_i = 0$.

- Step 2: Now that the test takers' classifications are known, the outcomes are simulated by the use of binomial distributions. If a test taker's classification is $z_i = 1$, she guesses at each question with success probability p_1 and one observes the test score which is the Binomial outcome $Y_i \sim \text{Binomial}(20, p_1)$. Otherwise if the classification is $z_i = 0$, she answers a question correctly with probability p_0 and one observes the test score $Y_i \sim \text{Binomial}(20, p_0)$.

Latent class models provide the flexibility of allowing unknown class assignments of observations and the ability to cluster observations with similar characteristics. In the true/false exam example, the fitted latent class model will pool one class of observations with a lower success rate and pool the other class with a higher success rate. This fitted model also estimates model parameters for each class, providing insight of features of each latent class.

13.4.2 A latent class model with two classes

This section builds on the previous section to describe the details of the model specification of a latent class model with two classes for the true/false exam example. The JAGS software is used for MCMC simulation and several inferences are described such as identifying the class for each test taker and learning about the success rate for each class.

Suppose the true/false exam has m questions and y_i denotes the score of observation i, $i = 1, \cdots, n$. Assume there are two latent classes and each observation belongs to one of the two latent classes. Let z_i be the class assignment for observation i and π be the probability of being assigned to class 1. Given the latent class assignment z_i for observation i, the score Y_i follows a Binomial distribution with m trials and a class-specific success probability. Since there are only two possible class assignments, all observations assigned to class 1 share the same correct success parameter p_1 and all observations assigned to class 0 share the same success rate parameter p_0. The specification of the data model is expressed as follows:

$$Y_i = y_i \mid z_i, p_{z_i} \sim \text{Binomial}(m, p_{z_i}), \quad (13.19)$$

$$z_i \mid \pi \sim \text{Bernoulli}(\pi). \quad (13.20)$$

In this latent class model there are many unknown parameters. One does not know the class assignment probability π, the class assignments $z_1, ..., z_n$, and the probabilities p_1 and p_0 for the two binomial distributions. Some possible choices for prior distributions are discussed in this section.

(a) The parameters π and $(1 - \pi)$ are the latent class assignment probabilities for the two classes. If additional information is available which indicates, for example, that 1/3 of the observations belong to class 1, then π is considered as fixed and set to the value of 1/3. If no such information is available, one can consider π as unknown and assign this parameter a prior distribution. A natural choice for prior on a success probability is a Beta prior distribution with shape parameters a and b.

(b) The parameters p_1 and p_0 are the success rates in the Binomial model in the two classes. If one believes that the test takers in class 1 are simply random guessers, then one fixes p_1 to the value of 0.5. Similarly, if one believes that test takers in class 0 have a higher success rate of 0.9, then one sets p_0 to the value 0.9. However, if one is uncertain about the values of p_1 and p_0, one lets either or both success rates be random and assigned prior distributions.

Scenario 1: known parameter values

We begin with a simplified version of this latent class model. Consider the use of the fixed values $\pi = 1/3$ and $p_1 = 0.5$, and a random p_0 from a uniform distribution between 0.5 and 1. This setup indicates that one believes strongly that one third of the test takers belong to the random-guessing class, while the remaining two thirds of the test takers belong to the knowledgeable class. One is certain about the success rate of the guessing class, but the location of the correct rate of the knowledgeable class is unknown in the interval (0.5, 1).

R The JAGS model script is shown below. One introduces a new variable `theta[i]` that indicates the correct rate value for observation `i`. In the sampling section of the JAGS script, the first block is a loop over all observations, where one first determines the rate `theta[i]` based on the classification value `z[i]`. The `equals` command evaluates equality, for example, `equals(z[i], 0)` returns 1 if `z[i]` equals to `0`, and returns 0 otherwise. This indicates that the rate `theta[i]` will either be equal to `p1` or `p0` depending on the value `z[i]`.

One should note in JAGS, the classification variable `z[i]` takes values of 0 and 1, corresponding to the knowledgeable and guessing classes, respectively. As π is considered fixed and set to 1/3, the variable `z[i]` is assigned a Bernoulli distribution with probability 1/3. To conclude the script, in the prior section the guessing rate parameter `p1` is assigned the value 0.5 and the rate parameter `p0` is assigned a Beta(1, 1) distribution truncated to the interval (0.5, 1) using `T(0.5, 1)`.

```
modelString<-"
model {
## sampling
for (i in 1:N){
   theta[i] <- equals(z[i], 1) * p1 + equals(z[i], 0) * p0
y[i] ~ dbin(theta[i], m)
}
for (i in 1:N){
   z[i] ~ dbern(1/3)
}
## priors
p1 <- 0.5
p0 ~ dbeta(1,1) T(0.5, 1)
}
"
```

One performs inference for `theta` and `p0` in JAGS by looking at their posterior summaries. Note that there are $n = 30$ test takers, each with an associated `theta` indicating the correct success rate of test taker `i`. The variable `p0` is the estimate of the correct rate of the knowledgeable class.

How are the correct rates estimated for different test takers by the latent class model? Before looking at the results, let's revisit the dataset as shown in Figure 13.13. Among the test takers with lower scores, it is obvious that test taker #6 with a score of 6 is likely to be assigned to the random-guessing class, whereas test takers #4 and #5 with a score of 13 are probably assigned to the knowledgeable class. Among test takers with higher scores, test takers #15 and #17 with respective scores of 20 and 19 are most likely to be assigned to the knowledgeable class, and test taker #24 with a score of 14 is also likely assigned to the knowledgeable class.

TABLE 13.2
Posterior summaries of the correct rate θ_i of six selected test takers.

Test Taker	Score	Mean	Median	90% Credible Interval
#4	13	0.553	0.500	(0.500, 0.876)
#5	13	0.555	0.500	(0.500, 0.875)
#6	6	0.500	0.500	(0.500, 0.500)
#15	20	0.879	0.879	(0.841, 0.917)
#17	19	0.878	0.879	(0.841, 0.917)
#24	14	0.690	0.831	(0.500, 0.897)

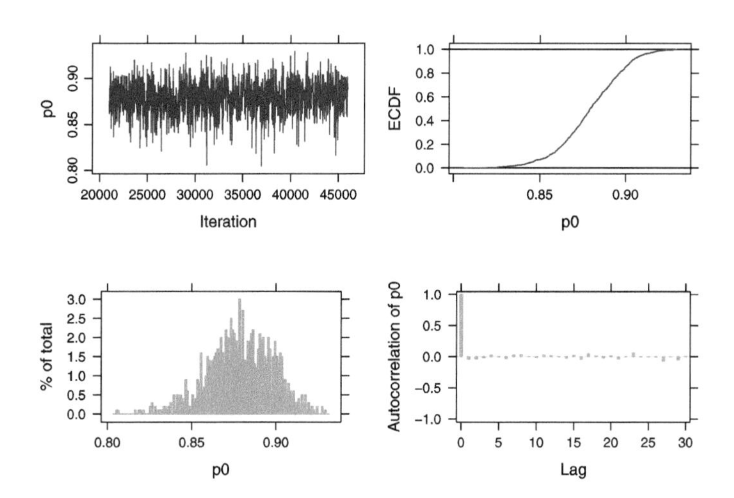

FIGURE 13.14
MCMC diagnostic plots for correct rate of the knowledgeable class, p_0.

The latent class model assigns observations to one of the two latent classes at each MCMC iteration, and the posterior summaries of `theta` provide estimates of the correct rate of each test taker. Table 13.2 provides posterior summaries for six specific test takers. The posterior summaries of the correct rate of test taker # 6 indicate that the model assigns this test taker to the random-guessing group and the posterior mean and median of the correct rate is at 0.5. Test takers #4 and #5 have similar posterior summaries and are classified as random-guessing most of the time with posterior mean of correct rate around 0.55. Test taker #24 has a higher posterior mean than the test takers #4 and #5. But with a posterior mean 0.69, the posterior probability for the true rate for #24 is somewhat split between random guessing and knowledgeable states. Test takers #15 and #17 are always classified as knowledgeable with posterior mean and median of correct rate around 0.88.

One also summarizes the posterior draws of p_0 corresponding to the success rate for the knowledgeable students. Figure 13.14 provides MCMC diagnostics of p_0. Its posterior mean, median, and 90% credible interval are 0.879, 0.879, and (0.841, 0.917). These estimates are very close to the correct rate of test takers #15 and #17. These test takers are always classified in the knowledgeable class and their correct rate estimates are the same as p_0.

Scenario 2: all parameters unknown

It is straightforward to generalize this latent class model relaxing some of the fixed parameter assumptions in Scenario 1. It was originally assumed that the class assignment parameter $\pi = 1/3$. It is more realistic to assume that the probability of assigning an individual into the first class π is unknown and assign this parameter a beta distribution with specific shape parameters. Here one assumes little is known about this classification parameter and so π is assigned a Beta$(1, 1)$, i.e. a uniform distribution on (0, 1). In addition, previously it was assumed that it was known that the success rate for the "guessing" group p_1 was equal to 1/2. Here this assumption is relaxed by assigning the success rate p_1 a uniform prior on the interval (0.4, 0.6). If one knows only that that the success rate for the "knowing" group is p_0 is larger than p_1, then one assumes p_0 is uniform on the interval $(p_1, 1)$.

R The JAGS script for this more general model follows. We introduce the parameter `q` as π, that is the class assignment parameter and assign it a beta distribution with parameters 1 and 1. The prior distributions for `p1` and `p0` are modified to reflect the new assumptions.

```
modelString<-"
model {
## sampling
for (i in 1:N){
   theta[i] <- equals(z[i], 1) * p1 + equals(z[i], 0) * p0
   y[i] ~ dbin(theta[i], m)
}
for (i in 1:N){
   z[i] ~ dbern(q)
}
## priors
p1 ~ dbeta(1, 1) T(0.4, 0.6)
```

```
p0 ~ dbeta(1,1) T(p1, 1)
q ~ dbeta(1, 1)
}
"
```

In Scenario 1, the posterior distributions of the correct rates `theta[i]` were summarized for all individuals. Here we instead focus on the classification parameters `z[i]` where `z[i] = 1` indicates a person classified into the random-guessing group. Figure 13.15 displays the posterior means of the z_i for all individuals. As expected, individuals #1 through #10 are classified as guessers and most individuals with labels 12 and higher are classified as knowledgeable. Individuals #11 and #24 have posterior classification means between 0.25 and 0.75 indicating some uncertainty about the correct classification for these people.

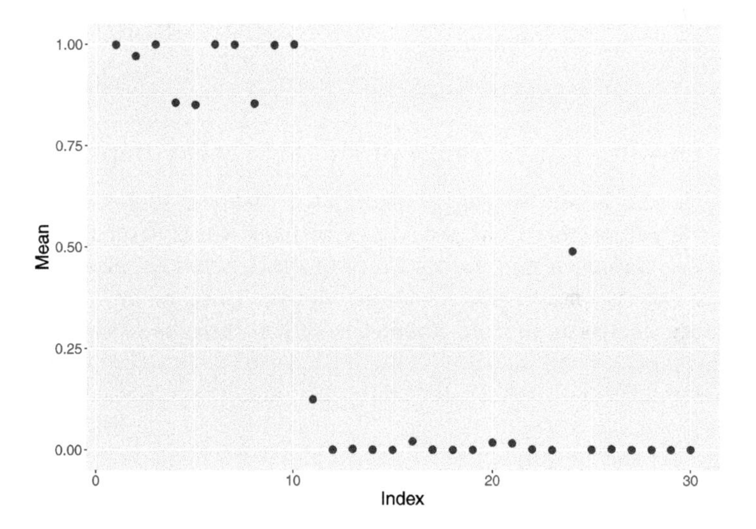

FIGURE 13.15
Posterior means of classification parameters $\{Prob(z_i = 1 \mid \{y_i\})\}$ for all test takers.

Figure 13.16 displays density estimates of the simulated draws from the posterior distributions of the class assignment parameter π and the rate parameters p_1 and p_0. As one might expect, the posterior distributions of p_1 and p_0 are centered about values of 0.54 and 0.89. There is some uncertainty about the class assignment parameter as reflected in a wide density estimate for π (q in the figure).

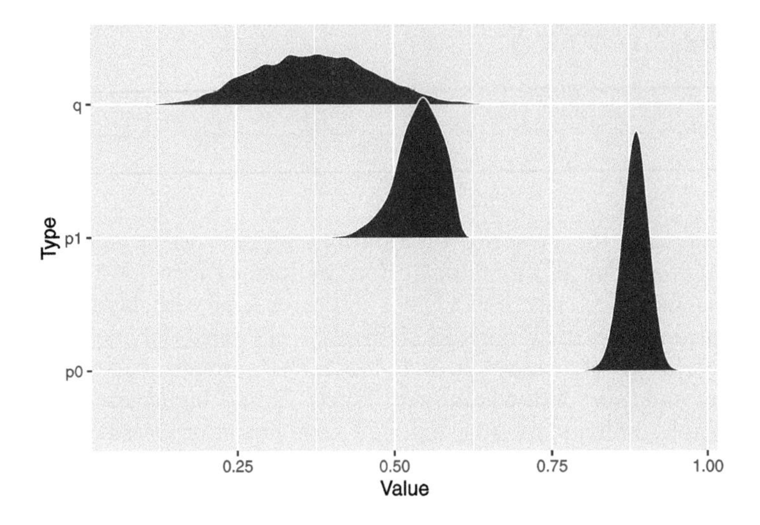

FIGURE 13.16
Posterior density plots of parameters π, p_1 and p_0.

13.4.3 Disputed authorship of the Federalist Papers

Returning to the *Federalist Papers* example of Section 13.2, the discussion focused on learning about the true rates of filler words for papers written by Alexander Hamilton and James Madison. But actually the true authorship of some of the papers was in doubt, and the primary task in Mosteller and Wallace (1963) was to learn about the true authorship of these disputed authorship papers from the data. This problem of disputed authorship can be considered a special case of latent data modeling where the latent variable is the authorship of a disputed paper. We describe how the Bayesian model of Section 13.2 can be generalized to learn about both the rates of a particular filler word and the identities of the authors of the papers of disputed authorship.

In our sample there are a total of 74 *Federalist Papers.* We assume that 49 of these papers are known to be written by Hamilton, 15 of the papers are known to be written by Madison, and the authorship of the remaining 10 papers is disputed between the two authors. We focus on the use of the filler word "can" in these papers. Let $\{(y_{1i}, n_{1i})\}$ denote the frequencies of "can" and total words in the Hamilton papers, $\{(y_{2i}, n_{2i})\}$ denote the frequencies and total words in the Madison papers, and $\{(y_i, n_i)\}$ denote the corresponding quantities in the disputed papers. As in Section 13.2, we assume $\{y_{1i}\}$ are Negative Binomial(p_{1i}, α_1) where $p_{1i} = \beta_1/(\beta_1 + n_{1i}/1000)$, and $\{y_{2i}\}$ are Negative Binomial(p_{2i}, α_2) where $p_{2i} = \beta_2/(\beta_2 + n_{2i}/1000)$.

The distribution of the frequencies $\{y_i\}$ is unknown (out of the total number of words $\{n_i\}$) since these correspond to the papers of disputed authorship.

Let z_i denote the unknown authorship of paper i among the disputed papers – if $z_i = 0$, the paper was written by Hamilton and if $z_i = 1$, the paper was written by Madison. If one knows the value of z_i, the distribution of the frequency y_i is known. If $z_i = 0$, then y_i is Negative Binomial(p_i, α_1) where $p_i = \beta_1/(\beta_1 + n_i/1000)$, and $z_i = 1$, then y_i is Negative Binomial(p_i, α_2) where $p_i = \beta_2/(\beta_2 + n_i/1000)$. To complete the model, one needs to assign a prior distribution to the latent authorship indicators $\{z_i\}$. It is assumed $z_i \sim$ Bernoulli(0.5) which means that z_i from the prior is equally likely to be 0 or 1.

The JAGS script for the disputed authorship problem is shown below. The data is structured so that `N1` papers are known to be written by Hamilton, `N2` papers are known to be written by Madison, and the authorship of the remaining `N3` papers are in doubt. The data includes the number of occurrences of the word "can" and the total number or words in each group of papers. Note that, as in Section 13.2, weakly informative priors are placed on the gamma priors for $\alpha_1, \beta_1, \alpha_2$ and β_2.

```
modelString = "
model{
for(i in 1:N1){
   p1[i] <- beta1 / (beta1 + n1[i] / 1000)
   y1[i] ~ dnegbin(p1[i], alpha1)
}
for(i in 1:N2){
   p2[i] <- beta2 / (beta2 + n2[i] / 1000)
   y2[i] ~ dnegbin(p2[i], alpha2)
}
for(i in 1:N3){
   theta[i] <- equals(z[i], 0) * alpha1 +
               equals(z[i], 1) * alpha2
   gamma[i] <- equals(z[i], 0) * beta1 +
               equals(z[i], 1) * beta2
   p[i] <- gamma[i] / (gamma[i] + n[i] / 1000)
   y[i] ~ dnegbin(p[i], theta[i])
   z[i] ~ dbern(0.5)
}
alpha1 ~ dgamma(.001, .001)
beta1 ~ dgamma(.001, .001)
alpha2 ~ dgamma(.001, .001)
beta2 ~ dgamma(.001, .001)
}
"
```

Using this script, a sample of 5000 draws were taken from the posterior distribution and Figure 13.17 displays posterior means of the classification parameters $z_1, ..., z_{10}$ for the ten disputed authorship parameters. Since $z_i = 1$

if the author is Madison, this graph is showing the posterior probability the author is James Madison for each paper. Note that most of these posterior means are located near 0.5, with the one exception of Paper 4 where the posterior probability of Madison authorship is 0.174. So really one has not learned much about the identity of the true author from this data.

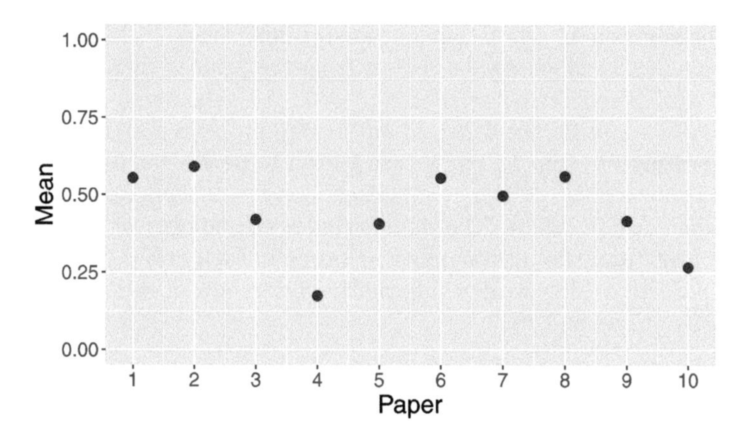

FIGURE 13.17
Posterior means of classification parameters for authorship problem using rates of the filler word "can".

But we have only looked at the frequencies of one particular filler word in our analysis. In a typical study such as the one done by Mosteller and Wallace (1963), a number of filler words are used. One can extend the analysis to include a number of filler words; the approach is outlined below and the implementation details are left to the end-of-chapter exercises.

Suppose y_{1i}^w denotes the number of occurrences of the word w in the i-th paper written by Hamilton. Similarly, y_{2i}^w denotes the word count of w in the i-th paper written by Madison and y_i^w denotes the word count of w in the i-th paper of disputed authorship. It is assumed that each word count follows a Negative Binomial distribution where the parameters of the distribution depend on the author and the word. So, for example, for a Hamilton paper, y_{1i}^w is distributed Negative Binomial(p_1^w, α_1^w) where $p_1^w = \beta_1^w / (\beta_1^w + n_{1i}/1000)$. For a Madison paper, y_{2i}^w is distributed Negative Binomial(p_2^w, α_2^w) where $p_2^w = \beta_2^w / (\beta_2^w + n_{2i}/1000)$. For a paper of disputed authorship, the count y_i^w will either be distributed according to one of the Negative Binomial distributions where the distribution depends on the value of the classification variable z_i.

A JAGS script can be written to fit this model with multiple filler words. In the script, one defines the matrix variable `y1` where `y1[i, j]` is defined to be the number of words of type j in the i-th paper of Hamilton. In a similar fashion one defines the matrices `y2` and `y` where `y2[i, j]` and `y[i, j]` denote respectively the counts of the j-th word of the i-th Madison and i-th disputed authorship paper. One will be learning about vectors $\alpha_1, \beta_1, \alpha_2, \beta_2$ where each vector has W values where W is the number of words in the study. As before `z[i]` denotes the classification variable where `z[i] = 1` denotes authorship of the i-th disputed paper by Madison. In an end-of-chapter exercise, the reader will be asked to implement the model fitting using a selection of filler words. One would anticipate that one would be able to discriminate between the two authors on the basis of a large group of filler words.

13.5 Exercises

1. **Federalist Papers Word Study**

   The frequencies of word use of Madison and Hamilton are stored in the data file `fed_word_data.csv`. Consider the counts of the word "there" in the 50 *Federalist Papers* authored by Hamilton. Suppose the count y_i in the i-th paper is distributed Poisson with mean $n_i\lambda/1000$ where n_i is the number of words in the paper and λ is the rate of the word "there" per 1000 words.

   (a) Assuming a weakly informative prior for λ, use JAGS to fit this Poisson sampling model.
   (b) Compute a 90% probability interval for the rate λ.
   (c) Consider a new essay with 1000 words. By simulating 1000 draws from the posterior predictive distribution, construct a 90% prediction interval for the number of occurrences of the word "there" in this essay.

2. **Federalist Papers Word Study (continued)**

   Instead of Poisson sampling, suppose the count of the word y_i "there" in the i-th *Federalist paper* is distributed negative binomial with parameters p_i and α, where $p_i = \beta/(\beta + n_i/1000)$ where n_i is the number of words in the paper and α/β is the rate of the word "there" per 1000 words.

   (a) Using a suitable weakly informative prior for α and β, use JAGS to simulate 1000 draws from the posterior distribution.
   (b) Construct a 90% interval estimate for the rate parameter α/β.
   (c) By simulating from the posterior predictive distribution, construct a 90% prediction interval for the number of uses of "there" in a new essay of 1000 words.
   (d) Compare your answers with the answers in Exercise 1 assuming Poisson sampling.

3. **Comparing Word Use**

   Using negative binomial sampling models, compare the average word use of Hamilton and Madison for the words "this", "on", "his", and "by". Suppose the mean rate per 1000 words is measured by α_1/β_1 and α_2/β_2 for Hamilton and Madison, respectively. For each word, construct a 90% interval estimate for the difference in use rates $D = \alpha_1/\beta_1 - \alpha_2/\beta_2$. By looking at the locations of these interval estimates, which words were more often used by Hamilton and which ones were more likely to be used by Madison? The data file is `fed_word_data.csv`.

4. **Comparing Word Use (continued)**

   As in Exercise 3, using negative binomial sampling models, compare the average word use of Hamilton and Madison for the words "this", "on", "his", and "by". If the mean rate per 1000 words is measured by α_1/β_1 and α_2/β_2 for Hamilton and Madison, respectively, suppose one is interested in comparing the rates using the ratio

$$R = \frac{\alpha_1/\beta_1}{\alpha_2/\beta_2}.$$

   Construct and graph 90% interval estimates for R for each word in the study.

5. **Basketball Shooting Data**

   Table 13.3 displays the number of free throw attempts FTA and the number of successful free throws FT for all the seasons of Isiah Thomas, a great basketball point guard who played in the National Basketball Association from 1982 to 1994. This data is contained in the file `nba_guards.csv` where the `Player` variable is equal to "THOMAS." Let p_j denote the probability of a successful free throw for the j-th season. Consider the quadratic logistic model

$$\log\left(\frac{p_j}{1-p_j}\right) = \beta_0 + \beta_1(x_j - 30) + \beta_2(x_j - 30)^2,$$

   where x_j is Thomas' age during the j-th season.

   (a) By using JAGS with a reasonable choice of weakly informative prior on the regression parameters, collect 5000 draws from the posterior distribution on $\beta = (\beta_0, \beta_1, \beta_2)$.

   (b) Construct a density estimate and a 90% interval estimate for the age $h_1(\beta)$ where Thomas attained peak performance.

   (c) Construct a density estimate and a 90% interval estimate for the probability p that Thomas makes a free throw at age 28.

TABLE 13.3
Free throw shooting data for the basketball player Isiah Thomas.

Age	FTA	FT	Age	FTA	FT
20	429	302	27	351	287
21	518	368	28	377	292
22	529	388	29	229	179
23	493	399	30	378	292
24	462	365	31	377	278
25	521	400	32	258	181
26	394	305			

6. **Basketball Shooting Data (continued)**

   The dataset `nba_guards.csv` contains the number of free throw attempts FTA and the number of successful free throws FT for all of the seasons played by fifteen great point guards in the National Basketball Association. Let p_{ij} denote the probability of a successful free throw of the i-th player during the j-th season. Suppose the probabilities $\{p_{ij}\}$ for the i-th player satisfy the quadratic model

   $$\log\left(\frac{p_{ij}}{1-p_{ij}}\right) = \beta_{i0} + \beta_{i1}(x_{ij} - 30) + \beta_{i2}(x_{ij} - 30)^2,$$

   where x_{ij} is the age of the i-th player during the j-th season and $\beta_i = (\beta_{i0}, \beta_{i1}, \beta_{i2})$ denotes the vector of regression coefficients for the i-th player.

   (a) Construct a hierarchical prior for the regression vectors $\beta_1, ..., \beta_{15}$ analogous to the one used for baseball hitters in the chapter.

   (b) Use JAGS to simulate a sample of 5000 draws from the posterior distribution of the β_j and also of the second stage prior μ_β.

   (c) For one player, consider the age $h_1(\beta)$ where he attained peak performance in free-throw shooting. Compare the posterior distributions of $h_1(\beta)$ using an individual logistic model and using the hierarchical model.

7. **Football Field Goal Kicking**

   The data file `football_field_goal.csv` contains data on field goal attempts for professional football kickers. Let y_j denote the response (success or failure) of a field goal attempt from x_j yards. One is interested in fitting the logistic model

   $$\log\left(\frac{p_j}{1-p_j}\right) = \beta_0 + \beta_1 x_j,$$

where p_j is the probability of a successful attempt. Figure 13.18 displays individual logistic fits for ten kickers in the 2005 season. These fits were found using weakly informative priors on the regression parameters β_0 and β_1 on individual fits for each player.

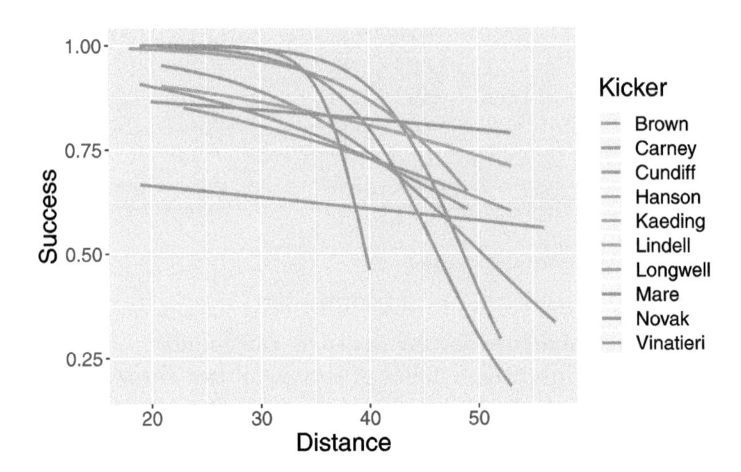

FIGURE 13.18
Individual logistic model fits for ten professional football kickers in the 2005 season.

(a) Looking at Figure 13.18, do you believe these individual fits of success probability against distance are suitable for all players? Explain.

(b) For the player Brown, assuming this logistic model and a weakly informative prior on the regression coefficients, use JAGS to simulate from the posterior distribution. From the output to construct a 90 percent interval estimate for the probability of success at 30 yards.

(c) Pool the data for all 10 players, and use JAGS to fit from the logistic model where the probability of success is a function of the distance. Use JAGS to simulate from the posterior and from the output construct a 90 percent interval estimate for the probability of success at 30 yards.

(d) Compare your answers to parts (b) and (c).

8. **Football Field Goal Kicking (continued)**

In the logistic model predicting success on a field goal attempt based on the distance in feet, suppose $\beta_i = (\beta_{i0}, \beta_{i1})$ denotes the regression vector for the logistic model on the i-th player.

(a) Write down a hierarchical prior for the ten regression vectors $\beta_1, ..., \beta_{10}$ similar to what was used for the baseball hitters in the chapter.

(b) Using JAGS, simulate a sample of 5000 from the posterior distribution of $\beta_1, ..., \beta_{10}$.

(c) Display the posterior means of the probabilities of success for all kickers as a function of distance similar to Figure 13.18.

(d) For the player Brown, construct a 90 percent interval estimate for the probability of success at 30 yards. Compare your answer to the individual fit (part (b) of Exercise 7) and the pooled data fit (part (c) of Exercise 7).

9. **Checking for Overdispersion**

In the hitter's career trajectory example in Section 13.3.4, it was assumed that the number of on-base events in season y_j was distributed binomial with a specific probability of success p_j where the $\{p_j\}$ satisfy a logistic quadratic model. If one views a scatterplot of the observed rates OB/PA against age for Chase Utley (Figure 13.7), one notices some variation about the fitted curve. It is natural to ask if the variability about this line is greater than one would predict from the binomial model.

(a) Following the example in Section 13.3.4, obtain a posterior sample from the posterior distribution of β using Utley's data and a weakly informative choice of prior distribution.

(b) Compute the posterior mean of β and obtain an estimate for the on-base probability for all of Utley's ages.

(c) Write a function to simulate one replicated sample of y_j^R from the posterior predictive distribution. Compute the sum of squares of the rates y_j^R/n_j about the fitted on-base probabilities.

(d) Using the function written in part (c), repeat this simulation 1000 times, obtaining 1000 sum of squares values from the posterior predictive distribution.

(e) By comparing the posterior predictive sample in part (d) with the observed sum of squares about the fitted curve, comment about the suitability of the binomial sampling model.

10. **Moby Dick Word Study**

Project Gutenberg offers a large number of free electronic books and the `gutenbergr` R package facilitates reading in these books into R for textual analysis. Use this package to download the famous novel *Moby Dick* by Herman Melville and collect all of the distinct words in this novel in a data frame.

(a) Divide the words of the novel into 1000-word groups and count the number of occurrences of a particular filler word in each group.

(b) Use a negative binomial distribution to model the counts of the filler word across groups.

(c) Consider the use of a different filler word, and use a Negative Binomial distribution to model the counts of this new word across 1000-word groups.

(d) By use of a suitable model, compare the rates (per 1000 words) of the two types of filler words. Construct a 90% interval estimate for the ratio of rates of use of the two words.

11. **An Outlier Model**

Suppose one observes a sample measurement data where there is a small possibility of an outlier. One observes y_i which is either Normal(μ, σ) with probability p or Normal($\mu, 3\sigma$) with probability $1-p$. Assume a weakly informative prior on μ and p is Beta(2, 20). Introduce latent class assignments where $z_i = 0$ or $z_i = 1$ depending if the observation is Normal(μ, σ) or Normal($\mu, 3\sigma$). Use JAGS with a script similar to used to Section 13.4 to fit this model. The dataset `darwin.csv` in the `ProbBayes` R package contains measurements on the differences of the heights of cross and self fertilized plants quoted by Fisher (1960). Compute the posterior probability that each observation is an outlier. Plot the measurements against these posterior outlier probabilities and comment on the pattern in the scatterplot.

12. **Another Latent Data Model**

Suppose n students are taking a multiple-choice exam of m questions. As in Section 13.4, suppose there are two types of students, the "guessing" students and the "knowledgeable" students who answer each question correctly with respective probabilities of p_0 and p_1. The following R code will simulate some data in this scenario where $n = 50$, $m = 20$, $p_0 = 0.4$, $p_1 = 0.7$ and the probability that a student is a guesser is $\pi = 0.2$ (p in the JAGS script).

```
set.seed(123)
p0 <- 0.40; p1 <- 0.70
m <- 20; n <- 50; p <- 0.2
z <- sample(0:1, size = m, prob = c(p, 1-p),
             replace = TRUE)
prob <- (z == 0) * p0 + (z == 1) * p1
y <- rbinom(m, size = n, prob)
```

By use of a latent class model similar to what was used in Section 13.4, simulate from the joint posterior distribution of all parameters.

Estimate values of p_0, p_1, and π from the posterior and compare these estimates with the "true" values of the parameters used in the simulation.

13. **Determining Authorship From a Single Word**

    In the dataset `federalist_word_study.csv`, the variable `Authorship` indicates the authorship of the *Federalist Papers* and the variable `Disputed` indicates the papers where the authorship is disputed. Following the work in Section 13.4.3 and the JAGS script, fit a latent variable model using the filler word "from". Using this particular word, examine the posterior probabilities of authorship for the ten papers of disputed authorship. Is this single word helpful for determining authorship for any of the papers? Repeat this analysis using the filler word "this".

14. **Determining Authorship From Multiple Words**

    Suppose one wishes to use all of the filler words "by", "from", "to", "an", "any", "may", "his", "upon", "also", "can", "of", "on", "there", and"this" to determine authorship of the ten disputed papers. Using the approach described at the end of Section 13.4.3, write a JAGS script to fit the latent variable model. Collect a posterior sample from the posterior distribution of the classification variables. Use the posterior means of the classification variables to determine authorship for each of the ten variables.

14

Appendices

14.1 Appendix A: The constant in the beta posterior

In the dining survey example of Chapter 7, it was claimed that the posterior $\pi(p \mid Y = 12) \sim \text{Beta}(15.06, 10.56)$ by recognizing that the product of the prior and likelihood was proportional to $p^{15.06-1}(1-p)^{10.56-1}$. Here it is shown that the normalizing constant of the density is $\frac{1}{B(15.06,10.56)}$. That is, it is shown that

$$\pi(p \mid Y = 12) = \frac{1}{B(15.06, 10.56)} p^{15.06-1}(1-p)^{10.56-1}. \quad (14.1)$$

The more general derivation of the normalizing constant is presented assuming the prior is $p \sim \text{Beta}(a, b)$ and the sampling density is $Y \sim \text{Binomial}(n, p)$. Using Bayes' rule,

$$\begin{aligned}
\pi(p \mid Y = y) &= \frac{\pi(p)f(y \mid p)}{f(y)} = \frac{\pi(p)f(y \mid p)}{\int_p \pi(p)f(y \mid p)dp} \\
&= \frac{\frac{\Gamma(a+b)}{\Gamma(a)\Gamma(b)}p^{a-1}(1-p)^{b-1}\binom{n}{y}p^y(1-p)^{n-y}}{\int_p \frac{\Gamma(a+b)}{\Gamma(a)\Gamma(b)}p^{a-1}(1-p)^{b-1}\binom{n}{y}p^y(1-p)^{n-y}dp} \\
&= \frac{\frac{\Gamma(a+b)}{\Gamma(a)\Gamma(b)}\binom{n}{y}p^{(a+y)-1}(1-p)^{(n+b-y)-1}}{\frac{\Gamma(a+b)}{\Gamma(a)\Gamma(b)}\binom{n}{y}\int_p p^{(a+y)-1}(1-p)^{(n+b-y)-1}dp} \\
&= \frac{p^{(a+y)-1}(1-p)^{(a+b-y)-1}}{\frac{\Gamma(a+y)\Gamma(n+b-y)}{\Gamma(n+a+b)}} \\
&= \frac{\Gamma(n+a+b)}{\Gamma(a+y)\Gamma(n+b-y)}p^{(a+y)-1}(1-p)^{(n+b-y)-1} \\
&\Rightarrow \; p \mid Y = y \sim \text{Beta}(a+y, n+b-y)
\end{aligned} \quad (14.2)$$

The key to the derivation is to recognize that

$$\int_p p^{(a+y)-1}(1-p)^{(n+b-y)-1}dp = \frac{\Gamma(a+y)\Gamma(n+b-y)}{\Gamma(n+a+b)}, \quad (14.3)$$

because of the fact that the beta distribution is proper, therefore it integrates to 1. That is,

$$\int_p \frac{\Gamma(n+a+b)}{\Gamma(a+y)\Gamma(n+b-y)} p^{(a+y)-1}(1-p)^{(n+b-y)-1} dp = 1. \tag{14.4}$$

14.2 Appendix B: The posterior predictive distribution

In Chapter 7, we considered the situation where $Y \sim$ Binomial(n, p) and the proportion $p \sim$ Beta(a, b). One observes $Y = y$ and one is interested in the posterior predictive distribution of the number of successes $\tilde{Y}$ in a future sample of size m. We provide a detailed derivation below, showing that this predictive mass function $f(\tilde{Y} \mid Y = y)$ is a special case of the beta-binomial distribution.

$$\begin{aligned}
f\tilde{Y} \mid Y = y) &= \int f(\tilde{Y}, p \mid Y = y) dp \\
&= \int f(\tilde{Y} \mid p, Y = y)\pi(p \mid Y = y) dp \\
&= \int_0^1 \binom{m}{\tilde{y}} p^{\tilde{y}}(1-p)^{m-\tilde{y}} \frac{\Gamma(a+b+n)}{\Gamma(a+y)\Gamma(b+n-y)} \\
& \quad p^{a+y-1}(1-p)^{b+n-y-1} dp \\
\texttt{[constants out} &= \binom{m}{\tilde{y}} \frac{\Gamma(a+b+n)}{\Gamma(a+y)\Gamma(b+n-y)} \\
\texttt{integral sign]} & \quad \int_0^1 p^{(\tilde{y}+a+y-1}(1-p)^{m-\tilde{y}+b+n-y-1} dp \\
\texttt{[Beta integral} &= \binom{m}{\tilde{y}} \frac{\Gamma(a+b+n)}{\Gamma(a+y)\Gamma(b+n-y)} \\
\texttt{integrates to 1]} & \quad \frac{\Gamma(\tilde{y}+a+y)\Gamma(m-\tilde{y}+b+n-y)}{\Gamma(a+m+b+n)} \\
\texttt{[Use } \Gamma(x+1) \texttt{ for} &= \frac{\Gamma(m+1)}{\Gamma(\tilde{y}+1)\Gamma(m-\tilde{y}+1)} \frac{\Gamma(a+b+n)}{\Gamma(a+y)\Gamma(b+n-y)} \\
x! \texttt{ when } x \texttt{ is integer]} & \quad \frac{\Gamma(\tilde{y}+a+y)\Gamma(m-\tilde{y}+b+n-y)}{\Gamma(a+m+b+n)}
\end{aligned} \tag{14.5}$$

14.3 Appendix C: Comparing Bayesian models using a mixture of priors

Chapter 7 considers the situation where the observation $Y \sim \text{Binomial}(n, p)$ and a mixture of beta priors of the form

$$\pi(p) = q\pi_1(p) + (1-q)\pi_2(p),$$

where π_1 is Beta(a_1, b_1), π_2 is Beta(a_2, b_2), and q is a constant between 0 and 1. After observing $Y = y$, we show that the posterior density can also be represented as a mixture of two beta distributions.

One can write the posterior density as

$$\begin{aligned}\pi(p \mid Y = y) &= \frac{\pi(p) f(Y = y \mid p)}{f(Y = y)} \\ &= \frac{\{q\pi_1(p) + (1-q)\pi_2(p)\} \binom{n}{y} p^y (1-p)^{n-y}}{f(Y = y)},\end{aligned}$$

where $f(Y = y)$ is the marginal density of Y evaluated at $Y = y$.

One finds the marginal density of Y by integrating out p from the joint density of (Y, p). By performing several calculations similar to the derivation in Appendix A, one obtains

$$\begin{aligned}f(y) &= \int_0^1 \{q\pi_1(p) + (1-q)\pi_2(p)\} \binom{n}{y} p^y (1-p)^{n-y} dp \\ &= \binom{n}{y} [qB(a_1 + y, b_1 + n - y) + (1-q)B(a_2 + y, b_2 + n - y)],\end{aligned}$$

where $B(a, b)$ is the beta function

$$B(a, b) = \frac{\Gamma(a)\Gamma(b)}{\Gamma(a+b)}.$$

If one substitutes the expression for $f(x)$ in the posterior density, one obtains

$$\begin{aligned}\pi(p \mid Y = y) &= \frac{\{q\pi_1(p) + (1-q)\pi_2(p)\} p^y (1-p)^{n-y}}{qB(a_1 + y, b_1 + n - y) + (1-q)B(a_2 + y, b_2 + n - y)} \\ &= \frac{qB(a_1 + y, b_1 + n - y)\pi_1(p \mid y) + (1-q)B(a_2 + y, b_2 + n - y)\pi_2(p \mid y)}{qB(a_1 + y, b_1 + n - y) + (1-q)B(a_2 + y, b_2 + n - y)} \\ &= q(y)\pi_1(p \mid y) + (1-q(y))\pi_2(p \mid y),\end{aligned} \tag{14.6}$$

where π_1 is a beta density with shape parameters $a_1 + y$ and $b_1 + n - y$, π_2 is a beta density with shape parameters $a_2 + y$ and $b_2 + n - y$, and $q(y)$ is the constant

$$q(y) = \frac{qB(a_1 + y, b_1 + n - y)}{qB(a_1 + y, b_1 + n - y) + (1 - q)B(a_2 + y, b_2 + n - y)}.$$

This shows that a mixture of beta densities is a conjugate density in that both the prior and posterior densities have the same mixture of beta functional forms.

Using the Deviance Information Criteria (DIC)

Chapter 12 describes the problem of choosing between a number of regression models. The deviance information criteria or DIC is a popular method for model selection. In a general Bayesian model, let $\pi(\theta)$ denote the prior density, $L(\theta \mid y)$ denote the likelihood, and $\pi(\theta \mid y)$ denote the posterior density. Define the deviation to be minus two times the log likelihood function

$$D(\theta) = -2 \log L(\theta \mid y). \tag{14.7}$$

After one observes the data y, one can summarize the model fit by computing the posterior expectation of the deviance

$$\bar{D} = \int D(\theta)\pi(\theta)\theta. \tag{14.8}$$

Generally as one chooses a model with more parameters, the value of $\bar{D}$ will decrease and so the value of $\bar{D}$ by itself is not useful in comparing models with different number of parameters. One has to balance the value of $\bar{D}$ with an additional term that measures the complexity of the model. One can measure model complexity by the effective number of parameters defined by the expected deviance minus the deviance evaluated at the posterior expectation:

$$p_D = \bar{D} - D(\bar{\theta}), \tag{14.9}$$

where $\bar{\theta}$ is the posterior mean.

One defines the DIC as the sum of $\bar{D}$ and the effective number of parameters.

$$DIC = \bar{D} + p_D. \tag{14.10}$$

When one has a number of plausible models, one computes the value of DIC from the simulated posterior sample for each model, and chooses the model with the smallest value of DIC.

Bibliography

[1] Albert, J. (2009), *Bayesian Computation with R*, 2nd edition, Springer, New York.

[2] Albert, J. (2019), ProbBayes: Probability and Bayesian Modeling, R package version 1.05.

[3] Berry, D. (1996), *Statistics: A Bayesian Perspective*, Duxbury Press, Belmont, CA.

[4] Cobb, G. (2015), Mere Renovation is Too Little Too Late: We Need to Rethink our Undergraduate Curriculum from the Ground Up, *The American Statistician*, 69:4, 226-282.

[5] Denwood, M. (2016), runjags: An R Package Providing Interface Utilities, Model Templates, Parallel Computing Methods and Additional Distributions for MCMC Models in JAGS, *Journal of Statistical Software*, 71, 9.

[6] Fisher, R. (1960), *Statistical Methods for Research Workers*, Oliver and Boyd, Edinburgh.

[7] Gelman, A., Carlin, J. B., Stern, H. S., Dunson, D. B., Vehtari, A., and Rubin, D. B. (2013), *Bayesian Data Analysis,* Chapman & Hall/CRC, Boca Raton, FL.

[8] Gelman, A., and Nolan, D. (2017), *Teaching Statistics: A Bag of Tricks*, Oxford University Press, New York.

[9] Grinstead, C. M., and Snell, J. L. (2012), *Introduction to Probability*, American Mathematics Society.

[10] Hoff, P. D. (2009), *A first Course in Bayesian Statistical Methods*, Springer: New York.

[11] Mohanty, S. P., S. Suresh Babu, and N. Sreekumaran Nair (2001), The Use of Arm Span as a Predictor of Height: A Study of South Indian women, *Journal of Orthopaedic Surgery* 9.1, 19-23.

[12] Mosteller, F., and Wallace, D. L. (1963), Inference in an Authorship Problem: A Comparative Study of Discrimination Methods Applied to the Authorship of the Disputed Federalist Papers, *Journal of the American Statistical Association*, 58, 302, 275-309.

[13] Plummer M. (2003), JAGS: A Program for Analysis of Bayesian Graphical Models Using Gibbs Sampling. In K. Hornik, F. Leisch, A. Zeileis (eds.), *Proceedings of the 3rd International Workshop on Distributed Statistical Computing* (DSC 2003).

[14] Plummer M., Best N., Cowles K., Vines K. (2006). Coda: Convergence Diagnosis and Output Analysis for MCMC, *R News*, 6(1), 7-11.

[15] R Core Team (2017), R: A Language and Environment for Statistical Computing, R Foundation for Statistical Computing, Vienna, Austria, `www.R-project.org`.

[16] Raftery, A. E. (1988), Inference for the Binomial N Parameter: A Hierarchical Bayes Approach, *Biometrika*, 75(2), 223-228.

[17] Wickham, H. (2016), *ggplot2: Elegant Graphics for Data Analysis*, Springer, New York.

Index

Note: The () symbol refers to a R function.